T0234878

Lecture Notes in Computer Science　　10261

Commenced Publication in 1973
Founding and Former Series Editors:
Gerhard Goos, Juris Hartmanis, and Jan van Leeuwen

More information about this series at http://www.springer.com/series/7407

Fengyu Cong · Andrew Leung
Qinglai Wei (Eds.)

Advances in Neural Networks – ISNN 2017

14th International Symposium, ISNN 2017
Sapporo, Hakodate, and Muroran, Hokkaido, Japan, June 21–26, 2017
Proceedings, Part I

 Springer

Editors
Fengyu Cong
Dalian University of Technology
Dalian
China

Qinglai Wei
Chinese Academy of Sciences
Beijing
China

Andrew Leung
City University of Hong Kong
Kowloon Tong
Hong Kong

ISSN 0302-9743 ISSN 1611-3349 (electronic)
Lecture Notes in Computer Science
ISBN 978-3-319-59071-4 ISBN 978-3-319-59072-1 (eBook)
DOI 10.1007/978-3-319-59072-1

Library of Congress Control Number: 2017941494

LNCS Sublibrary: SL1 – Theoretical Computer Science and General Issues

Printed on acid-free paper

This Springer imprint is published by Springer Nature
The registered company is Springer International Publishing AG
The registered company address is: Gewerbestrasse 11, 6330 Cham, Switzerland

Preface

The twin volumes of *Lecture Notes in Computer Science* constitute the proceedings of the 14th International Symposium on Neural Networks (ISNN 2017) held during June 21–26, 2017, in Sapporo, Hakodate, and Muroran, Hokkaido, Japan. Building on the success of the previous events, ISNN has become a well-established series of popular and high-quality conferences on the theory and methodology of neural networks and their applications. This year's symposium was held for the third time outside China, in Hokkaido, a beautiful island in Japan. As usual, it achieved great success. ISNN aims at providing a high-level international forum for scientists, engineers, educators, and students to gather so as to present and discuss the latest progress in neural network research and applications in diverse areas. It encouraged open discussion, disagreement, criticism, and debate, and we think this is the right way to push the field forward.

Based on the rigorous peer-reviews by the Program Committee members and reviewers, 135 high-quality papers from 25 countries and regions were selected for publication in the LNCS proceedings. These papers cover many topics of neural network-related research including intelligent control, neurodynamic analysis, memristive neurodynamics, computer vision, signal processing, machine learning, optimization etc. Many organizations and volunteers made great contributions toward the success of this symposium. We would like to express our sincere gratitude to City University of Hong Kong and Hokkaido University for their sponsorship, the IEEE Computational Intelligence Society, the International Neural Network Society, and the Japanese Neural Network Society for their technical co-sponsorship. We would also like to sincerely thank all the committee members for all their great efforts in organizing the symposium. Special thanks go to the Program Committee members and reviewers whose insightful reviews and timely feedback ensured the high quality of the accepted papers and the smooth flow of the symposium. We would also like to thank Springer for their cooperation in publishing the proceedings in the prestigious *Lecture Notes in Computer Science* series. Finally, we would like to thank all the speakers, authors, and participants for their support.

April 2017

Fengyu Cong
Andrew C.-S. Leung
Qinglai Wei

Organization

Honorary Chair

Shun'ichi Amari   RIKEN Brain Science Institute, Japan

General Chairs

Hidenori Kawamura   Hokkaido University, Japan
Jun Wang   City University of Hong Kong, SAR China

Advisory Chairs

Kunihiko Fukushima   Fuzzy Logic Systems Institute, Japan
Takeshi Yamakawa   Fuzzy Logic Systems Institute, Japan

Steering Chairs

Haibo He   University of Rhode Island, USA
Derong Liu   University of Illinois, Chicago, USA
Jun Wang   City University of Hong Kong, SAR China

Organizing Committee Chairs

Andrzej Cichocki   RIKEN Brain Science Institute, Japan
Min Han   Dalian University of Technology, China
Bao-Liang Lu   Shanghai Jiao Tong University, China
Masahito Yamamoto   Hokkaido University, Japan

Program Chairs

Fengyu Cong   Dalian University of Technology, China
Andrew C.-S. Leung   City University of Hong Kong, SAR China
Qinglai Wei   CAS Institute of Automation, China

Special Sessions Chairs

Long Cheng   CAS Institute of Automation, China
Satoshi Kurihara   University of Electro-Communications, Japan
Qingshan Liu   Huazhong University of Science and Technology, China
Tomohisa Yamashita   National Institute of Advanced Industrial Science
     and Technology, Japan
Nian Zhang   University of District of Columbia, USA

Tutorial Chairs

Hitoshi Matsubara   Future University Hakodate, Japan
Keiji Suzuki   Future University Hakodate, Japan

Workshop Chairs

Mianxiong Dong   Muroran Institute of Technology, Japan
Jay Kishigami   Muroran Institute of Technology, Japan
Yasuo Kudo   Muroran Institute of Technology, Japan

Publicity Chairs

Jinde Cao   Southeast University, China
Hisao Ishibuchi   Osaka Prefecture University, Japan
Zhigang Zeng   Huazhong University of Science and Technology, China
Huaguang Zhang   Northeastern University, China
Jun Zhang   South China University of Technology, China

Publications Chairs

Jin Hu   Chongqing Jiaotong University, China
He Huang   Soochow University, China
Xinyi Le   Shanghai Jiao Tong University, China
Yongming Li   Liaoning University of Technology, China

Registration Chairs

Shenshen Gu   Shanghai University, China
Hiroyuki Iizuka   Hokkaido University, Japan
Ka Chun Wong   City University of Hong Kong, SAR China

Local Arrangements Chairs

Takashi Kawakami   Hokkaido University of Science, Japan
Koji Nishikawa   Hokkaido University of Science, Japan

Secretaries

Miki Kamata   Hokkaido University, Japan
Ying Qu   Dalian University of Technology, China

Program Committee

Xuhui Bu   Henan Polytechnic University, China
Long Cheng   Chinese Academy of Sciences, China

Fengyu Cong   Dalian University of Technology, China
Ruxandra Liana Costea   Polytechnic University of Bucharest, Romania
Jisheng Dai   Jiangsu University, China
Wai-Keung Fung   Robert Gordon University, UK
Shenshen Gu   Shanghai University, China
Zhishan Guo   Missouri University of Science and Technology, USA
Zhenyuan Guo   Hunan University, China
Chengan Guo   Dalian University of Technology, China
Wei He   Beijing University of Science and Technology, China
Sanqing Hu   Hangzhou Dianzi University, China
Long-Ting Huang   Wuhan University of Technology, China
Min Jiang   Xiamen University, China
Danchi Jiang   University of Tasmania, Australia
Shunshoku Kanae   Fukui University of Technology, Japan
Rhee Man Kil   Korean Advanced Institute of Science and Technology,
                               South Korea
Chiman Kwan   Signal Processing, Inc., Singapore
Chi-Sing Leung   City University of Hong Kong, SAR China
Michael Li   Central Queensland University, Australia
Shoutao Li   Jilin University, China
Cheng Dong Li   Shandong Jianzhu University, China
Jie Lian   Dalian University of Technology, China
Jinling Liang   Southeast University, China
Meiqin Liu   Zhejiang University, China
Ju Liu   Shandong University, China
Wenlian Lu   Fudan University, China
Biao Luo   Chinese Academy of Sciences, China
Dazhong Ma   Northeastern University, China
Tiedong Ma   Chongqing University, China
Jinwen Ma   Peking University, China
Kim Fung Man   City University of Hong Kong, SAR China
Seiichi Ozawa   Kobe University, Japan
Sitian Qin   Harbin Institute of Technology at Weihai, China
Ruizhuo Song   Beijing University of Science and Technology, China
Qiankun Song   Chongqing Jiaotong University, China
John Sum   National Chung Hsing University, China
Weize Sun   Shenzhen University, China
Norikazu Takahashi   Okayama University, Japan
Christos Tjortjis   International Hellenic University, Greece
Kim-Fung Tsang   City University of Hong Kong, SAR China
Jun Wang   City University of Hong Kong, SAR China
Jian Wang   China University of Petroleum, China
Zhanshan Wang   Northeastern University, China
Jing Wang   Beijing University of chemical Technology, China
Shenquan Wang   Changchun University of technology, China
Dianhui Wang   La Trobe University, Australia

Xinzhe Wang	Dalian University of Technology, China
Zhuo Wang	Beihang University, China
Ding Wang	Chinese Academy of Sciences, China
Zhiliang Wang	Northeastern University, China
Qinglai Wei	Chinese Academy of Sciences, China
Xin Xu	National University of Defense Technology, China
Qinmin Yang	Zhejiang University, China
Xiong Yang	Tianjin University, China
Xu Yang	Beijing University of Science and Technology, China
Mao Ye	University of Electronic Science and Technology of China, China
Nian Zhang	University of the District of Columbia, USA
Chi Zhang	Dalian University of Technology, China
Jie Zhang	Chinese Academy of Sciences, China
Xiumei Zhang	Changchun University of Technology, China
Jie Zhang	Newcastle University, UK
Bo Zhao	Chinese Academy of Sciences, China
Dongbin Zhao	Chinese Academy of Sciences, China

Contents – Part I

Cognition Computation and Neural Networks

Contents – Part II

Signal, Image and Video Processing

Bio-signal and Medical Image Analysis

Clustering, Classification, Modeling, and Forecasting

Online Multi-threshold Learning
with Imbalanced Data Stream

Xufen Cai[1,3]($\boxtimes$), Min Yang[2], Rong Zhu[2]($\boxtimes$), Xiaoyan Li[3], Long Ye[1]($\boxtimes$),
and Qin Zhang[1]

[1] Department of Information Engineering,
Communication University of China, Beijing, China
{yelong,zhangqin}@cuc.edu.cn
[2] State Key Laboratory of Software Engineering, Computer School,
Wuhan University, Hubei, China
{yangmin,zhurong}@whu.edu.cn
[3] Institute of Computer Technology, CAS, Beijing, China
{xufen.cai,xiaoyan.li}@vipl.ict.ac.cn

Abstract. This paper addresses the imbalanced data problem in an
online fashion based on multi-threshold learning. The majority of existing works on processing large-scale imbalanced data stream assume a
prior distribution of data based on a training dataset, while we consider
a more challenging setting without any assumption of the prior, and
propose an online multi-threshold learning (OMTL) method by simultaneously learning multiple classifiers with different threshold based on F-
measure incremental updating. The proposed approach shows its potentials on recent benchmark datasets compared to previous cost-sensitive
and threshold fine-tuning based online learning algorithms.

Keywords: Online learning · Imbalanced data · Multi-threshold ·
F-measure

1 Introduction

Classification problems of imbalanced data are prevalent in real world, such as
medical fatal diseases diagnoses, finance fraud detection, security detection and
information retrieval, etc. Imbalanced training data probably lead to poor performance on the minority classes without a proper treatment [7]. The situation gets
even worsen when the loss of the minority is large [11], for example missing detection of cancer cells and failure in recognizing prohibited items. These real-life
machine learning problems can be more naturally viewed as online rather than
batch/offline learning, and online learning for imbalanced data pose a greater
challenge for data usually arrive sequentially and continuously in real-time, it is
impossible to store the whole subsequent large- scale data stream. In addition, as
the inherent complex characteristics of imbalanced sequential training data, it is
difficult to model the class distributions and achieving high classification accuracies for the minority classes without significantly jeopardizing the accuracies
of the minority on the fly.

© Springer International Publishing AG 2017
F. Cong et al. (Eds.): ISNN 2017, Part I, LNCS 10261, pp. 3–9, 2017.
DOI: 10.1007/978-3-319-59072-1_1

Despite the great effort and achievements on imbalanced data learning, online learning for imbalanced data is still at its early stage especially when compared with offline learning in this field.

A sampling based approach [12] was proposed to consider the properties of dataset as a whole and sample the instances to balance the data distribution. Subsequently various sampling methods have been developed [1,5,6,10] to specifically tackle the imbalanced learning problem. However, the embedded down-sampling step select a subset from all training data and then only the local properties of this subset is exploited. Obviously, a subset is not an ideal representation of the whole dataset and probably wastes valuable training data.

After that cost-sensitive based online classification techniques have been proposed and applied in various application domains [4]. Many studies cast the problem into cost-sensitive learning that assigns different costs to mistakes of different classes [3,9], these studies assume a given cost vector (or matrix) and modify conventional loss functions to incorporate the given cost vector/matrix. The issue with this approach is that Under the online learning scenario, the prior distribution of labels is unknown resulting in the uncertainty of reasonable cost vector/matrix and subsequently provide unfavorable classification perfomance.

Alternatively, in this paper, we proposes a threshold based online learning method for imbalanced data stream. The proposed framework simultaneously learns multiple classifiers with various thresholds. In particular, at each iteration, the prediction is made by a classifier which is selected according to the highest updated F-measure, which is a popular performance measure for imbalanced problem. The selection of the optimal classifier is adaptive and evolving according to the data stream. We emphasize that the proposed approach does not require any prior knowledge. Empirical studies demonstrate that the proposed algorithm is effective and outperform previous online learning algorithms for imbalanced data stream.

The rest of this paper is organized as follows. Section 2 proposes the Online Multi-threshold learning method for solving the online classification problem of imbalanced data. Section 3 demonstrates experiments to validate the proposed approach, fallowed by a detailed discussion of the results. Finally, a conclusion is provided in Sect. 4.

2   Online Multi-threshold Learning

In this section, we provide a multi-threshold based solution to the online classification for imbalanced data stream. We simultaneously maintain several classifiers with different threshold and briefly present how to update F-measure incrementally in an online fashion, the classifier with the highest F-measure is selected as the promising optimal classifier for the next incoming instance.

2.1   Notations and Problem

For clear description in the following, we first define some notations. Typically, for online learning, there are samples T in d dimension arriving in a sequence, denoted as $X = [x_1, x_2, \cdots x_T] \in \mathbb{R}^{d \times T}$ with their class labels $Y = [y_1, y_2, \cdots y_T]$, $y_t \in \gamma$, where $x_t \in \mathbb{R}^{d \times 1}$ is the feature representation of the t - th sample with $t \in [1, T]$. y_t is its class label, and γ is the space of classes. For online binary classification $\gamma \in \{-1, +1\}$; similarly, for online multi-class classification $\gamma \in \{1, 2, \cdots k\}$, which k is the number of classes. The problem to deal with in this paper is how to learn the optimal models Φ for online binary classification with imbalanced data stream X.

2.2 Online Binary Classification

Originally, for online binary classification, the learner processes an incoming instance $x_t \in X$ by a linear function f as follows:

$$f(x) = w^T x_t + b \tag{1}$$

where $w \in \mathbb{R}^{1 \times d}$s called a weight vector, $b \in \mathbb{R}^{1 \times 1}$ is a bias, the negative of the bias is sometimes called a threshold.

Then predicting its label $\hat{y}_t$ at each step t by

$$\hat{y}_t = \text{sgn}(f(x_t; w_t)) \in \gamma, \quad \gamma \in \{-1, +1\} \tag{2}$$

After the prediction, the true label $y_t \in \gamma$ is revealed and then the suffered loss $l_t(y_t, \hat{y}_t)$ between the predicted label $\hat{y}_t$ and the true label y_t is calculated by the corresponding loss function. When $l_t > 0$, the learner will update the learner: $w_{t+1} \leftarrow \nabla(w_t; (x_t, y_t))$. The learning objective is to minimize the cumulative mistake over the entire sequence of data examples.

2.3 Online Multi-threshold Learning

The typical online binary classification learner tries to train a good model with optimal weight vector w and bias/threshold b. Nevertheless, under the online learning scenario, it is impossible to select the optimal threshold for the classifier suitable for every imbalanced datasets without any prior knowledge. In this case, our proposed multi-threshold learning method attempts to train several models with different threshold instead, every single sample could find a promising optimal classifier with a particular threshold adaptively. First we map samples into a feature representation space with limited variation scale through non-linear transformation, like tanh function (ranges from -1 to 1) or *sigmoid* function (ranges from 0 to 1). In this paper, we utilize tanh function projecting samples into $[-1, 1]$ as follows:

$$f(x) = s(w^T x_t + b) \tag{3}$$

where $s(\cdot)$ is non-linear transformation and $s(x) = \frac{e^x - e^{-x}}{e^x + e^{-x}}$. After projection, we set a series of threshold $\tau^* \in (-1, 1)$, empirically, we set 19 thresholds as $\tau^* = \{-0.9, -0.8, -0.7, \cdots - 0.1, 0, 0.1 \cdots 0.6, 0.7, 0.8, 0.9\}$. Then we have a series of models Φ as follows:

$$\Phi(f^*) = s(w^T x_t + b) + \tau^* \tag{4}$$

Accordingly, predicting its label $\hat{y}_t$ at each step t by

$$\hat{y}_t = \text{sgn}(f^*(x_t; w_t)) \in \gamma, \quad \gamma \in \{-1, +1\} \tag{5}$$

The rest problem lies to how to choose a suitable classifier with promising performance among all the models for every incoming instance. Intuitively, we could utilize online criteria to measure the performance in terms of all models. But directly calculating the online measure by going through all examples is rather costly meanwhile requiring to store all predictions, which is not permitted in an online fashion. Thanks to the development of online measure metric optimization [2,14], the online F-measure can be calculated incrementally. We utilize F measure optimization method to select the promising optimal classifier with certain a threshold for the next incoming instance. More specifically, we could maintain several classifiers with different threshold simultaneously and calculate the F score online for each sample before selecting the classifier with the highest F score to produce the final prediction.

We calculate the online incrementally updating F-measure by

$$F(t) = \frac{2a_t}{c_t} \tag{6}$$

where $a_t = \sum_{t}^{T} y_t \hat{y}_t$ and $c_t = \sum_{t}^{T} y_t + \sum_{t}^{T} \hat{y}_t$, which a_t and c_t are updated by

$$
a_{t+1} = \begin{cases} a_t + 1, & if\ y_{t+1} = 1\ and\ \hat{y}_{t+1} > 0, \\ a_t, & otherwise; \end{cases}
$$
$$
c_{t+1} = \begin{cases} c_t + 2, & if\ y_{t+1} = 1\ and\ \hat{y}_{t+1} > 0, \\ c_t + 1, & if\ y_{t+1} = 1\ or\ \hat{y}_{t+1} > 0, \\ c_t, & if\ y_{t+1} = -1\ and\ \hat{y}_{t+1} < 0; \end{cases} \tag{7}
$$

3   Experiments and Discussion

In this section, we carry out extensive experiments to evaluate the performance of the proposed algorithm OMTL on various benchmark datasets from the LIBSVM repository of binary classification tasks[1].

3.1   The Datasets

To evaluate the performance, we used 3 imbalanced datasets with various imbalance ratios (the proportion of positive and negative samples) and significantly different dimension, as listed in Table 1. We construct imbalanced data for binary classification from multiclass datasets mnist, setting instances of the first class as positive and instances of the rest 9 classes as negative, denoted by mnist1all.

---

[1] http://www.csie.ntu.edu.tw/~cjlin/libsvmtools/datasets/.

Table 1. The discription of datasets

Datasets	size	dimension	Neg:Pos
codrna	59535	9	5.24
w8a	64700	300	32.5
mnist1all	64997	785	33.2

3.2 Experimental Design and Implementation

We compare the proposed OMTL algorithm with several state-of-the art online learning algorithms, namely the baseline OGD [15], the threshold fine-tuning method OFO [2], and two online cost-sensitive algorithms, which are PAUM [8] and CSOGD (the first type) [13]. To ensure a fair comparison, all algorithms adopt the same experimental setup to implement in MatLab R2015b and run in a Linux machine with 64 bit, $Intel\ Core\ i7 - 6700\ CPU@3.40\,GHz \times 8$ We run the proposed algorithm and all the competitors trained on 80% of each dataset and the rest 20% as the testing set. We conducting each experiment on 20 random permutations for each dataset. The final results are reported by averaging over these 20 runs. To examine the performance of using different loss functions, we investigate both the hinge loss and the logistic loss in the experiment and denote these two loss functions by suffixing -h and -l to the corresponding methods respectively.

3.3 Results

We compare testing performance on testing data to demonstrate the generalization ability of different online learning algorithms. Table 2 show the experimental results for each dataset. The classification results obtained using OMTL algorithm appear quite promising. For all datasets, the proposed OMTL method performs the best in terms of both hinge loss and logistic loss.

As can be observed from the testing performance, the OGD method shows the worst performance with the lowest F measure because it ignores imbalanced problem totally; Given the prior knowledge, cost-sensitive based online algorithms, CSOGD and PAUM, take imbalance ratio into account and outperform the baseline OGD method; The threshold fine-tuning based OFO method achieves better performance than online cost-sensitive algorithms. However, the improvement is not as significant as that of the proposed OMTL method; These results indicate that the proposed OMTL method improves the classification performance with large scale imbalanced data stream.

We also evaluate the changing of online F measure performance in terms of the training data of "w8a". The lelf picture of Fig. 1 demonstrates the online F-measure performance for all competitors; The right one illustrates the online F-measure performance for all models with different threshold of our proposed method.

Table 2. Evalutation of testing performance

Methods	codrna	w8a	mnist1all
	F-measure	F-measure	F-measure
OGD-h	0.6939 ± 0.0020	0.5732 ± 0.0094	0.8335 ± 0.0129
PAUM-h	0.7129 ± 0.0042	0.6615 ± 0.0068	0.8462 ± 0.0085
CSOGD-h	0.7132 ± 0.0035	0.6265 ± 0.0025	0.8462 ± 0.0085
OFO-h	0.7686 ± 0.0027	0.6655 ± 0.0011	0.8544 ± 0.0133
OMTL-h	$\mathbf{0.8148 \pm 0.0008}$	$\mathbf{0.6827 \pm 0.0028}$	$\mathbf{0.8846 \pm 0.0091}$
OGD-l	0.7012 ± 0.0022	0.4468 ± 0.0103	0.8362 ± 0.0164
PAUM-l	0.7129 ± 0.0042	0.6615 ± 0.0068	0.8462 ± 0.0085
CSOGD-l	0.7132 ± 0.0035	0.6265 ± 0.0025	0.8462 ± 0.0085
OFO-l	0.7751 ± 0.0025	0.5126 ± 0.0217	0.8583 ± 0.0166
OMTL-l	$\mathbf{0.8085 \pm 0.0027}$	$\mathbf{0.6713 \pm 0.0043}$	$\mathbf{0.8599 \pm 0.0170}$

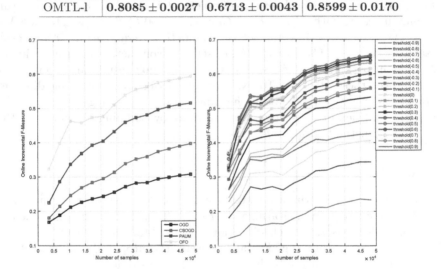

Fig. 1. Online performance for "w8a"

4   Conclusion

This work presents an F-measure incremental updating based online learning method for imbalance data stream. The proposed OMTL algorithm simultaneously trains multiple classifiers with various thresholds, and predicts by selecting a classifier based on the incremental updated F-measure determined by online performance of individual learners. Our promising results from extensive experiments validate the superior efficiency of our algorithm and its ability to improve online imbalanced problem.

Acknowledgments. The authors would like to acknowledge the funding supported by State Key Laboratory of Software Engineering, Computer School, Wuhan University, and research project number is SKLSE-2015-A-06, and also partially supported by the National Natural Science Foundation of China under the Project 61371191.

References

1. Bunkhumpornpat, C., Sinapiromsaran, K., Lursinsap, C.: DBSMOTE: density-based synthetic minority over-sampling technique. Appl. Intell. **36**(3), 664–684 (2012)
2. Busa-Fekete, R., Szörényi, B., Dembczynski, K., Hüllermeier, E.: Online F-measure optimization. In: Advances in Neural Information Processing Systems, pp. 595–603 (2015)
3. Elkan, C.: The foundations of cost-sensitive learning. In: International Joint Conference on Artificial Intelligence, vol. 17, pp. 973–978. Lawrence Erlbaum Associates Ltd. (2001)
4. Gao, J., Liu, X., Ooi, B.C., Wang, H., Chen, G.: An online cost sensitive decision-making method in crowdsourcing systems. In: Proceedings of the 2013 ACM SIGMOD International Conference on Management of Data, pp. 217–228. ACM (2013)
5. Han, H., Wang, W.-Y., Mao, B.-H.: Borderline-SMOTE: a new over-sampling method in imbalanced data sets learning. In: Huang, D.-S., Zhang, X.-P., Huang, G.-B. (eds.) ICIC 2005. LNCS, vol. 3644, pp. 878–887. Springer, Heidelberg (2005). doi:10.1007/11538059_91
6. He, H., Bai, Y., Garcia, E.A., Li, S.: ADASYN: adaptive synthetic sampling approach for imbalanced learning. In: 2008 IEEE International Joint Conference on Neural Networks (IEEE World Congress on Computational Intelligence), pp. 1322–1328. IEEE (2008)
7. He, H., Garcia, E.A.: Learning from imbalanced data. IEEE Trans. Knowl. Data Eng. **21**(9), 1263–1284 (2009)
8. Li, Y., Zaragoza, H., Herbrich, R., Shawe-Taylor, J., Kandola, J.: The perceptron algorithm with uneven margins. In: ICML, vol. 2, pp. 379–386 (2002)
9. Scott, C.: Surrogate losses and regret bounds for cost-sensitive classification with example-dependent costs. In: Proceedings of the 28th International Conference on Machine Learning (ICML 2011), pp. 153–160 (2011)
10. Stefanowski, J., Wilk, S.: Selective pre-processing of imbalanced data for improving classification performance. In: Song, I.-Y., Eder, J., Nguyen, T.M. (eds.) DaWaK 2008. LNCS, vol. 5182, pp. 283–292. Springer, Heidelberg (2008). doi:10.1007/978-3-540-85836-2_27
11. Sun, Y., Wong, A.K., Kamel, M.S.: Classification of imbalanced data: a review. Int. J. Pattern Recogn. Artif. Intell. **23**(04), 687–719 (2009)
12. Van Hulse, J., Khoshgoftaar, T.M., Napolitano, A.: Experimental perspectives on learning from imbalanced data. In: Proceedings of the 24th International Conference on Machine Learning, pp. 935–942. ACM (2007)
13. Wang, J., Zhao, P., Hoi, S.C.: Cost-sensitive online classification. IEEE Trans. Knowl. Data Eng. **26**(10), 2425–2438 (2014)
14. Ying, Y., Wen, L., Lyu, S.: Stochastic online AUC maximization. In: Advances in Neural Information Processing Systems, pp. 451–459 (2016)
15. Zinkevich, M.: Online convex programming and generalized infinitesimal gradient ascent (2003)

A Comparative Study of Machine Learning Techniques for Automatic Product Categorisation

Chanawee Chavaltada[1], Kitsuchart Pasupa[1(✉)], and David R. Hardoon[2]

[1] Faculty of Information Technology, King Mongkut's Institute of Technology Ladkrabang, Bangkok 10520, Thailand
Kitsuchart@it.kmitl.ac.th
[2] PriceTrolley Pte. Ltd., 573969 Singapore, Singapore

Abstract. The revolution of the digital age has resulted in e-commerce where consumers' shopping is facilitated and flexible such as able to enquire about product availability and get instant response as well as able to search flexibly for products by using specific keywords, hence having an easy and precise search capability along with proper product categorisation through keywords that allow better overall shopping experience. This paper compared the performances of different machine learning techniques on product categorisation in our proposed framework. We measured the performance of each algorithm by an Area Under Receiver Operating Characteristic Curve (AUROC). Furthermore, we also applied Analysis of Variance (ANOVA) to our results to find out whether the differences were significant or not. Naïve Bayes was found to be the most effective algorithm in this investigation.

Keywords: Product classification · Product categorisation · Machine learning

1 Introduction

The revolution of the digital age has resulted in e-commerce and purchasing of goods has shifted from buying at physical stores to buying from virtual outlets via online shopping where consumers are facilitated with shopping ease and flexibility such as having an ability to enquire about product availability and get instant response as well as having a flexibility to search for products using specific keywords while also being able to access a description and perform a call to action. In addition, intelligent search functions can provide consumers some suggested products that are relevant to the search keyword. Therefore, having an easy and precise search capability along with proper product categorisation through keywords allow potential customers to have an overall better shopping experience.

The United Nations Standard Products and Service (UNSPC) is a product and service taxonomy standard that was established according to the United

© Springer International Publishing AG 2017
F. Cong et al. (Eds.): ISNN 2017, Part I, LNCS 10261, pp. 10–17, 2017.
DOI: 10.1007/978-3-319-59072-1_2

Nations' Common Coding System (UNCCS) and the Dun & Bradstreet's Standard Product and Service Codes (SPSC) [1]. Furthermore, common products have also been categorised by product domain experts; however, these categorisation approaches have proven effective only if the number of products is small.

The challenge of large-scale, accurate and automated categorisation has motivated exploration of computerised approaches where machine learning is a natural avenue given a framework for a computer algorithm to learn from a set of data while continuously optimising the categorisation operation to reduce error, time and cost [2]. Supervised learning is a widely used type of machine learning that requires learning from a set of training data in order that the trained model will be efficient before it is used for an actual analysis. In this case, products can be classified into appropriate categories. However, a product dataset is usually represented as a corpus of documents that posseses an a priori text processing challenge to be overcome before a classification model can be developed [3]. Examples of text processing techniques are number removal, punctuation removal, stop word removal, conversion to lowercase, and tokenisation. Then, n-gram model is used for feature extraction, i.e., it counts the frequency of words that are subsequently vectorised for use in text classification [4]. The common classification techniques for document analysis include Naïve Bayes [5], Support Vector Machine [6], Artificial Neural Networks [7], Latent Discriminant [8] Regression and Logistic Regression [9].

In this paper, we focused comparing the performances of multiple machine learning methods on product catogorisation. In our experiment, we collected product name and category data from three online shopping websites. Prior to the classification process, the data were pre-processed with text processing techniques mentioned above and an n-gram model was used in feature extraction. Subsequently, classification models were built from popular techniques including NB, SVM, ANNs, and LR which were described in Sect. 2. The experiment framework is discussed in Sect. 3. In Sect. 4, we presented our results and finally conclude the paper with a discussion and conclusion in Sect. 5.

2   Methodology

In this paper, we followed the overall product classification methodology illustrated in Fig. 1.

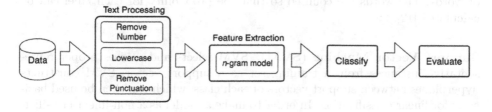

Fig. 1. Product classification processes.

2.1   Text Processing

Document data format is often done by converting the data into a compatible format for each respective process or text processing. This approach manipulates text into utility data. There are a number of text preprocessing techniques such as number removal, punctuation removal, conversion of letters to lowercase, and tokenisation. Usually, they are applied for information retrieval, information extraction and data mining.

2.2   Feature Extraction

A feature is an individual measurable attribute of an occurrence being observed [10]. This step is necessary for building effective algorithms. Effective features are discriminant, independent and informative. For extracting features from documents, count vectorisation is a good method. It counts word frequency. Here, we used an n-gram model as a linguistic probability model for predicting items in the same sequence order as that of a Markov model [11] and extracting features.

2.3   Classification

Classification is a data mining and supervised learning technique with an objective to predict an outcome by learning a statistical model of historical data attributes (also known as training data). Each classification method has different tuning parameters that affect the efficiency of the model.

Naïve Bayes (NB). This technique is based on Bayes' theorem with strong independence assumptions between features. It is usually used for text classification by calculating the probabilities of occurrences of items or posterior probabilities given that an occurrence and the previous occurrence are independent. Then the occurrence with highest probability is chosen. Moreover, some models such as text classification model has multiple labels, so a multinomial model has to be used. This model is used for prediction of frequency of corpus occurrences with an assumption that the length of the document is related to label according to Bayes' theorem. Therefore, each document represents a bag of words. The words are counted so that the probability for each label can be calculated [12].

Support Vector Machine (SVM). SVM selects and utilises proper representative instances from a training set as a support vector. SVM constructs hyperplanes between support vectors of each class, which can then be used basically for linear classification. In order to make a model as a non-linear classifier, a kernel function is applied. Kernel function maps input data in a lower dimensional feature space to a higher dimensional feature space. Some common kernel functions include Polynomial and Radial basis function (RBF) [13].

Artificial Neural Networks (ANNs). ANNs is a model of biological neural structure that receives an input through an axon into a cell body and send an output to the next neuron via a synapse. This process involves a lot of neurons that are connected in parallel and have an ability to learn from a mistake in order to improve themselves by adjusting the weight of each neuron. A neural networks model has three main type of layers. The first layer is an input layer for receiving data and sending them to next layer. The second layer consists of hidden layers that are responsible for computing and improving nodes. The performance and accuracy of a model depend on this the characteristics of this layer such the number of hidden layers and nodes. After the data were processed by the hidden layers, the output layer determines the answer by using an activated function for a specified problem.

Logistic Regression (LR). LR is a regression model where dependent attributes are categorical. Commonly, a regression model is used for analysing an event probability by using an expect value that affects event. There are two types of LR: Binary Logistic Regression and Multinomial Logistic Regression. The differences between both types are in the types of labels which are binary and multinomial, respectively. In the case that the labels are multiple values, we must use the multinomial logistic regression [14].

3 Experimental Framework

3.1 Data Collection

The data have been collected from three online shopping websites. It consisted of product names and categories. Details of each data are explained in Table 1.

Table 1. Details of dataset A, B and C.

Dataset	Product names	Categories
A	5,863	58
B	11,658	89
C	28,355	468

3.2 Data Preprocessing

The collected data were transformed to a structured format. This was done by applying text processing techniques on the product names. Punctuations and numbers were removed from the product names. Then, all of the letters were converted to lowercase. Then, we were able to extract features by using an n-gram model to transform the data into feature vectors for use in the models as shown in Fig. 1. The data were normalised by z-score. The product names were used as input data, but product categories were encoded into numerical data for use as labels, and the labels were used for predicting the target for classifier.

3.3 Experiment Setting

Data were split into two sets. Eighty and twenty percent of the data were used as a training set and a test set, respectively. They were pre-processed and feature extracted by methods explained in Sect. 2.1. Since all algorithms required parameter tuning, five-fold cross validation was applied to find optimal parameters for the best model on training set. The performance was evaluated by Area Under Receiver Operating Characteristic Curve (AUROC) measure that is more suitable for handling imbalance data than accuracy measure is and also more statistically consistent [15]. We evaluated the performances on different sets of features which were {1}-gram, {1, 2}-gram, {1, 2, 3}-gram and {1, 2, 3, 4}-gram. Subsequently, the parameters for each algorithm were varied as follows:

- ANNs: Hidden layers were {1, 2, 3}, and the number of neuron for each layer were {10, 20, 30, · · · , 100}
- SVM: C value was in range {$10^{-4}, 10^{-3}, · · · , 10^5, 10^6$}. We evaluated three types of kernel which are Linear, Polynomial, and Radial Basis Function (RBF). Degrees of Polynomial were in range 1–6 and Gaussian width range was {$10^{-6}, 10^{-5}, · · · , 10^5, 10^6$}.
- LR: Regularisation parameter was in the range of {$10^{-4}, 10^{-3}, · · · , 10^5, 10^6$}.

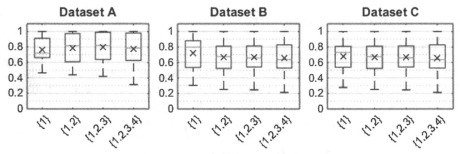

(a) A box plot of four sets of feature.

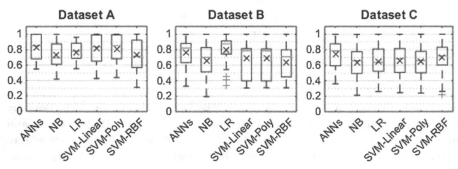

(b) A box plot of six algorithms.

Fig. 2. The box plots show average of AUROC across feature and algorithms for each datasets with 10 runs

Once the optimal parameters had been set, they were used to train a model which was later tested and evaluated on the test set. We ran the experiment 10 times, each with a different random split.

4   Results and Discussions

Figure 2(a) shows the average AUROC for each set of features across all considered classification techniques. Using the set of features $1, 2, 3$-gram gave the best performance for dataset A, while for the dataset B and C, using only the unigram ($n = 1$) set was needed for a good performance. However, from the results for dataset A, B and C, the determination of the best set of features was inconclusive at $p = 0.79, 0.35, 0.96$ respectively with one-way analysis of variance (ANOVA). Respectively, as determined by one-way analysis of variance (ANOVA, a statistical comparison technique that provides a capability to compare differences between means [16]).

Furthermore, we compared the performances of six different techniques on three datasets (averaged across four sets of features) as shown in Fig. 2(b). NB showed the highest average AUROC for sdataset A and C but for dataset B, the highest was LR. It was found that there were interactions between the sets of features and the algorithm used in this framework for every dataset–dataset A and B at $p < 0.01$ and C at $p < 0.05$ by two-way ANOVA. This means that

Table 2. Multiple comparison–it shows mean difference (MD) and its p-value. Bold face indicates statistically significant.

Algorithm 1	Algorithm 2	DataSet A		DataSet B		DataSet C	
		MD	p-value	MD	p-value	MD	p-value
NB	LR	**0.063**	**<0.05**	−0.025	0.889	**0.097**	**<0.01**
NB	ANNs	**0.091**	**<0.01**	**0.114**	**<0.01**	**0.104**	**<0.01**
NB	SVM Linear	0.013	0.990	**0.080**	**<0.01**	**0.083**	**<0.01**
NB	SVM Poly	0.025	0.843	**0.089**	**<0.01**	**0.095**	**<0.01**
NB	SVM RBF	**0.092**	**<0.01**	**0.134**	**<0.01**	0.045	0.086
LR	ANNs	0.029	0.746	**0.138**	**<0.01**	0.007	0.998
LR	SVM Linear	−0.050	0.166	**0.105**	**<0.01**	−0.013	0.967
LR	SVM Poly	−0.038	0.459	**0.106**	**<0.01**	−0.002	0.999
LR	SVM RBF	0.029	0.742	**0.159**	**<0.01**	**−0.052**	**<0.05**
ANNs	SVM Linear	**−0.079**	**<0.01**	−0.034	0.671	−0.021	0.817
ANNs	SVM Poly	**−0.067**	**<0.05**	−0.034	0.722	−0.009	0.995
ANNs	SVM RBF	<0.001	0.999	0.02	0.949	**−0.059**	**<0.01**
SVM Linear	SVM Poly	0.012	0.992	0.020	0.999	0.012	0.981
SVM Linear	SVM RBF	**0.079**	**<0.01**	0.054	0.172	−0.039	0.187
SVM Poly	SVM RBF	**0.067**	**<0.05**	0.052	0.204	**−0.051**	**<0.05**

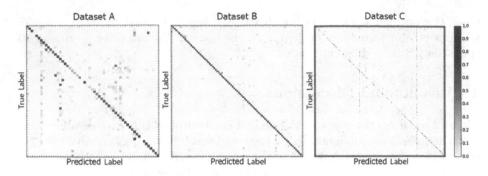

Fig. 3. Confusion matrix of NB on all three datasets.

there was a significant difference between at least one pair of means for each dataset; therefore, we subsequently conducted multiple comparison by two-way ANOVA on each dataset as illustrated in Table 2. Clearly, NB yielded a better performance than those of the others in 3/5 cases for dataset A, 4/5 cases for both dataset B, and C ($p < 0.01$). It can be seen, for dataset B, LR performances were significantly different better in 4/5 cases ($p < 0.01$), but inconclusive when comparing to NB ($p = 0.889$)–NB and LR are comparable in this case. ANNs were found to be worse than NB, SVM-Linear and SVM-Poly ($p < 0.05$) for dataset A, while they were worse than NB and LR for dataset B ($p < 0.01$), and they were worse than NB and SVM-RBF for dataset C ($p < 0.01$). It was inconclusive which algorithm was the worst after all. Moreover, the confusion matrices of NB on all three datasets in Fig. 3 show that it was the best algorithm in this framework.

5   Conclusion

This paper proposes a framework for automatic product categorisation. We evaluated and compared well-known machine learning techniques on three datasets obtained from the online websites and based on AUROC. We have found that the performance of NB was the best-statistically significant. Furthermore, it is inconclusive whether a set of proposed features was the best.

References

1. Ding, Y., Korotkiy, M., Omelayenko, B., Kartseva, V., Zykov, V., Klein, M., Schulten, E., Fensel, D.: GoldenBullet: automated classification of product data in e-commerce. In: Proceedings of the 5th International Conference on Business Information Systems (BIS 2002) (2002)
2. Simon, P.: Too Big to Ignore: The Business Case for Big Data. Wiley, Hoboken (2013)
3. Shankar, S., Lin, I.: Applying machine learning to product categorization. Technical report, Stanford University (2011)

4. Kozareva, Z.: Everyone likes shopping! multi-class product categorization for e-commerce. In: Proceedings of the 2015 Conference of the North American Chapter of the Association for Computational Linguistics: Human Language Technologies, pp. 1329–1333 (2015)
5. Zhang, H., Li, D.: Naïve bayes text classifier. In: Proceedings of the 2007 IEEE International Conference on Granular Computing (GRC 2007), p. 708 (2007)
6. Tong, S., Koller, D.: Support vector machine active learning with applications to text classification. J. Mach. Learn. Res. **2**, 45–66 (2001)
7. Wermter, S.: Neural network agents for learning semantic text classification. Inf. Retr. **3**(2), 87–103 (2000)
8. Wang, Z., Qian, X.: Text categorization based on LDA and SVM. In: 2008 International Conference on Computer Science and Software Engineering, vol. 1, pp. 674–677 (2008)
9. Cheng, W., Hüllermeier, E.: Combining instance-based learning and logistic regression for multilabel classification. Mach. Learn. **76**(2), 211–225 (2009)
10. Bishop, C.: Pattern Recognition and Machine Learning, vol. 128, 1st edn. Springer, New York (2006). pp. 1–58, ISSN 1613-9011
11. Jurafsky, D., Martin, J.H.: Speech and language processing. Int. Ed. **710**, 117–119 (2000)
12. Lewis, D.D.: Naive (Bayes) at forty: the independence assumption in information retrieval. In: Nédellec, C., Rouveirol, C. (eds.) ECML 1998. LNCS, vol. 1398, pp. 4–15. Springer, Heidelberg (1998). doi:10.1007/BFb0026666
13. Suykens, J.A.K., Vandewalle, J.: Least squares support vector machine classifiers. Neural Process. Lett. **9**(3), 293–300 (1999)
14. Yuth, K.: Principle and using logistic regression analysis for research. RMUTSV Res. J. **4**(1), 1–12 (2012)
15. Ling, X.C., Huang, J., Zhang, H.: AUC: a statistically consistent and more discriminating measure than accuracy. In: Proceedings of the 18th International Joint Conference on Artificial Intelligence (IJCAI 2003), vol. 3, pp. 519–524 (2003)
16. Viaene, S., Derrig, R.A., Baesens, B., Dedene, G.: A comparison of state-of-the-art classification techniques for expert automobile insurance claim fraud detection. J. Risk Insur. **69**(3), 373–421 (2002)

Bootstrap Based on Generalized Regression Neural Network for Landslide Displacement for Interval Prediction

Jiejie Chen[1(✉)], Zhigang Zeng[2], and Ping Jiang[3]

[1] College of Computer Science and Technnology,
Hubei Normal University, Huangshi 435002, China
chenjiejie118@gmail.com
[2] School of Automation, Huazhong University of Science and Technology,
Wuhan 430074, China
[3] Computer School, Hubei PolyTechnic University, Huangshi 435002, China

Abstract. A novel interval prediction (PIs) method, called bootstrap based on generalize neural network (Bootstrap-GRNN) for landslide displacement forecasting model is proposed. New algorithm contains B+1 GRNN and then divide two parts. The first part includes B GRNN to compute variance. The second part has one GRNN to get variance of errors. According to the interval prediction formula, we can get the corresponding interval prediction for landslide displacement with real case.

Keywords: Interval prediction · Bootstrap · Generalize neural network · Variance

1 Introduction

Landslide [1–3] as one of kind of complex geological disasters, and brings serious threat to human life and production environment. Landslides have complex causes, various and uncertainty effect factors, and bring very great difficultly to disaster prevention and control of project. Three gorges reservoir area which is located at upper reaches of Yangtze River in China is wide, large scale and has complicated geological conditions. It has frequent geological disasters, especially large-scale landslide widely occurred in geological history. The prediction of landslide disaster which mainly based on forecasting landslide sliding time, so that people can as soon as possible take measures and countermeasures for disaster prevention and reduction. These measures can be considered one of the most effective ways and means of decreasing landslide disaster.

Landslide is a dynamic change process of complex geological mechanics, which has been in disorderly, unsteady, unbalanced, uncertain and random state. And there exists a variety of nonlinear process, and it is difficult to use the traditional linear methods to solve. Therefore, artificial neural networks (ANNs) [4,5] and various nonlinear methods are applied in Landslide displacement. Currently, a large number of approaches have been widely proposed in solving the

© Springer International Publishing AG 2017
F. Cong et al. (Eds.): ISNN 2017, Part I, LNCS 10261, pp. 18–27, 2017.
DOI: 10.1007/978-3-319-59072-1_3

problem of the landslide forecast and prediction [4,5]. Most of these approaches are inclined to forecast specific value about displacement. Then, it is very hard to get exact value and has great deviation from actual value. Prediction interval prone to attain more accurate than the prediction precision. Interval prediction [6–11] can represent the dynamic characteristics of time series and the future development trend. To build interval prediction can get upper and lower bounds of the interval. Specific prediction results exist in the scope of the interval. There are four main methods of interval prediction Delta, Bayesian, Bootstrap, and Lower upper bound estimation (LUBE).

The paper has five sections. The first section is about introduction, methods and materials are listed in Sect. 2. Section 3 shows the analytical method about Bootstrap-GRNN [12]. A real case is proposed to illustrate the availability of our model 4. Finally, Sect. 5 presents conclusion about this paper.

2   Methods and Materials

2.1   Confidence Interval

It is often supposed that targets can be defined as:

$$S_i = Y_i + \epsilon_i \qquad (1)$$

where S_i is the ith measured target (totally n targets). The error with a zero expectation is noise ϵ_i. Then, it compute the value between true regression mean Y_i and measured value S_i. Supposed that errors are identically and independently distributed. Then, it defined

$$S_i - \hat{Y}_i = [Y_i - \hat{Y}_i] + \epsilon_i \qquad (2)$$

Confidence Interval (CIs) [6–11] solve with the uncertainty between the prediction true regression Y_i and $\hat{Y}_i$, and are based on the estimation of characteristics of the probability distribution $P(Y_i|\hat{Y}_i)$.

The measured value S_i and the true regression mean Y_i are statistically independent, the total variance comes from

$$\sigma_i^2 = \sigma_{\hat{Y}_i}^2 + \sigma_{\epsilon_i}^2 \qquad (3)$$

the measure of noise variance is $\sigma_{\epsilon_i}^2$, and $\sigma_{\hat{Y}_i}^2$ is computed from model parameter estimation errors and misspecification.

2.2   Interval Prediction

The coverage probability [6–11] is the important characteristic of PIs. Then, PI coverage probability (PICP) [6–11] is defined as counting the number of target values overlaped by the constructed PIs.

$$\text{PICP} = \frac{1}{N} \sum_{i=1}^{N} \delta_i \qquad (4)$$

where N is called the number of samples in the test set, and the lower bounds (LB) of is L_i and the upper bounds (UP) of is U_i for the ith PI, respectively. Theoretically, PICP need to approach to or larger than the nominal confidence level related to the PIs.

where

$$\delta_i = \begin{cases} 1, \text{ if } t_i \in [L_i, U_i] \\ 0, \quad \text{otherwise} \end{cases} \tag{5}$$

$$\text{NMPIW} = \frac{1}{N\eta} \sum_{i=1}^{N} (U_i - L_i) \tag{6}$$

2.3   Bootstrap

Efron [9–11] introduced the approach bootstrap to compute confidence intervals in the case that conventional method are not valid. Especially, when few data are obtainable, approximate large sample methods are not applicable. Bootstrap repeats resampling of the original sample set, and D_i^* (the index i represented for the i resampling) means B samples with random sample size of N were extracted from sample D_0. When M is large enough, it can provide an approximation for all possible values of the statistic T by repeated sampling from the D_i. Small sample data sets for statistical simulation. It can obtain unknown distribution and unknown parameter distribution, the implementation process was shown in Fig. 1.

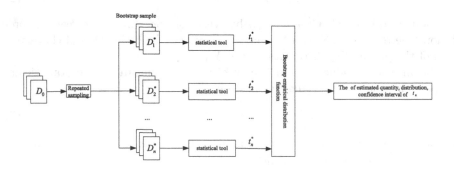

Fig. 1. Bootstrap

2.4   GRNNS

The Specht [12] introduced a variation of the radial basis neural network which named GRNN [12] to perform general (linear or nonlinear) regressions. It concludes the joint probability density function (PDF) of x and y with a training set. The pdf is derived from the data with no preconceptions about its form.

The $f(x, y)$ is the function of the known joint continuous probability density [12], including a vector random variable x and a scalar random variable y, the regression of y on X is defined as:

$$E[y|X] = \frac{\int_{-\infty}^{\infty} y f(X, y) dy}{\int_{-\infty}^{\infty} f(X, y) dy} \tag{7}$$

The estimator $f(X, Y)$ can be calculated, which is based on sample values X_i and Y_i of the random variables x and y. The number of sample of observations is n, and the dimension of the vector variable x is p:

$$\begin{aligned}
f(X, Y) = \ & \frac{1}{(2\pi)^{(p+1)/2} \sigma^{(p+1)}} \frac{1}{n} \\
& \times \sum_{i=1}^{n} \exp\left[-\frac{(X - X_i)^T (X - X_i)}{2\sigma^2}\right] \\
& \times \exp\left[-\frac{(Y - Y_i)^2}{2\sigma^2}\right].
\end{aligned} \tag{8}$$

An architecture of the GRNNS [12] is depicted in Fig. 2.

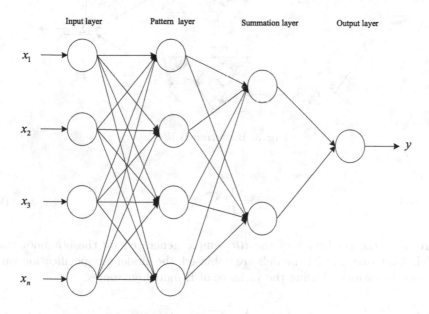

Fig. 2. Schematic diagram of a GRNNS architecture

3   Bootstrap-GRNN

3.1   Confidence Interval

Bootstrap-GRNN contains B+1 GRNN and then divide two parts. The first part includes B GRNN to compute variance $\sigma_{\hat{y}}^2$. The second part has one GRNN to get variance of errors $\sigma_{\epsilon_i}^2$. According to the interval prediction formula, we can get the corresponding interval prediction for landslide displacement.

In Fig. 3, it used Bootstrap to divide the original data into B parts, each part can be defined $D_{b=1}^B$. The estimate variance is decided by specification error. Under this assumption, the true regression $\hat{y}$ is estimated by averaging the point forecasts of $\hat{y}_i^b$ B models. (Each GRNN).

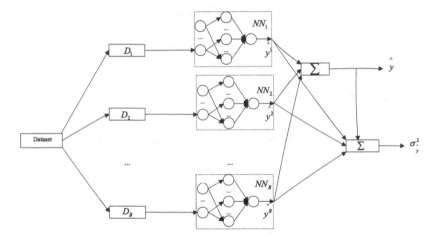

Fig. 3. Bootstrap B NN

$$\hat{y} = \frac{1}{B} \sum_{b=1}^B \hat{y}_i^b \tag{9}$$

where $\hat{y}_i^b$ is the prediction of the ith sample generated by the bth bootstrap model. Assuming that NN models are unbiased, the model misspecification variance can be estimated using the variance of B model outcomes.

$$\sigma_{\hat{y}_i}^2 = \frac{1}{B-1} \sum_{b=1}^B (\hat{y}_i^b - \hat{y}_i)^2 \tag{10}$$

According to two formula, the confidence interval can be defined:

$$\hat{y} \pm Z_{1-\alpha} \sqrt{\sigma_{\hat{y}_i}^2} \tag{11}$$

3.2   Cost Function

The one of character of Bootstrap-GRNN has only one regulation parameter (spread). Meanwhile, the learning of network depends entirely on data sample. And this trait point determines that the network can avoid the influence of the subjective assumption on the prediction results. The larger the value of spread, the more smooth function fitting. However, the value of approximation error will become larger, the more hidden neurons needed and the greater the calculation. On the contrary, the smaller the value of spread leads to more accurate the approximation of the function. The approximation process will not be smooth and present an over adaptation. The key of Bootstrap-GRNN is to find the proper spread. During application process, we will try to calculate different spread to obtain the proper value. Then, the CWC is introduced to solve with this problem.

$$C = \text{NMPIW}(1 + \delta(\text{PICP})e^{-\mu(\text{PICP}-\nu)}) \tag{12}$$

In the training process, $\delta(\text{PICP}) = 1$, in the testing process, $\delta(\text{PICP}) = 1$ is defined as:

$$\delta(\text{PICP}) = \begin{cases} 1, \text{PICP} < \nu \\ 0, \text{PICP} \geq \nu \end{cases} \tag{13}$$

4   Application of Landslide Prediction

4.1   Data

In this paper, we tried to present the Bootstrap-GRNN approach to derive meaningful nonlinear relationships between various parameters of one practical geotechnical problems. All experiments about this paper are based on MAT-LAB 2010b platform. The landslide is uncertainty, stability and complexity, so the formation of the landslide is extremely difficult to know. A case of Baishuihe landslide monitoring point ZG118 in Three Gorges Reservoir area in China is proposed to illustrate the availability of our model.

The important task of selecting the appropriate spread value and the best training samples to train GRNN. So parameter setting about Bootstrap-GRNN has two: decomposition number B and spread. The decomposition number B can decide the number of GRNN. Considering the length of data set and published papers, we choose 10 B in our real case. Then in this paper, it sets 11 GRNN, and 10 is for computing estimate variance, the left one is for computing error variance (Figs. 4 and 5).

4.2   Experiment and Results

Considering some other external factors rainfall and reservoir level, pearson cross correlation coefficients (PCC) and mutual information (MI) [4,5] are adopted to find the potential input variables. All data basis on the general rules and feature of landslide, which are divided into two groups to study. The first group includes

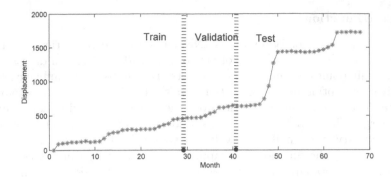

Fig. 4. Displacement

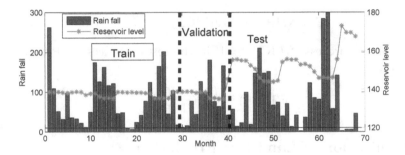

Fig. 5. Rainfall and reservoir

70% to set up the predicting model, and the rest recordings 30% to predict the displacement of landslide. Figures 6, 7, 8 and 9 are about interval prediction and confidence interval prediction based on Bootstrap-GRNN Fig. 6 is shown for training set about interval prediction, the minimum difference about UB and LB is 0.6391, the maximum difference about UB and LB is 1.8838, the mean about UB and LB is 1.3082, the standard deviations is 0.2661, and 11 points have not existed in interval. Figure 7 is shown for confidence interval, the minimum difference about UB and LB is 0.0107, the maximum difference about UB and LB is 0.6241, the mean about UB and LB is 0.3119, and 15 points have not existed in interval.

Figure 8 is shown for Prediction set about interval prediction, the minimum difference about UB and LB is 0.4927, the maximum difference about UB and LB is 2.7632, the mean about UB and LB is 1.5417, the standard deviations is 0.6965, and 5 points have not existed in interval. Figure 9 is shown for confidence interval, the minimum difference about UB and LB is 0.0480, the maximum difference about UB and LB is 1.7113, the mean about UB and LB is 0.3204, and 5 points have not existed in interval. And, the PICP and NMPIW of Training set are 68.06 and 65.41. And, the PICP and NMPIW of Training set are 75 and 77.09. In fact, the bigger PICP and smaller NMPIW are best results for PIs. So, we should make some ways to get lager PICP and smaller NMPIW.

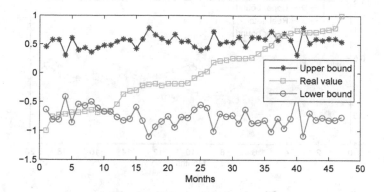

Fig. 6. PIs on training test

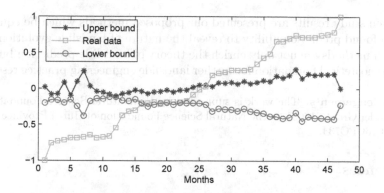

Fig. 7. CIs on training test

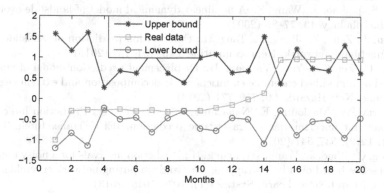

Fig. 8. PIs on prediction test

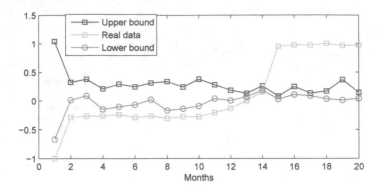

Fig. 9. CIs on prediction test

5   Conclusion

The given study results are presented our proposed methods which be equipped with profound prediction ability to reveal the nature of landslide evolution. Our proposed methods are not only enrich the theory of landslide prediction but also provide theoretical foundation for further landslide engineering practice research.

Acknowledgements. The work is supported by the Natural Science Foundation of China under Grant 61603129, the Natural Science Foundation of Hubei Province under Grant 2016CFC734.

References

1. Cubito, A., Ferrara, V., Pappalardo, G.: Landslide hazard in the Nebrodi Mountains (northeastern Sicily). Geomorphology **66**(1–4), 359–372 (2005)
2. Qin, S., Jiao, J., Wang, S.: The predictable time scale of landslides. Bull. Eng. Geol. Environ. **59**(4), 307–312 (2001)
3. Qin, S.Q., Jiao, J., Wang, S.: A nonlinear dynamical model of landslide evolution. Geomorphology **43**, 77–85 (2002)
4. Chen, J., Zeng, Z., Jiang, P., Tang, H.: Deformation prediction of landslide based on functional networtk. Neurocomputing **149**, 151–157 (2015)
5. Lian, C., Zeng, Z., Yao, W., Tang, H.: Displacement prediction model of landslide based on a modified ensemble empirical mode decomposition and extreme learning machine. Nat. Hazards **66**, 759–771 (2013)
6. Khosravi, A., Mazloumi, E., Nahavandi, S., Creighton, D.: Prediction intervals to account for uncertainties in travel time prediction. IEEE Trans. Intell. Transp. Syst. **12**(2), 537–547 (2011)
7. Sheng, C., Zhao, J., Wang, W., Leung, H.: Prediction intervals for a noisy nonlinear time series based on a bootstrapping reservoir computing network ensemble. IEEE Trans. Neural Netw. Learn. Syst. **24**(7), 1036–1048 (2013)
8. Khosravi, A., Nahavandi, S., Creighton, D., Atiya, A.: Lower upper bound estimation method for construction of neural network-based prediction intervals. IEEE Trans. Neural Netw. Learn. Syst. **24**(3), 337–346 (2011)

9. Efron, B.: Bootstrap methods: another look at the jackknife. Ann. Stat. **7**, 1–26 (1979)
10. Efron, B., Tibshirani, R.: Bootstrap methods for standard errors, confidence intervals, and other measures of statixtical accuracy. Stat. Sci. **1**(1), 54–77 (1986)
11. Efron, B.: The bootstrap and Markov chain Monte Carlo (2011)
12. Paojcic, J., Ibric, S., Djuric, Z., Milica, J., Owen, I.C.: An investigation into the usefulness of generalized regression neural network analysis in the developmen of level A in vitro-in vivo correlation. Eur. J. Pharm. Sci. **30**(3), 264–272 (2007)

Multi-task Learning with Cartesian Product-Based Multi-objective Combination for Dangerous Object Detection

Yaran Chen[1,2] and Dongbin Zhao[1,2(✉)]

[1] The State Key Laboratory of Management and Control for Complex Systems,
Institute of Automation, Chinese Academy of Sciences, Beijing 100190, China
[2] The University of Chinese Academy of Sciences, Beijing, China
dongbin.zhao@ia.ac.cn

Abstract. Autonomous driving has caused extensively attention of academia and industry. Vision-based dangerous object detection is a crucial technology of autonomous driving which detects object and assesses its danger with distance to warn drivers. Previous vision-based dangerous object detections apply two independent models to deal with object detection and distance prediction, respectively. In this paper, we show that object detection and distance prediction have visual relationship, and they can be improved by exploiting the relationship. We jointly optimize object detection and distance prediction with a novel multi-task learning (MTL) model for using the relationship. In contrast to traditional MTL which uses linear multi-task combination strategy, we propose a Cartesian product-based multi-target combination strategy for MTL to consider the dependent among tasks. The proposed novel MTL method outperforms than the traditional MTL and single task methods by a series of experiments.

Keywords: Dangerous object detection · Multi-task learning and convolutional neural network

1 Introduction

Nowadays, more and more people pay attention to driving safety. Dangerous object detection is an effective measure for improving driving safety which has been widely studied for several decades by many researchers. However, it is still challenging to accurately and promptly detect dangerous object.

Dangerous object detection aims to identify the potentially dangerous vehicles and pedestrians for drivers. According to input signals, dangerous object detection methods usually are divided into: general sensor-based methods and vision-based methods. Sensor-based methods mainly apply lasers and radars to sense surroundings and detect dangerous object. They have been widely used, thanks to the

D. Zhao—This work is supported by National Natural Science Foundation of China (NSFC) under Grants 61573353 and 61533017, and the National Key Research and Development Plan under Grant No. 2016YFB0101000.

F. Cong et al. (Eds.): ISNN 2017, Part I, LNCS 10261, pp. 28–35, 2017.
DOI: 10.1007/978-3-319-59072-1_4

great environmental perception capability. Autonomous cars such as Google Car and Baidu Car [1], use a rotating light detection and ranging (LIDAR) scanners to obtain the environment information and warn drivers about dangerous objects. However, these lidar sensors are too expensive to apply in a large scale. Compared with sensor-based methods, vision-based dangerous object detection is low cost and captures more traffic information, such as object distance, object categories and traffic signs [2]. In previous work, vision-based methods are usually formulated as an object detection problem and a distance prediction problem, which are dealt with using two independent models. The typical methods of object detection including faster R-CNN [5], and SSD (Single Shot MultiBox Detector) [6], can be used for object detection in autonomous driving. Object distance is generally measured by a RGB-D cameras LIDAR or radar [9].

In fact, vision-based object detection and distance prediction present prominent visual relationship. The objects far from the camera usually look small and cover few pixels of an image, while the closer ones are generally distributed in the near field of view and cover more pixels, shown as Fig. 1(a) and (b). In addition, Fig. 1(c) and (d) show that objects taken from different camera angles present different poses. Obviously, the visual relationship is very worthy to be exploited for detecting dangerous objects. However, it is much ignored in previous work which deals with object detection and distance prediction using two independent models. Therefore, simultaneously optimizing object detection and distance prediction in one model will probably improve the performance of dangerous object detection.

Fig. 1. The cars with different distances and poses

Multi-task learning (MTL) is a well-known method for simultaneously optimizing multiple tasks. MTL exploits shared information among multiple tasks to improve the performance of each other [3]. MTL has been widely applied in computer vision community: such as action recognition [14], pose estimation [10], face detection [11], facial landmark localization [12], and achieved great successes. MTL method generally linearly combines the objectives of multiple tasks to exploit the shared information and jointly optimizes the related tasks. However, it much ignores the correlations of multiple tasks.

In this paper, we propose a novel MTL method based on CNN to jointly optimize object detection and distance prediction. In order to facilitate distance prediction, it is formulated as a classification problem, through discretizing continuous distance. In the proposed MTL method, we propose a joint optimization objective according to the Cartesian product of object classes and distance categories. We prove that the proposed Cartesian product-based multi-task combination strategy outperforms the linear multi-task combination strategy in mathematics and experiments.

Our contributions are shown as follows. First, we use the MTL mechanism to dangerous object detection for exploiting the visual relationship between object detection and object distance prediction for the first time. Second, we propose a novel multi-task combination strategy based on the Cartesian product, and prove it outperforms the linearly combination strategy.

2   Multi-task Learning

Dangerous object detection deals with object detection and distance prediction. Object detection is usually expressed as a classification task. Namely we detect objects by classifying the proposed regions of images. It is difficult to accurately predicting continuous distance owing to the non-linear variation of the sight distance. Therefore, the distance prediction task is transformed into a classification problem. MTL is a popular technique for dealing with related multiple tasks. In this paper, we propose a novel MTL to jointly optimize the two classification problems by shared information.

2.1   Linear Multi-task Combination

Traditional MTL methods generally optimize multiple tasks by a linear multi-task combination strategy (LC-MTL). Namely the loss is a weighted linear combination of the multiple objective functions [12] shown as:

$$L_{c+d} = \alpha \cdot L_c + (1 - \alpha) \cdot L_d, \tag{1}$$

where L_c and L_d are the objective functions of the object detection task C and distance prediction task D, respectively. And α specifies the relative importance of each task and can be experimentally chosen.

Due to the powerful ability of representation learning, CNN has been widely used in multi-task learning, especially for the classification task. For dangerous object detection, through shared model parameters, CNN can jointly model the object detection C and distance prediction D. We use y_c to denote a class of objects, and $y_c \in \{c_1, c_2, ..., c_p\}_{1 \times p}$ where p represents the number of object classes. Similarly, y_d denotes a category of object distance, where $y_d \in \{d_1, d_2, ..., d_p\}_{1 \times q}$ and q is the number of object distance categories. For a given image $\mathbf{x} \in \Re_+^{m \times n}$, CNN simultaneously computes the probabilities of object recognition and distance classification: $p(y_c = c_i | \mathbf{x})$ the probability of the image $\mathbf{x}$ belonging to the i-th class of object and $p(y_d = d_j|\mathbf{x})$ the probability of the image $\mathbf{x}$ belonging to the j-th class of object distance.

A typical objective function of the classification with multiple categories is the cross entropy loss:

$$L_c = y_c \cdot log(p(y_c = c_i|\mathbf{x})). \tag{2}$$

Similarly, we get $L_d = y_d \cdot log(p(y_d = d_j|\mathbf{x}))$. Then the loss of the MTL (Eq. (1)) can be rewritten as:

$$L_{c+d} = y_c \cdot log(p(y_d|\mathbf{x})) + y_d \cdot log(p(y_d|\mathbf{x})), \tag{3}$$

where we ignore the constant α for simplification.

Through the MTL with the linear multi-task combination strategy, CNN can exploit the shared information for the related tasks from input images. However, it much ignores the dependence among multiple targets.

2.2   Cartesian Product-Based Multi-task Combination

To exploit the dependence among related targets, we propose a Cartesian product-based multi-task combination strategy (CP-MTL) to jointly optimize object detection and distance prediction. We denote the combined task based on the Cartesian product as $M = C \otimes D$, where $\otimes$ represents the Cartesian product operator. Concretely, we use $y_{c \otimes d} = y_c \otimes y_d$ as a category of M and $y_{c \otimes d} \in \{c_1 d_1, c_1 d_2, ...c_1 d_q, ..., c_i d_j, ..., c_p d_q\}_{1 \times pq}$, where pq is the number of the combined task category.

Then, the loss function of M is formulated as:

$$L_{c \otimes d} = y_{c \otimes d} \cdot log(p(y_{c \otimes d} = c_i d_j|\mathbf{x})). \tag{4}$$

Through taking the Cartesian product operator into Eq. (4), we can obtain:

$$l_{c \otimes d} = c_1 d_1 \cdot log(p(y_{c \otimes d} = c_1 d_1|\mathbf{x})) + c_1 d_2 \cdot log(p(y_{c \otimes d} = c_1 d_2|\mathbf{x})) + \\ \cdots + c_p d_q \cdot log(p(y_{c \otimes d} = c_p d_q|\mathbf{x})). \tag{5}$$

Equation (5) is the sum of pq entries, and each one contains a probability $p(y_{c \otimes d} = c_i d_j|\mathbf{x})$. It means that the image $\mathbf{x}$ belongs to c_i of task C and d_j of the task D. If the task C and D are completely independent, we can obtain:

$$p(y_{c \otimes d} = c_i d_j|\mathbf{x}) = p(y_c = c_i|\mathbf{x}) \cdot p(y_d = d_j|\mathbf{x}). \tag{6}$$

Then we take Eq. (6) into Eq. (5) and deduce that:

$$L_{c \otimes d} = L_c + L_d = L_{c+d}. \tag{7}$$

Compared Eqs. (7) and (3), we prove that if the two tasks are independent, the loss function of the traditional LC-MTL method is equal to the loss function of the proposed CP-MTL method. Otherwise, the CP-MTL method can exploit the dependency between two tasks, which is ignored by LC-MTL method. For dangerous object detection, the object detection task and object distance classification task are probably not independent, which may be more suitable for being modeled by the proposed CP-MTL model.

3   CP-MTL SSD Method

Dangerous object detection consists of object detection and distance prediction. Owing to the strong capability of learning representation, CNN-based object detection methods have achieved satisfactory performance. SSD is one of the art-of-the-state CNN-based object detection methods. It directly predicts object bounding boxes and object classes by sharing convolutional features, resulting a short detection time and high accurate. In this paper, we incorporate the proposed CP-MTL (Cartesian product-based combination multi-target) into the optimization objective of SSD to simultaneously optimize the object detection and distance classification tasks.

3.1   Model Architecture

Figure 2 shows the structure of the proposed CP-MTL SSD Method. It consists of multiple hierarchical convolutional layers, some default bounding boxes with different aspect ratios, and a number of detections. By the convolution operation, the hierarchical convolution layers can produce a lot of feature maps of different scales and resolutions for an input image. There are some default bounding boxes on these feature maps. For one default bounding box, the following detection consists of a full-connected classification layer and a regression layer, to regress the bounding box and classify the object category simultaneously. Due to the larger number of default bounding boxes, the model can produce a lot of detections of boxes. Through non-maximum suppression [8], the model will predicts the final boxes.

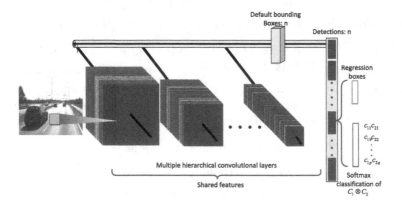

Fig. 2. The architecture of the proposed CNNVA.

CP-MTL is a variant of SSD. Although they seem similar, there is an essential difference between them. Namely CP-MTL optimizes the Cartesian product-based combination targets of object recognition and object distance classification, while SSD just only optimizes the target of object recognition.

3.2   Cartesian Product-Based Combination Targets

We propose a Cartesian product-based combination of object detection task and distance classification task to simultaneously optimize object detection and object distance prediction. Based on the sizes and shapes of objects, we classify objects into three categories: cars, vans and pedestrians, denoted as $\{c_1, c_2, c_3\}$. Due to the relationship between the distance and the object distance, we consider the distance category task from two dimensions: the vertical distance and the horizontal distance, shown as Fig. 3.

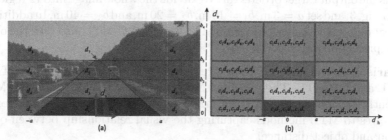

Fig. 3. (a) The image geographic division according to the distance and the visual angle. (b) The categories of the Cartesian product-based combination target, where (a) is mapped to the two dimensional plane (b) (Color figure online)

Figure 3(a) shows that the space is parted into 12 regions and 8 categories denoted as $\{d_1, d_2, \cdots, d_8\}$, due to the symmetry of vehicles. And the red one denotes the shortest vertical distance and the most dangerous category, followed by the yellow one, the green one, and the blue one. In Fig. 3(b), each region is a distance category and contains all the categories $\{c_1, c_2, c_3\}$ of C. Recognizing objects during a given distance category is much easier than recognizing them at all the range of distance.

4   Experiment

In this section, we comprehensively evaluate the proposed CP-MTL model on dangerous object detection task by comparing the proposed CP-MTL with the single task learning model (SSD) and the LC-MTL method with the linear multi-task combination strategy.

Dataset: KITTI dataset [4] contains more than 40,000 images which are collected by a car driving in European cities. About 16,000 images contain information of object positions. In the experiments, we randomly divide the 16,000 images into 3 parts: training set, testing set and validating set. Among them, the training set contains 12,000 images, the testing set contains 3000 images and the validating set contains 1000 images. All experimental configures are experimentally chosen according to the performances on the validating set.

Evaluation Metrics: In object detection, a common evaluation metrics is the average precision (AP). AP measures the comprehensive performance, including the recall rate and precision rate of object detection. The mAP is the mean value of the APs of different object categories.

Experimental Setup: In this study, we take SSD as the baseline model. It has 18 convolutional layers and 5 detectors. The early 13 layers are initialized by the Oxford VGG [7,13], and others are randomly initialized. There are five bounding boxes with different aspect ratios ($\{1, 2, 3, \frac{1}{2}, \frac{1}{3}\}$) at each position of 10-th, 15-th, 16-th, 17-th, and 18-th convolutional feature layers. The detection with multiple shapes, resolutions and scales, can deal with various objects with different shapes and sizes. The proposed models, whether the CP-MTL or the LC-MTL, are the variants of SSD. They have the same network architecture and configures with SSD. But the key difference is the output target of detector. We divide the whole image into 12 regions as shown in Fig. 3, and set $a = 3\,\text{m}, b_1 = 10\,\text{m}, b_2 = 20\,\text{m}$, and $b_3 = 40\,\text{m}$. In addition to the object detection, the proposed CP-MTL and the LC-MTL also take the object distance prediction into account.

Comparison Experiments: Table 1 reports the detection performance of SSD, LC-MTL and CP-MTL. The proposed MTL methods (LC-MTL and CP-MTL) consistently outperform the SSD on mAP and APs of each object category. It mainly owes to MTL methods capturing the visual relationship between object detection and object distance.

Compared with LC-MTL, the CP-MTL yields significant performance improvements in the mAP and APs of all object categories. In a sense, it is verified that the proposed the Cartesian product-based multi-task combination strategy outperforms the linear multi-task combination strategy. At the same time, we also note that the Cartesian product-based multi-task combination strategy increases the difficulty of multi-task learning due to the more detailed classification categories. Therefore, the proposed CP-MTL may require more data to be trained. Finally, we exhibit an example of real-time dangerous object detection on a video. Figure 4 shows four snapshots of the video at $t = 1\,\text{s}$, $t = 10\,\text{s}$, $t = 20\,\text{s}$ and $t = 30\,\text{s}$, respectively. Compared with other object detection systems, the proposed CP-MTL not only bounds the object in an image but also gives its danger level according to the predicted object distance, shown in Fig. 4. Moreover, the proposed CP-MTL based on a fast detection algorithm SSD can meet the real-time requirements of practical applications.

Table 1. The detection results with CP-MTL, LC-MTL and SSD

Method	mAP	AP (Cars)	AP (Pedestrians)	AP (Vans)
SSD	0.8104	0.8779	0.6741	0.8790
LC-MTL	0.8331	0.8933	0.8945	0.7113
CP-MTL	0.8405	0.8945	0.8980	0.7292

Fig. 4. Snapshots from video detection with CP-MTL model

5 Conclusion

We propose the CP-MTL algorithm for dangerous object detection in autonomous driving. Through Cartesian product-based multiple objectives combination, CP-MTL can simultaneously optimize object detection and object distance prediction to exploit the relationship between them. We mathematically prove that the proposed CP-MTL outperforms LC-MTL, when the two tasks are not independent. Also, we carry out systematic experiments to verify that the proposed method outperforms the state-of-art SSD object detection method and the traditional MTL method.

References

1. Bruch, M.: Velodyne HDL-64E lidar for unmanned surface vehicle obstacle detection. In: Proceedings of SPIE - The International Society for Optical Engineering, Florida, 05 April 2010
2. Chen, X., Kundu, K., Zhang, Z., Ma, H., Fidler, S., Urtasun, R.: Monocular 3D object detection for autonomous driving. In: The IEEE Conference on Computer Vision and Pattern Recognition (CVPR), June 2016
3. Evgeniou, A., Pontil, M.: Multi-task feature learning. Adv. Neural Inf. Process. Syst. **19**, 41 (2007)
4. Geiger, A., Lenz, P., Urtasun, R.: Are we ready for autonomous driving? The KITTI vision benchmark suite. In: Conference on Computer Vision and Pattern Recognition (CVPR) (2012)
5. Girshick, R.: Fast R-CNN. In: Proceedings of the IEEE Conference on Computer Vision, pp. 1440–1448 (2015)
6. Liu, W., Anguelov, D., Erhan, D., Szegedy, C., Reed, S., Fu, C.-Y., Berg, A.C.: SSD: single shot multibox detector. In: Leibe, B., Matas, J., Sebe, N., Welling, M. (eds.) ECCV 2016. LNCS, vol. 9905, pp. 21–37. Springer, Cham (2016). doi:10. 1007/978-3-319-46448-0_2
7. Lv, L., Zhao, D., Deng, Q.: A semi-supervised predictive sparse decomposition based on the task-driven dictionary learning. Cogn. Comput. (2016). doi:10.1007/ s12559-016-9438-0
8. Neubeck, A., Gool, L.V.: Efficient non-maximum suppression. In: International Conference on Pattern Recognition, pp. 850–855 (2006)
9. Xia, Y., Wang, C., Shi, X., Zhang, L.: Vehicles overtaking detection using RGB-D data. Sign. Proces. **112**, 98–109 (2015)
10. Yim, J., Jung, H., Yoo, B.I., Choi, C.: Rotating your face using multi-task deep neural network. In: Computer Vision and Pattern Recognition, pp. 676–684 (2015)
11. Zhang, C., Zhang, Z.: Improving multiview face detection with multi-task deep convolutional neural networks. In: IEEE Winter Conference on Applications of Computer Vision, pp. 1036–1041 (2014)
12. Zhang, Z., Luo, P., Chen, C.L., Tang, X.: Facial landmark detection by deep multi-task learning. In: European Conference on Computer Vision, pp. 94–108 (2014)
13. Zhao, D., Chen, Y., Lv, L.: Deep reinforcement learning with visual attention for vehicle classification. IEEE Trans. Cogn. Dev. Syst. (2016). doi:10.1109/TCDS. 2016.2614675
14. Zhou, Q., Wang, G., Jia, K., Zhao, Q.: Learning to share latent tasks for action recognition. In: IEEE International Conference on Computer Vision, pp. 2264–2271 (2013)

Collaborative Response Content Recommendation for Customer Service Agents

Cuihua Ma[1], Ping Guo[1,2(✉)], Xin Xin[1], Xiaoyu Ma[3], Yanjie Liang[3],
Shaomin Xing[3], Li Li[3], and Shaozhuang Liu[3]

[1] School of Computer Science and Technology,
Beijing Institute of Technology, Beijing 100081, China
{chma,xxin}@bit.edu.cn
[2] Laboratory of Graphics and Pattern Recognition,
Beijing Normal University, Beijing 100875, China
pguo@ieee.org
[3] Beijing Easemob Technology Co., Ltd., Beijing 100086, China
{jma,liangyj,xingshaomin,lili,stliu}@easemob.com

Abstract. The rapid development of artificial intelligence (AI) has
motivated extensive research on dialog system. Using dialog system to
automatize customer service is a common practice in many business
fields. In this paper, we investigate a novel task to recommend response
for customer service agents of each shop. A major challenge is the problem of data insufficiency for each shop. Meanwhile, we want to keep
the personalized information for shops with very different commodities.
To deal with such problems, we propose a LSTM (Long Short-Term
Memory) Neuron Tensor Network architecture to encode the common
features of all shops' data and model the personalized features of each
shop. Extensive experiments demonstrate that our method outperforms
four baseline methods evaluated by recall metric.

Keywords: Response recommendation · Dialogue system · Neural network

1 Introduction

The rapid progress of online transaction makes people's lives more convenient,
which requires many questions about commodity to be answered for customers.
This situation forces the fast development of customer service. The traditional
customer service system, such as airline booking [1] and restaurant reservation
[2], is slot-filling representation, while the research of response recommendation
for customer service is a novel application, which will relieve the workload of
customer service agents. And Fig. 1 represents our work that is aimed at recom-
mending response for customer service agents.

There is a challenge of data insufficiency for each shop in our dataset. The
largest number of dialogs from one shop is less than six thousand. In addition,
each shop has its own personalized features. How to collaboratively represent the

© Springer International Publishing AG 2017
F. Cong et al. (Eds.): ISNN 2017, Part I, LNCS 10261, pp. 36–43, 2017.
DOI: 10.1007/978-3-319-59072-1_5

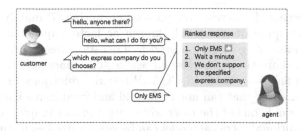

Fig. 1. The business scenario.

common features of all shops and keep the personalized features of each shop at the same time would be a problem worth solving, then a more targeted response would be recommended. So we need to find a personalized recommendation method which can resolve the problem of data insufficiency.

In this paper, we propose an effective method to address the aforementioned issues. Firstly, the common features of all shops' data are used to train a ranking model. Then, the personalized features are added to the model to recommend response with personalized information of specified shop. Finally, we will evaluate the effectiveness of our personalized recommendation method by calculating the recall of every shop. In our work, there are three primary contributions:

(1) We investigate a novel task – recommend response content for customer service agents. It settles the problem of data insufficiency and personalized response recommendation.
(2) From the aspect of model, we propose a LSTM Neuron Tensor Network (LNTN), which is a general architecture and need not prior knowledge and complicated syntactic analysis.
(3) The experimental results show that our recommendation method is effective. Especially, on average, 10@1 recall of our personalized method has a 6% obvious improvement compared with that of LSTM model with only 200 training samples.

2   Related Works

Related applications include:

Recommender System: Generally, recommender system recommends related items for users based on their rating information and usage frequency [3] on items, or according to relationships [4] among similar users. However, we recommend response for customer service according to the relevancy of the response to the question, rather than user-item rating information or frequency data.

Dialogue System: Normally, dialogue system [5] is realized by dialogue management [6], who treats dialogue act (DA) as the input to represent sentence. This type of data is represented using slot-filling. However, our data with the problem of data insufficiency isn't represented by DA.

Question Answering: Researches on question answering [7] mainly focus on the fact-based direct question commonly known as factoid question. Community based question answering (CQA), a branch of question answering, is a platform where users can share their expertise for askers. The characteristics of the data we used are similar with those in CQA. While the questions in our data are mostly short sentences that can not be factoid and need immediate response.

Retrieval model is one of the most widely used model in question answering system. The traditional retrieval model can be summarized as features and classifier, and mainly used in early CQA [8]. Features in this method are extracted according to rules of manual definition, but lack of good scalability. Then with the development of neural network, learning-based deep feature extraction comes into being. Especially for LSTM-based model [9], it is popular due to its capability of long term dependency memory. However, these methods cannot tackle personalized recommendation problem. In this paper, we will study how to personalize response recommendation for customer service.

3   Recommendation Framework

3.1   Problem Definition

The major task studied in this paper is to recommend response content for each shop's customer service agents. The dialogue system records the detailed chatting information between agents and customers, including shop ID, dialogue content etc. Our goal is to train a ranking model using these data. The framework of our customer service system is shown in Fig. 2. Suppose we have a dictionary $D = (w_{i=1}^{M} : e_{i=1}^{M})$, where w_i denotes the ith word and e_i denotes the embedding of ith word. M indicates the dictionary size. Then the segmentation of context can be transformed to the embedding representation by extracting corresponding word embedding from the dictionary. The model can use these embedding representation to calculate the relevancy of a context-response pair. The result would be used to recommend response.

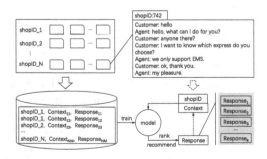

Fig. 2. Framework of our system.

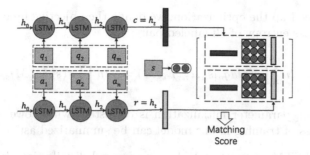

Fig. 3. Architecture of our LSTM Neuron Tensor Network (LNTN)

3.2 Tensor Network Model

Figure 3 shows a visualization of our LSTM Neuron Tensor Network (LNTN). Define $f(x) = \sigma(x)$, $g(x) = tanh(x)$, the LSTM architecture used in this paper is defined by the following equations:

$$inputgate : i_t = f(W_{xi}x_t + W_{hi}h_{t-1} + W_{ci}c_{t-1} + b_i)$$
$$forgetgate : f_t = f(W_{xf}x_t + W_{hf}h_{t-1} + W_{cf}c_{t-1} + b_f) \tag{1}$$
$$cell : \hat{c}_t = W_{xc}x_t + W_{hc}h_{t-1} + b_c$$
$$c_t = f_t \odot c_{t-1} + i_t \odot g(\hat{c}_t)$$
$$outputgate : o_t = f(W_{xo}x_t + W_{ho}h_{t-1} + W_{co}c_t + b_o)$$
$$hiddenstate : h_t = o_t \odot g(c_t)$$

Let matrixes q, a denote the embedding representation of the context sentence and response sentence respectively, vectors q_j, a_j indicate word embedding of the jth word in context and response respectively. Take q as the input of LSTM model, after a series of gate operations, then the final output of the model is sentence vector c for context q. The same applies to a to get r.

The traditional ways are to calculate their cosine distance. However, we need model the personalized features to measure their relevancy. A tensor is a geometric object that can be represented as a multi-dimensional array of numerical values. Following the Neural Tensor Network (NTN) [10] and Convolutional Neural Tensor Network (CNTN) [11], we place a tensor layer on top of the two LSTMs to model the relevancy of context and its response with its shop's personalized features.

3.3 Algorithm

We can optimize the model by minimizing the error between the observed value and the calculated relevance value. The cost function for our model is:

$$min\frac{1}{2}\sum_{i}^{I}\sum_{j}^{J}\sum_{k}^{K}(y_{ijk} - f_{ijk})^2 \tag{2}$$

In order to speed up the optimization, we define the objective function by minimizing the cross entropy of all labeled pairs:

$$L = -ln \prod_n^N p(flag_n|s_n, c_n, r_n, M) = -ln \prod_n^N \sigma(s_n^T(c_n^T M r_n)) \tag{3}$$

For training, parameter initialization is consistent with Lowe et al.'s work [9]. The process of training in our model can be summarized as:

(1) Initialize LSTM parameters W,b using normal distribution. Initialize personalized features s_i of shop i and core tensor M using a uniform distribution with values between -0.01 and 0.01
(2) Generate c_j, r_k of context j and response k from LSTM model respectively. Calculate relevancy y_{ijk} of context j from shop i and response k with outcome $y_{ijk} = \sum_i^I \sum_j^J \sum_k^K M_{ijk} s_{ie} c_{jl} r_{kl}$ for $e = 1, \cdots, d, l = 1, \cdots, h$, where d, h are dimensions. Generate observed value f_{ijk}
(3) Modify s_i and M based on gradient descent
(4) Update the neural network parameters based on backpropagation through time to reduce L.
(5) Iterate steps (2) to (4) until convergence.

3.4   Complexity Analysis

The number of parameters in LSTM is 600 k. The number of personal parameter is 10 for each shop which could be learned with a few data in the target shop. On the real-world dataset with more than 1300 k, the experiments are practical.

The main computation of gradient methods is evaluating the object function L and its gradients against variables. So the computation complexity is $O(N_F uv)$, where N_F is the number of training data, u is time of updating variables and v is the number of iterations. Since our algorithm will converge after 4 to 11 iterations, this indicates that the computational time of our method is linear with the number of training data. This complexity analysis shows that our proposed approach is very efficient.

4   Experimental Analysis

In this section, we conduct several experiments to analyze the recommendation quality of our proposed method.

4.1   Dataset

The primary source of data for this work is the multi-turn (3 is minimum) dialogue records for one month in customer service dialogue system. Each line of the data source may contains a lot of information. We only keep those entries which include shop information, contexts and responses. In experiments, we adopt 133 k dialogues for positive samples from 81 shops. In order to enrich training data, we require the number of dialogues for each shop is at least 1000. Similar to the work in [9], the data is processed to *(shopID, context, response, flag)* for test easily. To improve the robustness of the system, we consider the case of one positive example and nine negative examples.

4.2 Evaluation Metric and Baselines

We use metric Recall to evaluate the performance of the prediction. The recall is defined as: $Recall = \frac{TP}{TP+FN}$, where TP denotes the number of predicted positive examples who are actual positive examples, FN denotes the number of predicted negative examples who are actual positive examples. From the definition, we can see that a larger Recall value means a better performance.

A family of metrics used in our task is Recall $rc@k$. The agents are asked to select the k most likely responses from rc responses, and it is correct if the true response is among these k candidates.

We compare our framework with the following baselines:

1. TFIDF: Prediction is calculated by using cosine similarity of TFIDF between the context and response.
2. RNN: Prediction is calculated by using common features extracted through RNN without personalized features.
3. LSTM: Prediction is calculated by using common features extracted through LSTM without personalized features.
4. RNTN: Prediction is calculated by the Recursive Neuron Tensor Network (RNTN) with common features extracted through RNN and personalized features modeled by tensor.

Table 1. Average recall of all 81 shops

	TFIDF	RNN	LSTM	RNTN	LNTN
2@1	0.62	0.55	0.77	0.70	0.79
5@1	0.37	0.24	0.53	0.40	0.55
5@2	0.50	0.45	0.75	0.65	0.77
10@1	0.27	0.13	0.38	0.25	0.40
10@2	0.36	0.24	0.57	0.42	0.59
10@5	0.50	0.55	0.83	0.76	0.85

4.3 Performance

Table 1 presents the overall performances of our method and other methods. It is obvious that our personalized method outperforms other baseline methods. TFIDF shows the term frequency has some impact. The methods with RNN

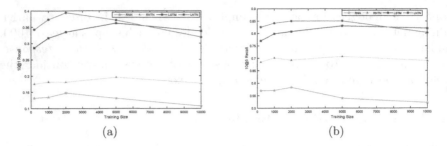

(a)   (b)

Fig. 4. Performance of methods with different training sizes in 20 shops, (a) is for 10@1 and (b) is for 10@5 recall of methods with different training sizes, respectively.

perform poor while the methods with LSTM perform well. Moreover, our LNTN method has a 2% improvement compared with LSTM model on average which proves that the personalized information works.

In order to verify the recommendation quality of our model for the shops with insufficient data, we uniformly and randomly select 20 shops from all 81 shops. In the 20 shops, we use different training sizes. 10@1 and 10@5 recall of different methods are shown in Fig. 4(a) and (b). Both results show that our method has excellent adaptability with less training size. When the training size is over 7000, the LNTN performance is not as good as LSTM. We guess it may be because we set the same iterations for methods with different training sizes, while the larger the training size is, the more iterations it will need.

Figure 5 shows the effect of neuron number on the performance. From Fig. 5(b) we can observe that the performance of 20 neurons is as good as that of 300 neurons. Both results shown in Fig. 5 indicate that when training size is small, it is a wise choice to select fewer neurons.

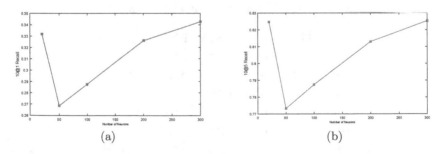

(a)   (b)

Fig. 5. Impact of different neuron numbers in our LNTN model with 200 training samples, (a) is for 10@1 and (b) is for 10@5 recall of methods with different neurons, respectively.

5   Conclusion

In this paper, we investigate a novel task to recommend response content for customer service agents. The experimental results demonstrate that the personalized information of each shop is important and our personalized method gets the best results compared with four baselines even in case of data insufficiency.

Acknowledgments. The work described in this paper was mainly supported by the National Nature Science Foundation of China (No. 61672100, 61375045), the Ph.D. Programs Foundation of Ministry of Education of China (No. 2013110112-0035), the Joint Research Fund in Astronomy under cooperative agreement between the National Natural Science Foundation of China and Chinese Academy of Sciences (No. U1531242), Beijing Natural Science Foundation (No. 4162054, 4162027), and the Excellent young scholars research fund of Beijing Institute of Technology.

References

1. Levin, E., Pieraccini, R., Eckert, W.: Learning dialogue strategies within the Markov decision process framework. In: Proceedings of the 1997 IEEE Workshop on Automatic Speech Recognition and Understanding, pp. 72–79 (1997)
2. Ondřej D., Filip J.: Sequence-to-Sequence Generation for Spoken Dialogue via Deep Syntax Trees and Strings (2016)
3. Ma, H., Liu, C., King, I., et al.: Probabilistic factor models for web site recommendation. In: 2011 International ACM SIGIR Conference on Research and Development in Information Retrieval, p. 274. ACM (2011)
4. Ma, H.: An experimental study on implicit social recommendation. In: 2013 International ACM SIGIR Conference on Research and Development in Information Retrieval, pp. 73–82 (2013)
5. Ondřej D., Filip J.: A context-aware natural language generator for dialogue systems. arXiv preprint arXiv:160807076 (2016)
6. Ge, W., Xu, B.: Dialogue management based on sentence clustering. In: 2015 Meeting of the Association for Computational Linguistics and the International Joint Conference on Natural Language Processing, pp. 800–805 (2015)
7. Blooma, M.J., Kurian, J.C.: Research issues in community based question answering. In: Pacific Asia Conference on Information Systems, Pacis 2011: Quality Research in Pacific Asia, Brisbane, Queensland, Australia, 7–11 July 2011. DBLP, p. 29 (2011)
8. Hong, L., Davison, B.D.: A classification-based approach to question answering in discussion boards. In: International ACM SIGIR Conference on Research and Development in Information Retrieval, SIGIR 2009, pp. 171–178 (2009)
9. Lowe, R., Pow, N., Serban, I., et al.: The ubuntu dialogue corpus: a large dataset for research in unstructured multi-turn dialogue systems. Computer Science (2016)
10. Socher, R., Chen, Q., Mannig, C.D., Ng, A.Y.: Reasoning with neural tensor networks for knowledge base completion. In: Advances in Neural Information Processing Systems (2013)
11. Qiu, X., Huang, X.: Convolutional neural tensor network architecture for community-based question answering. In: 2015 International Conference on Artificial Intelligence, pp. 1305–1311. AAAI Press (2015)

Text Classification Based on ReLU Activation Function of SAE Algorithm

Jia-le Cui[1], Shuang Qiu[1], Ming-yang Jiang[2], Zhi-li Pei[2(✉)],
and Yi-nan Lu[3]

[1] College of Mathematics, Inner Mongolia University for Nationalities,
Tongliao 028043, China
jialecui0228@163.com
[2] College of Computer Science and Technology,
Inner Mongolia University for Nationalities, Tongliao 028043, China
zhilipei@sina.com
[3] College of Computer Science and Technology,
Jilin University, Changchun 130012, China

Abstract. In order to solve the deep self-coding neural network training process, the Sigmoid function back-propagation gradient is easy to disappear, a method based on ReLU activation function is proposed for training the self coding neural network. This paper analyzes the performance of different activation functions and comparing ReLU with traditional Tanh and Sigmoid activation function and in Reuters-21578 standard for experiments on the test set. The experimental results show that using ReLU as the activation function, not only can improve the network convergence speed, and can also improve the accuracy.

Keywords: ReLU activation function · Self coding neural network · Text classification · Accuracy

1 Introduction

With the continuous development of information society, the problem of text classification has been the important content in the research of data mining and machine learning. Different algorithms and models of classification techniques have been proposed and widely applied in the Web document classification [1] and automatic abstract [2]. Stacked Auto-Encoder neural network [3] is a common unsupervised learning algorithm based on neural network algorithm. Each layer network is trained by the layer-by-layer greedy method and the whole model is optimized by back-propagation

This work is supported by National nature science fund project (61373067); Inner Mongolia autonomous region, 2013 annual "prairie talent project"; Autonomous region "higher school youth science and technology talents" (NJYT-14-A09); Inner Mongolia natural science foundation (2013MS0911); Jilin province science and technology development fund project (20140101195JC); Inner Mongolia autonomous region higher school science and technology research (NJZY16177); Inner Mongolia autonomous nature science fund project (2016MS0624).

F. Cong et al. (Eds.): ISNN 2017, Part I, LNCS 10261, pp. 44–50, 2017.
DOI: 10.1007/978-3-319-59072-1_6

algorithm to the objective function input value is equal to the output value. In recent years, many researchers have proposed many improved methods based on SAE algorithm to achieve more significant results. Wu [4] proposed SAE-PCA model improves the recognition rate in gesture recognition significantly than the traditional algorithms. Jiang [5] proposed a based on SAE-LBP web classifier combined with sparse auto-encoder and LBP neural network to improve the accuracy of the classification compared with the traditional BP neural network. However, these improved SAE algorithms use s-shaped growth curve (Sigmoid) and hyperbolic tangent (Tanh) function as the activation function in the process of learning characteristics. The convergence rate of the network using this kind of function is slow or Gradient diffusion leads to non-convergence [8]. More serious is the neuron saturation will lead to information loss, thereby affecting the classification results [7].

In order to solve these problems, it is popular in recent years by using unsaturated linear function ReLUs (Rectified Linear Units) as the activation function in the neural network. And ReLU is more prone to sparsity and reduce the interdependence between parameters and alleviate over-fitting problems [10]. Therefore, based on the advantages of ReLU function, this paper proposes a text classification method based on ReLU activation function for SAE algorithm and verifies by Reuters-21578 standard test set, and compared with other activation functions.

2   Related Work

2.1   Auto-encoder Neural Network (AENN) [3]

Self-coding neural network is an unsupervised learning model that makes the output target as close as possible to the input data itself [3]. The traditional self-coding neural network is divided into three layers, including input layer, hidden layer and output layer. The structure is shown in Fig. 1.

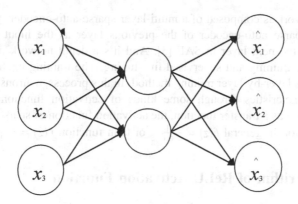

Fig. 1. Self coding neural network

The entire network model in the learning process is divided into two major steps:

(1) Encoding phase of the raw data from the input layer to the hidden layer:

$$h = g_{\theta_1}(x) = \sigma(w_1 x + b_1) \tag{1}$$

(2) Reconstruction from the hidden layer to the output layer of the original data decoding stage:

$$z = g_{\theta_2}(h) = \sigma(w_2 x + b_2) \tag{2}$$

It through the process of encoding and decoding, the original data x by the activation function mapping to the hidden layer of the feature compression expression h, and then use BP algorithm to reconstruct the original data information z to get hidden layer more useful and concise compression feature expression [11]. As a result, each training sample x_i is ultimately mapped to an implicitly compressed feature representation h_i and an output reconstructed feature representation z_i. Thus, the objective of the AE parameter model is to minimize the reconstruction error loss function. For self-coding neural networks, the function is generally defined as:

$$J(W,b) = \left[\frac{1}{m} \sum_{i=1}^{m} \left(\frac{1}{2} \left\| h_{w,b}(x^i) - y^i \right\|^2 \right) \right] + \frac{\lambda}{2} \sum_{l=1}^{n_l-1} \sum_{i=1}^{s_i} \sum_{j=1}^{s_j+1} \left(W_{ji}^{(l)} \right)^2 \tag{3}$$

Where m is the number of training samples, x^i and y^i represents the input and output of the sample i, $W_{ji}^{(l)}$ represents the weights of connections of the neuron i of layer l and the neuron j of layer $l+1$, b is the offset item.

2.2 Stacked Auto-encoder Neural Network (SAENN)

SAE neural network is composed of a multi-layer sparse auto-encoder. We regard the output of the sparse auto-encoder of the previous layer as the input of the sparse auto-encoder of the next layer in SAE [3]. And it's a depth network model through unsupervised pre-training and supervised fine-tuning. The training method of the this network model is layer-by-layer training method. In this process, neurons between each layer learn characteristics through some kinds of activation function, we see this activation function as a transfer function, the activation function of sparse auto-encoder is sigmoid function in general $f(z) = \frac{1}{1+e^{-z}}$, or tanh function $f(z) = \frac{e^z - e^{-z}}{e^z + e^{-z}}$.

3 SAE Algorithm of ReLU Activation Function

The activation function is at the core of the performance of the SAE network. If the activation function is chosen improperly, no matter what kind of learning method is used to learn the feature or improve the network structure can not achieve high learning accuracy, or even impossible to complete the classification task; the contrary, if you choose a better activation function can significantly improve the network performance [6].

In recent years, a new modifier linear activation function has been widely used in the Restricted Boltzmann Machine (RBM) [7] and Convolutional Neural Network (CNN) [6] in the filed of deep neural network simulating the neural system of the brain, gradually replace the Sigmoid activation function into the mainstream. The units produced by this function are called rectified linear units (ReLU) [10], and the function is defined as

$$rectifier(x) = \max(0, x) \tag{4}$$

If the output value of this function is less than 0, let it equal to 0; otherwise keep the original value of the same. Compared with the traditional Sigmoid function, ReLU has a certain sparse ability and is closer to the biological activation model. Because the gradient calculation is not used in the calculation and division of the exponential operation, ReLU computing faster and more accurate, better generalization. Krizheysky [6] pointed out that the convergence rate of ReLU can be increased by 6 times compared with tanh function.

Therefore, the ReLU activation function is introduced into the SAE algorithm to study the performance of the deep network.

4 Experimental Study

Based on the Reuters-21578 data set, this paper compares the performance of neural network using ReLU activation function and traditional activation in SAE algorithm. The experiment is completed in Matlab2012b, using the operating system WIN7 64-bit, CPU2.3 GHz, RAM8G.

4.1 Experimental Parameters Analysis

At present, the selection of SAE neural network structure is not perfect theoretical basis and often associated with specific application. In a reasonable range of SAE neural network will learn the higher-lever data more abstract features, but too many network layers may also reduce the performance of SAE, easily lead to over-fitting [9]. In [9], the numbers of the best hidden layer and the best hidden nodes in SAE neural network are determined experimentally. Therefore, in this paper, the number of neurons in input layer is 1000 and the number of nodes in output layer is 10.

In SAE neural network, two parameters (learning rate, momentum) of unsupervised pre-training phase have great influence on the classification performance. In this paper, the accuracy indexes of different parameters under the same activation function are tested under Reuters-21578 dataset as shown in figures. The parameters with highest accuracy are taken as the optimal parameters in the paper. The parameters are: Learning rate = 1.5, Momentum = 0.2 (Figs. 2 and 3).

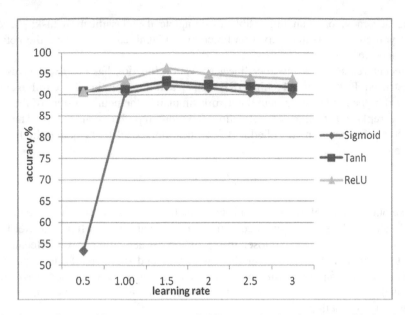

Fig. 2. Curve of recognition accurate rates of different activation functions with the change of learning rate

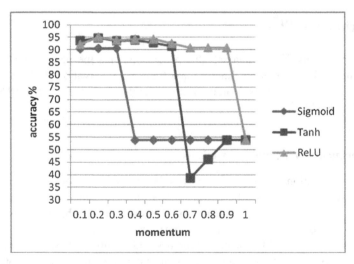

Fig. 3. Curve of recognition accurate rates of different activation functions with the change of momentum

However we noticed the precision of ReLU dropped drastically when the momentum is over 0.9. This is because the sparse ability of ReLU is to force some data to 0, which reduces the average performance of the network model. So we can improve the ability of ReLU [12], but we not described here in detail.

To avoid data duplication, the performance of the network with different activation functions was evaluated objectively. The averge of 5-fold cross validation (CV5) was used as the numerical result. In addition, the other parameters of the network are as follows: the number of iterations is 50, the number of sampers pre-training is 100.

4.2 Results Analysis

As is shown in the Table 1 the comparison between mean square training error and test sample error rate of the SAE algorithm for different activation functions (Sigmoid, Tanh, ReLU) in the dataset.

Table 1. Error rate of Reuters-21578 dataset in SAE algorithm with different activation function

Activation function	Training error	Test error
SAE (Sigmoid)	8.569%	46.625%
SAE (Tanh)	3.46%	24.063%
SAE (ReLU)	24.465%	10.938%

It can be seen from Table 1 that when using different activation functions as the transfer function in the SAE algorithm, both the mean square training error and final test sample error rate are different. Although the training error is the lowest in the network with Tanh as the activation function, it shows that the Tanh function is prone to over-fitting and lead to decrease generalization ability. The test error is the highest with Sigmoid function as the activation function. However, the ReLU function is not only the highest training error, but also the lowest test error, which shows that ReLU has sparse property better prevents the over-fitting and improves the accuracy of text classification.

5 Conclusion

In this paper, in order to solve the deep self-coding neural network training process, the Sigmoid function back-propagation gradient is easy to disappear, a method based on ReLU activation function is proposed for training the self coding neural network and experiments are carried on the Reuters-21578 dataset. The experimental results show that using ReLU as the activation function in depth self-coding neural network not only can prevent the over-fitting phenomenon, but also improve the accuracy of text classification. It also shows that the ReLU activation function with sparse nature is more suitable for classification and recognition in dealing with large data problems, and it has a stronger expression ability for models, which is important for the future research of text mining technology.

References

1. Deng, W., Wang, G., Hong, Z.: Weighting naïve Bayesian mail filtering based on rough set. Comput. Sci. **02**, 16–29 (2011)
2. Research on Automatic Summarization Based on Maximum Entropy. Kunming University of Science and Technology (2013)
3. Hinton, G.E., Salakhutdinov, R.R.: Reducing the dimensionality of data with neural networks. Science **313**(5786), 504–507 (2006)
4. Li, S.Z., Yu, B., Wu, W., et al.: Feature learning based on SAE-PCA net-work for human gesture recognition in RGBD images. Neuron Comput. **151**(2), 565–573 (2015)
5. Jiang, G., Gu, N., Zang, X.: Study on Web Classification based on SAE-LBP. Miniat. Microcomput. Syst. **04**, 223–276 (2016)
6. Krizhevsky, A., Sutskever, I., Hinton, G.E.: Image-net classification with deep convolutional neural networks. In: Advances in Neural Information Processing Systerms, pp. 1097–1105 (2012)
7. Glorot, X., Bordes, A., Bengio, Y.: Domain adaptation for large-scale sentiment classification: a deep learning approach. In: Proceedings of the 28th International Conference on Machine Learning, pp. 513–520 (2011)
8. Dahl, G.E., Sainnath, T.N., Hinton, G.E.: Improving deep neural networks for LVCSR using rectified linear units and dropout. In: Acoustics, Speech and Signal Processing (ICASSP), Piscataway, pp. 8609–8613 (2013)
9. Wang, Z., Ding, J.: Study on water body extraction method based on stacked self-coding. J. Comput. Application **35**(9), 2706–2709 (2015)
10. Glorot, X., Bordes, A., Bengio, Y.: Deep sparse rectifier networks. J. Mach. Learn. Res. **15**, 315–323 (2011)
11. Ge, S.S., Hang, C.C., Lee, T.H., et al.: Stable adaptive neural network control. Springer Publishing Company, Incorporated, Berlin (2010)
12. Xu, B., Wang, N., Chen, T., et al.: Empirical evaluation of rectified activations in convolutional network. arXiv preprint arXiv:1505.00853. (2015)

On Designing New Structures with Emergent Computing Properties

Daniela Danciu$^{(\boxtimes)}$ and Vladimir Răsvan

Department of Automation and Electronics, University of Craiova,
A.I. Cuza Street, No. 13, 200585 Craiova, Romania
{ddanciu,vrasvan}@automation.ucv.ro

Abstract. In this paper we continue the analysis of some structures with potentialities for Artificial Intelligence (AI) devices. We follow the suggestions of J.J. Hopfield about ensuring a sufficiently large number of equilibria which need to be asymptotically stable in some sense. Viewing AI devices as repetitive structures, we focus on those devices ensuring some stability properties of the equilibria from the design stage, pointing out that the so-called hyperstable blocks (in particular, the *triplet* connection) are suitable for this purpose. At the same time the possible number and localization of equilibria as well as their stability are discussed.

Keywords: Hyperstable structures · Triplet connection · Emergence properties · Asymptotic behavior · Hopfield neuron · Modified FitzHugh-Nagumo neuron

1 Introduction

In two previous papers [1,2] we have discussed the possibility of designing new devices for *Artificial Intelligence* (AI) based on networked that fulfil two basic requirements pointed out by a classic of the field [3,4]: emergence (i.e. existence of a quite large number of equilibria) and such global behavior properties which ensure that some of the aforementioned equilibria are attractors (the so-called *gradient behavior*).

In most cases, the approach for designing a AI device with feedback interconnections is as follows: it is "trained" to achieve some "useful goal" by modifying its coefficients (algorithmically) while the behavioral properties result *a posteriori*.

Our approach, within the framework of the *hyperstability theory* [5], will be different. We shall point out some structures which ensure: (a) the necessary qualitative properties from the beginning (e.g. global stability versus gradient behavior) as well as (b) the preservation of these properties with respect to some ways of interconnections. Among the hyperstable structures we shall take into account the most complex and the less known at the same time – the *triplet*. Hyperstability of triplets of Hopfield neurons is considered. Then some aspects

© Springer International Publishing AG 2017
F. Cong et al. (Eds.): ISNN 2017, Part I, LNCS 10261, pp. 51–59, 2017.
DOI: 10.1007/978-3-319-59072-1_7

concerning the localization and possible number of equilibria as well as their local stability are analyzed. The paper ends with suggestions concerning repetitive structures of triplet cells to achieve functional Artificial Neural Networks (ANNs).

2 Neural Networks Containing Feedback. Elementary Theory of Hyperstability

In contrast to feedforward AI networks (e.g. Perceptrons), Hopfield and others considered as useful and interesting the *objects containing feedback connections.* As mentioned in [6], feedback might be an instability generator and among the mechanisms that generate instability one should discover some hidden feedback. But it is quite well known that existence of unstable steady states is not incompatible with the existence of many stable steady states. Starting from the basic ideas of [3,4] we introduced in [1] two theses that we shall follow here also

- The performance of the neural networks as well as of other AI devices is conditioned by some "collective behavior" of many equilibria, among which the best is the *gradient like behavior.* Usually this condition is overlooked in the functional design stage and has to be checked *a posteriori.*
- The most efficient (and the most common) way of seeking the gradient like behavior is to make use of some Lyapunov like lemmas which are obtained from the Barbašin-Krasovskii-LaSalle invariance principle.

In order to seek for the solutions of the aforementioned problems we shall look for ensuring the *gradient behavior* by using the *theory of hyperstability* elaborated by Popov [5]. The basic notions of the theory can be found in short in [1]. The basic element of the theory is the so-called *Popov system* composed of some controlled dynamics equation and an integral index [1,5]. To a standard "block" of the control theory

$$\dot{x} = f(x, u(t), t), \quad v = g(x, u(t), t) \tag{1}$$

which is "square" (the number of the inputs equals the number of the outputs) one associates the Popov system

$$\dot{x} = f(x, u(t), t)$$
$$\eta(t_0, t) = \Re e\left(\int_{t_0}^{t} u^*(\tau)v(\tau)\mathrm{d}\tau\right) = \Re e\left(\int_{t_0}^{t} u^*(\tau)g(x(\tau), u(\tau), \tau)\mathrm{d}\tau\right). \tag{2}$$

Definition 1. *The block (1) is called hyperstable in the strict sense if system (2) is such i.e. there exist two Kamke-Massera functions $\rho : \mathbb{R}^+ \mapsto \mathbb{R}^+$ and $\psi : \mathbb{R}^+ \mapsto \mathbb{R}^+$ (also called K-functions) such that*

$$\rho(|x(t)|) \leq \eta(t_0, t) + \psi(|x(t_0)|), \quad \forall t \geq t_0 \tag{3}$$

along the solutions of (1).

The following two results are basic

Proposition 1. *The sum of two hyperstable Popov systems - defined by the Cartesian product of the two dynamics and the sum of the two indexes - is also hyperstable.*

Proposition 2. *Assume (1) is hyperstable in the sense of Definition 1. Then, among the free state transitions of*

$$\dot{x} = f(x, 0, t) \tag{4}$$

there is found the trivial equilibrium $x(t) \equiv 0$ which is stable in the sense of Lyapunov and all free state trajectories are globally bounded.

Remark that while Proposition 2 concerns a single block, Proposition 1 may concern block interconnections leading to the sum of the associated Popov systems. With respect to this we can make.

Statement 1. [1] *Those interconnections of hyperstable blocks to which one can associate sum systems generate hyperstable aggregated blocks.*

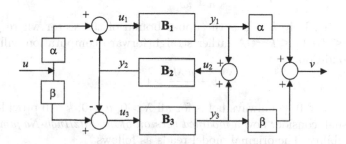

Fig. 1. The triplet interconnection (Source: [7]).

The most common such interconnections are the feedforward and the negative feedback interconnections [5]. However, in a less circulated paper [7], V.M. Popov introduced a new interconnection – of three blocks – which, at least in the case of SISO (Single Input Single Output) blocks turns to be independent of the aforementioned two (Fig. 1). This new interconnection was introduced within the context of real positive transfer function synthesis. Since real positiveness of the transfer function is necessary and sufficient for the hyperstability of the linear time invariant blocks, the connection with the theory of hyperstability is obvious.

We shall not insist any longer on this subject since this was done in a previous paper [1]. As stated there, we propose the triplet of hyperstable neurons as a basic cell for a possible new Cellular Neural Network. Among other advantages, this structure can give at users' disposal a stable structure with non-symmetric weight matrix.

3 Hyperstable Neurons

We shall discuss here briefly two types of artificial neurons for possible use in hyperstable interconnections.

A. The *standard Hopfield neuron* is described by

$$\dot{x} = -ax + wy + I, \quad y = f(\lambda x), \quad \lambda > 0 \tag{5}$$

where $f : \mathbb{R} \mapsto \mathbb{R}$ is a sigmoid i.e. sector restricted increasing function. Since the neurons are connected at the level of the variable y, this will be considered the output of the block. Moreover the "neuron block" itself is a feedback connection of a linear and a non-linear block containing a sector restricted nonlinear function. This last block is hyperstable [5] hence we need the hyperstability of the linear block

$$\dot{x} = -ax - wu(t), \quad v = \lambda x \tag{6}$$

where $a > 0$, $w \in \mathbb{R}$. As showed in [1,5], the necessary and sufficient condition for hyperstability is the fulfilment of the frequency domain inequality

$$\frac{1}{k} - \Re e \frac{(1 + \imath \omega q)w\lambda}{a + \imath \omega} > 0, \quad \forall \omega > 0 \tag{7}$$

for some $q \in \mathbb{R}$. Here $k > 0$ is the upper bound of the sector where f is confined i.e. $0 \le f(\sigma)\sigma \le k\sigma^2$. A rather straightforward computation will give the following condition

$$1 - \frac{w\lambda k}{a} > 0. \tag{8}$$

if $w > 0$. For $w < 0$, (7) always holds for all $\lambda > 0$, $a > 0$, $k > 0$ provided $q > 0$.

B. We shall consider now a *modified version of the FitzHugh-Nagumo neuron* for hyperstability. The original model reads as follows

$$\dot{u} = f(u) - v + I(t), \quad \tau \dot{v} = g(u) - v \tag{9}$$

where $f(u) = u - u^3/3$ and $g(u) = \delta u$ with $\delta = \delta_+$ for $u > 0$ and $\delta = \delta_-$ for $u < 0$. This model is much alike to that of a Van der Pol oscillator since $f(u)$ is a S-like function. For AI needs we shall consider a modified version where the S-function is substituted by a sigmoid to obtain

$$\dot{u} = -\phi(u) - v + I(t), \quad \tau \dot{v} = \delta u - v. \tag{10}$$

The original FitzHugh-Nagumo neurons are supposed to be interconnected as follows

$$\dot{u}_i = f(u_i) - v_i + \sum_1^N w_{ij}u_j + I_i(t), \quad \tau_i \dot{v}_i = \delta u_i - v_i, \quad i = \overline{1, N}. \tag{11}$$

From here we deduce the structure of the modified FitzHugh-Nagumo neuron

$$\dot{u} = -\phi(u) + wu - v + \mu(t), \quad \tau \dot{v} = \delta u - v, \quad y = u. \tag{12}$$

The structure being a feedback one, with a sector restricted nonlinear function, the approach is the same as in the case of the Hopfield neuron. The necessary and sufficient condition for hyperstability is the fulfilment of the frequency domain inequality

$$\frac{1}{k} + \Re e(1 + \imath\omega q)\gamma(\imath\omega) > 0, \quad \forall \omega > 0 \tag{13}$$

for some $q \in \mathbb{R}$. Here $\gamma(s)$ is the transfer function of the following linear block

$$\dot{u} = wu - v + \mu(t), \quad \tau\dot{v} = \delta u - v, \quad y = u \tag{14}$$

and is given by

$$\gamma(s) = \frac{1 + s\tau}{\delta - w + (1 - w\tau)s + \tau s^2}. \tag{15}$$

The simple conditions $w < \delta$, $w\tau < 1$ ensure $\gamma(s)$ has its poles with negative real part. Therefore the modified transfer function

$$\chi(s) = (1 + qs)\gamma(s) = \frac{(1 + sq)(1 + s\tau)}{\delta - w + (1 - w\tau)s + \tau s^2} \tag{16}$$

has its poles and its zeros in the left hand plane provided $q > 0$. It will thus be positive real provided $\Re e\chi(\imath\omega) \geq 0$ for some $q > 0$. A straightforward while tedious manipulation shows the following choices for $q > 0$:

- $w < 0$: if $\delta\tau < 1$, then $q > 0$ is good; if $\delta\tau > 1$ then take $q > -(w\tau^2)(\delta\tau - 1)^{-1}$;
- $w > 0$: if $\delta\tau < 1$ then take $q > (w\tau^2)(1 - \delta\tau)^{-1}$; if $\delta\tau > 1$ then take q between the two positive roots of the trinomial

$$(\delta\tau - 1)^2 q^2 + 2\tau^2(w(\delta\tau - 1) - 2(\delta - w))q + w^2\tau^4 = 0. \tag{17}$$

Therefore the modified FitzHugh-Nagumo neuron (12) is hyperstable for any $k > 0$ provided $w < \delta$, $w\tau < 1$; worth mentioning that the aforementioned conditions are fulfilled automatically for $w < 0$.

4   Localization and Stability for the Equilibria of a Triplet Cell of Hopfield Neurons

Consider the *triplet cell of Hopfield neurons* (TCH) described by the equations

$$\dot{x} = -Ax + Wy + I, \quad y = F(\Lambda x), \quad v = c^T y \tag{18}$$

where $A = diag(a_i)$, $\Lambda = diag(\lambda_i)$, $F(\Lambda x) = diag(f_i(\lambda_i x_i))$, $I = bu$, $b = c = \begin{bmatrix} \alpha & 0 & \beta \end{bmatrix}^T$, $\alpha, \beta \in \mathbb{R}_+$, $a_i > 0$, $\lambda_i > 0$, $\forall i = \overline{1,3}$ and the interconnections matrix W, induced by the triplet structure presented in Fig. 1, with the form

$$W = \begin{bmatrix} w_{11} & -w_{12} & 0 \\ w_{21} & w_{22} & w_{23} \\ 0 & -w_{32} & w_{33} \end{bmatrix}, \quad w_{ij} > 0, \quad \forall i, j = \overline{1,3}. \tag{19}$$

The input u and the output v of TCH are terminals which allow coupling the triplet cell with other similar structures within a more general *cell-based neural network*; we shall consider them as scalar real valued functions. The nonlinear sigmoid function of the Hopfield neuron of type $f(\lambda x) = 1/(1 + exp(-\lambda x))$, where the scaling factor $\lambda > 0$ may be different for each neuron, $f : \mathbb{R} \to (0, 1)$ is a continuous differentiable, monotonically increasing verifying, for $\lambda > 0$ sufficiently large,

$$\lim_{x \to \infty} f(x) = 1, \quad \lim_{x \to -\infty} f(x) = 0, \quad \lim_{|x| \to \infty} f'(x) = 0. \tag{20}$$

Due to these properties, the dynamics of the TCH evolves within the open cube $H = (0, 1)^3$; we define also the closed cube by $\bar{H} = [0, 1]^3$.

The main aim of this section is the study of the existence, number, location and stability properties of the equilibria for the TCH. The approach we shall use is inspired by a paper by Vidyasagar [8]. Due to the structure induced by triplet, (see (19) and Fig. 1) we shall not make use, as in [8], of the assumptions about W regarding either the "no self-interactions" ($w_{ii} = 0$) or "symmetrical inter-actions" ($w_{ij} = w_{ji}$). The Vidyasagar's approach is based on the relationships between the state x_{eq} and the output y_{eq} of an isolated neuron at equilibrium, as they result from the features of the intersection points of two plots. More precisely, by considering the equation of an isolated Hopfield neuron (5) and defining

$$g_1(x) = \frac{a}{w}x - \frac{I}{w}, \qquad g_2(x) = f(\lambda x) \tag{21}$$

the intersection points of $g_1(x)$ and $g_2(x)$ in Fig. 2, for $w > 0$ as well as for $w < 0$, allow deriving in [8] the following relationships for the equilibria pair (x_{eq}, y_{eq})

$$\text{case P:} \begin{cases} y_{eq} \to 0, \text{ if } x_{eq} < 0 \text{ (type } P_1) \\ y_{eq} \to 1, \text{ if } x_{eq} > 0 \text{ (type } P_2) \end{cases}, \quad \text{case Q:} \{ y_{eq} \in (0, 1), \text{ if } x_{eq} \to 0 \text{ (type } Q) \tag{22}$$

Accordingly, one can distinguish the following possible locations within $\bar{H}$ for the equilibria y_{eq} of the TCH when $\lambda_i \to \infty$:

[I] - Interior of H: $y_{eq} \in Int(H) \Leftrightarrow y_{eq}^i \in (0, 1), \ \forall \, i = \overline{1,3}$
[V] - Vertices of $\bar{H}$: $y_{eq} \to V(\bar{H}) \Leftrightarrow y_{eq}^i \to \{0, 1\}, \ \forall \, i = \overline{1,3}$
[F] - Faces of $\bar{H}$: $y_{eq} \to F(\bar{H}) \Leftrightarrow y_{eq}^i \to \{0, 1\}, \ y_{eq}^{j,k} \in (0, 1), \ \forall \, i, j, k = \overline{1,3}$
[E] - Edges of $\bar{H}$: $y_{eq} \to E(\bar{H}) \Leftrightarrow y_{eq}^{i,j} \to \{0, 1\}, \ y_{eq}^k \in (0, 1), \ \forall \, i, j, k = \overline{1,3}.$
$$\tag{23}$$

Case [I]: Equilibria within the interior of H. This case corresponds to points of type Q in (22). Consequently, by replacing $x = 0$ in the first equation of (18) we obtain

$$Wy + bu = 0. \tag{24}$$

Taking into account (19) it follows that $\det(W) > 0$ and thus, when $\lambda \to \infty$, the triplet of Hopfield neurons has an unique equilibrium defined by

$$y_{eq} = -W^{-1}bu \in Int(H). \tag{25}$$

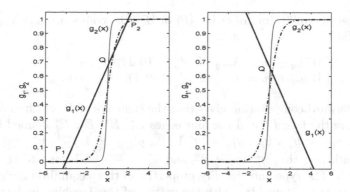

Fig. 2. Equilibria locations for the isolated Hopfield neuron (Source: [8]) as defined in (21)–(23). Remark: $\lambda \to \infty$ ensures the shape of sigmoid function to approach the shape of threshold function. Legend: $g_1(x)$ (thick line), $y = g_2(x)$ for: $\lambda = 2$ (dotted line), $\lambda = 8$ (thin line).

The Stodola criterion applied to the characteristic polynomial $p(s) = det(sI - W)$ gives that *the unique equilibrium of the TCH within the interior of H is unstable.*

Case [V]: Equilibria near the vertices V_i, $i = \overline{0,7}$ of the closed cube $\bar{H} = [0,1]^3$. Consider the vertices of $\bar{H}$: $V_0 = [0\,0\,0]^T$, $V_1 = [0\,0\,1]^T$, $V_2 = [0\,1\,0]^T$, $V_3 = [0\,1\,1]^T$, $V_4 = [1\,0\,0]^T$, $V_5 = [1\,0\,1]^T$, $V_6 = [1\,1\,0]^T$, $V_7 = [1\,1\,1]^T$. The existence and location of the TCH equilibria near the vertices of $\bar{H}$ follow from the fulfilment of the conditions of case P in (22) by the solutions of the equilibria equation derived from (18) for $\dot{x} = 0$

$$Wy + bu = Ax. \tag{26}$$

A straightforward computation leads to the following possible equilibria:

$$\begin{cases} y_{eq} \to \{V_2^*, V_3^*, V_6^*, V_7^*\} & \text{if } u \in \mathbf{I} = \left(\max\left\{\frac{w_{12}-w_{11}}{\alpha}, \frac{w_{32}-w_{33}}{\beta}\right\}, \min\left\{\frac{w_{12}}{\alpha}, \frac{w_{32}}{\beta}\right\}\right) \subset \mathbb{R}_+ \\ y_{eq} \to V_0^* & \text{if } u \leq 0. \end{cases}$$
$$\tag{27}$$

In order to evaluate the type and stability of these equilibria we take into account the properties (20) of the nonlinear functions $f_i(\cdot)$, $i = \overline{1,3}$ and apply the well-known *stability by the first approximation* method. Let x_{eq} be an equilibrium of the system (18) and $y_{eq} = F(\Lambda x_{eq})$. The linearized system of (18) in the equilibrium x_{eq} is

$$\dot{x}_L = A_L x_L, \quad A_L = -A + W J_F(x_{eq})\Lambda \tag{28}$$

where $J_F(x_{eq})$ is the Jacobian matrix $[\partial F/\partial x](x_{eq})$, a diagonal matrix. The conditions (20) applied to (28) give $A_L \to -A$. Consequently, $A_L = diag(-a_i)_{i=\overline{1,3}}$ and all its eigenvalues are negative. Thus, *all the equilibria located near the vertices defined by (27) are stable nodes.*

Cases [F and E]: Equilibria located near the faces and/or edges of $\bar{H} = [0,1]^3$. These two cases are described by the relations [F] and [E] in (23). In order y_{eq} be an equilibrium of either $[F]$ or $[E]$, the components of x_{eq} and y_{eq} have to fulfil

$$\begin{cases} W y_{eq} + bu = A x_{eq} & \text{if } y_{eq}^i \to \{0, 1\}, \; x_{eq}^i \in \mathbb{R} \setminus \{0\} \\ W y_{eq} + bu = 0 & \text{if } y_{eq}^i \in (0, 1), \; x_{eq}^i \to 0 \; . \end{cases} \tag{29}$$

A straightforward computation gives that the elements of $\bar{H}$ which may contain equilibria are the face F^* and the four edges E_1^*, E_2^*, E_3^*, E_4^* defined by: $y_F^* = [y_1^* \; 1 \; y_3^*]^T$, $y_{E_1}^* = [y_1^* \; 1 \; 0]^T$, $y_{E_2}^* = [y_1^* \; 1 \; 1]^T$, $y_{E_3}^* = [0 \; 1 \; y_3^*]^T$, $y_{E_4}^* = [1 \; 1 \; y_3^*]^T$. Worth mentioning that $y_1^* = \frac{w_{12} - \alpha u}{w_{11}}$ and $y_3^* = \frac{w_{32} - \beta u}{w_{33}}$ belong to the interval $(0, 1)$ if $u \in \mathbf{I}$. The type and stability properties of these equilibria are evaluated on the linearized system (28), with the entries of the Jacobian matrix $J_F(x_{eq})$ as follows

$$J_F^{\,i}(x_{eq}) = \begin{cases} \frac{y_i^*}{\lambda_i} & \text{if } y_{eq}^i \in (0, 1) \\ 0 & \text{if } y_{eq}^i \to \{0, 1\} \; . \end{cases} \tag{30}$$

An evaluation of the signs of A_L eigenvalues leads to the conclusion that *the TCH has*:

– *at least an equilibrium y_{eq} attaining the face F^*; this equilibrium is: (i) a stable node, provided $u > \max \left\{ \frac{w_{12} - a_1}{\alpha}, \frac{w_{32} - a_3}{\beta} \right\}$ or (ii) an hyperbolic equilibrium with the unstable manifold of dimension 2, otherwise;*
– *at least an equilibrium attaining each of the edges E_1^*, E_2^*, E_3^*, E_4^*; all these equilibria are: (i) stable nodes, provided $u > \max \left\{ \frac{w_{12} - a_1}{\alpha}, \frac{w_{32} - a_3}{\beta} \right\}$ or (ii) hyperbolic equilibria with the unstable manifold of dimension 1, otherwise.*

5   Conclusions

Considering the framework of *hyperstability theory* [5], this paper concerns the analysis of some structures with potentialities for AI devices. Our approach is, as far as we know, different from most approaches at least in the AI field. More specifically, we focus on those structures which ensure the necessary qualitative properties (e.g. global stability versus gradient behavior) from the beginning, i.e. from the design stage. Moreover, this approach also ensures the preservation of these qualitative properties with respect to some ways of interconnections.

Among the hyperstable structures, we focus on the *triplet* connection. First, there are derived conditions for hyperstablilty of both Hopfield and modified FitzHugh-Nagumo neurons – as component blocks. Then, the triplet cell of Hopfield neurons is studied from the points of view of equilibria localization and number as well as their local stability. The results of this contribution give reasons to further study the properties of repetitive hyperstable structures of triplet networks in order to eventually discover such emergent collective capabilities which allow them to achieve some functional ANNs.

References

1. Danciu, D., Răsvan, V.: On structures with emergent computing properties. A connectionist versus control engineering approach. In: Rojas, I., Joya, G., Catala, A. (eds.) IWANN 2015. LNCS, vol. 9094, pp. 415–429. Springer, Cham (2015). doi:10.1007/978-3-319-19258-1_35
2. Răsvan, V.: Reflections on neural networks as repetitive structures with several equilibria and stable behavior. In: Rojas, I., Joya, G., Cabestany, J. (eds.) IWANN 2013. LNCS, vol. 7903, pp. 375–385. Springer, Heidelberg (2013). doi:10.1007/978-3-642-38682-4_40
3. Hopfield, J.J.: Neural networks and physical systems with emergent collective computational abilities. Proc. Natl. Acad. Sci. USA **79**(1), 2554–2558 (1982)
4. Hopfield, J.J.: Neurons with graded response have collective computational properties like those of two-state neurons. Proc. Natl. Acad. Sci. USA **81**(5), 3088–3092 (1984)
5. Popov, V.M.: Hyperstability of Control Systems. Die Grundlehren der mathematischen Wissenschaften, vol. 204. Springer, Heidelberg (1973)
6. Neymark, Y.I.: Dynamical Systems and Controlled Processes (in Russian). Nauka, Moscow (1978)
7. Popov, V.M.: An analogue of electrical network synthesis in hyperstability (in Romanian). In: Proceedings of the Symposium on Analysis and Synthesis of Electrical Networks, no. III, p. 9.1. Power Institute of Romanian Academy (1967)
8. Vidyasagar, M.: Location and stability of the high-gain equilibria of nonlinear neural networks. IEEE Trans. Neural Netw. **4**(4), 660–671 (1993)

Fast Sparse Least Squares Support Vector Machines by Block Addition

Fumito Ebuchi[✉] and Takuya Kitamura[✉]

National Institute of Technology, Toyama College,
13 Hongo-machi, Toyama-shi, Toyama, Japan
h1612106@mailg.nc-toyama.ac.jp, kitamura@nc-toyama.ac.jp

Abstract. In this paper, we propose two fast feature selection methods for sparse least squares support vector training in reduced empirical feature space. In the first method, we select the training vectors as the basis vectors of the empirical feature space from the standpoint of the similarity. The complexity of the selection can be lower than that of the conventional method because we use the inner product values of training vectors without linear discriminant analysis or Cholesky factorization which are used by the conventional methods. In the second method, the selection method is forward selection by block addition which is a wrapper method. This method can decrease the size of the kernel matrix in the optimization problem. The selecting time can be shorter than that of the conventional methods because the computational complexity of the selecting basis vectors depends on the size of the kernel matrix. Using benchmark datasets, we show the effectiveness of the proposed methods.

Keywords: Empirical feature space · Least squares support vector machine · Pattern recognition

1 Introduction

Support vector machines (SVMs) [1,2] are powerful classifiers for pattern classification problems, and have been applied for several disciplines. As the advantage of SVMs, the solutions are sparse. Also, the generalization ability is high and it is easy to expand to non-linearly problems by the kernel methods [3]. However, the computational complexity of the standard SVMs may be too high because a quadratic programming problem must be solved in training SVMs. A least squares SVM (LS-SVM) [4] is one type of SVMs and can overcome this problem. The solutions of LS-SVMs can be obtained by solving a set of linear equations. Therefore, the computational complexity of LS-SVMs can be lower than that of the standard SVMs for the small problems. However, the solutions of the LS-SVM are not sparse because all training vectors become support vectors (SVs). A sparse LS-SVMs (SLS-SVMs) [5,6] in the reduced empirical feature space were proposed by Abe as the sparse models of LS-SVMs. In the SLS-SVMs, the solutions are obtained by solving LS-SVM in the empirical feature space [9] spanned by the linearly independent training vectors, which are

© Springer International Publishing AG 2017
F. Cong et al. (Eds.): ISNN 2017, Part I, LNCS 10261, pp. 60–70, 2017.
DOI: 10.1007/978-3-319-59072-1_8

selected by linear discriminant analysis (LDA) [7] or Cholesky factorization [8]. Then, the solutions are sparse because the basis vectors of the empirical feature space are defined as SVs. Using LDA or Cholesky factorization, the complexity of selection for the linearly independent training vectors may be large in large problems. Also, the generalization ability of SLS-SVMs may be worse than that of LS-SVM because the basis vectors of the empirical feature space are selected from the only standpoint of linearly independence. To overcome this problem, in an Improved SLS-SVM (ISLS-SVM) [10] which is proposed by Kitamura and Asano, the basis vectors of the empirical feature space are selected from the standpoints of classification and linearly independence. However, as with Abe's SLS-SVM, the computational complexity of ISLS-SVM is high.

In this paper, to reduce the computational complexity of ISLS-SVM, we propose two fast selection methods for the basis vectors of the empirical feature space. In the first method, we select the basis vectors based on the similarity among training vectors. First, as the ISLS-SVM, the training vectors are sorted based on the objective function value of the LS-SVM in each one-dimensional feature space spanned by each training vector. Then, we obtain phase angles between the basis vector corresponding to the smallest objective function value and other training vectors. We select the training vectors, whose phase angle is larger than the threshold value, as the candidates of the basis vector of the empirical feature space. We select another training vector based on the objective function values from the candidates of the basis vectors and iterate the above procedure. Thus, the similarities of the selected basis vectors are low. In this method, the basis vectors of the empirical feature space can be selected faster than LDA and Cholesky factorization because the computational complexity is $O(M^2)$. The second method is forward selection by Block Addition [11]. In this method, the basis vectors of the empirical feature space are selected from some subsets of training vectors. First, we obtain the objective function values of LS-SVM in each one-dimensional feature spaces spanned by each training vector. Then, based on these objective function values, all training vectors are divided equally into some subsets in increase order of the corresponding objective function values. Next, we generate the empirical feature space whose basis vectors are selected by the wrapper method from these subsets. Finally, we solve the LS-SVM in this generated empirical feature space. As with SLS-SVM, the basis vectors of the empirical feature space in the proposed method can be defined as the SVs.

In this paper, we describe the conventional SLS-SVM in the empirical feature space in Sect. 2. In Sect. 3, we show the proposed methods, and, in Sect. 4, we describe the effectiveness of the proposal method by computer experiment using the benchmark datasets. In Sect. 5, we deliver the conclusion.

2   Sparse Least Squares Support Vector Machine in the Sorted Empirical Feature Space

2.1   Least Squares Support Vector Machine

The decision function $D(\boldsymbol{x})$ is shown as follows:

$$D(\boldsymbol{x}) = \boldsymbol{w}^\top \boldsymbol{\phi}(\boldsymbol{x}) + b, \tag{1}$$

where x, w, $\phi(x)$ and b are the l-dimensional input vector, the n-dimensional weight vector, the mapping function into the n-dimensional feature space, and the bias term. We obtain w and b by solving the following optimization problem.

$$\min \quad \frac{1}{2}w^\top w + \frac{C}{2}\sum_{i=1}^{M}\xi_i^2, \tag{2}$$

$$\text{s.t.} \quad w^\top \phi(x_i) + b = y_i - \xi_i \tag{3}$$

$$\text{for} \quad i = 1, 2, ..., M.$$

Here, $y_i = 1$ or -1 if x_i belongs to Class 1 or 2, respectively. C is the margin parameter which determines the trade-off between the maximizing margin between classes and the minimizing error. ξ_i is the slack variable for $\phi(x_i)$. Introducing the Lagrange multipliers $\alpha = (\alpha_1, \alpha_2, ..., \alpha_M)$ into (2) and (3), we obtain the following unconstrained objective function.

$$\min \quad Q = \frac{1}{2}w^\top w + \frac{C}{2}\sum_{i=1}^{M}\xi_i^2 - \sum_{i=1}^{M}\alpha_i(w^\top \phi(x_i) + b - y_i + \xi_i). \tag{4}$$

We obtain α and b by solving (4) and define the decision function which is shown as follows:

$$D(x) = \sum_{i=1}^{M}\alpha_i \phi^\top(x_i)\phi(x) + b$$

$$= \sum_{i=1}^{M}\alpha_i K(x_i, x) + b. \tag{5}$$

Then, the decision function $D(x)$ needs all training vectors. Namely, all training vectors are SVs in LS-SVM and the LS-SVM doesn't have sparsity.

2.2 Sparse Least Squares Support Vector Machine in the Reduced Empirical Feature Space

The mapping function $h(x)$ which maps the input vector x into the N-dimensional empirical feature space is shown as follows:

$$h(x) = (K(x_1', x), ..., K(x_N', x))^\top, \tag{6}$$

where x' are N linearly independent training vectors, which are basis vectors of the empirical feature space. The decision function $D_e(x)$ in the empirical feature space is shown as follows:

$$D_e(x) = v^\top h(x) + b_e, \tag{7}$$

where v and b_e are the N-dimensional weight vector and bias term. We obtain v and b_e by solving the following optimization problem.

$$\min \quad \frac{1}{2}\boldsymbol{v}^{\top}\boldsymbol{v} + \frac{C}{2}\sum_{i=1}^{M}\xi_i^2, \tag{8}$$

$$\text{s.t.} \quad \boldsymbol{v}^{\top}\boldsymbol{h}(\boldsymbol{x}_i) + b_{\mathrm{e}} = y_i - \xi_i \tag{9}$$

$$\text{for} \quad i = 1, 2, ..., M.$$

In this problem, the solution can be obtained by solving the primal form because the computational complexity of primal form is smaller than that of dual form. From (6) and (7), the decision function $D_{\mathrm{e}}(\boldsymbol{x})$ need the linearly dependent training vectors. Hence, the solution is sparse. However, to obtain the linearly independent training vectors by LDA or Cholesky factorization, the computer complexity may be large in large problems. Moreover, there is no assurance that good training vectors from the standpoint of classification are selected as the basis vectors of the empirical feature space because the order of judge of linearly independence for the training vectors is random.

2.3   Sorting Method for Kernel Matrix Based on the Objective Function Value

In the ISLS-SVM [10], to select good training vectors as basis vectors of the empirical feature space, the order of judge of linearly independence for the training vectors is sorted from the standpoint of classification.

First, we select one training vector $\phi(\boldsymbol{x}_i)(i = 1, 2, ..., M)$. Then we generate the one-dimensional feature space spanned by $\phi(\boldsymbol{x}_i)$. We determine the following mapping function $g_i(\boldsymbol{x})$ into this one-dimensional feature space:

$$g_i(\boldsymbol{x}) = \frac{K(\boldsymbol{x}_i, \boldsymbol{x})}{\sqrt{K(\boldsymbol{x}_i, \boldsymbol{x}_i)}}. \tag{10}$$

Finally, we obtain the objective function value Q_i by solving (8), (9) which are replaced $\boldsymbol{h}(\boldsymbol{x})$ with $g_i(\boldsymbol{x})$. We obtain M objective function values Q_i ($i = 1, ..., M$). The training vector corresponding to the small objective function value is effective as the basis vector of the empirical feature space because the optimization problem of LS-SVM is the minimization problem. Then, we select the linearly independent training vectors in increasing order of the objective function values. However, in the conventional methods, the basis vectors of the empirical feature space are selected by LDA or Cholesky factorization which have a large computational complexity. The computational complexity of these algorithms is $O(M^3)$.

3   Fast Selection for Basis Vectors of the Empirical Feature Space

3.1   Fast Selection Using the Phase Angle

The phase angle between a training vector and other training vector is the degree of dissimilarity. In this method, we select the basis vectors of the empirical feature space such that the phase angles are smaller than the threshold value. First, as with the ISLS-SVM, we obtain the objective function value in the one-dimensional feature space spanned by each training vector. Then, based on these objective function values, the subscripts of the training vectors are sorted in ascending order. We determine the following the similarity between the training vector $\phi(x_p)$ and the training vector $\phi(x_q)$.

$$
\begin{aligned}
\cos\theta_{pq} &= \frac{\phi(x_p) \cdot \phi(x_q)}{||\phi(x_p)|| \cdot ||\phi(x_q)||} \\
&= \frac{K(x_p, x_q)}{\sqrt{K(x_p, x_p)}\sqrt{K(x_q, x_q)}}
\end{aligned}
\tag{11}
$$
$$
\text{for} \quad p = 1, ..., M - 1, q = p + 1, ..., M.
$$

If the value of $\cos\theta_{pq}$ is smaller than the threshold value β, the training vector $\phi(x_q)$ is selected as the basis vector of the empirical feature space. Then, the complexity of this selection is $O(M^2)$, which is lower than that of the conventional selection $O(M^3)$.

3.2 Fast Sparse Least Squares Support Vector Machines by Block Addition

In this section, we discuss the fast selection method for the basis vectors of the empirical feature space with block addition. The flow of the proposed method is showed in Fig. 1. In this method, we divide the training vectors into several sets. Let S be the SV candidate set. First, we sort the subscripts of the training vectors from the standpoint of the classification as with the ISLS-SVM. Then, we divide the training vectors equally into twenty sets $S^l(l = 1, ..., 20)$ in subscript order. We add S^1 to S and select the linearly independence training data from S by Cholesky Factorization or based on the degree of similarity. Then, we obtain the average recognition rate R^1 by five-fold cross-validation (CV). Let R_1 be R^1. Next, using the Cholesky Factorization or the degree of similarity, we select the linearly independent training vectors from S which is added to $S^l(l = 2, ..., 20)$ and we obtain the average recognition rate R_2^l by five-fold CV. Let R_2 be maximum value of R_2^l and add S^l to S. If $R_2 < R_1$, the iteration is terminated. Otherwise, we iterate the above procedure until $R_{i+1} > R_i$. In the following, we show the algorithm of our proposed method for the Cholesky Factorization or the degree of similarity.

Algorithm

Step 1 Sort the subscript of the all training vectors based on the objective function values.

Step 2 Divide the all training vectors equally into twenty sets $S^l(l = 1, ..., 20)$ in subscript order.

Step 3 Set $S \leftarrow S^1$, $F_l = 1(l = 2, ..., 20)$.

Step 4 Select the basis vectors of the empirical feature space from S by using the Cholesky Factorization or based on the degree of similarity.

Step 5 Solve the optimization problem (8), (9) to obtain the average recognition rate by five-fold CV. Let this average recognition rate be R_1.

Step 6 Set $i = 2$.

Step 7 Set $l = 2$.

Step 8 If $F_l = 1$, select the basis vectors of the empirical feature space from $S \leftarrow S + S^l$ by the Cholesky Factorization or based on the degree of similarity and go to Step 9. Otherwise, set $l = l + 1$ and go to Step 8.

Step 9 Solve the optimization problem of (8) and that of (9) to obtain the average recognition rate by CV. Let this average recognition rate be R_i^l.

Step 10 If $R_i^l < R_{i-1}$, set $F_l = 0$.

Step 11 If $l = 20$, go to step 12. Otherwise, set $l = l + 1$ and go to Step 8.

Step 12 Set $R_i = \max R_i^l$.

Step 13 If $R_i > R_{i-1}$, set $i = i + 1$, $S \leftarrow S + S^l$ ($l = \text{argmax} R_i^l$) and go to Step 7. Otherwise go to Step 14.

Step 14 Using (6), all training vectors are mapped into the empirical feature space whose basis vectors are the independent training vectors of the set S.

Step 15 Obtain v and b_e by solving the optimization problem of (8) and that of (9).

Step 16 Calculate the decision function (7).

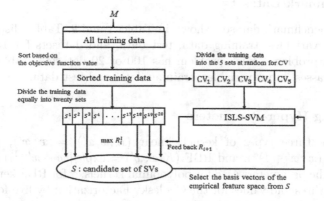

Fig. 1. The flow of the proposed method

4 Computer Experiment

In this section, we compare the proposed methods with the conventional ISLS-SVM using UCI benchmark datasets [4,12]. We measure the computational time using a computer (OS : Window 7 (64 bit), CPU : AMD Athlon (tm) II X 2 250 Processor, 4.00 GB memory). In this section, we refer to the ISLS-SVM as SLS-SVM and the proposed method as FSLS-SVM. To distinguish between the methods which use the Cholesky factorization and the phase angle, we add subscript P to the proposed method using the phase angle.

Table 1. Two-classes benchmark datasets

Data	Input	Training	Test	Sets
Banana	2	400	4900	100
B.cancer	9	200	77	100
Diabetes	8	468	300	100
German	20	700	300	100
Heart	13	170	100	100
Image	18	1300	1010	20
Ringnorm	20	400	7000	100
F.solar	9	666	400	100
Splice	60	1000	2175	20
Thyroid	5	140	75	100
Titanic	3	150	2051	100
Twonorm	20	400	7000	100
Waveform	21	400	4600	100

Table 2. Multi-classes benchmark datasets

Data	Class	Input	Training	Test
Iris	3	4	75	75
Numeral	10	12	810	820
Blood-cell	12	13	3097	3100
Thyroid(M)	3	21	3772	3428

4.1 Benchmark Datasets

We use the benchmark datasets shown in Tables 1 and 2. Table 1 lists the number of input variables, training data, test data, and datasets for 13 two-class classification problems. Each problem has 100 or 20 datasets. Table 2 lists the number of classes, input variables, training data, and test data.

4.2 Setting Hyper-Parameters

And, we use three types of kernel: linear $(K(\boldsymbol{x}, \boldsymbol{x}') = \boldsymbol{x}^{\mathsf{T}}\boldsymbol{x}')$, polynomial $(K(\boldsymbol{x}, \boldsymbol{x}') = (\boldsymbol{x}^{\mathsf{T}}\boldsymbol{x}' + 1)^d)$, and RBF $(K(\boldsymbol{x}, \boldsymbol{x}') = \exp(\gamma||\boldsymbol{x} - \boldsymbol{x}'||^2))$ kernel. We determined the parameters C, d for polynomial kernels, γ for RBF kernels, and η which is the threshold values for the Cholesky Factorization by five-fold CV. We selected C from $\{0.1, 1.0, 5.0, 10, 50, 100, 500, 10^3, 5\times 10^3, 10^4\}$, d from $\{2, 3, 4, 5\}$, γ from $\{0.1, 0.5, 1.0, 1.5, 3.0, 5.0, 10, 15, 20, 50, 100, 200\}$, and η which is the threshold value of Choleskey factorization from $\{10^{-2}, 10^{-3}, 10^{-4}, 10^{-5}, 10^{-6}\}$. We also set the threshold value $\beta = 0.99$ of the phase angle. Table 3 lists the hyper-parameters obtained by the five-fold CV.

4.3 Discussion

Table 4 shows the training time of each problem for the conventional method and the proposed methods. The shortest training time for each dataset is shown in bold. The shorter training time of the proposed methods than that of the SLS-SVM is marked with *. The training time of the SLS-SVM$_{\mathrm{P}}$ is shortest in

Table 3. Selected kernels and hyper-parameters by five-fold cross validation

	SLS-SVM			SLS-SVM$_P$		FSLS-SVM			FSLS-SVM$_P$	
Data	Kernel	C	η	Kernel	C	Kernel	C	η	Kernel	C
Banana	$\gamma = 100$	5	10^{-4}	$\gamma = 100$	5	$\gamma = 30$	5×10^3	10^{-5}	$\gamma = 50$	10^3
B.cancer	Linear	100	10^{-4}	$\gamma = 20$	1	$d = 3$	10^4	10^{-2}	$d = 4$	50
Diabetes	$d = 3$	10	10^{-3}	$d = 2$	500	$\gamma = 15$	100	10^{-4}	$\gamma = 15$	100
F.solar	$d = 3$	5	10^{-5}	$d = 5$	1	$\gamma = 5$	100	10^{-3}	$\gamma = 5$	100
German	Linear	5×10^3	10^{-3}	$\gamma = 3$	100	Linear	5×10^3	10^{-4}	Linear	5×10^3
Heart	Linear	10^3	10^{-2}	$\gamma = 50$	0.1	$\gamma = 1$	10^4	10^{-4}	$\gamma = 1$	10^4
Image	$\gamma = 200$	10^4	10^{-6}	$\gamma = 200$	10^4	$\gamma = 200$	10^4	10^{-4}	$\gamma = 200$	10^4
Ringnorm	$\gamma = 100$	0.1	10^{-2}	$\gamma = 100$	0.1	$\gamma = 100$	0.1	10^{-2}	$\gamma = 100$	0.1
Splice	$\gamma = 10$	100	10^{-2}	$\gamma = 10$	100	$\gamma = 3$	500	10^{-3}	$\gamma = 3$	500
Thyroid	$\gamma = 200$	1	10^{-2}	$\gamma = 200$	1	$\gamma = 200$	500	10^{-3}	$\gamma = 200$	500
Titanic	Linear	500	10^{-5}	Linear	5	$d = 2$	10	10^{-2}	$d = 2$	10
Twonorm	$\gamma = 1$	10	10^{-4}	$\gamma = 50$	1	Linear	1	10^{-3}	$\gamma = 5$	5
Waveform	$\gamma = 3$	10^4	10^{-4}	$\gamma = 3$	10^3	$\gamma = 50$	5	10^{-2}	$\gamma = 50$	5
Iris	$d = 2$	5×10^3	10^{-2}	$d = 2$	500	$\gamma = 5$	10^4	10^{-3}	$\gamma = 5$	10^4
Numeral	$\gamma = 15$	500	10^{-3}	$\gamma = 15$	500	$\gamma = 10$	5×10^3	10^{-4}	$\gamma = 10$	5×10^3
Blood-cell	$\gamma = 50$	100	10^{-3}	$\gamma = 50$	50	$\gamma = 50$	10^4	10^{-3}	$\gamma = 50$	5×10^3
Thyroid(M)	$\gamma = 200$	10^4	10^{-5}	$\gamma = 200$	10^4	$\gamma = 200$	10^4	10^{-6}	$\gamma = 200$	10^4

Table 4. Comparison of training time in second

Data	SLS-SVM	SLS-SVM$_\Gamma$	FSLS-SVM	FSLS-SVM$_P$
Banana	0.0942	*__0.0435__	*0.0680	*0.0828
B.cancer	0.0171	*__0.0088__	*0.0151	*0.0153
Diabetes	0.1795	*__0.0113__	*0.1243	*0.1239
F.solar	0.8096	*__0.0274__	*0.1103	*0.1096
German	1.0531	*__0.1008__	*0.1205	*0.1268
Heart	__0.0110__	0.0155	0.0177	0.0164
Image	9.6861	*__0.8837__	*1.7498	*2.1738
Ringnorm	0.0794	*0.0734	*__0.0727__	*0.0786
Splice	5.2503	*4.9179	*1.0425	*__1.0407__
Thyroid	0.0050	*__0.0039__	0.0081	0.0082
Titanic	0.0082	*__0.0008__	*0.0070	*0.0060
Twonorm	0.0889	*0.0727	*__0.0382__	*0.0792
Waveform	0.0802	*__0.0724__	0.0823	0.0836
Iris	0.001	*−	0.005	0.004
Numeral	1.279	*0.209	*__0.694__	0.773
Blood-cell	147.7	256.9	*__69.67__	*118.5
Thyroid(M)	275.8	*262.9	*__135.2__	*139.9

Table 5. Comparison of the average recognition rates and the standard deviation of the rate in percent

Data	SLS-SVM	SLS-SVM$_P$	FSLS-SVM	FSLS-SVM$_P$
Banana	**89.66 ± 0.42**	*89.61 ± 0.42	88.49 ± 0.60	88.97 ± 0.60
B.cancer	73.40 ± 4.69	*73.71 ± 4.44	***73.88 ± 4.70**	*73.84 ± 4.55
Diabetes	77.04 ± 1.55	***77.17 ± 1.66**	*76.82 ± 1.60	*76.82 ± 1.60
F.solar	66.71 ± 1.71	*66.55 ± 1.63	***66.80 ± 1.72**	*66.76 ± 1.73
German	75.66 ± 2.16	***75.85 ± 2.05**	*75.48 ± 2.32	*75.51 ± 2.23
Heart	84.00 ± 3.07	*83.97 ± 3.34	***84.01 ± 3.10**	*83.93 ± 3.07
Image	**95.10 ± 0.58**	*94.81 ± 0.50	93.67 ± 1.18	93.43 ± 1.26
Ringnorm	98.52 ± 0.10	*98.52 ± 0.10	***98.53 ± 0.10**	***98.53 ± 0.10**
Splice	**89.34 ± 0.70**	*89.34 ± 0.70	85.38 ± 0.67	85.34 ± 0.67
Thyroid	95.57 ± 1.96	*95.69 ± 1.93	*95.68 ± 2.13	***95.77 ± 2.07**
Titanic	77.32 ± 1.13	***77.36 ± 1.15**	*77.02 ± 1.85	77.03 ± 1.84
Twonorm	97.55 ± 0.16	*97.41 ± 0.15	***97.62 ± 0.12**	*97.54 ± 0.17
Waveform	**90.29 ± 0.43**	***90.29 ± 0.42**	*89.91 ± 0.46	*89.91 ± 0.46
Iris	92.00	*92.00	*96.00	***98.67**
Numeral	99.39	*99.39	***99.76**	***99.76**
Blood-cell	**93.42**	93.39	93.00	93.00
Thyroid(M)	**94.89**	***94.89**	94.81	94.81

the eight problems which are almost small. In three large problems, the training time of the FSLS-SVM is shortest in the three problems. And, the training time of the FSLS-SVM is shorter than that of SLS-SVM in thirteen problems because the size of kernel matrix in training of FSLS-SVM is smaller than that of SLS-SVM. The training time of the SLS-SVM$_P$ is shorter than that of the FSLS-SVM$_P$ in fourteen problems. In large problems, the training time of the FSLS-SVM$_P$ is shorter than that of the SLS-SVM$_P$. Because Cholesky Factorization has complexity $O(\frac{1}{6}M^3 - \frac{1}{3}M)$, the complexity are reduced by dividing the training vectors into subsets. And, the complexity of LS-SVM in the empirical feature space depend on the number of the basis vector of the empirical feature space. Hence, the complexity of FSLS-SVM depend on the number of dimensions of the empirical feature space and the number of iterations. Therefore, our proposed methods are effective from the standpoint of the computational complexity.

Table 5 shows the average recognition rates and standard deviations of each problems for the conventional methods and the proposed methods. The maximum average recognition rate for each data set is shown in bold. The average recognition rates of the proposed methods, which differ significantly from that of the SLS-SVM in Welch's t-test (5%) or is higher than that of the SLS-SVM, is marked with *. The recognition rates of the FSLS-SVM do not differ significantly

Table 6. Comparison of the numbers of SVs

Data	SLS-SVM	SLS-SVM$_P$	FSLS-SVM	FSLS-SVM$_P$
Banana	174	304	*34	*44
B.cancer	**9**	179	19	19
Diabetes	108	*101	*48	*49
F.solar	71	87	*34	*35
German	**20**	304	*20	40
Heart	**13**	169	14	14
Image	907	*446	*175	*179
Ringnorm	400	400	*40	*40
Splice	977	977	*121	*121
Thyroid	90	136	*15	*15
Titanic	**3**	10	8	8
Twonorm	240	400	*23	*34
Waveform	392	400	*43	*43
Iris	14	23	*6	*6
Numeral	457	*455	*61	*67
Blood-cell	1933	3059	*271	*462
Thyroid(M)	2327	2823	*376	*376

from that of the SLS-SVM in almost the problems. And, the recognition rates of the FSLS-SVM is highest in six problems. However, for Splice dataset, the recognition rate of the FSLS-SVM is too worse than that of the SLS-SVM because the number of the basis vectors of the empirical feature space is insufficient. The recognition rates of the FSLS-SVM$_P$ do not differ significantly from that of the SLS-SVM$_P$ in eleven problems. Given these facts, there is little performance degradation from the standpoint of the generalization ability in the proposed method.

Table 6 shows the numbers in SVs of each problem for the conventional and the proposed methods. The minimum number of the SVs for each dataset is shown in bold. The number of SVs of the proposed method which is smaller than that of the SLS-SVM is marked with *. The number of SVs of the FSLS-SVM is smaller than that of SLS-SVM in fourteen problems. The number of SVs of the FSLS-SVM$_P$ is smaller than of SLS-SVM$_P$ in sixteen problems. For the SLS-SVM$_P$, the all training vectors become SVs in two problems. Although it can be improved by changing the threshold value. Thus, we can confirm that the solutions of the proposed methods are sparse.

5   Conclusion

In this paper, we presented two fast selection methods for the basis vectors of the empirical feature space. In the first method, by obtaining and using cosine values among the training vectors, we can fast-select the basis vectors of the empirical feature space. In the second method, we introduced the block addition to selection for basis vectors of the empirical feature space. As mentioned above, in large problems, the training time of the proposed methods is much smaller than that of the conventional methods. In small problems, the training time of the proposed methods were almost the same as that of the conventional methods. And, the generalization ability of the proposed methods was almost the same as that of the conventional method. Furthermore, the sparsity improved by the proposed methods. Therefore, our proposed method is more effective than the conventional SLS-SVM.

References

1. Vapnik, V.N.: Statistical Learning Theory. Wiley, New York (1998)
2. Abe, S.: Support Vector Machines for Pattern Classification. Advances in Pattern Recognition. Springer, London (2010)
3. Schlkopf, B.: The kernel trick for distances. In: Proceedings of the Neural Information Processing Systems 13 (NIPS 2000), pp. 301–307 (2000)
4. Suykens, J.A.K., Vandewalle, J.: Least squares support vector machine classifiers. Neural Process. Lett. **9**(3), 293–300 (1999)
5. Abe, S.: Sparse least squares support vector training in the reduced empirical feature space. Pattern Anal. Appl. **10**(3), 203–214 (2007)
6. Kitamura, T., Sekine, T.: A novel method of sparse least squares support vector machines in class empirical feature space. In: Huang, T., Zeng, Z., Li, C., Leung, C.S. (eds.) ICONIP 2012. LNCS, vol. 7664, pp. 475–482. Springer, Heidelberg (2012). doi:10.1007/978-3-642-34481-7_58
7. Mika, S., Rätsh, G., Weston, J., Schölkopf, B., Müller, K.R.: Fisher discriminant analysis with kernels. In: Proceedings of the IEEE Workshop on Neural Networks for Signal Processing IX, pp. 41–48 (1999)
8. Zdenek, D., Tomas, K., Martin, M., Alexandros, M.: Cholesky decomposition of a positive semidefinite matrix with known kernel. Appl. Math. Comput. **213**(13), 6067–6077 (2001)
9. Xiong, H., Swamy, M.N.S., Ahmad, M.O.: Optimizing the kernel in the empirical feature space. IEEE Trans. Neural Netw. **16**(2), 460–474 (2005)
10. Kitamura, T., Asano, K.: Sparse LS-SVM in the sorted empirical feature space for pattern classification. In: Proceedings of the International Conference on Neural Information Processing, pp. 549–556 (2015)
11. Nagatani, T., Ozawa, S., Abe, S.: Fast variable selection by block addition and block deletion. J. Intell. Learn. Syst. Appl. **2**(4), 200–211 (2010)
12. UCI machine learning repository. http://archive.ics.uci.edu/ml/datasets.html

Construction and Analysis of Meteorological Elements Correlation Network

Cui-juan Fang[1], Feng-jing Shao[1(✉)], Wen-peng Zhou[2],
Chun-xiao Xing[1], and Yi Sui[1]

[1] College of Computer Science and Technology,
Qingdao University, Qingdao 266071, China
sfj@qdu.edu.cn
[2] Institute of Science and Technology Information of Qingdao,
Qingdao 266003, China

Abstract. Analysis of the correlation between meteorological elements could help find climate changing patterns. In this paper, the time series of meteorological elements, such as pressure, temperature and humidity, are converted to a correlation network, in which nodes represent the correlation relation (state) between the two meteorological elements and edges represent the transformation between different states. By analyzing the topological properties of the correlation network (degree, strength, path, etc.), the correlation patterns between meteorological elements could be found. Empirical studies of Weifang with 9 years climate observation data show that the correlation network has a power-law distribution and sub-seasonal characteristics. The correlation between temperature and pressure are more strongly negative and it did not change significantly with the year went. The correlation shows a seasonal variation that more negative correlation in summer and the spring as follows.

Keywords: Correlation network · Time series · Topological properties · Meteorological factors

1 Introduction

Climate system is a complex system. Exploring its complexity has important theoretical significance and practical value. An important objective of studying meteorological networks is to understand the interdependent nature of meteorological elements. In recent years, meteorological networks that describe the correlation between climatic factors have become a research hotspot.

Guolin Feng and Lei Zhou constructed the temperature fluctuation network by using the Chinese temperature data and the coarseness method. They also constructed stochastic and chaotic fluctuation network for comparing with the temperature network [2]. For analyzing the regional characteristics of the temperature evolving in China, Zhou used the average temperature to construct temperature fluctuation network [3]. Qin et al. [4] constructed a small - world network and scale - free network of meteorological station. Tsonis et al. [6] found the community of climate network can reflect various global climate characteristics.

© Springer International Publishing AG 2017
F. Cong et al. (Eds.): ISNN 2017, Part I, LNCS 10261, pp. 71–80, 2017.
DOI: 10.1007/978-3-319-59072-1_9

However, researches mentioned above lacked the internal correlation of climates elements and its dynamic mechanism, such as how different climate elements correlated with time. The purpose of this paper is to study the internal correlation changes of meteorological elements, to analyze the meteorological change pattern.

2 The Method of Constructing a Meteorological Correlation Network

Let $<p_1, p_2, \ldots, p_n>$, $Q = <q_1, q_2, \ldots, q_n>$ two sequences with N elements. $P(s, l)$ means sub sequence of P starting from s-*th* and the length of it is l. Pearson coefficient is used for indicating the correlation between two sequences:

$$coor(P, Q, s, l) = \frac{\sum_{i=s}^{l+s-1} (p_i - \bar{p})(q_i - \bar{q})}{\sqrt{\sum_{i=s}^{l+s-1} (p_i - \bar{p})^2} \sqrt{\sum_{i=s}^{l+s-1} (q_i - \bar{q})^2}} \tag{1}$$

where $\bar{p}$ is average mean of p, $1 \le s \le n$, $1 \le l \le n$.

According to the value of $coor(P, Q, s, l)$, we use symbols P, p, U, n, N to describe different correlation between the two sequences. F is a function from $[-1, 1]$ to $\{N, n, U, p, P\}$, where

$$F(coor) = \begin{cases} P & (0.8 < coor(P, Q, s, l) \le 1, \text{Strong positive correlation}) \\ p & (0.3 < coor(P, Q, s, l) \le 0.8, \text{Weak positive correlation}) \\ U & (-0.3 < coor(P, Q, s, l) \le 0.3, \text{No significant correlation}) \\ n & (-0.8 < coor(P, Q, s, l) \le -0.3, \text{Weak negative correlation}) \\ N & (-1 < coor(P, Q, s, l) \le -0.8, \text{Strong negative correlation}) \end{cases} \tag{2}$$

$CS_l = <F(coor(P, Q, 1, l)), F(coor(P, Q, 2, l)), \ldots, F(coor(P, Q, n-l+1, l))>$ is a sequence with length $n-l+1$, where each element indicates consecutive period with l length of symbol sequence. We call each element as the state of the correlation. For example, NNN indicates a strong negative correlation between two sub sequences.

Then, we take each element of CS_l as a node and transformation between nodes as edge. And the weight of edge is how many times of this transformation occurs. Formal description as follows: $CS_l(s)$ is a sequence, starting from s with l length. Nodes set is $N = \{CS_l(s) | s = 1, 2, 3, \ldots, n-l+1\}$; edges set is $E = \{CS_l(s), CS_l(s') | s' - s = 1\}$. The weight is how many times of this transformation occurs.

3 The Topological Properties of Meteorological Correlation Network

3.1 Degree

Degree is simple but important attribute in node properties. The correlation network we have established is directed, therefore the node degrees in this network includes out-degree and in-degree.

Take this network for example, the in-degree of one node explains direct conversion from other states to this state, and the out-degree of it represents a conversion from this state to another state. Edges are linked by time sequences. Therefore except the first and the last node, all other nodes' in-degree and out-degree must be equal. We choose the out-degree, namely a sequence directly converts to another. The degree of a node indicates the degree of its short-range correlation with other nodes. The larger the degree of a node, the more the node directly converters to its neighbor node.

3.2 Strength

Node strength means the total weight of its edges. The strength is defined as follows:

$$s_i = \sum_{j \in N_i} \omega_{ij} \tag{3}$$

N_i: a set that contains those nodes connecting and pointing to node i.
ω_{ij}: The weight of the edge from node i to node j.

Strength reflects the importance of the node in the network. It not only takes account of all neighboring nodes connected into account, but also considers compact degree between neighboring nodes and it [7].

In order to clearly observe the relationship between each node and the entire network, we define strength ratio of node as follows:

$$S_i = \frac{\sum\limits_{j \in N_i} \omega_{ij}}{\sum\limits_{i \in N, j \in N_i} \omega_{ij}} \tag{4}$$

We count continuous frequency of correlation symbol, and the results are in the following form:

$$cs\{1 : x_1, 2 : x_2, 3 : x_3, 4 : x_4, \ldots\}, cs \in \{N, n, U, P, p\}$$

Number represents continuous frequency of correlation symbol, $x_i (i = 1, 2, 3, 4 \ldots)$ is the occurrences number of continuous frequency. For example, when $cs = N$, occurrences number of N is x_1, the occurrences number of NN is x_2, the occurrences number of NNN is x_3.

3.3 Betweenness Centrality

The betweenness centrality of node i is defined as:

$$BC_i = \sum_{s \neq i \neq t} \frac{n_{st}^i}{g_{st}} \qquad (5)$$

g_{st} is the number of shortest path from node s to node t. n_{st}^i is the number of shortest path which from node s to node t and through the node i.

From the aspect of control information transmission, the higher BC, the important of the node is. Although in actual network, the transmission frequency of nodes is not the same, and not all transmission of nodes is based on the shortest path. The BC of the nodes still approximate depicts the influence of information flow on the network For the correlation network, understanding the central nodes can better know the changing process of double variables correlation.

3.4 Clustering Coefficient

The correlation network defined in this paper is a weighted network. Weighted clustering coefficient is adopted.

$$C^w(i) = \frac{1}{s_i(k_i - 1)} \sum_{j,k} \frac{(w_{ij} + w_{ik})}{2} a_{ij} a_{jk} a_{ki}, \qquad (6)$$

$C^w(i)$: the weighted clustering coefficient of node i.
s_i: the strength of node i; k_i: the degree of node i.
w_{ij}, w_{ik}: the weight of the edge $(i,j), (i,k)$.
$a_{ij} a_{jk} a_{ki}$: Whether the node i,j,k constitute a triangle. If $a_{ij} a_{jk} a_{ki} = 1$, they would constitute a triangle; If not, they did not constitute a triangle.

$0 \leq C \leq 1$. $C = 0$ If and only if all the nodes in the network are isolated nodes. $C = 1$ If and only if the network is a global coupled.

Weighted clustering coefficient is a statistical parameter which depicts gathered properties between neighbor nodes in complex network, and the higher the clustering coefficient is, the closer the degree of association is between adjacent nodes.

The clustering coefficient of the whole network C is the average of the clustering coefficients C_i^w of all the nodes i.

4 The Empirical of Meteorological Correlation Network

This article selects Weifang on January 1, 2006 to November 30, 2015, the temperature, humidity and pressure data. Because some data are missing, we actually access to 3590 sets of data.

For the symbol sequence CS_l, we take $l = 3$. It indicates that the symbol sequence P, Q is 3 consecutive days state. At that time, there are 125 states theoretically.

Using temperature and pressure data, we finally create a weighted directed network which contains 112 nodes, 405 edges.

4.1   Correlation Analysis

By computing the correlation coefficient of temperature and pressure, we found that the negative correlation samples are 2519, positive correlation samples are 1062, and completely irrelevant samples are 3. Through Eqs. 1 and 2, we got the symbol sequence. After counting the number of symbol sequence, we found that the symbol "N" turns up 1634 times, accounting for more than 45.6%; meanwhile, "n" appears 654 times, accounting for more than 18.2%. The correlation illustrating temperature and pressure is more likely to be negative one. Through experiments, we can find the correlation between humidity and pressure, humidity and temperature is also more likely to be negative one.

At the same horizontal plane, the air temperature changes is an important cause of the pressure change. When air cooling, air shrinkage, density, and weight per unit area under the air column increasing, the pressure would rise. As a result, the cold air, is always coupled with an increase in air pressure. But usually the relationship between temperature and air pressure is also impacted by other factors, such as wind speed, location, etc. Thus in the correlation statistics of temperature and pressure, negative correlation usually occurred.

4.2   The Degree of Node

In the correlation network, the maximum out-degree of nodes is 5. It shows that the kinds of state conversion is no more than 5. There are 41 nodes with the maximum out-degree. The minimum out-degree of nodes is 1. In fact, nodes with minimum means that only one state could be transformed to, indicating stable transformation.

For example, there is only one state which state pPn could convert to, that is npp. We list those stable transformation in Table 1. The conclusion can be drawn that those transformation are stable but rare changing pattern.

Table 1. Nodes with minimum out-degree

Nodes with minimum out-degree	Nodes being pointed to	Number of transformation
nPn	Pnn	1
PNP	NPU	
UNp	NpP	
pnP	npp	
Npn	pnn	
PUP	UPn	
NpN	pNN	
Pnp	npp	

4.3 The Distribution of Edge Weight

As is depicted in Fig. 1, the number of edges in the network decreases as the weight of the edge increases, showing a power-law distribution. In this figure, the weight of most edges is very small, only a small part of the edge of the weight value is larger. In the temperature and pressure correlation network, the maximum edge weight is 388 and the minimum is 1. In humidity-pressure and humidity-pressure correlation network, the number of edges and the weight of the edge show same character.

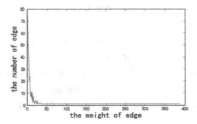

Fig. 1. The distribution of weight

4.4 Strength

The strength and the strength ratio of the temperature and pressure correlations network describe the degree of correlation states. The bigger the strength and the strength ratio, the more important the state is in the correlation network. The probability of occurrence and the probability of conversion to other states is larger.

From Table 2, the strength of node *NNN* and *NNn* are higher, which indicating that the strong negative correlation of states *NNN* and *NNn* for temperature and pressure is 575, 217. The cumulative distribution of the two nodes is 22.05%.

Table 2. Strength ration and strength of each node (Descending order)

Rank (R)	1	2	3	4
Node	NNN	NNn	nNN	nnn
Strength	575	217	206	129
Strength ratio (S)	16.03%	6.05%	5.74%	3.58%
Rank (R)	5	6	...	112
Node	nNn	UNN	...	pNP
Strength	96	70	...	1
Strength ratio (S)	2.66%	1.95%	...	0.03%

Using logarithmic scales on both the horizontal and vertical axes of the strength ratio(S) and rank(R), The linear regression equation is obtained:

$$y = -0.8655x = 2.66833, R^2 = 0.8322.$$

Therefore, the distribution of correlation sequence is following power-law distribution. As is shown in Fig. 2:

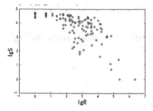

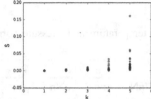

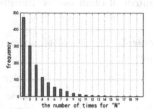

Fig. 2. The logarithm relationship between the point strength and its ranking

Fig. 3. The relationship between degree and strength

Fig. 4. The relationship between the number of "N" consecutive occurrences and the frequency

As is shown in Fig. 3: the strength of the node increases with the degree increasing in the network.

Through the frequency of the occurrence of continuous sequence in the correlation statistics, we found that strong negative correlation composition of the sequence length up to 19, and the frequency of "N" decreases with the increasing of consecutive times, as shown in Fig. 4.

Table 3. Edges whose weight rank in the top 24 in the network (Descending order)

Rank	1	2	3	4	5	6	7	8	9	10	11	12
Source	NNN	NNN	nNN	NNn	nnN	NNn	nNN	NnN	Nnn	nnn	NNN	UNN
Target	NNN	NNn	NNN	Nnn	nNN	NnN	NNn	nNN	nnn	nnN	NNU	NNN
weight	388	129	127	82	80	70	62	61	52	51	46	45
Rank	13	14	15	16	17	18	19	20	21	22	23	24
Source	Nnn	nnn	UnN	NNn	nNn	nnN	NnN	Unn	UUN	nNn	NnU	NNn
Target	nnN	nnn	nNN	NnU	Nnn	nNn	nNn	nnn	UNN	NnN	nUU	Nnp
weight	44	41	38	35	33	33	29	27	26	26	26	26

Table 3 lists the edge whose weight was ranked in top 23. Maximum weight value is 388, which appears in the NNN to its own ring. Those indicate that temperature and pressure show a strong negative correlation. It is found that those nodes whose weight rank in top 10 were composed by 'N' and 'n'. And 'U' first appeared in 11th, 'p' first appeared in 24th, 'P' first appeared in 38th. Again, the temperature and pressure show a strong negative correlation.

4.5 Betweeness Centrality

There are some nodes whose BC is bigger than others', which means any two states convert more likely to through these nodes, and these nodes are important for the control of other nodes' transitions (Table 4).

Table 4. The BC of all nodes in temperature and pressure correlation network (descending order)

Order	1	2	3	4
Node	UnN	pnn	nNn	pUn
BC	0.0549	0.0462	0.0442	0.0427
Order	5	6	7..	8
Node	nnp	PpU	nnP	PUU
BC	0.0410	0.0406	0.0401	0.0398
Order	9	10	…	112
Node	NNn	…	…	UUU
BC	0.0391	…	…	0

Theoretically those nodes with bigger BC have larger control power. We use temperature and pressure correlation network to analyze the highest BC rank of the nodes.

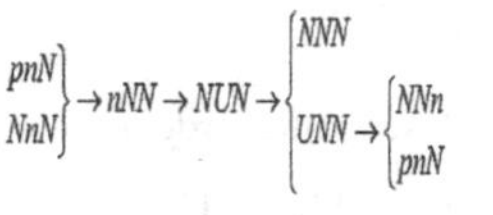

$$\left.\begin{matrix} pnN \\ NnN \end{matrix}\right\} \to nNN \to NUN \to \left\{ \begin{matrix} NNN \\ UNN \to \left\{ \begin{matrix} NNn \\ pnN \end{matrix}\right. \end{matrix}\right.$$

Fig. 5. The conversion pattern of network

$$\left.\begin{matrix} pnN \\ NnN \end{matrix}\right\} \to nNN \to NUN \to \left\{ \begin{matrix} NNN \\ UNN \to \left\{ \begin{matrix} NNn \\ pnN \end{matrix}\right. \end{matrix}\right.$$

with boxes *UnN*, *pnn*, *nNn*

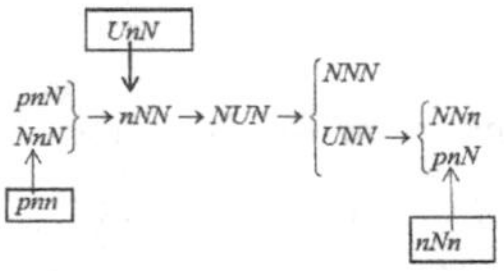

Fig. 6. The relationship between node (state) who has higher BC and conversion pattern of network

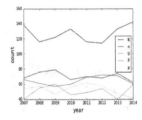

Fig. 7. The total number of characters with

In the network, we assume that any two nodes are named as node *i* and node *j*, and the weight of the edge represents the conversion frequency from node *i* to node *j*. From one state frequently converts to another state, we called them conversion pattern. In the network, the performance of the conversion pattern is from one node to another node, and the weight of the edge is big.

The conversion pattern in this network is as Fig. 5. It can be observed from Fig. 5 that there are more negative correlations in the correlation network and a few states with weakly or non-correlated symbols. It indicates that there are a lot of negative correlation between temperature and pressure, and a few positive correlation to a negative correlation. In Fig. 6, the node in the box with the higher BC, other nodes (states) from those nodes(states) convert to the nodes in the conversion pattern.

4.6   The Impact of the Season

First of all, we count the frequency of each character after symbolization changes with time. The results are shown in Fig. 7. It can be observed that in the eight years from 2007 to 2014, the correlation between temperature and pressure is a strong negative one, and the number of correlations varies little with the changing of year.

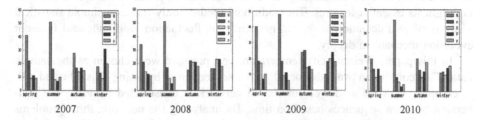

Fig. 8. Different season character statistical figure in 2007–2010

What can be observed from Fig. 8 is that the summer temperature and pressure show more negative correlation, followed by spring. In the autumn and winter, the correlation performance are relatively uniform, and there are no particularly significant correlation characteristics.

In order to further verify the different fluctuations of the temperature and pressure correlations in different seasons, we constructed the network for four seasons and found that the average clustering coefficient in the whole network is 0.007819, and the mean clustering coefficients of the two-variable correlation in spring, summer, fall and winter network respectively are 0.202, 0.208, 0.2179 and 0.1019. The average network coefficients of four seasons are respectively 20 times larger than those of the integrated network. Table 5 lists the nodes those clustering coefficients in top 4 in the four season networks.

Table 5. The nodes those clustering coefficients in top 4 in the four seasons networks

Season	Rank			
	1	2	3	4
Spring	NnN	Nun	UPU	pUP
Summer	NnN	Unp	NUN	UNn
Fall	pUp	ppP	nUn	NnN
Winter	UNN	UnU	NnN	nNN

From Table 5, we can observe that there are differ in nodes who has higher clustering coefficient between spring and summer network. In the spring and summer, showing a high negative correlation and in the autumn and winter, the relative performance of more uniform. Studying the clustering coefficient of the temperature and pressure network will provide us some reference in the future studies of the changing of temperature and pressure correlation [7].

5　Results and Discussion

The fluctuation of time series bivariate is a nonlinear, unsteady complex system, therefore it is difficult to reveal the fluctuation mechanism by using the existing traditional analysis methods. But it provides us a new way of thinking and methods in the field of physical and economic research. Using coarseness method, which symbol the coefficient correlation, abandoning the details of smaller level, is more conducive to highlight its essential features. The traditional method only makes a simple statistical analysis of its fluctuation state, but not study its fluctuation amplitude and inherent evolution mechanism deeply.

In this paper, Weifang city temperature and pressure were chosen as the sample data. The correlation was abstracted into one sequence by using Eqs. (1) and (2). A complex network model of correlation network is constructed by using the conversion between sequences based on time. By analyzing the network, three problems are solved, which are bivariate correlation fluctuation statistic, variation rule and evolution mechanism.

References

1. Tsonis, A.A., Roebber, P.J.: The architecture of the climate network. Phys. A Stat. Mech. Appl. **333**(4), 497–504 (2004)
2. Zhou, L., Zhi, R., Feng, A.X., Gong, Z.Q.: Topological analysis of temperature networks using bipartite graph model. Acta Phys. Sin. (Chin. Ed.) **59**(9), 6689–6696 (2010)
3. Gong, Z.-Q., Zhi, R., Zhou, L., Feng, G.-L.: Study on the regional characteristics of the temperature changes in china based on complex network. Acta Phys. Sin. (Chin. Ed.) **58**(10), 7351–7358 (2009)
4. Qin, K., Li, D.Y., Hu, X.L.: Research on weather data mining based on complex network (Chinese Edition). In: CCCN (2006)
5. Palu, M., Hartman, D., Hlinka, J., Vejmelka, M.: Discerning connectivity from dynamics in climate networks. Nonlinear Process. Geophys. **18**(5), 751–763 (2011)
6. Wang, X.F., Li, X., Chen, G.R.: The Theories and Application of Complex Network (Chinese Edition), p. 10. Tsinhhua University Press, Beijing (2006)
7. Tsonis, A.A., Wang, G., Swanson, K.L., Rondrigues, F.A., Costa, L.F.: Community structure and dynamics in climate networks. Clim. Dyn. **37**(5–6), 933–940 (2011)
8. Zhou, L., Gong, Z.Q., Zhi, R., Feng, G.L.: An approach to research the topology of chines temperature sequence based on complex network. Acta Phys. Sin. (Chin. Ed.) **57**(11), 7380–7389 (2008)
9. Gao, X.Y., An, H.Z., Fang, W.: Research on fluctuation of bivariate correlation of time series based on complex networks theory. Acta Phys. Sin. (Chin. Ed.) **61**(9), 1321–1323 (2012)

Classifying Helmeted and Non-helmeted Motorcyclists

Atsushi Hirota[1]([✉]), Nguyen Huy Tiep[2], Le Van Khanh[2], and Natsuki Oka[1]

[1] Kyoto Institute of Technology, Kyoto, Japan
hirota@ii.is.kit.ac.jp, nat@kit.ac.jp
[2] Panasonic R&D Center Vietnam Co., Ltd., Hanoi, Vietnam

Abstract. Riding a motorcycle without a helmet can cause serious injury. Although a crackdown on traffic violations is a good way to stop this unsafe practice, it is not realistic to manually find and arrest riders who do not wear helmets where there are numerous motorcycle riders, as in Vietnam. In consideration of this situation, we developed an automatic detection system for riders who are not wearing a helmet using deep learning. The proposed method's accuracy, precision, recall, and F-measure in classifying motorcyclists into helmeted and non-helmeted are 0.966, 0.957, 0.936, and 0.946, respectively. The quality of the classification was higher than in previous work which did not use deep learning. As with other image-processing systems using deep learning, our system achieved state-of-the-art performance. This system will reduce not only the number of motorcycle riders not wearing a helmet, but also the manual work of arresting illegal riders.

Keywords: Convolutional neural network · Classification · Helmet

1 Introduction

There is a very large number of motorcycles in use in several countries; for example, in Vietnam, Malaysia, and India. In the case of Vietnam, this is because cars are extremely expensive in terms of average annual income, and public transportation is insufficient. There is no subway yet available, and construction plans for a subway have been postponed many times. Considering this situation, it is expected that a very large number of motorcycles will continue to be used into the foreseeable future. Unfortunately, some riders do not wear helmets. Riding motorcycles without a helmet is not only illegal in Vietnam, but can also cause severe injury. To force motorcycle riders wear helmets, a enforcement of penalties for traffic violations is desirable. However, it is impractical to manually identify and arrest riders not wearing helmets from among the large number of active riders. Therefore, automatic detection of non-helmeted motorcyclists is required. In this paper, we report the implementation of a system capable of identifying riders without a helmet in real time using deep neural networks.

Visual object tracking via deep neural networks has been studied previously [1]. Compared with conventional methods using SIFT [2] or HOG [3],

© Springer International Publishing AG 2017
F. Cong et al. (Eds.): ISNN 2017, Part I, LNCS 10261, pp. 81–86, 2017.
DOI: 10.1007/978-3-319-59072-1_10

deep neural networks have achieved higher levels of performance [5,6]. We adopted deep learning because of its advantages, which are as follows.

- Fast execution: Although it takes an appreciable amount of time to train the network, once learning has been completed, it operates at high speed.
- High precision.
- Situation independence: Conditions such as weather, day or night, etc. can be disregarded.

2   Overall System Configuration

Our system consists of an IP camera for recording traffic, a motorcycle detector for identifying motorcycles and their riders, a preprocessing system for resizing and cropping the upper bodies of motorcycle riders in images, a classifier to categorize riders into helmeted and non-helmeted, and a report system for recording the images of non-helmeted riders with the corresponding time stamps. Figure 1 shows the information flow of the system. We constructed the motorcycle detector, the preprocessor, and the classifier. This paper focuses on the classifier, which is described in detail in Sect. 3.

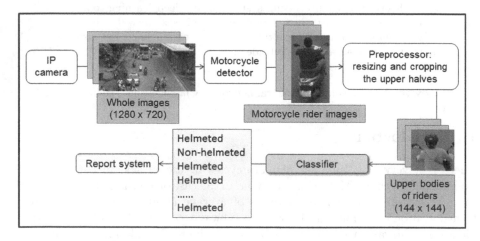

Fig. 1. Information flow of the overall system, which consists of an IP camera, a motorcycle detector, a preprocessor, a classifier, and a report system.

As for the IP camera settings, the shutter speed must be fast enough to capture images clearly. We captured the back views of riders in order to read their motorcycles' license plates. The size of the whole image was 1280 × 720 to allow recognition of the numbers on the plates. The motorcycle detector successfully cut out almost all motorcycle rider images from the entire image using deep learning technology. Details of the motorcycle detector will be reported in another paper. The sizes of the clipped images vary depending on the situation. In order to fix the images' sizes and improve classification accuracy, we resized them to 144×288 after which the system cropped the upper halves of the resized images. We call these the resizing and cropping processes *preprocessing*. Preprocessed images are sent as input to the classifier, where each image is classified into helmeted or non-helmeted. The classifier is explained in detail in the next section. In the final report system, if the result of the classification is non-helmeted, the system records the rider's image and the time when the motorcycle passed the detector site.

3 Rider Classifier

This section describes how to classify images of riders into helmeted and non-helmeted.

3.1 Network Design

Each motorcycle rider image is classified using deep learning technology. The features are computed by propagating a mean-subtracted 144×144 RGB image in the forward direction through three sets of convolutional layers, a pooling layer, a normalization layer, and three fully connected layers. A final softmax layer outputs the probability distribution as to whether a rider is wearing a helmet or not. The system processes each frame in a batch. Each convolutional hidden layer convolves k filters of kernel size $h \times h$ with stride s and pad p with the input image and applies a rectified linearity. In each pooling layer, max pooling is done over a 3×3 region with stride 2 without any padding. The hyper-parameters of the convolutional layers are shown in Table 1.

Table 1. Details of the convolutional layers.

Layer name	Kernel size $h \times h$	Stride s	Pad p	Number of filters k
conv1	11×11	4	0	64
conv2	5×5	1	2	128
conv3	3×3	1	1	256

To implement the classifier, we used *Caffe*[1], which was developed by the Berkley Vision Learning Center, as a frame work for deep learning.

3.2 Training

Images outputted from the motorcycle detector include those in which motorcycles and riders are not properly cropped. We thus manually eliminated such improper images and obtained 3,984 helmeted and 1,838 non-helmeted rider

---

[1] http://caffe.berkeleyvison.org/.

images. We divided these into training data and validation data randomly. The ratio of the training set and the validation set was 4:1. We trained a 13-layer classifier network with the training data. It took about 30 min to train the network with a single GPU.

4   Results

4.1   Classifier Performance

We evaluated the classification performance of the classifier by cross validation. It should be noted that the performance evaluation of the classifier is based on the assumption that the motorcycle detector performed ideally. This is valid because inappropriate images were manually excluded as described in Sect. 3.2. In this test, we calculated the accuracy, precision, recall, and F-measure with the data set of 3,984 helmeted and 1,838 non-helmeted rider images described in Sect. 3.2. Table 2 shows the result. The performance of our classifier was better than in previous work [4]. Waranusast et al. [4] showed an accuracy of 0.74 in the classification of helmeted and non-helmeted riders using a k-nearest neighbor (kNN) classifier. Note that we used different data from that in the previous work.

Table 2. Classification result.

Measure	Value
Accuracy	0.966
Precision	0.957
Recall	0.936
F-measure	0.946

4.2   Execution Speed

The processing speed of the three components of the motorcycle detector, preprocessor, and classifier (see Fig. 1) was 20.2 frames per second (fps) on average. Hence, the system functions in real time.

5   Discussion

5.1   Typical Errors

There were two typical error sources. Firstly, when a motorcycle rider wore a black helmet whose color was similar to the hair color of most Asians, the classifier sometimes judged them as non-helmeted riders. Secondly, in the case of two people riding a motorcycle together where one wore a helmet and the other did not, the classifier tended to judge them as helmeted riders. There were not many images of these kinds, so these two types of errors would be reduced by gathering more data to train the classifier network.

5.2   Importance of Preventing False Positives

Table 3 shows all the possible cases in the classification. In this work, it is important to reduce the number of false positives (FPs) because, if the system judges a rider who actually wears a helmet as a non-helmeted rider, the rider is erroneously reported as having violated traffic rules. Consequently, it is necessary to check all the images that were judged as positive by the system and remove false positives manually. In general, decreasing FPs tends to increase the number of false negatives (FNs). In other words, if we try to reduce the number of errors in which a rider who actually wears a helmet is judged to be a non-helmeted rider, the other error of judging a rider who does not wear a helmet to be a helmeted rider will occur more often. However, FN errors are not as critical, as the system then fails to report some illegal riders but still poses an adequate deterrent against traffic violations. Therefore, we prefer to decrease the number FP errors even though the number of FN errors will then increase. In implementation, we used the following condition to reduce the number of FPs:

$$\begin{cases} if\ T \leq 5 \\ \quad if\ N \leq 4\ then\ helmeted \\ otherwise \\ \quad if\ N \leq T/2\ then\ helmeted \end{cases}$$

where T represents how many times a rider is classified, and N designates how many times he/she is classified as a non-helmeted rider. Note that each rider is classified multiple times. This is because, as long as a motorcycle continues to be detected by the motorcycle detector, it is classified every time it is detected in order to increase accuracy. Using the above condition, we reduce the number of FPs. We determined the hyper-parameters in the above condition by trial and error.

Table 3. All possible classification cases.

		Actual result	
		Helmeted	Non-helmeted
System prediction	Helmeted	True negative (TN)	False negative (FN)
	Non-helmeted	False positive (FP)	True positive (TP)

6   Conclusion

We are developing a system to detect motorcyclists who are not wearing a helmet in traffic images. This paper focused on the classifier of riders into helmeted and non-helmeted. The performance cross validation of the classifier revealed that it outperformed previous work which did not use deep learning. We will develop license plate recognition software in the report system (see Fig. 1), which is necessary in order to arrest illegal riders. Once the entire system has been completed, it will reduce the instances of motorcycle riders not wearing helmets and will reduce the manual work required to arrest illegal riders.

References

1. Girshick, R., Donahue, J., Darrell, T., Malik, J.: Rich feature hierarchies for accurate object detection and semantic segmentation. In: Proceedings of the IEEE Conference on Computer Vision and Pattern Recognition, pp. 580–587 (2014)
2. Lowe, D.G.: Distinctive image features from scale-invariant keypoints. Int. J. Comput. Vis. **60**(2), 91–110 (2004)
3. Dalal, N., Triggs, B.: Histograms of oriented gradients for human detection. In: 2005 IEEE Computer Society Conference on Computer Vision and Pattern Recognition, pp. 886–893 (2005)
4. Waranusast, R., Bundon, N., Timtong, V., Tangnoi, C., Pattanathaburt, P.: Machine vision techniques for motorcycle safety helmet detection. In: 28th International Conference on Image and Vision Computing, New Zealand, pp. 35–40 (2013)
5. Ciregan, D., Meier, U., Schmidhuber, J.: Multi-column deep neural networks for image classification. In: Proceedings of the IEEE Conference on Computer Vision and Pattern Recognition, pp. 3642–3649 (2012)
6. Krizhevsky, A., Sutskever, I., Hinton, G.E.: Imagenet classification with deep convolutional neural networks. In: Advances in Neural Information Processing Systems, pp. 1097–1105 (2012)

Dominant Set Based Density Kernel and Clustering

Jian Hou[1,2(✉)] and Shen Yin[3]

[1] College of Engineering, Bohai University, Jinzhou 121013, China
dr.houjian@gmail.com
[2] ECLT, Università Ca' Foscari Venezia, 30124 Venezia, Italy
[3] Research Institute of Intelligent Control and Systems,
Harbin Institute of Technology, Harbin 150001, China

Abstract. The density peak based clustering algorithm has been shown to be a potential clustering approach. The key of this approach is to isolate and identify cluster centers by estimating the local density of data appropriately. However, existing density kernels are usually dependent on user-specified parameters evidently. In order to eliminate the parameter dependence, in this paper we study the definition of dominant set, which is a graph-theoretic concept of a cluster. As a result, we find that the weights of data in a dominant set provides a non-parametric measure of data density. Based on this observation, we then present an algorithm to estimate data density without parameter input. Experiments on various datasets and comparison with other density kernels demonstrate the effectiveness of our algorithm.

Keywords: Density peak · Clustering · Dominant set · Density kernel

1 Introduction

Data clustering has wide application in pattern recognition, image analysis and fault diagnosis [14,15]. In the past decades, much efforts were devoted to data clustering and various clustering algorithms have been proposed. Unfortunately, in applying these algorithms to real clustering tasks, there are still many problems to be solved. The k-means-like algorithm, NCuts [12] and the general spectral clustering algorithms [17] uses as input the number of clusters and their results rely on the parameters heavily. In addition, these algorithms tend to generate clusters of spherical shapes, and the results are also influenced by cluster center initialization. The DBSCAN [5], AP [1] and DSets [10] algorithms are able to determine the number of cluster automatically. However, all these three algorithms have their own problems. The DBSCAN algorithm depends on two parameters *Eps* and *MinPts* for density estimation. Generally, the other density based clustering algorithms are also dependent on user-specified density

---

J. Hou—This work is supported in part by the National Natural Science Foundation of China under Grant No. 61473045 and by China Scholarship Council.

F. Cong et al. (Eds.): ISNN 2017, Part I, LNCS 10261, pp. 87–94, 2017.
DOI: 10.1007/978-3-319-59072-1_11

parameters. The AP algorithm must be fed the preference value of each data, which impacts on clustering results significantly. The DSets algorithm are built on the pairwise similarity matrix of the data to be clustered and is parameter independent in itself. However, in the case that data are represented as feature vectors, the estimation of data similarity usually introduces one or more similarity parameters, which have been found to impact on the clustering results [7,8]. Besides, both the AP and DSets algorithms have the tendency to generate spherical clusters only. These observations show that it is still necessary to explore new clustering algorithms, although there are already a vast amount of clustering algorithms in the literature.

Our work in this paper is on the basis of the density peak (DP) based clustering algorithm proposed in [11]. The DP algorithm is based on the assumption that cluster centers are density peaks and they are relatively far from each other. With the local density ρ_i of each data i and the distance δ_i to the nearest neighbor with higher density to represent the data in a decision graph, it is found that the cluster centers are with both high ρ and high δ, whereas the non-center data are with either small ρ or small δ. As a result, the cluster centers are isolated from non-centered data and it is relatively easy to differentiate between two kinds of data. By assuming that the label of one data is the same as that of its nearest neighbor with higher density, all the non-center data can be grouped into clusters sequentially.

Local density calculation is the key of the DP algorithm as it determines ρ, δ and then the cluster centers. In [11] the authors use cut-off and Gaussian kernels, both of which involve a cut-off distance d_c. Although [11] presents an empirical method to calculate the range of d_c, we have found that the clustering results vary significantly with different d_c's in this range. In order to solve this problem, in this paper we present a non-parametric density kernel based on the DSets algorithm. One important feature of the DSets algorithm is that each data in a cluster is assigned a weight. Our study of the dominant set definition shows that this weight reflects the relationship of the data with all the others, and can be viewed as a measure of the local density. By calculating the pairwise data similarity matrix properly, all the data can be included in one single cluster, and therefore the weights of all the data can be obtained with the DSets algorithm. We show that this process can be accomplished independent of user-specified parameters. The effectiveness of our algorithm is demonstrated in experiments and comparisons with other density kernels.

2   Density Peak Clustering

2.1   The DP Algorithm

The DP algorithm is proposed based on the following observations. First, cluster centers are usually the density peaks in the neighborhood. This means that compared with non-center data, cluster centers have relatively large local density ρ. Second, in practice few data are with the same local density, therefore the distance δ of one data to its nearest neighbor with higher density is usually

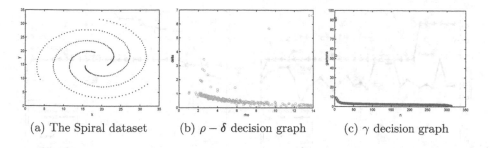

(a) The Spiral dataset   (b) $\rho - \delta$ decision graph   (c) γ decision graph

Fig. 1. The Spiral dataset and two decision graphs with the Gaussian kernel.

not large. In contrast, cluster centers are surrounded by data with lower density, therefore their δ's are relative large. In summary, the cluster centers usually have both large ρ's and large δ's, whereas non-center data have either small ρ's or small δ's. This difference between cluster centers and non-center data makes it possible to isolate and identify cluster centers from non-center data. Then based on the assumption that the label of one data is the same as that of its nearest neighbor with higher density, the non-center data can be grouped into clusters. Although this assumption has no theoretic ground, it is consistent with human intuition and works well in experiments.

From the above description we see that the key of the DP algorithm is the calculation of local density. While the local density can be estimated in different ways, existing approaches usually involve user-specified parameters, which may influence the density values and then the clustering results. For example, the cutoff kernel and Gaussian kernel is used in [11] to calculate the local density, and both kernels involve the cutoff distance d_c. The cutoff kernel measures the density by the number of data in the neighborhood of radius d_c, and the Gaussian kernel uses d_c as the decay parameter. After the local density ρ's are calculated, the distance δ_i is obtained by

$$\delta_i = \min_{j \in S, \rho_j > \rho_i} d_{ij}. \tag{1}$$

With the ρ's and δ's of all the data, we use a $\rho - \delta$ decision graph to illustrate the relationship of the cluster centers and non-center data in the $\rho - \delta$ space. Taking the Spiral dataset [3] for example, we show the $\rho - \delta$ decision graph in Fig. 1(b). For space reason, here we use only the Gaussian kernel and d_c is calculated by including 1.6% of all the data in the neighborhood.

It is evident in Fig. 1(b) there are three data with both large ρ's and large δ's, and they are presented as the outliers of the set of data. Obviously these three data are the centers of the three clusters. Considering that identifying cluster centers with the $\rho - \delta$ decision graph involves two thresholds, [11] then proposes to use $\gamma = \rho\delta$ as the single criterion of cluster center selection. We sort the data in the decreasing order according to their γ's and obtain the γ decision graph in Fig. 1(c), where the three cluster centers with large γ's can be recognized relatively easily.

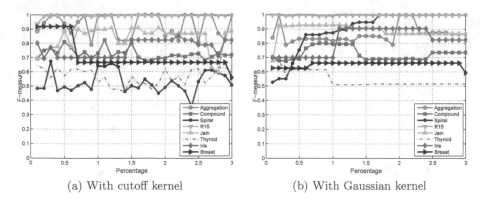

(a) With cutoff kernel   (b) With Gaussian kernel

Fig. 2. The influence of the percentage in calculating d_c on the clustering results.

In general, while the $\rho - \delta$ decision graph and γ's decision graph are helpful in identifying cluster centers, it is still quite difficult to find out the correct cluster centers automatically. In this paper we assume the number of clusters, N, is determined beforehand, and use the N largest γ's to identify the cluster centers.

2.2   The Problems

In both the cutoff and Gaussian kernels the parameter d_c needs to be specified, and it is suggested in [11] to determine d_c such that 1% to 2% of all data are included in the neighborhood on average. However, we have found that with both kernels, different values of d_c in this range causes significant variance in the clustering results. In addition, the best results may not be obtained with d_c in this range. In fact, we show how the clustering results vary with the percentages used to calculate d_c in Fig. 2, where we use F-measure to evaluate the clustering results. Eight datasets, namely Aggregation [6], Compound [16], Spiral, R15 [13], Jain [9] and three UCI datasets Thyroid, Iris and Breast, are used in experiments.

Figure 2 indicates that the percentage and the parameter d_c has a significant influence on the clustering results. Unfortunately, we cannot arrive at any useful conclusion as to the appropriate range of d_c from Fig. 2. In addition, it is not clear how the clustering results are correlated with d_c. Consequently, it is very difficult to determine the appropriate d_c.

3   Our Algorithm

In order to solve the parameter dependence problem of existing density kernels, in this paper we present a non-parametric density by making use of the nice properties of the DSets algorithm. In this section we firstly introduce the dominant set definition, and then present in details how the DSets algorithm can be used to calculate the local density.

3.1   Dominant Set

In order to derive the non-parametric density kernel based on the dominant set, we firstly present the definition of dominant set briefly. The details of the definition can be found in [10].

We use S to denote the set of data for clustering, and $A = (a_{ij})$ to represent the pairwise similarity matrix. With D as a non-empty subset of S and $i \in D$, $j \notin D$, we measure the relationship of j and i by

$$\phi_D(i, j) = a_{ij} - \frac{1}{|D|} \sum_{k \in D} a_{ik}, \tag{2}$$

with $|D|$ denoting the number of data in D. The we define

$$w_D(i) = \begin{cases} 1, & \text{if } |D| = 1, \\ \sum_{l \in D \setminus \{i\}} \phi_{D \setminus \{i\}}(l, i) w_{D \setminus \{i\}}(l), & \text{otherwise.} \end{cases} \tag{3}$$

With this key variable and $W(D) = \sum_{i \in D} w_D(i)$, the formal definition of dominant set can be presented as follows. A subset D such that $W(T) > 0$, for all non-empty $T \subseteq D$ is called a dominant set if

1. $w_D(i) > 0$, for all $i \in D$.
2. $w_{D \cup \{i\}}(i) < 0$, for all $i \notin D$.

In [10] the authors show that a dominant set can be extracted with game dynamics, e.g., replicator dynamics, developed in evolutionary game theory. Specifically, we use $x \in R^n$ to denote the weights of the data, which can be obtained by replicator dynamics. In this paper we adopt the more efficient dynamics proposed in [2]. It is shown that this weight vector corresponds to the weighted characteristic vector x^D of a dominant set D, which is defined as

$$x_i^D = \begin{cases} \frac{w_D(i)}{W(D)}, & \text{if } i \in D, \\ 0, & \text{otherwise} \end{cases} \tag{4}$$

In other words, after we obtain the weight vector, the data with positive weights form a dominant set. In extracting a dominant set, the weights of all the data for clustering can be initialized to $\frac{1}{n}$. The dominant sets can be obtained sequentially in a peeling-off manner [10].

3.2   Non-parametric Density Kernel

From the last subsection we observe that in a dominant set, each data i is assigned a weight equaling to $\frac{w_D(i)}{W(D)}$. On the other hand, Eq. (3) indicates that $w_D(i)$ measures the similarity between i and the other data, and a large $w_D(i)$ means that i has a high overall similarity with other data. It is evident that if i is in the central area of a dominant set, then it is likely that $w_D(i)$ is large and

i has a large weight. In contrast, one data i in the border area of a dominant set tends to have a small weight. Since the weights of data in a dominant set can be used to differentiate between the data in central and border areas, they can be treated as the data density in the DP algorithm. In this sense, we can make use of the dominant sets algorithm to calculate the density, and therefore treat dominant set as a density kernel.

However, in applying this density kernel to the DP algorithm, there are still two problems to be solved. First, while the dominant set extraction uses as input only the pariwise similarity matrix and no parameters are involved, the calculation of data similarity usually introduces parameters. For example, the commonly used similarity measure $s(i,j) = exp(-d(i,j)/\sigma)$ introduces the parameter σ. Second, in the case that there are more than one clusters in the dataset and the dynamics proposed in [2] are used, there will be some data with zero weights. These data with identical density will influence the clustering of non-center data negatively. By studying the definition of dominant set, in the following we show how to solve these two problems.

The definition of $w_D(i)$ in Eq. (3) indicates that a large $w_D(i)$ corresponds to large similarities between i and other data. Then the dominant set definition states that each data in a dominant set has a positive $w_D(i)$. This is equivalent to saying that each data in a dominant set are similar to all the others. As a result, the dominant set definition imposes a high requirement on the internal similarity in a dominant set. With a fixed dataset, the variance of σ results in the change of similarity value. A small σ leads to small similarity values, which further result in a large amount of small dominant sets. In contrast, a large σ corresponds to large similarity values and then a small number of large dominant sets. By adopting a sufficiently large σ, we can group all the data into a dominant set, and therefore assign non-zero weights to all the data. Although σ influences the similarity values, it does not change the magnitude ordering of these similarity values, and therefore has no influence on the ordering of data weights. Consequently, the value of σ does not impact on the DP clustering results, only if all data are assigned positive weights.

In practice, if σ is too large, many large similarity values may become identical due to limited digits after decimal. Therefore we use the following algorithm to determine the σ and generate the density used in the DP algorithm. With $\bar{d}$ denoting the average of pairwise distances, we build a list composed of $\bar{d}, 10\bar{d}, 50\bar{d}, 100\bar{d}, 200\bar{d}, \cdots$. Given a dataset, we assign σ with the values in the list from small ones to large ones, until all the resulted data weights are greater than zero.

4   Experiments

We test the proposed density kernel in experiments on the eight datasets, and compare the results with those from the cutoff kernel and Gaussian kernel. In addition, we also compare with some other algorithms, including k-means, DBSCAN, NCuts, AP and SPRG [17]. With k-means, NCuts and SPRG, we set

Table 1. Clustering results (F-measure) comparison on eight datasets.

	k-means	NCuts	DBSCAN	AP	SPRG	DP-cutoff	DP-Gaussian	Ours
Aggregation	0.83	0.99	0.80	0.82	0.73	0.99	0.99	0.94
Compound	0.68	0.70	0.88	0.77	0.64	0.82	0.69	0.81
Spiral	0.35	0.58	1.00	0.35	0.37	0.64	1.00	1.00
R15	0.82	0.99	0.77	0.54	0.93	0.99	0.99	0.95
Jain	0.79	0.63	0.87	0.57	0.86	0.90	0.87	1.00
Thyroid	0.83	0.64	0.68	0.52	0.97	0.55	0.51	0.72
Iris	0.89	0.93	0.77	0.93	0.87	0.70	0.90	0.78
Breast	0.96	0.64	0.87	0.82	0.97	0.67	0.66	0.69
Average	0.77	0.76	0.83	0.66	0.79	0.78	0.83	0.86

Table 2. Clustering results (Jaccard index) comparison on eight datasets.

	k-means	NCuts	DBSCAN	AP	SPRG	DP-cutoff	DP-Gaussian	Ours
Aggregation	0.64	0.98	0.67	0.71	0.49	0.98	0.99	0.87
Compound	0.46	0.46	0.84	0.69	0.42	0.71	0.47	0.74
Spiral	0.20	0.30	1.00	0.20	0.20	0.39	1.00	1.00
R15	0.65	0.99	0.42	0.25	0.83	0.96	0.99	0.86
Jain	0.53	0.42	0.91	0.29	0.63	0.71	0.65	1.00
Thyroid	0.64	0.40	0.57	0.29	0.90	0.29	0.29	0.58
Iris	0.69	0.79	0.59	0.77	0.66	0.51	0.73	0.60
Breast	0.87	0.39	0.78	0.56	0.89	0.48	0.40	0.54
Average	0.59	0.59	0.72	0.47	0.63	0.63	0.69	0.77

the required number of clusters as ground truth and report average results of 10 runs. With DBSCAN, the $MinPts$ is set as 2, manually selected from 1 to 10, and Eps is determined based on $MinPts$ [4]. For AP, the required preference value is manually selected to be $p_{min} + 9.2step$, where $step = (p_{max} - p_{min})/10$ and $[p_{min}, p_{max}]$ is the preference value range calculated with the method proposed by the authors of [1]. In DP algorithm with the cutoff and Gaussian kernel, the percentage of data used to calculate d_c is set as 1.1% and 2.0%, respectively, both of which are manually selected from 1, 1.1, 1.2, $\cdots$, 2.0. The comparison of these algorithms are presented in Tables 1 and 2, where F-measure and Jaccard index are used to evaluate the clustering results. The comparison indicates that in terms of average clustering quality, our non-parameter kernel performs better than the cutoff and Gaussian kernels, and our algorithm also outperforms some other algorithms with carefully selected parameters.

5   Conclusions

In this paper we present a non-parametric density kernel to be used in the density peak based clustering algorithm. We study the dominant set definition and propose to treat the extraction of dominant set as a density kernel which is independent of parameters. We compare with the cutoff and Gaussian kernels in the DP algorithm and also some other clustering algorithms to illustrate the effectiveness of the proposed density kernel. One problem with the proposed density kernel is the relatively high computation load involved in similarity calculation and density calculation, which will be studied in our future work.

References

1. Brendan, J.F., Delbert, D.: Clustering by passing messages between data points. Science **315**, 972–976 (2007)
2. Bulo, S.R., Pelillo, M., Bomze, I.M.: Graph-based quadratic optimization: a fast evolutionary approach. Comput. Vis. Image Underst. **115**(7), 984–995 (2011)
3. Chang, H., Yeung, D.Y.: Robust path-based spectral clustering. Pattern Recognit. **41**(1), 191–203 (2008)
4. Daszykowski, M., Walczak, B., Massart, D.L.: Looking for natural patterns in data: Part 1. Density-based approach. Chemom. Intell. Lab. Syst. **56**(2), 83–92 (2001)
5. Ester, M., Kriegel, H.P., Sander, J., Xu, X.W.: A density-based algorithm for discovering clusters in large spatial databases with noise. In: International Conference on Knowledge Discovery and Data Mining, pp. 226–231 (1996)
6. Gionis, A., Mannila, H., Tsaparas, P.: Clustering aggregation. ACM Trans. Knowl. Discov. Data **1**(1), 1–30 (2007)
7. Hou, J., Gao, H., Li, X.: DSets-DBSCAN: a parameter-free clustering algorithm. IEEE Trans. Image Process. **25**(7), 3182–3193 (2016)
8. Hou, J., Liu, W., Xu, E., Cui, H.: Towards parameter-independent data clustering. Pattern Recognit. **60**, 25–36 (2016)
9. Jain, A.K.: Data clustering: user's dilemma. In: Perner, P. (ed.) MLDM 2007. LNCS, vol. 4571, pp. 1–1. Springer, Heidelberg (2007). doi:10.1007/978-3-540-73499-4_1
10. Pavan, M., Pelillo, M.: Dominant sets and pairwise clustering. IEEE Trans. Pattern Anal. Mach. Intell. **29**(1), 167–172 (2007)
11. Rodriguez, A., Laio, A.: Clustering by fast search and find of density peaks. Science **344**, 1492–1496 (2014)
12. Shi, J., Malik, J.: Normalized cuts and image segmentation. IEEE Trans. Pattern Anal. Mach. Intell. **22**(8), 167–172 (2000)
13. Veenman, C.J., Reinders, M., Backer, E.: A maximum variance cluster algorithm. IEEE Trans. Pattern Anal. Mach. Intell. **24**(9), 1273–1280 (2002)
14. Yin, S., Gao, H., Qiu, J., Kaynak, O.: Descriptor reduced-order sliding mode observers design for switched systems with sensor and actuator faults. Automatica **76**, 282–292 (2017)
15. Yin, S., Gao, H., Qiu, J., Kaynak, O.: Fault detection for nonlinear process with deterministic disturbances: a just-in-time learning based data driven method. IEEE Trans. Cybern. (2016). doi:10.1109/TCYB.2016.2574754
16. Zahn, C.T.: Graph-theoretical methods for detecting and describing gestalt clusters. IEEE Trans. Comput. **20**(1), 68–86 (1971)
17. Zhu, X., Loy, C.C., Gong, S.: Constructing robust affinity graphs for spectral clustering. In: IEEE International Conference on Computer Vision and Pattern Recognition, pp. 1450–1457 (2014)

Web Content Extraction Using Clustering with Web Structure

Xiaotao Huang$^{(\boxtimes)}$, Yan Gao$^{(\boxtimes)}$, Liqun Huang$^{(\boxtimes)}$, Zhizhao Zhang,
Yuhua Li, Fen Wang, and Ling Kang

Huazhong University of Science and Technology,
Wuhan City, Hubei Province, China
{huangxt,wangfen,kling}@hust.edu.cn,
kevin_gao1212@163.com, huanglq2002cn@163.com,
245442801@qq.com, 1491378629@qq.com

Abstract. Web content extraction is an essential part of data preprocessing in web information system. An algorithm for web content extraction based on clustering with web structure is proposed. The whole process can be divided in two steps. In the first step, clustering with the web pages collected from different websites. During this processing, similarity measurement of web page based on dynamic programming of weight is used. First, the web page is parsed to DOM tree; second, the weight is assigned to every node according to the position of the node and the amount of nodes in same depth and the depth of the DOM tree; third, calculating the similarity of two pages according to the given formula. When the first step is finished, web pages with similar structure would be divided into a set. In the second step, pages in the same set are compared and the same parts of pages will be removed, thus the remain is the web content. Experiments show that the proposed algorithm works with great effectiveness and accuracy.

Keywords: Web content extraction · Similarity · DOM tree · Cluster

1 Introduction

With the rapid development of internet which result into the situation that huge amount of HTML pages is created on World Wide Web, more and more complex components in web page makes research on web page difficult, so web content extraction is an essential part of data preprocessing in web information system. Many researchers have paid a lot of effort in extracting content from HTML documents. Many approaches have been proposed, all of which can be divided into four categories [1].

Content Extraction Based on HTML Template. There are two methods to get HTML template. The one is that extracting the template from web page set in same structure. Always, pages from same website share the same HTML template, so the template can be made by artificial for the web pages with same structure. The another is that generalizing the characteristics from different pages and then extracting general abstract template. The accuracy of this kind method is well, but manual service is required.

© Springer International Publishing AG 2017
F. Cong et al. (Eds.): ISNN 2017, Part I, LNCS 10261, pp. 95–103, 2017.
DOI: 10.1007/978-3-319-59072-1_12

Content Extraction Based on Heuristic Rules. Heuristic rules are created by analyzing the HTML source code and generalize the characteristics of the real content and noisy, then formulate the heuristic rules. Xiong et al. has proposed an extraction algorithm of Chinese HTML content based on similarity [2]. However, this method only suits for the pages that contain long text and the tag of every paragraph are same. Chang Yaohong has proposed a method based on tags, which are not applicable to the unknown pages because of various layout of web pages from different websites [3].

Content Extraction Based on Vision Division. It is easy to find that a web page can be visually divided into several areas. So this kind of algorithm divide the page into several areas according to tag <style> and CSS source files and then find out the content area. Microsoft Research Asia has put forward an algorithm called VIPS which needs a large amount of calculation because it needs to analyze numerous CSS source files containing much source code [4–6]. Some other improved algorithms based on VIPS and TVPS (Table and vision based page segmentation) also show that they need much computing even if the accuracy has been improved.

Content Extraction Based on Machine Learning. In this kind of algorithm, the content layout, tags and text features are counted and the results are used to build and train model. Aanshi Bhardwaj put forward a novel approach to extract content. The ratio of the amount of words under a node and the amount of the subtrees is calculated, however, this method cannot be used to pages with little content because of the low accuracy [7]. Many researchers extracted the content by analyzing the amount and density of punctuation [8–12]. They didn't take the situation into consideration that noisy data are showed as text with punctuation just like real content.

It is necessary to propose a content extraction algorithm that needs less artificial participation, that costs less computing resources, whose accuracy doesn't vary dramatically according to different websites.

To calculate the web page similarity, some researchers have proposed a method that the similarity is calculated through calculating the similarity of tree route which is a path from root node to a leaf node and a DOM tree can be described by a set of tree route [13]. This method will cost less time and system resources and is easy to be implemented. But there are two defects, a. the duplicated paths would result in much unnecessary calculations; b. there are several optimum matching paths to one path. In order to solve those problems, Liao et al. proposed an improved algorithm of web structure similarity based on tree path matching which defined the sequence similarity and position similarity of tree path and which cut the costs of time and improved the accuracy, however it did not take the hierarchy into consideration [14]. Some other researchers advised that the similarity can be calculated through counting the nodes and links in a DOM tree [15, 16]. However, this method only paid attention to the nodes and links, did not take the hierarchy into consideration, which resulted into low accuracy.

2 CECWS Algorithm

According to the discussion above, we know that the same parts of two pages from same website are the noisy data. Thus, for a set of web pages whose structure are similar to each other, we can extract the content through comparing the pages and then remove the same parts, the remain would be the content. We call algorithm CECWS.

However, in actual applications, the data collection is made up of many pages collected from different websites. In order to divide the pages from same website into same set, the cluster algorithm will be used.

2.1 Select Cluster Algorithm

It is easy to find that the cluster result has the following characteristics: the number of the sets is uncertain, high density of the cluster, the distance between clusters is large.

The cluster algorithm could be divided into five kinds including partitioning methods, hierarchical methods, density-based methods, grid-based methods and model-based methods. K-MEANS is the representative algorithm for partitioning methods, however, the number of the cluster (K) need to be assigned at the beginning, so it is not applicable. Density-based methods have two parameters (radius and minimum number) which have to be assigned at beginning, the result would be inaccurate if the two parameters are assigned inappropriately. Grid-based methods usually are used for multidimensional data. Model-based methods need to build the model according to characteristics of target dataset, the model determines the accuracy of the result. Hierarchical methods would cost too much resources.

Compared with those five kinds methods, Canopy have many advantages, the logic is simple and it is easy to convergence. Usually it is used to pre-process the data quickly and extensively. We think that it is the best algorithm to cluster the dataset.

2.2 Web Page Structure Similarity Measurement Method

To cluster the dataset, web page structure similarity measurement is required. A web page structure similarity measurement method based on dynamic programming on weight is proposed.

The DOM tree show in Fig. 2 is the parsing result of the web page show in Fig. 1. According to the discussion above, node a, c and d are static node, while node e is dynamic node which is generated by script.

According to the features discussed above, a structure similarity measurement method based on average distributed weight. The process can be described as follow:

- Assume that the weight of the whole DOM tree is 1, the amount of the nodes whose depth is 1 is N, thus, the weight that every node get is 1/N;
- Distribute the weight the node got equally to his child nodes.
- Distribute the weight iteratively utile reaching the leaf node.

- As for leaf node x and y, if x equals to y, the similarity of x and y is the weight they got, if not, the similarity is 0. As for non-leaf node x and y, if x equals to y, the similarity is sum of their child nodes' similarity, if not, the similarity is 0.
  Define: As for non-leaf node x and y, if the tag name, element set value and the amount of child nodes of two nodes are same, x = y, otherwise, x ≠ y.
- The similarity of two DOM trees equals to the similarity of their root nodes.

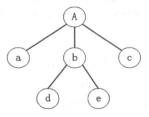

Fig. 2. DOM tree

Fig. 1. Web page structure

This method works well for most web pages, but in the situation that node b and c in two DOM trees are same correspondingly while node a are different, the similarity value using this method will be large enough to identify that two pages have the same template which is obviously wrong.

In order to solve this problem, a structure similarity measurement method based on dynamic programming on weight is proposed (SMDPW). We find that static areas always lay at the top and bottom of page while the content areas always lay at the center. In other words, the closer to the ends of page the areas lay at, the more likely it is static areas. Thus, the weight is distributed by dynamic programming, the closer to the ends of page the areas lay at, the bigger the weight is. This distribution strategy is only applicable to the nodes at depth 1. The other nodes get the weight which is calculated by distributing their parent nodes' weight equally.

In summary, the calculation process can be described as:

$$\text{sim}(P_1, P_2) = sim(P_1 X_{11}, P_2 X_{11}) \tag{1}$$

$$\text{sim}(P_1 X_{nm}, P_2 X_{nm}) = \sigma \times \left[\sum_{i=1}^{Num(n+1)} sim\left(P_1 X_{n+1,i}, P_2 X_{n+1,i}\right) \right] \times Lp_n(m) \tag{2}$$

$$Lp_n(m) = \begin{cases} 1, & n = 1 \\ \dfrac{\left|\frac{Num(n)\,+\,1}{2} - m\right| + 1}{\sum_{m=1}^{Num(n)} \left(\left|\frac{Num(n)\,+\,1}{2} - m\right| + 1\right)}, & n = 2 \\ \dfrac{1}{Num(n)}, & n > 2 \end{cases} \tag{3}$$

$$\sigma = \begin{cases} 1, & P_1X_{nm} = P_2X_{nm} \\ 0, & \text{other} \end{cases} \tag{4}$$

Where P is a DOM tree, P = $(X_{11}, X_{21}, X_{22}, \ldots, Xnm)$; n is the height of the tree, the depth of root node is 1; X11 is the root node; m is the sequence number of the node; Xnm is mth node at depth n; P_iX_{nm} is the node Xnm in tree Pi; $Num(n)$ is the amount of the node at depth n.

The algorithm SMDPW (P1, P2, N) can be described as following:

```
BEGIN
  IF P₁X₁₁ ≠ P₂X₁₁
    RETURN 0;
  IF Num(n + 1) = 0
    RETURN 1;
  IF n=1
    Lp₁(1) = 1;
  IF n=2
    Lp₂(m) = ...
  IF n>2
    Lpₙ(m) = 1/Num(n)
  FOR i = 1:Num(n)
    sim = SMDPW (P₁X₂ᵢ, P₂X₂ᵢ,n+1)
    sum+= Lp(2) × sim
  ENDFOR;
  RETURN sum;
END
```

$$Lp_2(m) = \frac{\left|\frac{Num(n)+1}{2}-m\right|+1}{\sum_{m=1}^{Num(n)}\left(\left|\frac{Num(n)+1}{2}-m\right|+1\right)}$$

$$Lp_n(m) = \frac{1}{Num(n)}$$

2.3 Content Extraction

As described in Sect. 2.1, the last step in our algorithm is to extract the content through comparing the pages and then remove the same parts, the remain would be the content. In order to remove the same parts from two web pages, we parse the web page to DOM tree first, then compare the corresponding nodes, the nodes will be removed if they are equal, otherwise, their brother nodes will be compared.

3 Experimental Result

We collect 5 group of news pages for our experiments, each group has 100 pages collected from 5 Chinese websites. We examine our algorithm using one group at a time. We implement the experiments programmed with java on a PC with 3.2 GHZ Core I5 2processors, 4 GB RAM and Win7 OS.

3.1   Performance Metrics

In our experiments, we use the precision P, the recall R and the harmonic mean F to evaluate our algorithm for every page. They are calculated as follows:

$$P = \frac{Length(C_M \cap C_E)}{Length(C_E)} \tag{5}$$

$$R = \frac{Length(C_M \cap C_E)}{Length(C_M)} \tag{6}$$

$$F = 2PR/(P + R) \tag{7}$$

Where C_M is the content of pages extracted by manual, C_E is the content of pages extracted by machine using our algorithm. Then we calculate the mean respectively among all pages:

3.2   Result

Since the threshold of Canopy in our algorithm is a previously determined value, there is a tradeoff between how high the threshold should be in order to balance precision, recall, and harmonic mean. Fig. 3 and Table 1 shows the change of R, P, F as the threshold increases.

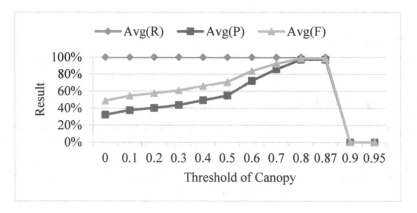

Fig. 3. Avg(P), Avg(R) and Avg(F) as the threshold increases from 0.1 to 0.95

In order to compare the performance of our algorithm with other approaches, we implement three other approaches. A, the approach based on page structure clustering proposed by Liao et al. [14]. B, a novel approach proposed by Aanshi and Veenu [7]. C, an approach based on layout similarity proposed by Yang et al. [1]. The results are presented in Table 2.

Table 1. Result of our algorithm when threshold is 0.8

	R	P	F
Yahoo!	100.00%	97.24%	98.60%
SOHU	100.00%	95.62%	97.76%
NETEASY	100.00%	96.51%	98.22%
CNN	100.00%	92.14%	95.91%
Average	100.00%	95.38%	97.62%

Table 2. Comparison between the three alternative algorithm and our algorithm

	Algorithm	Avg(R)	Avg(P)	Avg(F)
Yahoo!	Liao	67.63%	92.26%	78.05%
	Aanshi et al.	86.24%	91.25%	88.67%
	Yang et al.	97.01%	**95.63%**	**96.32%**
	Our	**100.00%**	93.24%	**96.50%**
SOHU	Liao	97.36%	92.47%	94.85%
	Aanshi et al.	76.64%	91.25%	83.31%
	Yang et al.	94.27%	90.53%	92.36%
	Our	**100.00%**	**95.62%**	**97.76%**
NETEASY	Liao	90.24%	**98.28%**	94.09%
	Aanshi et al.	97.34%	92.14%	94.67%
	Yang et al.	83.24%	92.73%	87.73%
	Our	**100.00%**	96.51%	**98.22%**
CNN	Liao	83.24%	97.01%	89.60%
	Aanshi et al.	95.24%	**97.24%**	**96.23%**
	Yang et al.	95.41%	89.37%	92.29%
	Our	**100.00%**	92.14%	95.91%
Average	Liao	84.62%	**95.01%**	89.15%
	Aanshi et al.	88.87%	92.97%	90.72%
	Yang et al.	92.48%	92.07%	92.17%
	Our	**100.00%**	94.38%	**97.10%**

3.3 Discussion

Why does R keep 100% no matter how the threshold changes? It is because that the content is extracted through comparing two pages who have similar structures and removing the same parts. In this process, the threshold can only affect the cluster result, in other words, inaccurate cluster result would result into the situation that noisy data cannot be removed exactly, which has no affection on the content. Thus, R would keep 100% no matter how the threshold changes, P would be low if the cluster result is inaccurate.

Why does R decreases rapidly when threshold is bigger than 0.87? It is because that the threshold bigger than 0.87 is too big to cluster a page into any set. According to the formula (3), in a three parts web page, the weight which is assigned to content is 20%, thus the minimum similarity between two pages (if they have same HTML template) is 80%; in a four parts web page, the minimum similarity is 82%. Besides, there will be some same parts in content, for the dataset in our experiments, the maximum similarity is 87%. Thus, if the threshold is bigger than 0.87, there will be no similar pages, every page in dataset will be divided into a single set, which result into that R and P are 0 as shown in Fig. 4-1. This is also the reason that threshold should better be set at 0.8.

Table 2 shows the performance of the other three algorithms and our algorithm. First, the results show that our algorithm achieves the best precision and the best harmonic mean. Second, our algorithm works stably, Third, the recall is good enough even if it is not the best.

4 Conclusion

We propose an algorithm for web content extraction based on clustering with web structure in which similarity measurement of web page based on dynamic programming of weight is used. Experimental results show that our algorithm works with great effectiveness and accuracy compared with three other algorithms.

In future, we plan to adopt our algorithm for Web data mining applications.

Acknowledgements. This work is supported by National Natural Science Foundation of China under grants 61572221.

References

1. Yang, L., Li, X., Geng, G.: Study of web pages content extraction based on layout similarity. Appl. Res. Comput. **32**(9), 2581–2586 (2015)
2. Xiong, Z., Zhang, H., Lin, M.: An extraction algorithm of Chinese HTML content based on similarity. J. Southwest Univ. Sci. Technol. **25**(1), 80–84 (2010)
3. Chang, Y., Zheng, Y., Chen, Y.: Content extraction technique for web pages based on HTML-tags. J. Comput. Eng. Des. **31**(24), 5187–5191 (2010)
4. Cai, D., Yu, S., Wen, J., et al.: VIPS: a vision- based page segmentation algorithm (2003)
5. Cai, D., Yu, S., Wen, J.-R., Ma, W.-Y.: Extracting content structure for web pages based on visual representation. In: Zhou, X., Orlowska, M.E., Zhang, Y. (eds.) APWeb 2003. LNCS, vol. 2642, pp. 406–417. Springer, Heidelberg (2003). doi:10.1007/3-540-36901-5_42
6. Mehta, R., Mitra, P., Karnick, H.: Extracting semantic structure of web document using content and visual information. In: Proceedings of the 14th Special Interest Tracks and Posters of International Conference on World Wide Web, pp. 928–929, ACM Press, New York (2005)
7. Aanshi, B., Veenu, M.: A novel approach for content extraction from web pages. In: Proceedings of 2014 RAECS UIET, pp. 6–8. Panjab University, Chandigarh (2014)
8. Peng, Q., Wang, Q., Li, Y., Zhang, J., et al.: Content extraction from chinese web pages based on punctuations distribution. In: International Conference on Computer Science and Service System, pp. 1351–1355 (2012)
9. Guo, Y., Tang, H., Song, L., et al.: ECON: an approach to extract content from web news page. In: International Asia-Pacific Web Conference, pp. 314–320 (2010)

10. Yang, Q., Yang, M.: A method of webpage content extraction based on point density. J. Intell. Comput. Appl. **5**(4), 42–44 (2015)
11. Lin, S., Chen, J., Niu, Z.: Combining a segmentation-like approach and a density-based approach in content extraction. Tsinghua Sci. Technol. **17**(3), 256–264 (2012)
12. Xiong, Z., Lin, X., Zhang, Y., et al.: Content extraction method combining web page structure and text feature. Comput. Eng. **17**(3), 256–264 (2013)
13. Joshi, S., Agrawal, N., Krishnapuram, R., Negi, S.: A bag of paths model for measuring structural similarity in Web documents. In: Proceedings of the 9th ACM SIGKDD International Conference on Knowledge Discovery and Data Mining (SIGKDD), pp. 577–582. ACM Press, Washington (2003)
14. Liao, H., Yang, Y., Jia, Z., et al.: An improved web structure similarity based on matching algorithm of tree paths. J. Jilin Univ. **50**(6), 1199–1203 (2012)
15. Joshi, S., Agrawal, N., Krishnapuram, R., Negi, S.: A bag of paths model for measuring structural similarity in web documents. In: SIGKDD 2003 (2003)
16. Cruz, I.F., Borisov, S., Marks, M.A., Webb, T.R.: Measuring structural similarity among web documents: preliminary results. In: Hersch, R.D., André, J., Brown, H. (eds.) EP/RIDT-1998. LNCS, vol. 1375, pp. 513–524. Springer, Heidelberg (1998). doi:10.1007/BFb0053296

Optimal KD-Partitioning for the Local Outlier Detection in Geo-Social Points

Teerawat Kumrai[1], Kyoung-Sook Kim[2(✉)], Mianxiong Dong[1], and Hirotaka Ogawa[2]

[1] Muroran Institute of Technology, Muroran, Japan
15096013@mmm.muroran-it.ac.jp, mx.dong@csse.muroran-it.ac.jp
[2] Artificial Intelligence Reserach Center,
National Institute of Advanced Industrial Science and Technology, Tokyo, Japan
{ks.kim,h-ogawa}@aist.go.jp

Abstract. Coupling social media with geographic location has boosted the worth of understanding the real-world situations. In particular, event detection based on clustering algorithms or bursty detection aims to find more specific topics that represent real-world events from geo-tagged social media. However, it is also necessary to identify unusual and seemingly inconsistent patterns in data, namely outliers. For example, it is difficult to obtain social media posted by residents of the places where a disaster is happening for quite some while. In this paper, we focus on a problem in partitioning a space to find a meaningful local outlier pattern by using a genetic algorithm (GA). We first describe a model of local patterns based on spatio-temporal neighbors and a normal distribution test. Then we propose our optimization process to maximize the number of patterns. Finally, we show results of the performance simulation with a real dataset related to a landslide disaster.

Keywords: Geo-social media · Spatio-temporal analysis · Outlier detection · KD-tree partitioning · Genetic algorithm

1 Introduction

In recent years, the analysis of geo-tagged social media (in short, geo-social media) is being emphasized to capture and predict real-world situations in emergency management, as evidenced by experiences from recent natural disasters such as the Tsunami and earthquake in Japan, and Hurricanes in the USA (Sandy and Katrina) and Haiti (Fay, Gustav, Hannah, and Ike). In [1], many Twitter messages were concentrated in the path of Hurricane Sandy and the tweets related to flooding showed a similar pattern to the path. We can find those messages reflect human experiences of the storm and allow us to obtain local information. Working with geo-location data in social media raises the need for the enhanced spatio-temporal data analysis to automatically discover potentially useful patterns and knowledge.

© Springer International Publishing AG 2017
F. Cong et al. (Eds.): ISNN 2017, Part I, LNCS 10261, pp. 104–112, 2017.
DOI: 10.1007/978-3-319-59072-1_13

There have been proposed many techniques and tools of spatio-temporal data mining: spatio-temporal clustering, hotspot detection, outlier detection, co-occurrence pattern discovery, and so on [2]. In particular, spatio-temporal clustering methods have been used to find more specific topics that represent real-world events, which unfold over space and time. For example, TwitterStand [3] captures tweets related to breaking news from noise by implementing online clustering. In [4,5], localized events within a small geographic area, such as public events and emergency situations, are extracted by using clustering techniques. Also, a space-time scan statistic approach is introduced in [6] to detect hotspots as topic events within a dataset across both space and time. They present spatio-temporal regions to depict disaster events from Twitter data. However, unusual and seemingly inconsistent patterns in data, namely outliers, sometimes represent useful information about abnormal situations in many applications, such as network traffic monitoring, credit card fraud detection, and outbreak of disease.

In this paper, we propose a genetic algorithm based approach that can detect spatio-temporal outliers with a multi-objective optimization function. First, we divide a spatio-temporal domain into several small areas and measure the keyword impartance of each area. Even though there are many definitions of outliers, we assume that a spatio-temporal outlier of a keyword in geo-social media has a different pattern of the local impartance measures with respect to the surrounding neighbors. Here, we classify patterns into 2 categories: L-pattern that is lower bound and H-pattern that is outbound. However, the problem is that we get a different result depending on the size and way of sub-space partitioning. In this paper, we focus on an optimal partitioning problem in a spatio-temporal domain as an NP-complete problem [8]. In order to find the optimal way to partition a space, we employ a genetic algorithm (GA) that is capable of making a global search and requires a shorter processing time than other meta-heuristics algorithms such as particle swarm optimization and ant colony optimization. The main contribution of this paper is summarized as follows:

- We apply a k-dimensional (KD) tree to divide a spatio-temporal domain into sub-cells.
- Two outlier patterns of a keyword are formulated by the local importance measures among the surrounding sub-cells.
- We investigate the optimal solution of partitioning to maximize the number of patterns.
- The simulation results are used to evaluate our approach.

The organization of this paper is as follows. Section 2 defines a problem of outlier pattern detection and Sect. 3 describes our genetic algorithm for the optimal solution of KD tree partitioning. Then, a simulation result of the proposed algorithm is described in Sect. 4. Finally, we conclude this paper with future work in Sect. 5.

2   Problem Statement

This section states the problem to partition a spatio-temporal domain and find outlier patterns. Let gp be a geo-social point as a tuple of (s, t, B), where

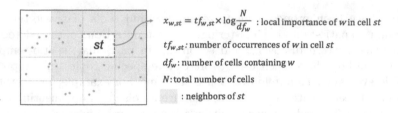

$x_{w,st} = tf_{w,st} \times \log\frac{N}{df_w}$: local importance of w in cell st

$tf_{w,st}$: number of occurrences of w in cell st

df_w: number of cells containing w

N: total number of cells

    : neighbors of st

Fig. 1. TF-IDF based on the two-dimensional grid partitioning (Color figure online)

$s = (lat, lon)$ is a geographical coordinate of latitude (lat) and longitude (lon), t is a timestamp, and B is a bag of words. Given a set of geo-social points, $GP = \{gp_1, gp_2, \cdots, gp_n\}$, in a spatio-temporal bounding box, $ST = ([lon_{min}, lon_{max}], [lat_{min}, lat_{max}], [time_{min}, time_{max}])$, consisting of three bounding intervals of longitudes, latitudes, and timestamps to cover all elements in GP, we examine the local importance measurement of a keyword, w, by applying tf-idf (term frequency-inverse document frequency) weighting scheme. Instead of document-based tf-idf, we divide ST into non-overlapping small cells, and each cell is regarded as a document to calculate tf-idf as shown in Fig. 1, i.e.; $ST = \{st_1, st_2, \ldots, st_N\}$. For each keyword, we scan the local importance score (x) of each cell. Then we identify the outlier pattern of each cell with respect to the surrounding neighbors. In this study, we simply use the normal distribution test (Z-test) [9] to detect outliers because our interest is a way how to partition ST. Given the local importance scores of cells, the Z value of a target cell is calculated as $Z = (\overline{x} - \mu_0)/(\sigma/\sqrt{n})$, where $\overline{x}$ is the average of local importance scores of queen contiguity neighbors (blue rectangles in Fig. 1) of the target cell, μ_0 are the mean value of the normal distribution, σ is the standard deviation of the local importance scores, and n is the number of neighbors including the target cell. If the Z values is more than the $Z_{\alpha/2}$, we count the target cell as one pattern of outbound outlier, called H-pattern. On the other hand, if the Z values of the target cell is less than the $-Z_{\alpha/2}$, we count it as one pattern of lower bound, called L-pattern.

Now we turn the optimization problem of the space partitioning. The reason why we use the space-partitioning approach instead of the distance metric is that the partitioning is suitable to apply a large dataset such as geo-social media. However, it leads to different results of the outlier detection depending on the size and shape of cells. In this paper, we take account of kd-tree [10] to divide ST as a three-dimensional (3D) space into small disjoint cells. The kd-tree is like a hybrid spatial grid and binary search tree and supports an efficient processing of range and nearest neighbor searches for high-dimensional point data comparing to other space partitioning methods. An example of 2-dimensional space partitioning by using kd-tree is showed in Fig. 2. From a subset of geo-social points containing keyword w, $GP_w \in GP$, we randomly select m-number of sampling points and create a kd-tree with them. Each node in the kd-tree represents one subdivision of ST. Then we determine whether each cell has an outlier pattern such as H- and L-pattern by the Z-test. However, the results have a critical issue

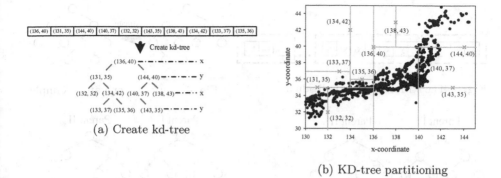

(a) Create kd-tree

(b) KD-tree partitioning

Fig. 2. 2-dimensional pattern example

about dependency of the samples. In order to find the optimal partition of space, we here employ a genetic algorithm (GA) as a well-known meta-heuristic approach that solves an optimization problem. It has already used for the outlier detection in several researches. In [11], two detection algorithms are proposed for high dimensional data based on the data distribution. Also, a fitness function based GA approach is introduced to detect outliers in [12]. Unlike the existing methods that focus on the detection of outlier objects, we consider a spatio-temporal outlier pattern with respect to the neighbors' point pattern.

3   Genetic Optimization Process

This section describes the process to seek the Pareto-optimal kd-tree partitioning for the outlier pattern detection. In order to operate GA, we define our objective function as follows:

$$U = \omega_1 HP + \omega_2 LP = \omega_1 \sum_{i=1}^{N} p_i^H + \omega_2 \sum_{i=1}^{N} p_i^L, \tag{1}$$

where HP and LP are the sums of H-pattern and L-pattern, ω_1 and ω_2 are the weighting factors of the number of two patterns, respectively, and N is the total number of sub-cells. The p_i^H and p_i^L is each outlier indicator of H-pattern and L-pattern of i-th sub-cell in ST as follows:

$$p_i^H = \begin{cases} 1, & if \ Z_i \geq Z_{\alpha/2} \\ 0, & otherwise \end{cases}, \qquad p_i^L = \begin{cases} 1, & if \ Z_i \leq -Z_{\alpha/2} \\ 0, & otherwise \end{cases}$$

In other words, we want to partition ST to get the maximum number of two outlier patterns. We assume if we get more patterns, then the areas would be smaller. Also we need more cells to obtain candidates where a certain unusual event happens. In Eq. (1), the function is dependent on two weight factors: ω_1 and ω_2 ($\omega_1 + \omega_2 = 1$). However, the finding the best value for the weights ω_1 and

Fig. 3. The structure of an individual

Fig. 4. Mutation operator example

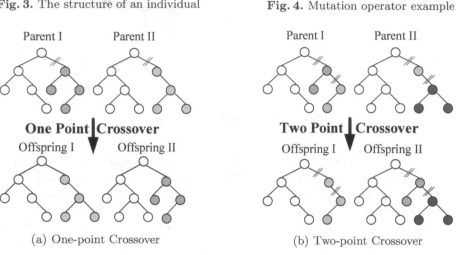

Fig. 5. Crossover operator example

ω_2 are very difficult. Thus, we consider the two objectives as a multi-objective optimization problem as well as maximizing the number of patterns.

The algorithm executes its optimization process to adjust the 3D kd-tree partitioning until the number of the generation reaches its maximum number of generations. After GA is finished, the set of solution and objective values are provided to select one of the solutions to use in 3D kd-tree partitioning for the detection of spatio-temporal outlier patterns. For the GA process, we define the population which consists of M individuals. Each individual i in the population represents by multiple segments, which is a set of the 3D coordinates of (x, y, t) in ST for kd-tree. The number of nodes in the kd-tree represents by the number of multiple segments in each individual. Figure 3 shows an example of the individual structure. Initially, the population will be generated by random coordinates in ST. Then, a fitness value of each individual is calculated by using Eq. (1). After that, the GA selects a pair of individuals who have the highest fitness values as parents by a selection operator (e.g., binary tournament and random selection). Then, selected parents reproduce two offspring by a crossover operator (e.g., one-point crossover, two-point crossover, and simulated binary crossover). The crossover operator operates with crossover rate. In this study, we use one-point crossover [15] and two-point crossover to implement the algorithm. An example of one-point and two-point crossover operator in our algorithm is showed in Figs. 5a and b, respectively.

The offspring can be evolved by a mutation operator (e.g., bit-flip mutation, uniform mutation, and polynomial mutation). The mutation operator randomly selects a node in kd-tree with a mutation rate. Also, the operator randomly decides to add or delete nodes. An example of mutation operator of the algorithm is shown as Fig. 4. The GA repeats these operators until the number of the offspring achieves size N. Then, this set of the offspring will be combined with the set of population. Finally, the algorithm selects the best M individuals from $M + N$ individuals by a selection operator as a new population for the next generation. After the fitness value error is satisfied or the maximum limit of the number of generations is found, the optimization processes of the GA are terminated.

4   Performance Simulation

This section shows the performance of our GA-based partitioning to maximize the number of H-patterns and L-patterns. We set up a simulation with a dataset of samples of Twitter messages that contains keyword 'landslide' related to 2014 Hiroshima landslides.

4.1   Simulation Configurations

The number of the sample data we used is 33,030 consisting of geographical coordinates (x and y) and timestamps (t). We calculated *tf-idf* values of keyword 'landslide' in each tweet. As mentioned in Sect. 2, we applied a normal distribution test (Z-test) to find the optimal number of H and L patterns. Two-tailed Z test with α is considered by 0.05. Therefore, the decision rule to find the number of H or L patterns is $Z \leq -1.960$ or $Z \geq 1.960$, respectively. The simulation parameters of GA were set as: 250 max generations, 100 population size, 0.9 crossover rate, and $1/n$ mutation rate. We evaluated our GA to find the optimal partition with the maximum number of H-patterns and L-patterns. The maximum number of kd-tree partitioning depends on the size of an individual in the Hiroshima dataset. We consider three cases of the individual size as shown in Table 1. There are two crossover operators are compared: one-point crossover and two-point crossover. The operators are compared in term of convergence.

Table 1. The size of an individual

Individual size	1% of Hiroshima	5% of Hiroshima	10% of Hiroshima
The number of coordinates	330	1652	3303

4.2 Simulation Results

In this simulation results, we show a performance metric that represents how individuals obtained from one algorithm outperform the individuals from another, called C-metric [16]. The C-metric between algorithm A and B is represented by $C(A, B)$ and calculated by $C(A, B) = |\{b \in B | \exists a \in A : a \succ b\}| / |B|$, where operator $\succ$ denotes the dominating (e.g., individual a dominates individual b is represented by $a \succ b$). Thus, there is at least one individual in A that dominated all individuals in B, if $C(A, B) = 1$. On the other hand, there is no individual in B that dominated by individual in A, if $C(A, B) = 0$.

Table 2. C-metric

Individual Size	1% of Hiroshima	5% of Hiroshima	10% of Hiroshima
C(OPX,TPX)	0.23	1.00	0.00
C(TPX,OPX)	0.02	0.41	1.00

Table 2 shows the C-metric at generation 200. The result shows that the one-point crossover (OPX) operator contribute in GA to better non-dominated frontier than the two-point crossover (TPX) operator in the case of the number of a partition that is 1% and 5% of the sample data. However, the number of a partition is 10% of the sample data, the OPX operator contributes in GA to better non-dominated frontier than the TPX operator, because there is at least one individual from the TPX operator dominated all individuals from the OPX operator, and there is no individual from the TPX operator that dominated by individual in the OPX operator.

Generation distance (GD) represents how fast an algorithm can optimize the solution. The minimum Euclidean distance from the non-dominated individuals to the Utopian point in objective space (U) at the end of each generation is calculated. It is measured as follows: $GD = \min_{i \in \mu} \sqrt{\sum_{k=1}^{n} (x_i[k])^2}$. Figure 6 shows the generation distance of the individuals at the end of each generation by using one-point crossover and two-point crossover. Figure 6a, b, and c show the generation distance in scenario 1%, 5%, and 10% of the individual size, respectively. Figure 6a shows that OPX contributes to increase the generation distance faster than TPX at the 100 and 150 generation. On the other hand, TPX contributes to increase the generation distance faster than OPX at the 50 and 200 generation. Figure 6c shows that OPX contributes to increase the generation distance faster than TPX at the 50 and 150 generation. On the other hand, TPX contributes to increase the generation distance faster than OPX at the 100 and 200 generation. However, OPX contributes to increasing the generation distance faster than TPX for 5% of the Hiroshima data as shown in Fig. 6b.

Figure 7a and b confirm the maximum number of the H-patterns and L-patterns by using one-point crossover (OnePointXover) and two-point crossover (TwoPointXover). The figures explain that the TwoPointXover is able to find the

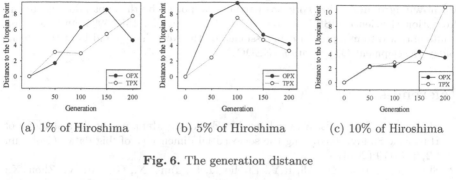

(a) 1% of Hiroshima   (b) 5% of Hiroshima   (c) 10% of Hiroshima

Fig. 6. The generation distance

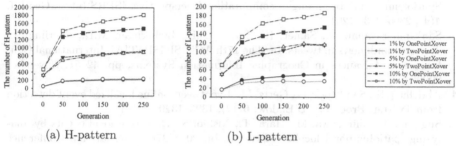

(a) H-pattern   (b) L-pattern

Fig. 7. The number of patterns

maximum number of H-patterns while maximizing the number of L-patterns in the case of 10% of the individual size better than the OnePointXover. However, in the 5% case, the OnePointXover is able to find the maximum number of H-patterns better than the TwoPointXover, but the TwoPointXover found the maximum number of L-patterns better than the OnePointXover. On the other hand, in the 1% case, the TwoPointXover is able to find the maximum number of H-patterns better than the OnePointXover, but the OnePointXover found the maximum number of L-patterns better than the TwoPointXover.

5   Conclusion

In this paper, we investigated the genetic algorithm for the outlier pattern detection in a spatio-temporal domain. We considered a normal distribution test (Z-test) of the *tf-idf* values of keyword in a subcell with respect to the surrounding neighbors. The simulation results show that the genetic algorithm is able to find an optimal kd-tree partitioning with the maximum number of the H and L outlier patterns. Also, the two-point crossover operator is able to find appropriate sets of the kd-tree partitioning for maximizing the number of the patterns in the both of H and L-patterns better than the one-point crossover operator in the case of the large number of partition. In future work, we will proceed experiments with large volumes of geo-social media and reduce the complexity of the GA considering a parallel processing to enhance our method.

Acknowledgments. This work was partially supported by the Japan Society for the Promotion of Science (JSPS) KAKENHI under Grant No. 15K15995, and based on results obtained from a project commissioned by the New Energy and Industrial Technology Development Organization (NEDO).

References

1. Shelton, T., Poorthuis, A., Graham, M., Zook, M.: Mapping the data shadows of Hurricane Sandy: uncovering the sociospatial dimensions of 'big data'. Geoforum **52**, 167–179 (2014)
2. Shekhar, S., Jiang, Z., Ali, R.Y., Eftelioglu, E., Tang, X., Gunturi, V., Zhou, X.: Spatiotemporal data mining: a computational perspective. ISPRS Int. J. Geo-Inf. **4**(4), 2306–2338 (2015)
3. Sankaranarayanan, J., Samet, H., Teitler, B.E., Lieberman, M.D., Sperling, J.: TwitterStand: news in tweets. In: The 17th ACM SIGSPATIAL International Conference on Advances in Geographic Information Systems, pp. 42–51, November 2009
4. Abdelhaq, H., Sengstock, C., Gertz, M.: Eventtweet: online localized event detection from Twitter. Proc. VLDB Endow. **6**(12), 1326–1329 (2013)
5. Sugitani, T., Shirakawa, M., Hara, T., Nishio, S.: Detecting local events by analyzing spatiotemporal locality of tweets. In: 2013 27th International Conference on Advanced Information Networking and Applications Workshops (WAINA), pp. 191–196, March 2013
6. Cheng, T., Wicks, T.: Event detection using Twitter: a spatio-temporal approach. PloS One **9**(6), e97807 (2014)
7. Kulldorff, M.: A spatial scan statistic. Commun. Stat.-Theory Methods **26**(6), 1481–1496 (1997)
8. LeFevre, K., DeWitt, D.J., Ramakrishnan, R.: Mondrian multidimensional k-anonymity. In: Proceedings of the 22nd International Conference on Data Engineering (ICDE 2006), pp. 25–25, April 2006
9. Bamnett, V., Lewis, T.: Outliers in Statistical Data. Wiley, Hoboken (1994)
10. Bentley, J.L.: Multidimensional binary search trees used for associative searching. Commun. ACM **18**(9), 509–517 (1975)
11. Aggarwal, C.C., Yu, P.S.: Outlier detection for high dimensional data. ACM SIGMOD Rec. **30**(2), 37–46 (2001)
12. Raja, P.V., Bhaskaran, V.M.: An effective genetic algorithm for outlier detection. Int. J. Comput. Appl. **38**(6), 30–33 (2012)
13. Kumrai, T., Ota, K., Dong, M., Champrasert, P.: An incentive-based evolutionary algorithm for participatory sensing. In: Global Communications Conference (GLOBECOM), pp. 5021–5025, December 2014
14. Deb, K.: Multi-objective Optimization Using Evolutionary Algorithms, vol. 16. Wiley, Hoboken (2001)
15. Poli, R., Langdon, W.B.: Schema theory for genetic programming with one-point crossover and point mutation. Evol. Comput. **6**(3), 231–252 (1998)
16. Zitzler, E., Thiele, L.: Multiobjective evolutionary algorithms: a comparative case study and the strength Pareto approach. IEEE Trans. Evol. Comput. **3**(4), 257–271 (1999)

V2G Demand Prediction Based on Daily Pattern Clustering and Artificial Neural Networks

Junghoon Lee and Gyung-Leen Park[✉]

Department of Computer Science and Statistics,
Jeju National University, Jeju City, Republic of Korea
{jhlee,glpark}@jejunu.ac.kr

Abstract. This paper presents how to manage the power consumption history in a microgrid, clusters days according to their time series patterns, and develops a prediction model for next day demand. Daily consumption patterns, each of which consists of quarter-hourly records, are grouped into 6 clusters, taking advantage of the dynamic time warping method in measuring the similarity between all feasible pairs of days. We select 3 main parameters for the cluster prediction of the next day, namely, month, day-of-week, and day-high temperature given by the weather forecast. For machine learning, learning patterns are generated after joining tables of power consumption, weather archive and day-group association on a daily basis. The next step builds an artificial neural network model using well-known open software. The model shows the accuracy of 67%, making it possible to estimate next day behavior, select the best demand model, and estimate power demand for vehicle-to-grid trades.

Keywords: Smart grid · Power demand estimation · Consumption pattern clustering · Dynamic time warping · Artificial neural network

1 Introduction

In modern power networks called the smart grid, EVs (Electric Vehicles) are making transportation as a part of the energy grid [1]. Having electric batteries, they can play a role of energy storage when they do not move. This feature brings a completely new energy service model called V2G (Vehicle-to-Grid), which makes EVs absorb over-produced electricity and emit when the system load is high [2]. As the two periods, namely, peak-load and low-load intervals, have different price rates, the EV owners can get economic benefits by buying cheap and selling high. Moreover, considering that the power facility is provisioned to match the peak demand in a target community, the V2G service can suppress the construction of a new power plant, as it can reduce or shift the energy peak.

This research was supported by Korea Electric Power Corporation through Korea Electrical Engineering & Science Research Institute. (Grant number: R15XA03-62).

© Springer International Publishing AG 2017
F. Cong et al. (Eds.): ISNN 2017, Part I, LNCS 10261, pp. 113–119, 2017.
DOI: 10.1007/978-3-319-59072-1_14

However, nation-wide or community-wide energy trade is not mature yet mainly due to the lack in the availability of bidirectional-flow batteries or attractive reward policies for selling electricity. Meanwhile, a microgrid having its own autonomous power system can be an active buyer for EV-stored electricity. During the peak time, it possibly does not use the expensive grid-supplying energy but takes cheap EV-stored electricity which has been bought much cheaper at an off-peak rate. For the V2G trade in which electricity flows from EVs to a microgrid, EVs must be plugged-in to the microgrid. Hence, the microgrid needs to decide not only how much electricity it wants to buy at a specific time instant but also from which EV it will buy much before the actual electricity injection [3]. Hence, the demand prediction is indispensable.

The energy consumption pattern is different microgrid by microgrid, according to how it is equipped, how many people will gather most, and the like. A microgrid can be an airport, a shopping mall, an office building, and the like. Hence, the prediction model must be different for each autonomous grid, and must be elaborately built based on its previous consumption history [4]. Here, many energy venders offer consumers with such records, making it possible to build an efficient energy consumption schedule and select an appropriate price plan [5]. In addition, recently, open software is widely available for processing a massive volume of big data, allowing even non-experts to conduct an intelligent data analysis.

In this regard, this paper analyzes the power consumption pattern in a campus of our university and builds a forecast model, aiming at the development of an efficient and convenient V2G service building block [6]. The process begins with cooking raw energy consumption files downloaded from the utility company and then groups each day according to the consumption shape by a hierarchical clustering method. After that, we make a prediction model which estimates to which group the next day will belong, mainly taking advantage of machine learning and big data handling techniques [7].

2   Data Processing

As shown in Fig. 1, our energy company provides a time series of power consumption records on quarter-hourly basis along with graphic display, after user authentication (http://pccs.kepco.co.kr). A record consists of timestamp, consumption amount (kwh), maximum consumption rate (kw), lagging reactive power ($kVarh$), leading reactive power ($kVarh$), lagging power factor, and leading power factor for each the 15-minute interval.

The set of records for a day are downloaded as a single file formatted by the HTML syntax. We download 366 records dating from 2015-7-1 to 2016-6-30. A C language program parses the set of files one by one, extracting those fields necessary in each record and standardizing the timestamp representation, and then converts to a series of SQL statements. The SQL script is moved to and executed on the MySQL server running on a Linux machine. Figure 2 shows that 35,136 records are currently inserted in the predefined *PowDemand* table having those fields described previously.

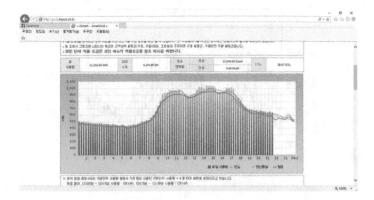

Fig. 1. Power consumption information

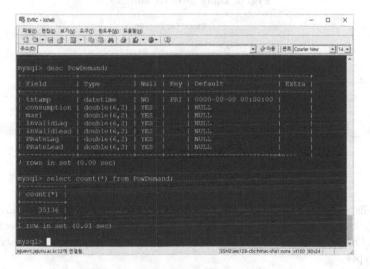

Fig. 2. Database table definition

To begin with, as the daily peak is very important to the power purchase plan and energy cost, Fig. 3 plots daily high consumptions for the whole period. We can see a seasonal effect as well as sharp decrease in weekends. In addition, at the beginning of the spring semester, the peak increases substantially, possibly due to the construction of new buildings.

Next, it is possible to retrieve information from this table in an R workspace, one of the most widely used statistics packages nowadays [8]. We take the timestamp and power consumption fields for all days by an R command and split them into 366 series, each of which includes 24 × 4 consecutive values. We implement an R script which compares the similarity between each series (366 × 365 pairs) according to the dynamic time warping method to identify the cluster of days. It takes quite a long time to calculate the distance between any two points belonging to the respective

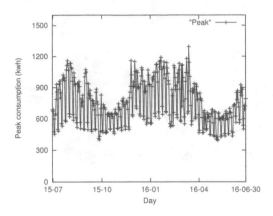

Fig. 3. Daily peak dynamics

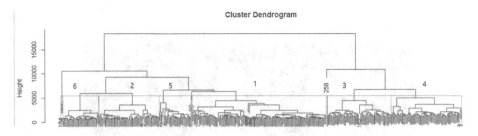

Fig. 4. Clustering result

series. Indispensably, the consumption pattern of a school will be different day by day according to whether it is during the semester or vacation as well as it is a weekday or weekend. Other microgrids also have different daily consumption curves according to their types.

The clustering result is shown in Fig. 4 and the number of days in each group is shown in Table 1. From the hierarchy created from the similarity measure, we can find the proper number of clusters for the target data set, with quite intuitive observation. Actually, there are outlier days. Some of them have corrupted data containing many 0's in their records. A day has extraordinarily large power consumption and we cannot know the reason. Most days belong to Group 1 and this group embraces weekdays during the semester. Group 2 includes semester weekdays and vacation weekends. Group 3 and Group 4 includes vacation weekdays and Group 4 shows much higher consumption. Group 5 includes some of vacation weekends and Group 6 some of semester weekends. It is not possible to discover such factors capable of perfectly classifying each group.

Figure 5 plots 6 curves, each of which is randomly selected day from a single group. Those curves have common features. First of all, the consumption stays low during the night and increases according to the beginning of work hours. How sharply the consumption increases decides the group. Group 4 shows the

Table 1. Clustering results

Group	Number of days
1	122
2	55
3	58
4	65
5	29
6	33

highest demand during the work hours. Miscellaneously, the demand slightly decreases around noon, or lunch time.

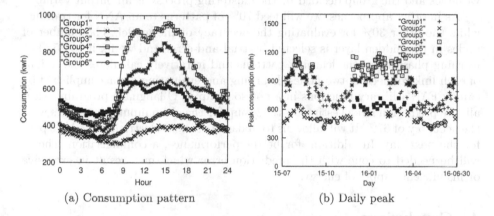

(a) Consumption pattern   (b) Daily peak

Fig. 5. Consumption statitics

In addition, Fig. 5 shows the daily peak and the group the day belongs to. The peak of each day is marked by a symbol representing its group. Group 3 and Group 4 are intensively found during the summer and winter vacation. Even though the number of students staying at the campus is small, the heating and cooling equipment spends quite much energy. If the cut-down of the peak load is the main goal, the arrangement of V2G trades must focus on this period, which is highly likely to suffer from the lack in the number of EVs participating in the trade. It is possible to assign a better reward or run a membership program. Semester weekends show low peak, hence the V2G trade is not so urgent. Semester weekdays can benefit from a better availability of EVs.

3   Data Fusion and Group Prediction

Assume that respective prediction models are developed for each group. Then, the V2G scheduler must know which group the next day will belong to and how much electricity is desirably purchased according to the predicted demand. The decision criteria can be weather forecast and temporal information such as month and day-of-week. The month field can account for the seasonal feature and whether the next day is in semester or vacation. First, we continuously retrieve the weather archive of our region. It comprises temperature, wind speed, insolation, and the like. The records are published on an hourly basis. In addition, the grouping result associates a day having a set of parameter instances with a specific group.

Machine learning is one of the most efficient and commonly-used methods for prediction. Particularly, artificial neural networks (ANN) can trace a non-linear behavior of a given time series based on the principle of learn-by-example [9]. Moreover, we can find versatile libraries of ANN such as FANN (Fast ANN). The first step is to make a set of learning patterns consisting of input and output variables. Here, we select month, day-of-week, and temperature as input variables and the group decided by the clustering process as an output variable.

As in other approaches, we will feed 70% of patterns to an ANN for training, while the other 30% for evaluating the accuracy of the model. The number of nodes in the hidden layer is selected by trial-and-error and we fix it to 25. The learning phase takes the learning patterns and iteratively calculates the weight of each link connecting two nodes. Learning and evaluating are accomplished by calling FANN library functions in a not so complex C language program. After all, the prediction model estimates the group for the given input parameters with the accuracy of 67%. It will allow us to make an efficient electricity purchase plan for the next day. In addition, for better performance, a compensation scheme will be needed to cope with the prediction error which may result in excessive or insufficient supply of energy.

4   Conclusions

In this paper, we have obtained the energy consumption records in our campus, grouped each day according to its consumption pattern, and developed a prediction model telling us which group the next day will belong to. Built on top of the combination of MySQL, R, and FANN, the hierarchical clustering process has found 6 groups with the dynamic time warping method. Then, FANN takes month, day-of-week, temperature forecast as input and the expected group as output to make a prediction model for a non-linear series, achieving the accuracy of 67%. This model makes it possible to develop a group-by-group energy demand model and thus build an efficient V2G service. In the whole steps, we have demonstrated the powerfulness and ease-to-use of open software and artificial intelligence. The smart grid will be enriched by abundant big data and sophisticated computational intelligence for high-level energy efficiency.

As future work, we are planning to combine more data streams such as charger status monitoring, energy consumption, wind and solar energy generation, and the like, for the sake of building an energy-efficient and eco-friendly green energy network in a city level [10].

References

1. Ramchrun, S., Vytelingum, R., Rogers, A., Jennings, N.: Putting the 'Smarts' into the smart grid: a grand challenge for artificial intelligence. Commun. ACM **55**(4), 89–97 (2012)
2. Bayram, I., Shakir, M., Abdallah, M., Qaraqe, K.: A survey on energy trading in smart grid. In: IEEE Global Conference on Signal and Information Processing, pp. 258–262 (2014)
3. Lee, J., Park, G.: A heuristic-based electricity trade coordination for microgrid-level V2G services. Int. J. Veh. Des. **69**(1/2/3/4), 208–223 (2015)
4. Lee, J., Park, G., Cho, Y., Kim, S., Jung, J.: Spatio-temporal analysis of state-of-charge streams for electric vehicles. In: 4th ACM/IEEE International Conference on Information Processing in Sensor Networks, pp. 368–369 (2015)
5. Goiri, I., Le, K., Nguyen, T., Guitart, J., Torres, J., Bianchini, R.: GreenHadoop: leveraging green energy in data-processing frameworks. In: Proceedings of Eurosys (2012)
6. Aman, S., Simmhan, Y., Prasanna, V.: Holistic measures for evaluating prediction models in smart grids. IEEE Trans. Knowl. Data Eng. **27**(2), 475–488 (2015)
7. Duerden, C., Shark, L.-K., Hall, G., Howe, J.: Prediction of granular time-series energy consumption for manufacturing jobs from analysis and learning of historical data. In: Annual Conference on Information Science and Systems, pp. 625–630 (2016)
8. Brunsdon, C., Comber, L.: An Introduction to R for Spatial Analysis & Mapping. SAGE Publication Ltd, Thousand Oaks (2015)
9. Nissen, S.: Neural Network Made Simple. Software 2.0 (2005)
10. Karnouskos, S., Ilic, D., Silva, G.: Assessment of an enterprise energy service platform in a smart grid city pilot. In: IEEE International Conference on Industrial Informatics, pp. 24–29 (2013)

An Arctan-Activated WASD Neural Network Approach to the Prediction of Dow Jones Industrial Average

Bolin Liao$^{(\boxtimes)}$, Chuan Ma, Lin Xiao, Rongbo Lu, and Lei Ding

College of Information Science and Engineering,
Jishou University, Jishou 416000, China
mulinliao8184@163.com

Abstract. Accurate prediction of the stock market index is a very challenging task due to the highly nonlinear characteristic of financial time series. For this reason, a single hidden-layer feed-forward neural network, activated by the arctan function, is proposed and investigated for predicting the Dow Jones Industrial Average. Then, a weights and structure determination (WASD) method is exploited to train the proposed arctan-activated neural network (termed arctan-activated WASD neural network). The relatively optimal weight and structure could be obtained by the presented WASD method. Numerical experiments are carried out based on huge amounts of historical data. The experimental results demonstrate the effectiveness and superior abilities of the arctan-activated WASD neural network for predicting the Dow Jones Industrial Average.

Keywords: Arctan · WASD neural network · Prediction · Dow Jones Industrial Average

1 Introduction

The stock market is a barometer of the country's economic situation. It reflects a country's capital mobility and expectations of the future economic trends [1,2]. Because the stock market interacts with many factors such as political events, general economic conditions and the expectations of traders, it possess the characteristics of non-linearity, discontinuities and high-frequency polynomial components [3]. In addition, with the help of modern technology, the rapid data processing of the stock market has led to very rapid fluctuations of stock market index. As a result, numerous banks, financial institutions, investors and stock brokers must buy and sell stocks in the shortest possible time, even within a few hours between buying and selling [4]. The topic has attracted more and more attention as well as research enthusiasm. Many researchers have devoted much effort to the predictability of the stock market and put forward many methods, such as the basic analysis, time series prediction and machine learning methods [5]. Among them, the artificial neural network (ANN) method is considered

© Springer International Publishing AG 2017
F. Cong et al. (Eds.): ISNN 2017, Part I, LNCS 10261, pp. 120–126, 2017.
DOI: 10.1007/978-3-319-59072-1_15

to be the best predictive method with a high level of validity in the field of stock market index forecasting [6]. However, some of the key points of the ANN structure should be carefully investigated. Because the choice of input variables directly affects the prediction accuracy, the definition of the optimal set of ANN input variables is one of the main problems in ANN. Number of neurons in the hidden layer, which often needs to be adjusted, is another key point of ANN. Unfortunately, there is no definite method to determine the optimal number of neurons in the hidden layer. Therefore, researchers usually employ the trial and error method to achieve this purpose [1]. However, the trial and error method is time-consuming, it does not apply to modern society that the time is money.

The Dow Jones Industrial Average, also termed the Dow Jones or simply the Dow, is a stock market index that shows how thirty large-cap companies located in the United States have traded during a trading cycle in the stock market. The Dow Jones Industrial Average is calculated by adding the prices of the thirty component stocks in the average and dividing by a divisor. Although the Dow Jones Industrial Average is compiled to evaluate the performance of the industrial department within the United States economy, the index's performance continues to be affected by not only economic reports, but also by political events such as war and terrorism, as well as by natural disasters that could potentially lead to economic harm. It is of great significance to make accurate prediction. In this paper, we propose a arctan-activated weights and structure determination (WASD) neural network to predict the Dow Jones Industrial Average. The WASD neural network comes from the error back propagation (BP) neural network. Theoretically, it can approximate any nonlinearity continuous function with arbitrary accuracy as a universal approximator [7]. In addition, the arctan-activated WASD neural network overcomes the weaknesses of classical neural networks, such as the presence of relatively slow convergence and local minima. Through illustrative experiments, the efficacy and the superiority of the proposed arctan-activated WASD neural network are well-verified.

The remainder of this paper is organized as follows. Section 2 provides a single hidden-layer feed-forward neural network, which is activated by the arctan function. Then, we describe experiments that used arctan-activated neural network to predict the return of Dow Jones Industrial Average, and we use arctan-activated WASD neural network and support vector machine (SVM) network for comparative analysis in Sect. 3. Finally, Sect. 4 presents conclusions.

2   Arctan-Activated WASD Neural Network and Theoretical Analysis

In this section, a multi-input arctan-activated WASD neural network is constructed and analyzed theoretically.

2.1   Arctan-Activated WASD Neural Network

The arctan-activated WASD neural network consists of K input layer neurons, an output neuron, and M hidden neurons. In this paper, we denote by w_{km}

the weight connecting the kth input-layer neuron (with $k = 1, 2, \ldots, K$) and the mth hidden-layer neuron (with $m = 1, 2, \ldots, M$), which is chosen randomly within interval (χ_1, ς_1) to make the period of the arctan function different. In addition, we denote by u_m the weight connecting the mth hidden-layer neuron and the output-layer neuron, which is determined in the next section by adopting the weights direct determination method [8,9]. Without loss of generality, the threshold b_m of the mth hidden layer neuron is randomly chosen in the interval (χ_2, ς_2), and the bias of the input-layer and output-layer neurons are all set to be zero simply.

Accordingly, we obtain the output of the mth hidden layer neuron as

$$\varphi_m = \arctan\left(\sum_{k=1}^{K}(x_k w_{k,m})\right) - b_m, \tag{1}$$

where x_k denotes the kth input-layer neuron. The output of the arctan-activated WASD neural network can be further described as

$$y = \sum_{m=1}^{M}(u_m \varphi_m). \tag{2}$$

The above equality means that the objective function is just a weighted combination of numerous arctan functions.

2.2   Weights Direct Determination Method

Hidden layer output matrix can be expressed as

$$H = \begin{bmatrix} \varphi_{0,0} & \varphi_{0,1} & \varphi_{0,2} & \cdots & \varphi_{0,M-1} \\ \varphi_{1,0} & \varphi_{1,1} & \varphi_{1,2} & \cdots & \varphi_{1,M-1} \\ \varphi_{2,0} & \varphi_{2,1} & \varphi_{2,2} & \cdots & \varphi_{2,M-1} \\ \vdots & \vdots & \vdots & \ddots & \vdots \\ \varphi_{N-1,0} & \varphi_{N-1,1} & \varphi_{N-1,2} & \cdots & \varphi_{N-1,M-1} \end{bmatrix} \in \mathbb{R}^{N \times M}, \tag{3}$$

where N is the number of samples, M is the number of neurons in the hidden layer, and $\varphi_{n,m}$ is the nth day's the mth hidden layer neuron output value. The connection weights of hidden layer neurons to output layer neurons can be directly determined as follows

$$u = (H^T H)^{-1} H^T \delta = H^+ \delta,$$

where H^+ is the pseudo-inverse of the output matrix H of the hidden layer, and δ is the column vector of the learning sample of Dow Jones Industrial Average.

2.3   Structure Determination Method

With regard to the arctan-activated WASD neural network, we define the following mean square error (MSE)

$$\text{MSE} = \frac{1}{N}\sum_{n=0}^{N-1}\left(\delta_n - \sum_{m=0}^{M-1} u_m \varphi_{n,m}\right)^2, \tag{4}$$

where δ_n is the nth day's actual value of the opening index of the Dow Jones. The training error E_{tra} and the internal check error E_{val} are calculated according to Eq. (4). The introduction of weighted error in the neural network can effectively prevent the occurrence of over-fitting phenomenon. Therefore, we define the weighted error of the arctan-activated WASD neural network as

$$E_{\text{tol}} = \alpha E_{\text{tra}} + (1 - \alpha)E_{\text{val}}, \tag{5}$$

where α is the proportion of training samples to the total number of samples. The weighted error E_{tol} is involved in the subsequent optimal structure determination process of the neural network.

First, we set the maximum number of resets. The hidden layer neurons are added one by one. When the weighted error no longer drops, the newly-added hidden layer neuron are reset (the weights of the input layer neurons to newly-added hidden layer neuron and the threshold of the newly-added hidden layer neuron). If the number of resets has reached the maximum number of resets and it still can not make the weighted error drop, then the structure is the optimal structure.

3 Experimental Verification

The Dow Jones Stock Average, also known as US30, is the worlds most influential and widely used stock index. It is based on the New York Stock Exchange listed on the part of the representative of the company stock as the preparation of the four stock price index composition. This paper uses 3748 valid data of the period from April 8, 2002 to November 11, 2016. The data includes the opening index, the highest index, the lowest index, the day's trading volume, and the closing index. In this paper, we set the weights of the input layer to the hidden layer and the thresholds of the hidden layer between $[-1, 1]$. In order to eliminate the impact of latitude on the results, the data need to be normalized. In this paper, we set the normalization interval as $[-1, 1]$.

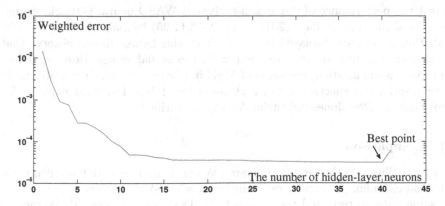

Fig. 1. Weighted error generated by arctan-activated WASD neural network

3.1 Training

In this paper, based on a lot of experiments, we use the first 2898 data for training and the last 850 data for internal verification (i.e., $\alpha = 2898/3748$). In addition, we set the maximum number of resets as 1300. To illustrate the relationship among the training error and M (i.e., the number of hidden-layer neurons), the corresponding result is displayed in Fig. 1. It is seen from Fig. 1 that by adding the hidden-layer neurons one by one, the weighted error first drops and then starts rise. Clearly, the number of hidden layer neurons at 40 is a turning point. Therefore, the optimal number of hidden layer neurons is approximately 40.

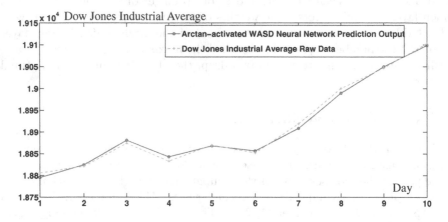

Fig. 2. Prediction results Arctan-activated WASD neural network

3.2 Prediction

To test the performance of the arctan-activated WASD neural network, we use the data of the next 10 days (2016.11.14–2016.11.25) to make predictions. The prediction results are displayed in Fig. 2. From this figure, on can observe that the predicted results are in agreement with the actual values. Besides, before the inverse normalization, the value of MSE is $1.781e - 06$, which is quite small. Consequently, the effectiveness of arctan-activated WASD neural network for predicting the Dow Jones Industrial Average is verified.

3.3 Comparison

In order to illustrate that arctan-activated WASD neural network has advantages in the stock index prediction, we compare it with SVM network. They all use the same data to test, and then predict the Dow Jones Industrial Average of 10 days. The actual errors of the two approaches are shown in Fig. 3. From

Fig. 3, one can see that the absolute value of the actual error generated by arctan-activated WASD neural network is less than 20. However, the value of actual error generated by SVM network is gradually increased and the maximum actual error is more than 120. Therefore, the superior performance of arctan-activated WASD neural network for predicting the Dow Jones Industrial Average is demonstrated once again.

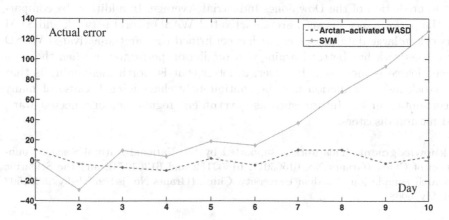

Fig. 3. Actual errors generated by arctan-activated WASD neural network and SVM network

In order to avoid accidental errors, we repeat the experiment ten times, and compare the best results and average results of ten experiments. The results are shown in Table 1. From Table 1, one can observe that the arctan-activated WASD neural network prediction error is one order of magnitude lower than the SVM network prediction error from the best angle or from the average point of view. In addition, the training time of arctan-activated WASD neural network is less than half of the SVM network. As a result, it can be concluded that the prediction performance of arctan-activated WASD neural network is superior to the SVM network.

Table 1. Numerical results of arctan-activated WASD (AWASD) neural network and SVM network

Models	AWASD best	AWASD average	SVM best	SVM average
MSE	1.91e−06	7.65e−06	2.0571e−05	2.0571e−05
CPU time	27.509 s	27.207 s	74.0209 s	74.3352 s

4   Conclusion

WASD neural network is a new feedforward neural network, which directly determines the weight and structure and reduces the complexity of the iterative process. The numerical experiments have demonstrated that the arctan-activated WASD neural network can effectively realize the self-determination of the optimal weights and the optimal structure, and show a very good accuracy in the prediction of the Dow Jones Industrial Average. In addition, by comparing the performance between arctan-activated WASD neural network and SVM network in large data processing, it has confirmed that arctan-activated WASD neural network has better learning and prediction performance when the nonlinear degree is increased. However, a detail that is worth mentioning is that the stock index prediction may be continuously challenging because of many uncertainties in the future, such as government regulation, unexpected wars, and natural disasters.

Acknowledgment. This work is supported by the National Natural Science Foundation of China (Grants No. 61563017, 61503152 and 61363073), and the Scientific Research Foundation of Jishou University, China (Grants No. jsdxxcfxbskyxm201508 and Jdy16008).

References

1. Mustafa, G., Mehmet, Ö., Aslı, B., Ayşe, T.D.: Integrating metaheuristics and artificial neural networks for improved stock price prediction. Expert Syst. Appl. **44**, 320–331 (2016)
2. Perwej, Y., Perwej, A.: Prediction of the Bombay Stock Exchange (BSE) market returns using artificial neural network and genetic algorithm. J. Intell. Learn. Syst. Appl. **4**, 108–119 (2012)
3. Hadavandi, E., Shavandi, H., Ghanbari, A.: Integration of genetic fuzzy systems and artificial neural networks for stock price forecasting. Knowl.-Based Syst. **23**, 800–808 (2010)
4. Bonde, G., Khaled, R.: Stock price prediction using genetic algorithms and evolution strategies. In: Proceedings of the 2012 International Conference on Genetic and Evolutionary Methods, pp. 10–15 (2012)
5. Prasanna, S., Ezhilmaran, D.: An analysis on stock market prediction using data mining techniques. Int. J. Comput. Sci. Eng. Technol. **4**, 49–51 (2013)
6. Sureshkumar, K., Elango, N.: Performance analysis of stock price prediction using artificial neural network. Glob. J. Comput. Sci. Technol. **12**, 18–26 (2012)
7. Zhang, Y., Guo, D., Luo, Z., Zhai, K., Tan, H.: CP-activated WASD neuronet approach to Asian population prediction with abundant experimental verification. Neurocomputing **198**, 48–57 (2016)
8. Li, S., You, Z., Guo, H., Luo, X., Zhao, Z.: Inverse-free extreme learning machine with optimal information updating. IEEE Trans. Cybern. **46**, 1229–1241 (2015)
9. Zhang, Y., Yin, Y., Guo, D., Yu, X., Xiao, L.: Cross-validation based weights and structure determination of Chebyshev-polynomial neural networks for pattern classification. Pattern Recogn. **47**, 3414–3428 (2014)

State Estimation for Autonomous Surface Vehicles Based on Echo State Networks

Zhouhua Peng[1,2]($\boxtimes$), Jun Wang[2], and Dan Wang[1]

[1] School of Marine Engineering, Dalian Maritime University,
Dalian 116026, People's Republic of China
zhpeng@dlmu.edu.cn, dwangdl@gmail.com
[2] Department of Computer Science, City University of Hong Kong,
Kowloon Tong, Hong Kong
jwang.cs@cityu.edu.hk

Abstract. This paper investigates the state estimation for autonomous surface vehicles in the presence of unknown dynamics and unmeasured states. The unknown dynamics comes from parametric model uncertainty, unmodelled hydrodynamics, and external disturbances caused by wind, waves and ocean currents. A nonlinear adaptive observer is proposed based on echo state networks, which are used to approximate the unknown dynamics using input-output data. By using the proposed observer, the unmeasured states and unknown dynamics can be simultaneously estimated in real time. The stability of the observer is analyzed via Lyapunov analysis. The proposed observer can be used in various motion control scenario, such as target tracking, trajectory tracking, path following, formation control, and even sideslip angle identification, not only for fully-actuated marine vehicles but also for under-actuated marine vehicles.

Keywords: Echo state network · State estimation · Unknown dynamics · Fully-actuated marine vehicles · Under-actuated marine vehicles

1 Introduction

In applications, the position information of marine surface vehicles can be easily obtained by using cheap Global Navigation Satellite System; however, the

The work of Z. Peng was supported in part by the National Natural Science Foundation of China under Grant 51579023, and in part by the Hong Kong Scholars Program under Grant XJ2015009, and in part by the China Post-Doctoral Science Foundation under Grant 2015M570247, and in part by High Level Talent Innovation and Entrepreneurship Program of Dalian under Grant 2016RQ036.
The work of J. Wang was supported in part by the Research Grants Council of the Hong Kong Special Administrative Region, China, under Grant 14207614, and in part by the National Natural Science Foundation of China under Grant 61673330.
The work of D. Wang was supported in part by the National Natural Science Foundation of China under Grant 61673081, and in part by the Fundamental Research Funds for the Central Universities under Grant 3132016313, and in part by the National Key Research and Development Program of China under Grant 2016YFC0301500.

© Springer International Publishing AG 2017
F. Cong et al. (Eds.): ISNN 2017, Part I, LNCS 10261, pp. 127–134, 2017.
DOI: 10.1007/978-3-319-59072-1_16

velocity information may not be readily measured. Therefore, the development of a stable observer to achieve a state estimation is essential for a marine vehicle. In the literature, a variety of stable observers have been proposed, for instances, wave-filtering observer [1–3], high-gain observer [4,6,8], linear observer [5], neural network-based observers [7,10], disturbance observer [9]. It is worth mentioning that the observers proposed in [4–6,8] are not able to identify the uncertainty. In addition, the model parameters in [1–3,9] are assumed to known. These may limits their applications.

In this paper, we aim to address the state estimation problem of autonomous surface vehicles (ASVs) in the presence of unknown dynamics, and unmeasured surge velocity, sway velocity, and yaw rate. The unknown dynamics stems from parametric model uncertainty, unmodelled hydrodynamics, and external disturbances such as wind, waves and ocean currents. A nonlinear adaptive observer is developed where echo state networks are employed to approximate the unknown kinetics using measured position-yaw information. With the proposed observer, the unmeasured states and unknown dynamics can be simultaneously estimated in real time. The stability of the echo-state-network-based observer is analyzed based on linear matrix inequality and Lyapunov theory. The proposed observer can be used in various applications, such as target tracking [14,15], trajectory tracking [17–19], path following [16,22,23], formation control [21], and sideslip angle estimation, not only for fully-actuated marine vehicles but also for under-actuated marine vehicles.

The paper is organized as follows: The problem formulation and preliminaries are introduced in Sect. 2. The echo-state-network-based observer design and analysis is presented in Sect. 3. Conclusions are drawn in Sect. 4.

2   Preliminaries and Problem Formulation

2.1   Echo State Network

Echo State Network (ESN) is widely used to model dynamical systems [13], and consists of a hidden layer and a memoryless output layer. Let $X \in \mathbb{R}^n$ be a reservoir state, $U \in \mathbb{R}^k$ be an input, $Y \in \mathbb{R}^l$ be its output vector, $W_u \in \mathbb{R}^{n \times k}$ and $W_x \in \mathbb{R}^{n \times n}$ be weight matrices, $b \in \mathbb{R}$ be a leaking decay rate, $c \in \mathbb{R}$ be a time constant, and $\sigma(\cdot)$ be an activation function vector. Then, the dynamical equation of an ESN with k inputs, n neurons in the hidden layer, and l neurons in the memoryless output layer, is expressed as

$$\dot{X} = \frac{1}{c} \{-bX + \sigma(W_u U + W_x X)\}, \tag{1}$$

and the output vector

$$Y = g(W^T X), \tag{2}$$

where $W \in \mathbb{R}^{n \times l}$ represents the output weight matrix and $g(\cdot)$ denotes the output activation function.

ESN holds the universal approximation property of recurrent neural networks. That is, given an arbitrary positive number ε^*, a continuous function $f(\xi), \mathbb{R}^n \to \mathbb{R}^l$ can be approximated by an ESN as [13]

$$f(\xi) = W^T \sigma_f(\xi) + \varepsilon, \forall \xi \in \Omega \subset \mathbb{R}^n, \tag{3}$$

where Ω is a sufficiently large compact set and ε is the approximation error satisfying $\|\varepsilon\| \leq \varepsilon^*$. Besides, there exists positive constants $W^* \in \mathbb{R}$ and $\sigma_f^* \in \mathbb{R}$ such that $\|W\|_F \leq W^*$ and $\|\sigma_f(\xi)\| \leq \sigma_f^*$.

The optimal weight W is given by

$$W = \arg \min_{\hat{W} \in \mathbb{R}^{n \times l}} \{\sup_{\xi \in \Omega} f(\xi) - \hat{W}^T \sigma_f(\xi)\}, \tag{4}$$

where $\hat{W}$ is an estimation of W.

2.2 Problem Formulation

In marine practice, the position, altitude and velocity information of a marine surface vehicle are described in body-fixed and earth-fixed reference frames. Let $\eta = [x, y, \psi]^T \in \mathbb{R}^3$ be a position-yaw vector in the earth-fixed frame, and $\nu = [u, v, r]^T \in \mathbb{R}^3$ be a velocity vector in the body-fixed frame. Then, the *kinematics* and *kinetics* of ASVs can be expressed by [11]

$$\dot{\eta} = R(\psi)\nu, \tag{5}$$

and

$$M\dot{\nu} = -C(\nu)\nu - D(\nu)\nu + g(\nu, \eta) + \tau_w(t) + \tau, \tag{6}$$

where $M = M^T \in \mathbb{R}^{3 \times 3}$ is an inertial matrix; $C(\nu) = -C(\nu) \in \mathbb{R}^{3 \times 3}$ is a Coriolis matrix; $D(\nu) \in \mathbb{R}^{3 \times 3}$ is a damping matrix; $g(\nu, \eta)$ represents the unmodelled hydrodynamics; $\tau_w = [\tau_{wu}, \tau_{wv}, \tau_{wr}]^T \in \mathbb{R}^3$ is a disturbance vector; $\tau = [\tau_u, \tau_v, \tau_r]^T \in \mathbb{R}^3$ is a control input; $R(\psi) \in \mathbb{R}^{3 \times 3}$ is defined as

$$R(\psi) = \begin{bmatrix} \cos\psi & -\sin\psi & 0 \\ \sin\psi & \cos\psi & 0 \\ 0 & 0 & 1 \end{bmatrix}. \tag{7}$$

In this paper, we aim to develop an adaptive observer for ASVs (5) and (6) using the position-yaw information of η.

3 ESN-Based Observer Design and Analysis

The vehicle dynamics contains parametric model uncertainty, unmodelled dynamics, and external disturbances caused by wind, waves and ocean currents. Besides, the velocity information of ASVs may not be available. In this paper, an ESN-based observer is developed to estimate the unknown dynamics, and unmeasured velocity information, simultaneously.

To facilitate the observer design, rewrite the Eqs. (5) and (6) as

$$\begin{cases} \dot{\eta} = R(\psi)\nu, \\ \dot{\nu} = M^{-1}[\tau - f(\cdot)], \end{cases} \tag{8}$$

where $f(\cdot) = C(\nu)\nu + D(\nu)\nu + g(\nu, \eta) - \tau_w(t)$.

Because the velocity information ν is not available, $f(\cdot)$ cannot be directly reconstructed by ESN. Here, the recorded output data η and input data τ are used to reconstruct the unknown function $f(\cdot)$.

Lemma 1. Given a positive constant ε^*, there exists a set of bounded weights $W \in R^{n \times 3}$, such that the continuous function $f(\cdot)$ can be approximated by an ESN as

$$f(\cdot) = W^T \sigma_f(\xi) + \varepsilon(\xi), \tag{9}$$

using the input vector $\xi = [\eta^T(t), \eta^T(t - t_d), \eta^T(t - 2t_d), \tau^T]^T$ with $t_d > 0$, and $\|\varepsilon(\xi)\| \leq \varepsilon^*$ provided there exists a suitable basis of activation function $\sigma(\cdot)$ over a compact set Ω.

In implementations, the incremental information $\Delta_1 = R(\psi)[\eta(t) - \eta(t - t_d)]$ and $\Delta_2 = R(\psi(t - t_d))[\eta(t - t_d) - \eta(t - 2t_d)]$ can be used as NN inputs for reducing the input dimension of ESN. When choosing $\xi = [\Delta_1^T, \Delta_2^T, \tau^T]^T$ or $\xi = [\Delta_1^T - \Delta_2^T, \tau^T]^T$, the input dimension of ESNs is reduced to nine or six, respectively. The ESN can be replaced by static neural network [20].

Let $\hat{\eta} = [\hat{x}, \hat{y}, \hat{\psi}]^T$ and $\hat{\nu} = [\hat{u}, \hat{v}, \hat{r}]^T$ be the estimates of η and ν. Then, an ESN-based full-state observer is designed as

$$\begin{cases} \frac{d\hat{\eta}}{dt} = -K_1 \tilde{\eta} + R(\psi)\hat{\nu}, \\ \frac{d\hat{\nu}}{dt} = -K_2 R^T(\psi)\tilde{\eta} + M^{-1}[-\hat{W}^T \sigma_f(\xi) + \tau], \end{cases} \tag{10}$$

and $\hat{W}$ is updated as follows

$$\dot{\hat{W}} = \Gamma[\sigma_f(\xi)\tilde{\eta}^T R(\psi) - k_W \hat{W}], \tag{11}$$

where $\tilde{\eta} = \hat{\eta} - \eta$; $K_1 \in \mathbb{R}^{3 \times 3}$ and $K_2 \in \mathbb{R}^{3 \times 3}$ are design parameters; $\Gamma \in \mathbb{R}$ and $k_W \in \mathbb{R}$ are positive constants.

Letting the velocity estimation error be $\tilde{\nu} = \hat{\nu} - \nu$ and recalling (8) and (10), it follows that

$$\begin{cases} \frac{d\tilde{\eta}}{dt} = -K_1 \tilde{\eta} + R(\psi)\tilde{\nu}, \\ \frac{d\tilde{\nu}}{dt} = -K_2 R^T(\psi)\tilde{\eta} + M^{-1}[-\tilde{W}^T \sigma_f(\xi) + \varepsilon], \end{cases} \tag{12}$$

where $\|\varepsilon\| \leq \varepsilon^*$ with ε^* being a positive constant. Letting $\tilde{X} = [\tilde{\eta}^T, \tilde{\nu}^T]^T$, the Eq. (12) can be further put into

$$\begin{cases} \frac{d\tilde{X}}{dt} = A\tilde{X} + B[-\tilde{W}^T \sigma_f(\xi) + \varepsilon], \\ \tilde{\eta} = C_0 \tilde{X}, \end{cases} \tag{13}$$

where

$$A = \begin{pmatrix} -K_1 & R(\psi) \\ -K_2 R^T(\psi) & 0_3 \end{pmatrix}, B = \begin{pmatrix} 0_{3\times3} \\ M^{-1} \end{pmatrix}, C_0 = (I_3, 0_3). \qquad (14)$$

In order to eliminate the dependence of A on ψ, a block-diagonal transformation $Z = T\tilde{X}$ with

$$T = \mathrm{diag}(R^T(\psi), I_3), \qquad (15)$$

is applied to (13), and it follows that

$$\begin{cases} \dot{\tilde{X}} = T^T A_0 T \tilde{X} + B\{-\tilde{W}^T \sigma_f(\xi) + \varepsilon\}, \\ \tilde{\eta} = C_0 \tilde{X}, A_0 = \begin{pmatrix} -K_1 & I_3 \\ -K_2 & 0_{3\times3} \end{pmatrix}. \end{cases} \qquad (16)$$

Using a mapping $Z = T\tilde{X}$, it follows that

$$\begin{cases} \dot{Z} = (A_0 + rS_T)Z + B[-\tilde{W}^T \sigma_f(\xi) + \varepsilon], \\ \dot{\tilde{W}} = \Gamma[\sigma_f(\xi)\tilde{\eta}^T R(\psi) - k_W \hat{W}], \end{cases} \qquad (17)$$

where $S_T = \mathrm{diag}(S^T, 0_3)$ and

$$S = \begin{pmatrix} 0 & -1 & 0 \\ 1 & 0 & 0 \\ 0 & 0 & 0 \end{pmatrix}. \qquad (18)$$

By exploiting the minimal and maximal bound of r, the stability of (17) is analyzed by concerning the following simultaneous inequalities

$$\begin{cases} A_0^T P + PA_0 + Q + PBB^T P + \ell FF^T + \bar{r}^*(S_T^T P + PS_T) \le 0, \\ A_0^T P + PA_0 + Q + PBB^T P + \ell FF^T - \bar{r}^*(S_T^T P + PS_T) \le 0, \end{cases} \qquad (19)$$

where $F = C^T - PB$; Q is a positive definite matrix; $\ell \in \mathbb{R}$ is a positive constant; $\bar{r}^*$ is an upper bound for r satisfying $|r| \le \bar{r}^*$; The stability property of the error dynamics (17) is stated as follows.

Theorem 1. If there exists a positive definite matrix P satisfying (19) and the control parameters are selected as

$$\frac{k_W}{2} - \frac{\sigma_f^{*2}}{2\ell} > 0, \qquad (20)$$

the error dynamics (17) is input-to-state stable, and all error signals in the closed-loop observer system are uniformly ultimately bounded.

Proof. Consider the Lyapunov function as

$$V = \frac{1}{2}\left\{ Z^T PZ + \Gamma^{-1}\mathrm{tr}(\tilde{W}^T \tilde{W}) \right\}. \qquad (21)$$

Using the inequalities

$$
\begin{cases}
-k_W \operatorname{tr}(\tilde{W}^T \hat{W}) \le -\frac{k_W}{2}\|\tilde{W}\|_F^2 + \frac{k_W}{2}\|W\|_F^2, \\
-Z^T E \tilde{W}^T \sigma_f(\xi) \le \frac{\ell}{2} Z^T E E^T Z + \frac{\tilde{W}^T \sigma_f(\xi)\sigma_f^T(\xi)\tilde{W}}{2\ell}, \\
Z^T P B \varepsilon \le \frac{1}{2} Z P B B^T P Z + \frac{1}{2}\varepsilon^2,
\end{cases}
\tag{22}
$$

it follows from (19) that the time derivative of V is

$$
\dot{V} \le -\lambda_{\min}(Q)\|Z\|^2 - \left(\frac{k_W}{2} - \frac{\sigma_f^{*2}}{2\ell}\right)\|\tilde{W}\|_F^2
$$

$$
+ \frac{k_W\|W\|_F^2}{2} + \frac{\|\varepsilon\|^2}{2}.
\tag{23}
$$

Letting $c = \min\left\{\lambda_{\min}(Q), \frac{k_W}{2} - \frac{\sigma_f^{*2}}{2\ell}\right\} > 0$ and noting that $c > 0$ under the condition (20), one has

$$
\dot{V} \le -\frac{c\|E\|^2}{2} - \left(\frac{c\|E\|^2}{2} - \frac{k_W\|W\|^2}{2} - \frac{\|\varepsilon\|^2}{2}\right),
\tag{24}
$$

where $E = [Z^T, \tilde{W}^T]$.

Noting that

$$
\|E\| \ge \frac{\sqrt{k_W}\|W\|}{\sqrt{c}} + \frac{\|\varepsilon\|}{\sqrt{c}} \ge \sqrt{\frac{k_W\|W\|^2 + \|\varepsilon\|^2}{c}},
\tag{25}
$$

renders

$$
\dot{V} \le -\kappa_3(\|E\|)
\tag{26}
$$

with $\kappa_3(s) = \frac{c}{2}s^2$, it follows that the system (17) is input-to-state stable [12].

Choosing $\kappa_1(s) = \frac{\lambda_{\min}(P)}{2}s^2$ and $\kappa_2(s) = \frac{\lambda_{\max}(P)}{2}s^2$ with $P = \operatorname{diag}\{P, \Gamma^{-1}\}$, there exists a class $\mathcal{KL}$ function β such that

$$
\begin{aligned}
\|E\| &\le \beta(\|E(0)\|, t) + \kappa_1 \circ \kappa_2\left(\frac{\sqrt{k_W}}{\sqrt{c}}\|W\|\right) + \kappa_1 \circ \kappa_2\left(\frac{1}{\sqrt{c}}\|\varepsilon\|\right), \\
&\le \beta(\|E(0)\|, t) + \kappa^W(\|W\|) + \kappa^\varepsilon(\|\varepsilon\|), \\
&\le \beta(\|E(0)\|, t) + \kappa^W(W^*) + \kappa^\varepsilon(\varepsilon^*),
\end{aligned}
\tag{27}
$$

where the the gain functions given by

$$
\kappa^W(s) = \sqrt{\frac{\lambda_{\max}(P)}{\lambda_{\min}(P)}}\frac{\sqrt{k_W}}{\sqrt{c}}s, \kappa^\varepsilon(s) = \sqrt{\frac{\lambda_{\max}(P)}{\lambda_{\min}(P)}}\frac{1}{\sqrt{c}}s.
\tag{28}
$$

Remark 1. Note that the vehicle model (6) becomes an under-actuated configuration when setting $\tau_v = 0$. Therefore, the developed observer can be applied to both under-actuated and fully-actuated marine surface vehicles.

4 Conclusions

In this paper, an echo-state-network-based observer is developed for state estimation of autonomous surface vehicles with unknown dynamics, and unmeasured surge velocity, sway velocity, and yaw rate. By using the developed echo-state-network-based observer, the unmeasured states and unknown dynamics can be simultaneously estimated, using the measured position-yaw information only. Based on Lyapunov analysis, the error signals in the closed-loop estimation system are proved to be uniformly ultimately bounded.

References

1. Fossen, T., Grovlen, A.: Nonlinear output feedback control of dynamically positioned ships using vectorial observer backstepping. IEEE Trans. Control Syst. Technol. **6**(1), 121–128 (1998)
2. Fossen, T., Strand, J.: Passive nonlinear observer design for ships using Lyapunov methods: full-scale experiments with a supply vessel. Automatica **35**(1), 3–16 (1999)
3. Ihle, I., Skjetne, R., Fossen, T.: Output feedback control for maneuvering systems using observer backstepping. In: Proceedings of the 13th Mediterrean Conference on Control and Automation Intelligent Control (2005)
4. Tee, K., Ge, S.: Control of fully actuated ocean surface vessels using a class of feedforward approximators. IEEE Trans. Control Syst. Technol. **14**(4), 750–756 (2006)
5. Zhang, L., Jia, H., Qi, X.: NNFFC-adaptive output feedback trajectory tracking control for a surface ship at high speed. Ocean Eng. **38**(13), 1430–1438 (2011)
6. Dai, S.L., Wang, M., Wang, C., Li, L.: Learning from adaptive neural network output feedback control of uncertain ocean surface ship dynamics. Int. J. Adapt. Control Signal Process. **28**(3–5), 341–365 (2012)
7. Peng, Z., Wang, D., Liu, H.H., Sun, G., Wang, H.: Distributed robust state and output feedback controller designs for rendezvous of networked autonomous surface vehicles using neural networks. Neurocomputing **115**, 130–141 (2013)
8. He, W., Yin, Z., Sun, C.: Adaptive neural network control of a marine vessel with constraints using the asymmetric barrier Lyapunov function. IEEE Trans. Cybern. (2016). doi:10.1109/TCYB.2016.2554621
9. Peng, Z., Wang, D., Wang, J.: Cooperative dynamic positioning of multiple marine offshore vessels: a modular design. IEEE/ASME Trans. Mechatron. **31**(3), 1210–1221 (2016)
10. Peng, Z., Wang, D., Shi, Y., Wang, H., Wang, W.: Containment control of networked autonomous underwater vehicles with model uncertainty and ocean disturbances guided by multiple leaders. Inf. Sci. **316**(20), 163–179 (2015)
11. Fossen, T.: Handbook of Marine Craft Hydrodynamics and Motion Control. Wiley, Hoboken (2011)
12. Sontag, E.D., Wang, Y.: On characterizations of the input-to-state stability property. Syst. Control Lett. **24**(5), 351–359 (1995)
13. Sun, G., Li, D., Ren, X.: Modified neural dynamic surface approach to output feedback of MIMO nonlinear systems. IEEE Trans. Neural Netw. Learn. Syst. **26**(2), 224–236 (2015)

14. Peng, Z., Wang, D.: Robust adaptive formation control of underactuated autonomous surface vehicles with uncertain dynamics. IET Control Theory A. **5**(12), 1378–1387 (2011)

15. Peng, Z., Wang, D., Chen, Z., Hu, X., Lan, W.: Adaptive dynamic surface control for formations of autonomous surface vehicles with uncertain dynamics. IEEE Trans. Control Syst. Technol. **21**(2), 513–520 (2013)

16. Xiang, X., Lapierre, L., Jouvencel, B.: Smooth transition of AUV motion control: from fully-actuated to under-actuated configuration. Robot. Auton. Syst. **67**, 14–22 (2015)

17. Chen, M., Ge, S.S., How, B.V.E., Choo, Y.S.: Robust adaptive position mooring control for marine vessels. IEEE Trans. Control Syst. Technol. **21**(2), 395–409 (2013)

18. Yin, S., Xiao, B.: Tracking control of surface ships with disturbance and uncertainties rejection capability. IEEE/ASME Trans. Mechatron. (2016). doi:10.1109/TMECH.2016.2618901

19. Cui, R., Yang, C., Li, Y., Sharma, S.: Adaptive neural network control of auvs with control input nonlinearites using reinforcement learning. IEEE Trans. Syst. Man Cybern. Syst. (2016). doi:10.1109/TSMC.2016.2645699

20. Peng, Z., Wang, D., Zhang, H., Sun, G.: Distributed neural network control for adaptive synchronization of uncertain dynamical multiagent systems. IEEE Trans. Neural Netw. Learn. Syst. **25**(8), 1508–1519 (2014)

21. Peng, Z., Wang, J., Wang, D.: Containment maneuvering of marine surface vehicles with multiple parameterized paths via spatial-temporal decoupling. IEEE/ASME Trans. Mechatron. (2016). doi:10.1109/TMECH.2016.2632304

22. Zheng, Z., Sun, L.: Path following control for marine surface vessel with uncertainties and input saturation. Neurocomputing **177**, 158–167 (2016)

23. Xiang, X., Yu, C., Zhang, Q., Xu, G.: Path-following control of an AUV: fully actuated versus under-actuated configuration. Mar. Technol. Soc. J. **50**(1), 34–47 (2016)

Using Neural Network Formalism to Solve Multiple-Instance Problems

Tomáš Pevný[1,2]([✉]) and Petr Somol[1,3]

[1] Cisco Systems, Charles Square 10, Prague, Czech Republic
pevnak@gmail.com
[2] Faculty of Electrical Engineering, Czech Technical University,
Prague, Czech Republic
[3] UTIA, Czech Academy of Sciences, Prague, Czech Republic

Abstract. Many objects in the real world are difficult to describe by means of a single numerical vector of a fixed length, whereas describing them by means of a set of vectors is more natural. Therefore, *Multiple instance learning* (MIL) techniques have been constantly gaining in importance throughout the last years. MIL formalism assumes that each object (sample) is represented by a set (bag) of feature vectors (instances) of fixed length, where knowledge about objects (e.g., class label) is available on bag level but not necessarily on instance level. Many standard tools including supervised classifiers have been already adapted to MIL setting since the problem got formalized in the late nineties. In this work we propose a neural network (NN) based formalism that intuitively bridges the gap between MIL problem definition and the vast existing knowledge-base of standard models and classifiers. We show that the proposed NN formalism is effectively optimizable by a back-propagation algorithm and can reveal unknown patterns inside bags. Comparison to 14 types of classifiers from the prior art on a set of 20 publicly available benchmark datasets confirms the advantages and accuracy of the proposed solution.

1 Motivation

The constant growth of data sizes and data complexity in real world problems has increasingly put strain on traditional modeling and classification techniques. Many assumptions cease to hold; it can no longer be expected that a complete set of training data is available for training at once, models fail to reflect information in complex data unless a prohibitively high number of parameters is employed, availability of class labels for all samples can not be realistically expected, and particularly the common assumption about each sample to be represented by a fixed-size vector seems to no longer hold in many real world problems.

Multiple instance learning (MIL) techniques address some of these concerns by allowing samples to be represented by an arbitrarily large set of fixed-sized vectors instead of a single fixed-size vector. Any explicit ground truth information (e.g., class label) is assumed to be available on the (higher) level of samples

© Springer International Publishing AG 2017
F. Cong et al. (Eds.): ISNN 2017, Part I, LNCS 10261, pp. 135–142, 2017.
DOI: 10.1007/978-3-319-59072-1_17

but not on the (lower) level of instances. The aim is to utilize unknown patterns on instance-level to enable sample-level modeling and decision making. Note that MIL does not address the Representation Learning problem [3]. Instead it aims at better utilization of information in cases when ground truth knowledge about a dataset may be granular and available on various levels of abstraction only.

From a practical point of view MIL promises to (i) save ground truth acquisition cost – labels are needed on sample-level, i.e., on higher-level(s) of abstraction only, (ii) reveal patterns on instance level based on the available sample-level ground truth information, and eventually (iii) achieve high accuracy of models through better use of information present in data.

Despite significant progress in recent years, the current battery of MIL tools is still burdened with compromises. The existing models (see next Sect. 2 for a brief discussion) clearly leave open space for more efficient utilization of information in samples and for a clearer formalism to provide easily interpretable models with higher accuracy. The goal of this paper is to provide a clean formalism bridging the gap between the MIL problem formulation and classification techniques of neural networks (NNs). This opens the door to applying latest results in NNs to MIL problems.

2   Prior Art on Multi-instance Problem

The pioneering work [10] coined *multiple-instance* or *multi-instance* learning as a problem where each sample b (called *bag* in the following) consists of a set of instances x, i.e., $b = \{x_i \in \mathcal{X} | i \in \{1, \ldots, |b|\}\}$, equivalently $b \in \mathcal{B} = \cup_{k>1}\{x_i \in \mathcal{X} | i \in \{1, \ldots, k\}\}$ and each instance x can be attributed a label $y_x \in \{-1, +1\}$, but these instance-level labels are not known even in the training set. The sample b is deemed positive if at least one of its instances had a positive label, i.e., label of a sample b is $y = \max_{x \in b} y_x$. Most approaches solving this definition of MIL problem belong to *instance-space paradigm*, in which the classifier is trained on the level of individual instances $f : \mathcal{X} \mapsto \{-1, +1\}$ and the label of the bag b is inferred as $\max_{x \in b} f(x)$. Examples of such methods include: Diverse-density [16], EM-DD [22], MILBoost [21], and MI-SVM [2].

Later works (see reviews [1,11]) have introduced different assumptions on relationships between labels on the instance level and labels of bags or even dropped the notion of instance-level labels and considered only labels on the level of bags, i.e., it is assumed that each bag b has a corresponding label $y \in \mathcal{Y}$, which is for simplicity assumed to be binary, i.e., $\mathcal{Y} = \{-1, +1\}$ in the following. Most approaches solving this general definition of the problem follow either the *bag-space paradigm* and define a measure of distance (or kernel) between bags [12, 13,17] or the *embedded-space paradigm* and define a transformation of the bag to a fixed-size vector [5,6,20].

Prior art on neural networks for MIL problems is scarce and aimed for *instance-space paradigm*. Reference [18] proposes a smooth approximation of the maximum pooling in the last neuron as $\frac{1}{|b|} \ln \left(\sum_{x \in b} \exp(f(x)) \right)$, where $f(x) : \mathcal{X} \mapsto \mathbb{R}$ is the output of the network before the pooling. Reference [23] drops the requirement on smooth pooling and uses the maximum pooling function in the last neuron. Both approaches optimize the L_2 error function.

Due to space limits, the above review of the prior art was brief. The Interested reader is referred to [1,4,11] for a more thorough discussion of a problem and algorithms.

3 Neural Network Formalism

The proposed neural network formalism is intended for a general formulation of MIL problems introduced in [17]. It assumes a non-empty space $\mathcal{X}$ where instances live with a set of all probability distributions $\mathcal{P}^{\mathcal{X}}$ on $\mathcal{X}$. Each bag corresponds to some probability distribution $p_b \in \mathcal{P}^{\mathcal{X}}$ with its instances being realizations of random a variable with distribution p_b. Each bag b is therefore assumed to be a realization of a random variable distributed according to $P(p_b, y)$, where $y \in \mathcal{Y}$ is the bag label. During the learning process each concrete bag b is thus viewed as a realization of a random variable with probability distribution p_b that can only be inferred from a set of instances $\{x \in b | x \sim p_b\}$ observed in data. The goal is to learn a discrimination function $f : \mathcal{B} \mapsto \mathcal{Y}$, where $\mathcal{B}$ is the set of all possible realizations of distributions $p \in \mathcal{P}^{\mathcal{X}}$, i.e., $\mathcal{B} = \{x_i | p \in \mathcal{P}^{\mathcal{X}}, x_i \sim p, i \in \{1, \ldots l\}, l \in \mathbb{N}\}$. This definition includes the original used in [10], but it also includes the general case where every instance can occur in positive and negative bags, but some instances are more frequent in one class.

The proposed formalism is based on the *embedded-space* paradigm representing bag b in an m-dimensional Euclidean space $\mathbb{R}^m$ through a set of mappings

$$(\phi_1(b), \phi_2(b), \ldots, \phi_m(b)) \in \mathbb{R}^m \qquad (1)$$

with $\phi : \mathcal{B} \mapsto \mathbb{R}$. Many existing methods implement embedding function as

$$\phi_i = g\left(\{k(x, \theta_i)\}_{x \in b}\right), \qquad (2)$$

where $k : \mathcal{X} \times \mathcal{X} \mapsto \mathbb{R}_0^+$ is a suitably chosen distance function, $g : \cup_{k=1}^{\infty} \mathbb{R}^k \mapsto \mathbb{R}$ is the pooling function (e.g. minimum, mean or maximum), and finally $\Theta = \{\theta_i \in \mathcal{X} | i \in \{1, \ldots, m\}\}$ is the dictionary with instances as items. Prior art methods differ in the choice of aggregation function g, distance function k, and finally in the selection of dictionary items, Θ. A generalization was recently proposed in [6] defining ϕ using a distance function (or kernel) over the bags $k : \mathcal{B} \times \mathcal{B} \mapsto \mathbb{R}$ and dictionary Θ containing bags rather instances. This generalization can be seen as a crude approximation of kernels over probability measures used in [17].

The computational model defined by (1) and (2) can be viewed as a neural network sketched in Fig. 1. One (or more) lower layers implement a set of distance functions $\{k(x, \theta_i)\}_{i=1}^m$ (denoted in Fig. 1 in vector form as $k(x, \theta)$) projecting each instance x_i from the bag $\{x_i\}_{i=1}^m$ from the input space $\mathbb{R}^d$ for $\mathbb{R}^m$. The pooling layer implementing the pooling function g produces a single vector $\bar{x}$ of the same dimension $\mathbb{R}^m$. Finally subsequent layers denoted in the figure as

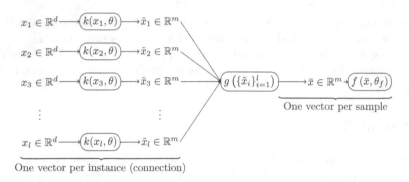

Fig. 1. Sketch of the neural network optimizing the embedding in embedding-space paradigm.

$f(\bar{x})$ implement the classifier that already uses a representation of the bag as a feature vector of fixed length m. The biggest advantage of this formalism is that with a right choice of pooling function $g(\cdot)$ (e.g. mean or maximum) all parameters of the embedding functions $k(x, \theta)$ can be optimized by the standard back-propagation algorithm. Therefore embedding at the instance-level (layers before pooling) is effectively optimized while requiring labels only on the bag-level. This mechanism identifies parts of the instance-space $\mathcal{X}$ with the largest differences between probability distributions generating instances in positive and negative bags with respect to the chosen pooling function. This is also the most differentiating feature of the proposed formalism to most prior art, which typically optimizes embedding parameters θ_i regardless of the labels.

The choice of a pooling function depends on the type of the MIL problem. If the bag's label depends on a single instance, as it is the case for the instance-level paradigm, then the maximum pooling function is appropriate, since its output also depends on a single instance. On the other hand if a bag's label depends on properties of all instances, then the mean pooling function is appropriate, since its output depends on all instances and therefore it characterizes the overall distribution.

Remark: the key difference of the above approach to the prior art [23] is in performing pooling *inside the network* as opposed to after the last neuron or layer as in the cited reference. This difference is key to the shift from instance-centric modeling in prior art to bag-centric advocated here. However the proposed formalism is general and includes [23] as a special case, where instances are projected into the space of dimension one ($m = 1$), pooling function g is set to maximum, and layers after the pooling functions are not present (f is equal to identity).

4   Experimental Evaluation

The evaluation of the proposed formalism uses publicly available datasets from a recent study of properties of MIL problems [7], namely *BrownCreeper,*

CorelAfrican, CorelBeach, Elephant, Fox, Musk1, Musk2, Mutagenesis1, Muta-genesis2, Newsgroups1, Newsgroups2, Newsgroups3, Protein, Tiger, UCSB-BreastCancer, Web1, Web2, Web3, Web4, and *WinterWren*. The supplemental material [8] contains equal error rate (EER) of 28 MIL classifiers (and their variants) from prior art implemented in the MIL matlab toolbox [19] together with the exact experimental protocol and indexes of all splits in 5-times repeated 10-fold cross-validation. Therefore the experimental protocol has been exactly reproduced and results from [8] are used in the comparison to prior art.

The proposed formalism has been compared to those algorithms from prior art that has achieved the lowest error on at least one dataset. This selection yielded 14 classifiers for 20 test problems, which demonstrates diversity of MIL problems and difficulty to choose suitable method. Selected algorithms include representatives of instance-space paradigm: *MIL Boost* [21], *SimpleMIL, MI-SVM* [2] with Gaussian and polynomial kernel, and prior art in Neural Networks (denoted *prior NN*) [23]; bag-level paradigm: k-nearest neighbor with *citation* distance [20] using 5 nearest neighbors; and finally embedded-space paradigm: *Miles* [5] with Gaussian kernel, Bag dissimilarity [6] with *minmin, meanmin, meanmean, Hausdorff*, and Earth-moving distance (EMD), *cov-coef* [8] embedding bags by calculating covariances of all pairs of features over the bag, and finally *extremes* and *mean* embedding bags by using extreme and mean values of each feature over instances of the bag. All embedded space paradigm methods except Miles used a logistic regression classifier.

The proposed MIL neural network consists of a single layer of rectified linear units (ReLu) [14] with transfer function $\max\{0, x\}$, followed by a mean-pooling layer and a single linear output unit. The training minimized a hinge loss function using the Adam [15] variant of stochastic gradient descend algorithm with mini-batch of size 100, maximum of 10 000 iterations, and default settings. L1 regularization on weights of the network was used to decrease over-fitting. The topology had two parameters — the number of neurons in the first layer defining the dimension of bag representation, m, and the strength of the L1 regularization, λ. Suitable parameters were found by estimating equal error rates by five-fold cross-validation (on training samples) on all combinations of $k \in \{2, 4, 8, 12, 16, 20\}$ and $\lambda \in \{10^{-7}, 10^{-6}, \ldots, 10^{-3}\}$ and using the combination achieving the lowest error. The prior art of [23] was implemented and optimized exactly as the proposed approach with the difference that the max pooling layer was *after* the last linear output unit.

Figure 2 summarizes results in critical difference diagram [9] showing the average rank of each classifier over the problems together with the confidence interval of corrected Bonferroni-Dunn test with significance 0.05 testing whether two classifiers have equal performance. The critical diagram reveals that the classifier implemented using the proposed neural net formalism (caption *proposed NN*) achieved overall the best performance, having the average rank 4.3. In fact, Table 1 shows that it provides the lowest error on nine out of 20 problems. Note that the second best, Bag dissimilarity [6] with minmin distance and prior art in NN [23], achieved the average rank 6.4 and was the best only on three and one problems respectively.

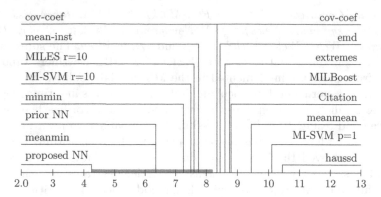

Fig. 2. Critical difference diagram shows average rank of each method over 20 problems. The thick black line shows the confidence interval of corrected Bonferroni-Dunn test with significance 0.05 testing whether two classifiers have equal performance.

Table 1. Average equal error rate of the proposed NN formalism on training and testing set and average equal error rate on the testing set of the best prior art for the given problem. Abbreviations of the prior art are as introduced in Sect. 4.

	Error of NN on		Prior art	
	Training set	Testing set	Error	Algorithm
BrownCreeper	0	**5.0**	11.2	MILBoost
CorelAfrican	2.6	**5.5**	11.2	minmin
CorelBeach	0.2	**1.2**	17	extremes
Elephant	0	**13.8**	16.2	minmin
Fox	0.4	**33.7**	36.1	meanmin
Musk1	0	17.5	**12.8**	Citation
Musk2	0	**11.4**	11.8	Hausdorff
Mutagenesis1	7.5	**11.8**	16.9	cov-coef
Mutagenesis2	14.9	**10.0**	17.2	emd
Newsgroups1	0	42.5	**18.4**	meanmean
Newsgroups2	0	35	**27.5**	prior NN
Newsgroups3	0	37.5	**31.2**	meanmean
Protein	2.5	**7.5**	15.5	minmin
Tiger	0	20.0	**19**	MILES
UCSBBreastCancer	0	25	**13.6**	MI-SVM g
Web1	0	40.6	**20.9**	MILES
Web2	0	28.1	**7.1**	MI-SVM p
Web3	0	25	**13.6**	MI-SVM g
Web4	0	18.8	**1.5**	mean-inst
WinterWren	0	5.9	**2.1**	emd

Exact values of EER of the best algorithm from the prior art and that of the proposed NN formalism is summarized in Table 1. From the results it is obvious that the proposed neural network formalism have scored poorly on problems with a large dimension and a small number of samples, namely Newsgroups and Web (see Table 1 of [7] for details on the data). The neural network formalism has easily overfit to the training data, which is supported by zero errors on the training sets.

5   Conclusion

This work has presented a generalization of neural networks to multi-instance problems. Unlike the prior art, the proposed formalism embeds samples consisting of multiple instances into vector space, enabling subsequent use with standard decision-making techniques. The key advantage of the proposed solution is that it simultaneously optimizes the classifier and the embedding. This advantage was illustrated on a set of real-world examples, comparing results to a large number of algorithms from the prior art. The proposed formalism seems to outperform the majority of standard MIL methods in terms of accuracy. It should be stressed though that results were compared to those published by authors of survey benchmarks; not all methods in referred tests may have been set in the best possible way. However, as many such cases would be very computationally expensive, the proposed formalism becomes competitive also due to its relatively modest computational complexity that does not exceed that of a standard 3-layer neural network. The proposed formalism opens up a variety of options for further development. A better and possibly more automated choice of pooling functions is one of the promising ways to improve performance on some types of data.

Acknowledgements. This work has been partially supported by Czech Science Foundation project 15-08916S.

References

1. Amores, J.: Multiple instance classification: review, taxonomy and comparative study. Artif. Intell. **201**, 81–105 (2013)
2. Andrews, S., Tsochantaridis, I., Hofmann, T.: Support vector machines for multiple-instance learning. In: Becker, S., Thrun, S., Obermayer, K. (eds.) Advances in Neural Information Processing Systems 15, pp. 577–584. MIT Press (2003)
3. Bengio, Y., Courville, A., Vincent, P.: Representation learning: a review and new perspectives. arXiv preprint arXiv:1206.5538v2 (2012)
4. Carbonneau, M.-A., Cheplygina, V., Granger, E., Gagnon, G.: Multiple instance learning: a survey of problem characteristics and applications. arXiv preprint arXiv:1612.03365 (2016)
5. Chen, Y., Bi, J., Wang, J.Z.: Miles: multiple-instance learning via embedded instance selection. IEEE Trans. Pattern Anal. Mach. Intell. **28**(12), 1931–1947 (2006)

6. Cheplygina, V., Tax, D.M., Loog, M.: Multiple instance learning with bag dissimilarities. Pattern Recogn. **48**(1), 264–275 (2015)
7. Cheplygina, V., Tax, D.M.J.: Characterizing multiple instance datasets. In: Feragen, A., Pelillo, M., Loog, M. (eds.) SIMBAD 2015. LNCS, vol. 9370, pp. 15–27. Springer, Cham (2015). doi:10.1007/978-3-319-24261-3_2
8. Cheplygina, V., Tax, D.M.J., Loog, M.: Supplemental documents to characterizing multiple instance datasets
9. Demšar, J.: Statistical comparisons of classifiers over multiple data sets. J. Mach. Learn. Res. **7**(Jan), 1–30 (2006)
10. Dietterich, T.G., Lathrop, R.H., Lozano-Pérez, T.: Solving the multiple instance problem with axis-parallel rectangles. Artif. Intell. **89**(1), 31–71 (1997)
11. Foulds, J., Frank, E.: A review of multi-instance learning assumptions. Knowl. Eng. Rev. **25**(01), 1–25 (2010)
12. Gärtner, T., Flach, P.A., Kowalczyk, A., Smola, A.J.: Multi-instance kernels. In: ICML, vol. 2, pp. 179–186 (2002)
13. Haussler, D.: Convolution kernels on discrete structures, Raport instytutowy, Universityof California at Santa Cruz (1999)
14. Jarrett, K., Kavukcuoglu, K., Ranzato, M., LeCun, Y.: What is the best multi-stage architecture for object recognition? In: 2009 IEEE 12th International Conference on Computer Vision, pp. 2146–2153, September 2009
15. Kingma, D., Ba, J.: Adam: a method for stochastic optimization. arXiv preprint arXiv:1412.6980 (2014)
16. Maron, O., Lozano-Pérez, T.: A framework for multiple-instance learning. In: Jordan, M.I., Kearns, M.J., Solla, S.A. (eds.) Advances in Neural Information Processing Systems 10, pp. 570–576. MIT Press (1998)
17. Muandet, K., Fukumizu, K., Dinuzzo, F., Schölkopf, B.: Learning from distributions via support measure machines. In: Advances in Neural Information Processing Systems, pp. 10–18 (2012)
18. Ramon, J., De Raedt, L.: Multi instance neural networks (2000)
19. Tax, D.M.J., Cheplygina, V.: MIL, A Matlab Toolbox for Multiple Instance Learning, version 1.2.1, June 2016. http://prlab.tudelft.nl/david-tax/mil.html
20. Wang, J., Zucker, J.-D.: Solving the multiple-instance problem: a lazy learning approach. In: Proceedings of the Seventeenth International Conference on Machine Learning, ICML 2000, pp. 1119–1126. Morgan Kaufmann Publishers Inc., San Francisco (2000)
21. Zhang, C., Platt, J.C., Viola, P.A.: Multiple instance boosting for object detection. In: Weiss, Y., Schölkopf, B., Platt, J.C. (eds.) Advances in Neural Information Processing Systems 18, pp. 1417–1424. MIT Press (2006)
22. Zhang, Q., Goldman, S.A.: EM-DD: an improved multiple-instance learning technique. In: Dietterich, T.G., Becker, S., Ghahramani, Z. (eds.) Advances in Neural Information Processing Systems 14, pp. 1073–1080. MIT Press (2002)
23. Zhou, Z.-H., Zhang, M.-L.: Neural networks for multi-instance learning. In: Proceedings of the International Conference on Intelligent Information Technology, vol. 182. Citeseer (2002)

Many-Objective Optimisation of Trusses Through Meta-Heuristics

Nantiwat Pholdee[1], Sujin Bureerat[1], Papot Jaroenapibal[2],
and Thana Radpukdee[1(✉)]

[1] Aircraft Multidisciplinary Optimisation Research Unit,
Department of Mechanical Engineering, Faculty of Engineering,
Khon Kaen University, Khon Kaen 40002, Thailand
tthanar@gmail.com
[2] Department of Industrial Engineering, Faculty of Engineering,
Khon Kaen University, Khon Kaen 40002, Thailand

Abstract. A truss is one of the most used engineering structures in daily life due to several advantages. A process for truss optimisation is usually set to minimise its mass while structural safety constraints are imposed. This design problem always leads to structures with less reliability since the solution is generally on the borderline of structural failure. Such a phenomenon can be alleviated by adding effects of all possible load cases with safety factors to design constraints. Alternatively, the design problem should be many-objective optimisation assigned to optimise mass and reliability indicators for all load cases. This paper is the first attempt to study such a design process. A number of many-objective meta-heuristics are employed to solve the test problems for many-objective truss optimization where their performances are compared.

Keywords: Many-objective optimisation · Truss design · Meta-heuristics · Evolutionary computation · Constrained optimisation

1 Introduction

Over the last few decades, Meta-heuristics (MHs), also known as Evolutionary Algorithms (EAs), havearguably been the mostprominent optimisation toolswidely applied for solving various practical optimisation problems. The advantages of MHs are global optimisation capability without requirement of function derivatives, therefore, they can deal with almost any kind of function and design variables [1–4]. In addition, they are very effective when dealing with multi-objective optimisation problems as they can explore a Pareto front within one optimisation run [2–4]. The methods are simple to understand, code, and implement. However, a lack of search consistency and convergence of MHs due to their random search is still an issue, particularly when the number of objective functions is more than three, which is usually called "many objective optimisation" [5]. Also, a large-scale design problem (a problem with a great many of design variables) may cause difficulty in using MHs [6, 7]. As a result, numerous MHs on solving many-objective optimisation problems have been developed in the past few years [5–14].

© Springer International Publishing AG 2017
F. Cong et al. (Eds.): ISNN 2017, Part I, LNCS 10261, pp. 143–152, 2017.
DOI: 10.1007/978-3-319-59072-1_18

Development of MHs for design problems with many objectives can be done in several ways such as: improving the search convergence by means of modifying the selection and/or updating mechanisms [10, 12–14], reducing the spaces of objective functions [5, 8], and a preference approach [9, 11]. Although many-objective MHs have been proposed successfully, researchers in the field are usually focused on implementing them for the unconstrained test functions. Therefore, the performance of MH algorithms for real engineering design with many objectives and design constraints needs to be investigated.

A truss is one of the most applied structures amid various engineering applications which require a lightweight design, such as bridges, roofs, electric towers, billboard structures, industrial structures etc. Under working conditions, the truss is subject to several mechanical phenomena such as stress, buckling, and vibration [1, 4, 15, 16]. Over the years, there has been a great amount of work related to truss optimisation studies. Design problems can be single-objective [16–18] or multi-objective [1, 2, 4]. Design variables can be categorised as topology, shape, and size where they can be determined separately [16, 18] or simultaneously [1, 2, 4, 17]. In cases of statically indeterminate trusses, stress and local buckling constraints can cause a non-convex feasible region which is often difficult to solve. Truss optimisation with natural frequency constraints also poses the same difficulty. It has been shown that using MHs for solving such problems can be successful [1, 2, 4, 16–19]. A typical single-objective truss optimisation is traditionally set to minimise mass subject to stress, displacement, buckling, and/or natural frequency constraints. It has been found that the obtained design results are too unreliable to be used. This is due to the fact that the optimum solution will lie on the constraint boundaries; the borderline between failure and safety. One way to prevent such a problem is by adding all possible load cases to the design problem. The factor of safety is also employed to increase design reliability. An alternative approach is to perform multi-objective optimisation to minimise mass and at the same time maximise structural reliability [1, 2, 4]. One indicator for truss reliability is structural compliance [1, 2] where minimising such a parameter is equivalent to maximising static structural global stiffness. Due to many load cases being applied for the real-world truss design, there are thus many compliance indicators. Other reliability indicators can be buckling factor and natural frequencies maximisation. The structural natural frequencies can be regarded as dynamic structural stiffness. The aforementioned objectives lead the design problem to many-objective optimisation. To our knowledge, such a design problem type is yet to be investigated in the literature. The baseline or benchmark results for future investigation need to be established.

Therefore, this paper presents a comparative performance study of various recently proposed many-objective meta-heuristic algorithms on solving many-objective optimisation of trusses. Five six-objective constrained optimisation of trusses are proposed where the employed MHs include multi-objective evolutionary algorithm based on decomposition (MOEA/D) [20], an improved two-archive algorithm (Two_Arch2) [10], a preference-inspired co-evolutionary algorithm using goal vectors (PICEA-g) [11], a knee point driven evolutionary algorithm for many-objective optimization (KnEA) [12], and KnEA with an approximate efficient non-dominated sorting approach (KnEA-A-ENS) [12, 13]. The C-indicator as detailed in [4, 21] is applied as a performance indicator of the methods.

2 Many-Objective Optimisation of Truss Structure

In this study, the many-objective optimisation problem for a truss structure can be stated as follows:

$$\min_{\mathbf{x}} \mathbf{f}(\mathbf{x}) = \{f_1, f_2, f_3, f_4, f_5, f_6\}$$

subjected to

$$\frac{|d_{max}|}{d_{all}} \leq 1, \; \frac{\sigma_{max}}{\sigma_{all}} \leq 1, \; \frac{P_i}{P_{cr,i}} \leq 1$$

$$A_{min} \leq A_i \leq A_{max} \quad \text{m}^2 \; x_{min} \leq x_i \leq x_{max} \quad \text{m}, \; y_{min} \leq y_i \leq y_{max} \quad \text{m}$$

where $\mathbf{f}(\mathbf{x})$ is a vector of objective functions to be minimised. The functions f_1, f_2, f_3, are structural mass, compliance and the ratio of the maximum compressive load to the critical buckling load (P_{maxs}/P_{cr}), respectively. The functions f_4, f_5, and f_6 are the first lowest three modes of natural frequencies. $\mathbf{x}^T = \{x_1, \ldots, x_n\}$ is the vector of design variables including the cross-sectional areas of the truss elements (A_i) and nodal positions in x-direction (x_i) and y-direction (y_i). The variables d_{max} and σ_{max} are the maximum nodal displacement and the maximum stress developed in the structure respectively, while d_{all} and σ_{all} are the allowable limits on nodal displacement and stress constraints. P_i is the compressive load in the i-th element of the truss with the corresponding buckling load limit while $P_{cr,i}$. A_{min} and A_{max} are the lower and upper limits of element cross-sectional areas respectively.

For bars subjected to compressive forces, the critical buckling load $P_{cr,i}$ is determined as:

$$P_{cr,i} = \begin{cases} A_i \frac{\pi^2 E}{(l_i/r_i)^2} & ; if \; \frac{l_i}{r_i} < \sqrt{\frac{2\pi^2 E}{\sigma_{all}}} \\ A_i \sigma_{all} \left\{ 1 - \frac{\sigma_{all}}{4\pi^2 E} \left(\frac{l_i}{r_i}\right)^2 \right\} & ; otherwise \end{cases}$$

where L_i, A_i and r_i are the effective length, cross-sectional area and cross-section radius of gyration of the i-th element, respectively. E is the Young's modulus. Truss structures used in this study are detailed as follows:

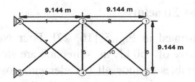

Fig. 1. Schematic of the planar 10-bar truss structure

2.1 Planar 10-Bar Truss Structure

The schematic drawing of the truss structure is illustrated in Fig. 1 [1, 16, 22]. The structure is subject to non-structural mass of 454 kg at each free node and external force P_1 and P_2 where the force P_1 is acted on nodes 2 and 4 while the force P_2 is acted on nodes 1 and 3. The force $P_1 = -5,000$ N and $P_2 = 5,000$ N are applied in y-direction. The design variables include all bar element cross-sectional areas where A_{min} and A_{max} are set to be 0.645×10^{-4} m^2 and 0.005 m^2 respectively. d_{all} and σ_{all} are set to be 0.05 and 68.95×10^6 N/m^2 while material density and the modulus of elasticity are 2770.0 kg/m^3 and 6.98×10^{10} N/m^2 respectively.

2.2   Planar 37-Bar Truss Structure

The schematic diagram of the truss structure is illustrated in Fig. 2 [1, 22]. The structure is subject to non-structural mass of 10 kg and external force of $-3,000$ N in y-direction at each free node of the lower chord. The elements of the lower chord were set as bar elements with an unchanged cross-sectional area of 40 cm^2 while the others were set as bar elements with initial cross-sectional area 1 cm^2. The design variables include all bar element cross-sectional areas (except the lower chord bar elements) and y-direction of nodal positions of the upper chord. The design variables are treated to have a symmetrical structure with respect to the y axis. The cross-sectional areas are assigned as the first 14 elements of the design vector while the nodal positions are assigned as the 15th–19th elements of the design vector. A_{min} and A_{max} are set to be 1×10^{-4} m^2 and 10×10^{-4} m^2 respectively, while the y-direction nodal positions of the upper chord nodes are limited to $0.1 \leq Y_i \leq 3$ m (changes in Y_i are symmetric to the y axis). d_{all} and σ_{all} are set to be 0.05 and 400×10^6 N/m^2 while the material density and the modulus of elasticity are 7800.0 kg/m^3 and 2.1×10^{11} N/m^2 respectively.

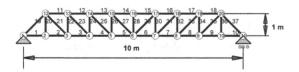

Fig. 2. Schematic of the planar 10-bar truss structure

2.3   Spatial 72-Bar Truss Structure

The truss structure is illustrated in Fig. 3 [16, 22]. Four Non-structural masses of 2270 kg and the external force of -50 kN in z-direction are attached and applied to the top nodes. The design variables include all bar element cross-sectional areas which were divided into 16 groups according to Table 1. A_{min} and A_{max} are set to be 0.645×10^{-4} m^2 and 0.003 m^2, respectively. d_{all} and σ_{all} are set to be 0.05 and 68.95×10^6 N/m^2 while the material density and the modulus of elasticity are 2770.0 kg/m^3 and 6.98×10^{10} N/m^2 respectively.

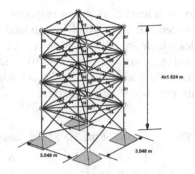

Fig. 3. Schematic of the spatial 72-bar truss structure

Table 1. Element grouping adopted for the 72-bar truss structure

Element groups	Element numbers in the groups	Element groups	Element numbers in the groups	Element groups	Element numbers in the groups	Element groups	Element numbers in the groups
1	1–4	5	19–22	9	37–40	13	55–58
2	5–12	6	23–30	10	41–48	14	59–66
3	13–16	7	31–34	11	49–52	15	67–70
4	17–18	8	35–36	12	53–54	16	71–72

2.4 Fifty-Two Bar Dome Truss (52barTruss)

The truss structure is illustrated in Fig. 4 [22]. The structure has non-structural masses of 50 kg at each free node. External forces of −5,000 N, −3,000 N and −2,000 N are

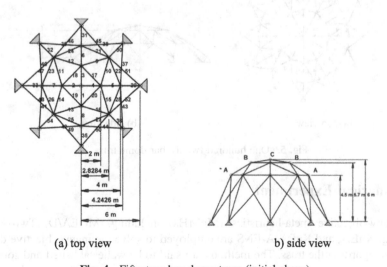

(a) top view   b) side view

Fig. 4. Fifty-two bar dome truss (initial shape)

acted in z-direction on all nodes in level C, B, and A respectively. The design variables include all bar element cross-sectional areas which are divided into 8 groups according to Table 2 and all free nodes which are allowed to move ± 2 m in all directions in a symmetrical manner. A_{min} and A_{max} are set to be 1×10^{-4} m^2 and 0.001 m^2 respectively. d_{all} and σ_{all} are set to be 0.05 and 400×10^6 N/m^2 while the material density and the modulus of elasticity are 7800.0 kg/m^3 and 2.1×10^{11} N/m^2 respectively.

Table 2. Element groups of the 3D 52 bar dome truss

Element group	1	2	3	4	5	6	7	8
Element number in group	1–4	5–8	9–16	17–20	21–28	29–36	37–44	45–52

2.5   One Hundred Twenty Bar Dome Truss (120barTruss)

The schematic drawing of the truss structure is illustrated in Fig. 5 [1, 22]. The structure has non-structural masses and external forces as 3000 kg and -150 kN in z-direction at all nodes at the level height of 7 m, 500 kg and -100 kN in z-direction at all nodes at the level height of 5.85 m, and 100 kg -75 kN in z-direction at the rest. The design variables include all bar element cross-sectional areas which are divided into 7.

Groups as shown in Fig. 5a. A_{min} and A_{max} are set to be 1×10^{-4} m^2 and 0.01293 m^2 respectively. d_{all} and σ_{all} are set to be 0.05 and 400×10^6 N/m^2 while the material density and the modulus of elasticity are 7971.810 kg/m^3 and 2.1×10^{11} N/m^2 respectively.

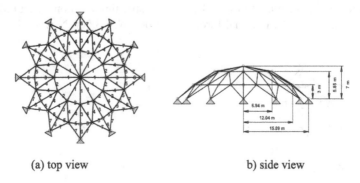

(a) top view   b) side view

Fig. 5. One hundred twenty bar dome truss

3   Numerical Experiment

Five many-objective meta-heuristics (MnMHs) including, MOEA/D, Two-Arch2, PICEA-g, KnEA, and KnEA-A-ENS are employed to solve the many-objective design problems regarding the truss. The methods are said to be wellestablished and some are currently considered the best optimisers amid solving unconstrained many objective test problems. The MnMHs and their optimisation parameter settings used in this study (details of notations can be found in the corresponding references of each method) are detailed as follows:

- MOEA/D: The code was coded by Pholdee and Bureerat [2] based on reference [20]. The number of neighbouring weight vectors, crossover and mutation probabilities being 6, 1.0, and 0.1 respectively.
- Two-Arch2: Used the code from Wang et al. [10]. The crossover probability and mutation probability are set to be 1 and 0.1, respectively.
- PICEA-g: employed the code from Wang et al. [11]. All optimization parameters such as the simulated binary crossover (SBX) parameter, type of crossover, probability of crossover between a pair of individuals, probability of internal crossover, etc. are set as default values from [11].
- KnEA: used the default code and parameter setting from Zhang et al. [12].
- KnEA-A-ENS: used the default code and parameter setting from Zhang et al. [12, 13].

Each optimiser is applied to uncover a Pareto optimal front amid the problems as detailed in Sect. 2 for 10 optimisation runs. For all design problems, the population number is set to 50, while the number of generations is set at 200. The Pareto archive number is set at 50. For the optimisers employing different population size, their search processes are terminated with the total number of function evaluations (FEs) equal to 50×200 FEs. Also, as the majority of the MnMHs used in this study are box-constrained many-objective optimisers, to deal with the constrained problems, the penalty function technique which was efficient for truss design with natural frequencies constraints in [23] is used.

4 Results and Discussion

After performing 10 optimisation runs of five many-objective optimisers on solving five truss many-objective optimisation test problems, the search performances are evaluated based on the C-indicator [4, 21]. The C-indicator compares each pair of optimisers from using the non-dominated fronts obtained. Such an indicator compares two particular non-dominated fronts and can be defined as:

$$C(A, B) = \frac{|\{b \in B; \exists a \in A : a \quad \text{dominate} \quad \text{or} \quad \text{equal} \quad \text{to} \quad b\}|}{|B|}$$

where $|B|$ is the total number of members in the set B. If $C(A, B) = 1$, it means all members in B are dominated by or equal to some members in A. If $C(A, B) > C(B, A)$, it implies that the front A is better than the front B or vice-versa.

Figure 6 shows the box-plots of the C-values comparing all pairs of all optimisers for the test problems 1–5. In the figure, the upper and lower lines show the maximum and minimum C-values while the middle line shows median of the C-values from 10 optimisation runs. The box-plot at row i and column j give the results of C(optimiser_i,

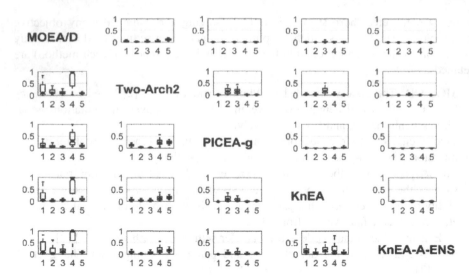

Fig. 6. Show box plot of C-value for all design problems

optimiser_j). The quantitative assessment is given in Tables 3, 4, 5, 6 and 7 where each value on row i and column j is the average value of C (optimiser_i, optimiser_j). For example in Table 3 (test problem 1), average C (MOEA/D, Two_arch2) = 0.0466 while average C (Two_arch2, MOEA/D) = 0.2481. This means Two_arch2 is better than MOEA/D. Tables 3, 4, 5, 6 and 7 show the average C-values of all optimisers on solving truss problems 1–5 respectively. From the tables, it is revealed that KnEA-A-ENS is the best optimiser for all design problems. KnEA is the second best optimiser for the planar 37-bar truss structure and the 120-bar dome truss. For the spatial 72-bar truss structure and 52-bar dome truss, the second best performers are Two-Arch2 and PICEA-g respectively.

Table 3. Mean C-values for planar 10-bar truss structure

MOEA/D	0.0466	0.0028	0.0110	0.0022
0.2481	Two-Arch2	0.0100	0.0210	0.0048
0.1168	0.1200	PICEA-g	0.0104	0.0032
0.2049	0.0706	0.0030	KnEA	0.0084
0.2875	0.0926	0.0054	0.1498	KnEA-A-ENS

Table 4. Mean C-values for planar 37-bar truss structure

MOEA/D	0.0120	0.0040	0.0000	0.0000
0.2296	Two-Arch2	0.1594	0.0236	0.0000
0.1086	0.0332	PICEA-g	0.0018	0.0000
0.0828	0.0468	0.1224	KnEA	0.0004
0.1418	0.0218	0.0574	0.0804	KnEA-A-ENS

Table 5. Mean C-values for spatial 72-bar truss structure

MOEA/D	0.0250	0.0036	0.0264	0.0030
0.1116	Two-Arch2	0.1488	0.1838	0.0334
0.0442	0.0230	PICEA-g	0.0052	0.0006
0.0632	0.0348	0.0758	KnEA	0.0140
0.1498	0.0616	0.0654	0.1946	KnEA-A-ENS

Table 6. Mean C-values for fifty-two bar dome truss

MOEA/D	0.0426	0.0014	0.0096	0.0000
0.7344	Two-Arch2	0.0160	0.0468	0.0092
0.4320	0.2526	PICEA-g	0.0658	0.0126
0.7110	0.1526	0.0148	KnEA	0.0334
0.7728	0.1882	0.0212	0.2124	KnEA-A-ENS

Table 7. Mean C-values for one hundred twenty bar dome truss

MOEA/D	0.1094	0.0070	0.0092	0.0048
0.1006	Two-Arch2	0.0076	0.0102	0.0046
0.1014	0.2374	PICEA-g	0.0274	0.0064
0.0982	0.1738	0.0342	KnEA	0.0078
0.0966	0.1702	0.0380	0.0698	KnEA-A-ENS

5   Conclusions

The test problems for 6-objective optimisation of trusses are proposed. Five many-objective meta-heuristics are implemented to solve the various problems. Comparative results show that KnEA-A-ENS gives the best results for all five test problems. The design results of many-objective optimisation of trusses, in practice, provide the optimal choices for better decision making. Future work will be directed to the development of a new powerful meta-heuristic for solving this new type of truss design problems. More objective functions can be added if the trusses are subject to multiple load cases.

Acknowledgements. The authors are thoroughly grateful for the financial support provided by the KKU Engineering Research Fund, the Faculty of Engineering, Khon Kaen University.

References

1. Pholdee, N., Bureerat, S.: Hybrid real-code population-based incremental learning and approximate gradients for multi-objective truss design. Eng. Opt. **64**, 1032–1050 (2013)
2. Pholdee, N., Bureerat, S.: Hybridisation of real-code population-based incremental learning and differential evolution for multiobjective design of trusses. Inform. Sci. **23**, 136–152 (2013)

3. Bureerat, S., Srisomporn, S.: Optimum plate-fin heat sinks by using a multi-objective evolutionary algorithm. Eng. Opt. **42**, 305–343 (2010)

4. Noilublao, N., Bureerat, S.: Simultaneous topology, shape, and sizing optimisation of plane trusses with adaptive ground finite elements using MOEAs. Math. Probl. Eng. **838102**, 9 (2013). doi:10.1155/2013/838102

5. Bandyopadhyay, S., Mukherjee, A.: An algorithm for many-objective optimization with reduced objective computations: a study in differential evolution. IEEE T. Evolut. Comput. **19**, 400–413 (2015)

6. Ma, X., Liu, F., Qi, Y., Wang, X., Li, L., Jiao, L., Yin, M., Gong, M.: A multiobjective evolutionary algorithm based on decision variable analyses for multiobjective optimization problems with large-scale variables. IEEE Trans. Evol. Comput. **20**, 275–298 (2016)

7. Zhang, X., Tian, Y., Cheng, R., Jin, Y.: A decision variable clustering-based evolutionary algorithm for large-scale many-objective optimization. IEEE Trans. Evol. Comput. (2017, in press)

8. Murata, T., Taki, A.: Examination of the performance of objective reduction using correlation-based weighted-sum for many objective knapsack problems. In: 10th International Conference on Hybrid Intelligent Systems (HIS), pp. 175–180. IEEE press, New York (2010)

9. Cheng, R., Jin, Y., Sendhoff, B.: A reference vector guided evolutionary algorithm for many-objective optimization. IEEE Trans. Evol. Comput. **20**, 773–791 (2016)

10. Wang, H., Jiao, L., Yao, X.: Two_Arch2: an improved two-archive algorithm for many-objective optimization. IEEE Trans. Evol. Comput. **19**, 524–541 (2015)

11. Wang, R., Purshouse, R.C., Fleming, P.J.: Preference-inspired coevolutionary algorithms for many-objective optimization. IEEE Trans. Evol. Comput. **17**, 474–494 (2013)

12. Zhang, X., Tian, Y., Jin, Y.: A knee point-driven evolutionary algorithm for many-objective optimization. IEEE Trans. Evol. Comput. **19**, 761–776 (2015)

13. Zhang, X., Tian, Y., Jin, Y.: Approximate non-dominated sorting for evolutionary many-objective optimization. Inform. Sci. **369**, 14–33 (2016)

14. Deb, K., Jain, H.: Handling many-objective problems using an improved NSGA-II procedure. In: WCCI 2012 IEEE World Congress on Computational Intelligence. IEEE press, New York (2012)

15. Kaveh, A., Ghazaan, M.I.: Hybridized optimization algorithms for design of trusses with multiple natural frequency constraints. Adv. Eng. Softw. **79**, 137–147 (2015)

16. Bureerat, S., Pholdee, N.: Optimal truss sizing using an adaptive differential evolution algorithm. J. Comput. Civil Eng. **30**, 04015019 (2015)

17. Dede, T., Ayvaz, Y.: Combined size and shape optimization of structures with a new meta-heuristic algorithm. Appl. Soft Comput. **28**, 250–258 (2015)

18. Kaveh, A., Sheikholeslami, R., Talatahari, S., Keshvari-Ilkhichi, M.: Chaotic swarming of particles: a new method for size optimization of truss structures. Adv. Eng. Softw. **67**, 136–147 (2014)

19. Kaveh, A., Bakhshpoori, T., Afshari, E.: An efficient hybrid particle swarm and swallow swarm optimization algorithm. Comput. Struct. **143**, 40–59 (2014)

20. Qingfu, Z., Hui, L.: MOEA/D: a multiobjective evolutionary algorithm based on decomposition. IEEE Trans. Evol. Comput. **11**, 712–731 (2007)

21. Zitzler, E., Deb, K., Thiele, L.: Comparison of multiobjective evolutionary algorithms: empirical results. Evol. Comput. **8**, 173–195 (2000)

22. Pholdee, N., Bureerat, S.: Comparative performance of meta-heuristic algorithms for mass minimisation of trusses with dynamic constraints. Adv. Eng. Softw. **75**, 1–13 (2014)

23. Kaveh, A., Zolghadr, A.: Truss optimization with natural frequency constraints using a hybridized CSS-BBBC algorithm with trap recognition capability. Comput. Struct. **102–103**, 14–27 (2012)

Clustering with Multidimensional Mixture Models: Analysis on World Development Indicators

Leonard K.M. Poon[✉]

Department of Mathematics and Information Technology,
The Education University of Hong Kong, Hong Kong, China
kmpoon@eduhk.hk

Abstract. Clustering is one of the core problems in machine learning. Many clustering algorithms aim to partition data along a single dimension. This approach may become inappropriate when data has higher dimension and is multifaceted. This paper introduces a class of mixture models with multiple dimensions called pouch latent tree models. We use them to perform cluster analysis on a data set consisting of 75 development indicators for 133 countries. We further propose a method that guides the selection of clustering variables due to the existence of multiple latent variables. The analysis results demonstrate that some interesting clusterings of countries can be obtained from mixture models with multiple dimensions but not those with single dimensions.

Keywords: Multidimensional clustering · Pouch latent tree models · Mixture models · World development indicators · Clustering variables selection

1 Introduction

Clustering [8] is a core problem in machine learning. Many clustering algorithms aim to partition data along a single dimension [2,16]. To handle data with higher dimensions, *feature selection* and *subspace clustering* approaches are often adopted. The former approach selects a subset of relevant features on data in which a clustering can be found [5,14]. The latter approach considers dense regions as clusters and tries to identify all dense subspaces (with reduced dimension) for partitioning the data [9,11]. Both approaches partition data along only a single dimension, in the sense that each data point belongs to at most one partition.

The above approach becomes inappropriate when data is multifaceted and multiple meaningful clusterings can be obtained. Suppose we want to cluster countries into different groups. We may partition them based on their land sizes and populations, systems of government, income levels, levels of freedom, etc. To obtain clusterings on different aspects, one may perform cluster analysis on data sets with different attributes. However, sometimes one may not know

© Springer International Publishing AG 2017
F. Cong et al. (Eds.): ISNN 2017, Part I, LNCS 10261, pp. 153–160, 2017.
DOI: 10.1007/978-3-319-59072-1_19

which aspect of data will yield to meaningful clusterings and sometimes the attributes in different aspects are interdependent. Hence it is more appropriate to perform cluster analysis that produces clustering along multiple dimensions simultaneously.

In our previous work [12,13], we propose a class of probabilistic graphical models called *pouch latent tree models* (PLTMs) for multidimensional clustering. The models are similar to Gaussian mixture models. However, they can contain multiple latent variables and hence can produce multiple clusterings.

In this paper, we present the results of a cluster analysis on countries based on the world development indicators. The indicators are statistics provided by the World Bank about the development and human lives for different countries. The data set we used includes 75 indicators relevant to risk management in the context of development for 133 countries. The data obviously represent different aspects of countries and our study aims to show the usefulness of multidimensional clustering. Due to the existence of multiple latent variables, we propose a method that guides the selection of clustering variables. Before we show the results, we review model-based clustering and introduce PLTMs.

2   Model-Based Clustering

Gaussian mixture models (GMMs) are commonly used in model-based clustering for numeric data [10]. GMMs assume that the population is be made up from a finite number of clusters. Suppose a variable Y is used to indicate this cluster, and variables $\boldsymbol{X}$ represent the attributes in the data. The variable Y is referred to as a *latent* (or unobserved) variable, and the variables $\boldsymbol{X}$ as *manifest* (or observed) variables. The manifest variables $\boldsymbol{X}$ is assumed to follow a mixture distribution

$$P(\boldsymbol{x}) = \sum_y P(y)P(\boldsymbol{x}|y),$$

where $P(\boldsymbol{x}|y)$ is known as the *component distribution* and in GMMs is assumed to be a multivariate Gaussian distribution $\mathcal{N}(\boldsymbol{x}|\boldsymbol{\mu}_y, \boldsymbol{\Sigma}_y)$, with mean vector $\boldsymbol{\mu}_y$ and covariance matrix $\boldsymbol{\Sigma}_y$ conditional on the value of Y.

3   Pouch Latent Tree Models

There is only one single latent variable in GMMs and hence can produce one clustering. To allow having multiple clusterings, we previously propose *pouch latent tree models* (PLTMs) [12,13]. A PLTM is a tree-structured probabilistic graphical model, where each internal node represents a latent variable, and each leaf node represents a set of manifest variables. All the latent variables are discrete, while all the manifest variables are continuous. A leaf node, also called *pouch node*, may contain a single manifest variable or several of them. An example is shown in Fig. 2.

In PLTMs, the dependency of a discrete latent variable Y on its parent $\Pi(Y)$ is characterized by a conditional discrete distribution $P(y|\pi(y))$. Let $\boldsymbol{W}$ be the variables of a pouch node with a parent node $Y = \Pi(W)$. We assume that, given a value y of Y, W follows the conditional Gaussian distribution $P(w|y) = \mathcal{N}(w|\mu_y, \Sigma_y)$ with mean vector μ_y and covariance matrix Σ_y. Denote the sets of pouch nodes and latent nodes by $\mathcal{W}$ and $\mathcal{Y}$, respsectively. The whole model defines a joint distribution over all observed variables X and latent variables Y

$$P(x, y) = \prod_{W \in \mathcal{W}} P(w|\pi(W)) \prod_{Y \in \mathcal{Y}} P(y|\pi(Y)) \tag{1}$$

Given a model structure m, the parameters can be estimated by the EM algorithm [4]. To learn the model structure, we use a greedy search that aims to maximize the BIC score [15]: $BIC(m|\mathcal{D}) = \log P(\mathcal{D}|m, \theta^*) - \frac{d(m)}{2} \log N$, where $\mathcal{D}$ is the data set, θ^* are the parameters estimated by the EM algorithm, $d(m)$ is the number of parameters in the model, and N is the data size. Interested readers are referred to [13] for details of the learning algorithm.

After we have learned a PLTM, we can partition data using each of the latent variables Y. Each data point d can be classified to one of the states of Y by computing the probability $P(y|d)$ based on the joint distribution (Eq. 1).

4   Analysis on World Development Indicators

Here we present the results of a cluster analysis on world development indicators using PLTM aiming to show its effectiveness for multidimensional clustering.

4.1   Data Set

In our experiment, we used the data set called World Development Report (WDR) 2014 provided by the World Bank.[1] The data set includes 75 indicators relevant to risk management in the context of development for 133 countries. The indicators are grouped into seven categories, namely key indicators of development, selected risk indicators, selected indicators related to risk management at the household level, enterprise sector level, financial sector level, macroeconomy level, respectively, and natural disasters and climate change indicators. For some indicators, the data set includes multiple values at different time periods. Some statistics are not available for some countries. In summary, the data set has 93 attributes and 133 samples with 15% of missing data.

4.2   Empirical Comparison

We included three methods based on GMMs for comparison in our experiment. The first method is mclust [6], which is an implementation of the parsimonious Gaussian mixture models [1]. The second method is the GS method [7]. It models the data using a collection of independent GMMs, each on a distinct subset

---

[1] http://data.worldbank.org/data-catalog/world-development-report-2014.

Table 1. Comparison of methods on the World Development Report 2014 data set. The table shows the numbers of latent variables (#LV), numbers of parameters (dim), and BIC scores of the models obtained. It also shows the NMI and number of clusters (#C) of the clustering closest to the given classifcation.

Method	#LV	dim	BIC	NMI	#C
mclust	1	1109	−51965	0.41	6
GS model	40	2125	(−37422)	0.52	4
PLTM	28	1043	**−46706**	**0.62**	4

of attributes. The third method is PLTMs. The first method produces unidimensional clusterings, whereas the other two method produce multidimensional clusterings. Since mclust and the GS method cannot handle missing data, we impute the missing data using the R package mice [18] before training them.

Table 1 shows the results obtained by the three methods. The mclust model contains one latent variable, whereas the GS model and PTLM contains 40 and 28 latent variables, respectively. In terms of model complexity, PLTM has the lowest number of parameters. This happens even though PLTM has more latent variables than mclust model and it has connections between latent variables unlike GS model.

To evaluate the model quality, we compute the BIC score of the models. We use the completed data as the test data set for consistency. The parameters of PLTM were re-estimated on the complete data after learning the model structure on the incomplete data. This should not be unfair to other methods since mclust and GS method used the same test data set for training while PTLM optimized its the structure using a data set different from the test data set.

The BIC scores in Table 1 show that PLTM has a higher quality than the mclust model. The BIC of GS model is even higher. However, this was possibly due to spurious clusters [10]. Those clusters have component distributions with very small variance and hence can attain very high likelihood on data. This can be seen from the fact that although the smallest variance in the data is 0.32, the smallest scale of variance of the component distribution in the GS model is much smaller at 2.8×10^{-16}.

The WDR includes a classification of countries based on four income levels, namely low, lower middle, upper middle, and high. The classification is used as a class variable for evaluating the clusterings given by the models. To evaluate the similarity between the partition given by a latent variable Y and the class variable C, we use the normalized mutual information $NMI(C;Y)$ [17]: $NMI(C;Y) = \frac{MI(C;Y)}{\sqrt{H(C)H(Y)}}$, where $MI(C;Y)$ is the mutual information between C and Y and $H(V)$ is the entropy of a variable V [3].

Table 1 shows the NMI attained by the three methods. Among the multiple clusterings given by GS method and PLTM, only the ones with the highest NMI are reported. The result shows that PLTM performed best in recovering the classification based on income levels.

4.3   Selection of Clustering Variables

Figure 2 shows the PLTM obtained from the WDR data set. The latent nodes are represented by the oval nodes. Each of them produces a partition of data. They partition the data along different facets of data as can be seen from the different attributes connected to them. For example, the latent variable Y_1 is connected to three attributes, namely gross national income per capita (gni_pc), PPP gross national income per capita (ppp_gni_pc), and worldwide government indicator (worldwide_government_indicator). The three observed variables are put inside a pouch node meaning that they are not independent conditionally on Y_1. The partition given by Y_1 happens to be the one closest to the classification based on income level given by WDR.

The PLTM obtained contains 28 latent variables and thus provides 28 ways to partition data. One issue arising from multidimensional clustering is how to select clustering among those available. In practice, there may not be any reference clustering for selection as we do in the previous subsection. Therefore, we propose a method for selecting clustering variables below.

Due to the model structure, each latent variable partitions data based on a different subset of attributes. To quantify this, we compute the NMI between a latent variable and each of the attributes.[2] After obtaining a vector of NMI values for each latent variable, we normalize them such that each one has unit magnitude. We then cluster the variables using hierarchical clustering.

The clustering of variables can help us look for a clustering of interest. We illustrate the idea using the PLTM obtained as an example. Figure 1 shows the hierarchical clustering result. We see that some latent variables (e.g. Y_1, Y_{21}–Y_{25}) are closer to each other, while some latent variables (e.g. Y_6, Y_7, Y_{16}, Y_{27}) are further away from the others. We cut off the tree at the red horizontal line in Fig. 1. There are four groups of latent variable below the line and they are indicated by different colors in Fig. 2. The grouping of variables is consistent with the model structure. It shows which latent variables partition data along a similar subset of

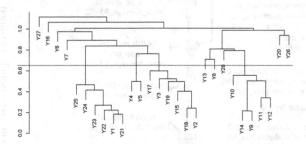

Fig. 1. Hierarchical clustering on latent variables based on the subset of attributes on which they partition the data. We cut off the tree at the red horizontal line. (Color figure online)

---

[2] The NMI can be computed using the empirical distribution after discretizing the continuous attributes.

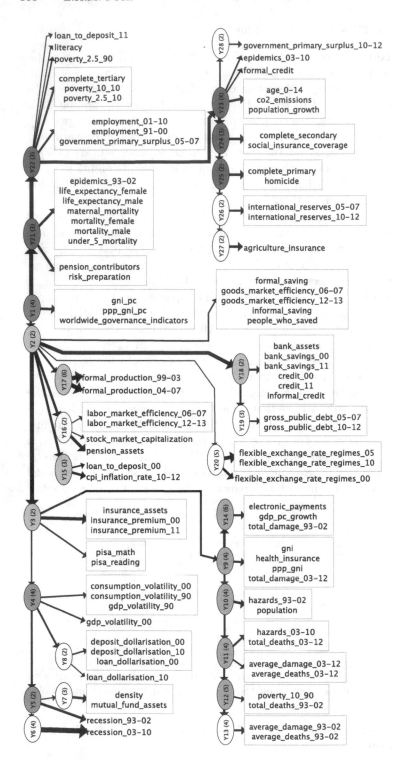

Fig. 2. Pouch latent tree models obtained from the World Development Report 2014 data set. Latent nodes are represented by oval nodes. Each of them produces a partition of data. Pouch nodes with multiple observed variables are shown as text in rectangular border, whereas those with single observed variables are shown as text without borders. The width of an edge indicates the strength of probabilistic dependency in terms of NMI between two nodes. The latent variable Y_1 (blue) yields a clustering of countries closest to the income level classification given by the report. The colors of latent nodes indicate their grouping. (Color figure online)

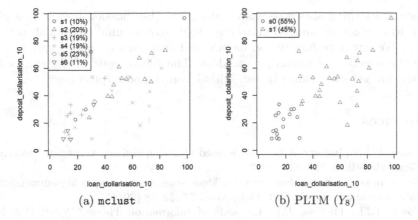

(a) `mclust`   (b) PLTM (Y_8)

Fig. 3. Clusterings along two dollarization attributes obtained from unidimensional clustering method `mclust` and multidimensional clustering method PLTM. PLTM partitions data neatly along this facet while `mclust` does not.

attributes. On the other hand, the ungrouped latent variables partition data along a relatively distinct subset of attributes.

We now take a look at the four groups of latent variables. The first group, Y_1 and Y_{21}–Y_{25}, depends on attributes related to general development level, such as GNI, life expectancy, mortality and poverty. The second group, Y_2, Y_3, Y_{15}, Y_{17} and Y_{18}, depends mainly on attributes related to the financial sector. The third group, Y_9–Y_{12} and Y_{14}, is mainly related to deaths and damages from natural disasters. The fourth group of variables Y_4 and Y_5 partition data based on number of recessions and GDP and household volatility.

The ungrouped variables also partition data based on meaningful subsets of attributes. For example, Y_8 partitions data based on dollarization, Y_{19} partitions based on flexibility of exchange rate, Y_{20} partitions based on gross public debt, and Y_{26} partitions based on international reserves.

The grouping of latent variables allows us to have an overview on the aspects of attributes from which we obtain a clustering. We can look for the attributes in which we are interested and select the latent variable connected to it to partition the data. As an example, suppose we are interested in dollarization. We can use the latent variable Y_8 in PLTM for clustering. Figure 3(b) shows the different countries projected on two dollarization attributes. The countries are classified into two groups neatly by Y_8. For comparison, we show the clustering obtained by `mclust` in Fig. 3(a). The comparison shows that the multidimensional clustering method PLTM can obtain some meaningful clusterings that cannot be obtained by the unidimensional clustering method `mclust`.

5 Conclusion

In this paper, we introduce a class of multidimensional mixture models called pouch latent tree models and use them for cluster analysis on world development indicators. PLTM is shown to recover the given classification better. It is also shown to produce meaningful clusterings that another unidimensional method cannot. We illustrate how to use hierarchical clustering on latent variables to guide the selection of clustering variables. The source code of algorithms for PLTMs can be found online: https://github.com/kmpoon/pltm-east.

References

1. Banfield, J.D., Raftery, A.E.: Model-based Gaussian and non-Gaussian clustering. Biometrics **49**(3), 803–821 (1993)
2. Bouveyrona, C., Brunet-Saumard, C.: Model-based clustering of high-dimensional data: a review. Comput. Stat. Data Anal. **71**, 52–78 (2014)
3. Cover, T.M., Thomas, J.A.: Elements of Information Theory. Wiley, Hoboken (2006)
4. Dempster, A.P., Laird, N.M., Rubin, D.B.: Maximum likelihood from incomplete data via the EM algorithm. J. R. Stat. Soc. Ser. B (Methodol.) **39**(1), 1–38 (1977)
5. Dy, J.G., Brodley, C.E.: Feature selection for unsupervised learning. J. Mach. Learn. Res. **5**, 845–889 (2004)
6. Fraley, C., Raftery, A.E., Murphy, T.B., Scrucca, L.: MCLUST version 4 for R: normal mixture modeling for model-based clustering, classification, and density estimation. Department of Statistics, University of Washington, Technical report (2012)
7. Galimberti, G., Soffritti, G.: Model-based methods to identify multiple cluster structures in a data set. Comput. Stat. Data Anal. **52**, 520–536 (2007)
8. Jain, A.K., Murty, M.N., Flynn, P.J.: Data clustering: a review. ACM Comput. Surv. **31**(3), 264–323 (1999)
9. Kriegel, H.P., Kröger, P., Zimek, A.: Clustering high dimensional data: a survey on subspace clustering, pattern-based clustering, and correlation clustering. ACM Trans. Knowl. Discov. Data **3**(1), 1–58 (2009)
10. McLachlan, G.J., Peel, D.: Finite Mixture Models. Wiley, New York (2000)
11. Parsons, L., Haque, E., Liu, H.: Subspace clustering for high dimensional data: a review. ACM SIGKDD Explor. Newsl. **6**(1), 90–105 (2004)
12. Poon, L.K.M., Zhang, N.L., Chen, T., Wang, Y.: Variable selection in model-based clustering: to do or to facilitate. In: Proceedings of the 27th International Conference on Machine Learning, pp. 887–894 (2010)
13. Poon, L.K.M., Zhang, N.L., Liu, T., Liu, A.H.: Model-based clustering of high-dimensional data: variable selection versus facet determination. Int. J. Approx. Reason. **54**(1), 196–215 (2013)
14. Raftery, A.E., Dean, N.: Variable selection for model-based clustering. J. Am.Stat. Assoc. **101**(473), 168–178 (2006)
15. Schwarz, G.: Estimating the dimension of a model. Ann. Stat. **6**(2), 461–464 (1978)
16. Shirkhorshidi, A.S., Aghabozorgi, S., Wah, T.Y., Herawan, T.: Big data clustering: a review. In: Murgante, B., et al. (eds.) ICCSA 2014. LNCS, vol. 8583, pp. 707–720. Springer, Cham (2014). doi:10.1007/978-3-319-09156-3_49
17. Strehl, A., Ghosh, J.: Cluster ensembles – a knowledge reuse framework for combining multiple partitions. J. Mach. Learn. Res. **3**, 583–617 (2002)
18. van Buuren, S., Groothuis-Oudshoorn, K.: MICE: multivariate imputation by chained equations in R. J. Stat. Softw. **45**(3), 1–67 (2011)

Logic Calculation Based on Two-Domain DNA Strand Displacement

Xiaobiao Wang, Changjun Zhou, Xuedong Zheng,
and Qiang Zhang[✉]

Key Laboratory of Advanced Design and Intelligent Computing (Dalian
University), Ministry of Education, Dalian 116622, China
zhangq@dlu.edu.cn

Abstract. DNA strand displacement technology has become one of the most commonly used in biological computing technology. In this paper, we design a calculation model of the basic logic unit based on two-domain DNA strand displacement and logical relation, including AND, OR logic gates. The calculation process is simple and easy to understand, because of the unified single strand structure. The process of the reaction is more thorough and more easy to control. The model is used to construct a converter of a four-bit binary into BCD code. The whole reaction process can be programmed and simulated in the software Visual DSD, and the result also verifies the correctness of the design of the basic logic calculation model.

Keywords: DNA strand displacement · Logic gate · Calculation model

1 Introduction

With the development of large data in recent years, the pressure of traditional computers is growing; more researchers gradually pay close attention to the biological computation. The biological computation is composed of science, biology, medicine and so on [1–3]. A lot of theoretical and computational models has been raised [4–6]. DNA strand displacement technology [7] is one of the most commonly used and the most important technical means for molecular computing.

Since Adleman [1] using DNA to solve a 7-city of Hamiltion path problem, DNA computation is used for solving a wide variety of application. In recent years, the research of the logic circuits and neural network is popular [8–11]. In 2010, Cardelli [11] proposed a structure of two-domain strand to investigate the computing power of a restricted class of DNA strand displacement structures. In 2011, Winfree et al. [2] achieved a four-neuron Hopfield associative memory experimental by using cascade the DNA strand displacement. In 2013, Zhang et al. [8] proposed the calculation model of logic AND, OR gate, and detected the experimental results by gel electrophoresis.

In this paper, we introduced the Luca Cardelli's join and fork gate model based on two-domain DNA strand displacement [11], and design AND, OR gate calculation model. We analyze and verify the correctness of the model by the software of Visual DSD [12]. Finally, on the basis of the established logic basic calculation model, we design a four-binary conversion BCD code computation model and verify it.

© Springer International Publishing AG 2017
F. Cong et al. (Eds.): ISNN 2017, Part I, LNCS 10261, pp. 161–169, 2017.
DOI: 10.1007/978-3-319-59072-1_20

2 Methods and Materials

2.1 Two-Domain DNA Strand Displacement

The technology of strand displacement is used to simulation of chemical reaction and the Petri net. In general, single strand is signal strand, double strand is gate structure; once signals and gates are mixed together, they will automatically react without further intervention until the gate or signal is completely exhausted. In this work, we construct the model structure by using two-domain strand structure [11].

In the structure of two-domain strand model, the single strand is composed of two parts, toehold (short) domain and recognition (long) domain. Double strand is gate structure and composed of the top strand and the bottom strand. The top strand of double strand is broken into segments by the nicks; the bottom strand is a complete strand generally. When the short toehold domain binding between a double strand and a single strand, if the long domain of single strand and double strand meet the principle of base pairing, the long domain of the single strand will gradually replace the top strand of the double strand by branch migration and replace the corresponding strand, otherwise no replacement reaction. As shown in Fig. 1, it is the strand displacement process. Here the single strand can only react with double strand.

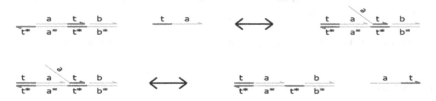

Fig. 1. DNA strand displacement.

2.2 AND Gate and OR Gate Based on Two-Domain

As we know, the most basic logical relationship is AND, OR, and NOT. In the logic circuit, the basic logic gate is AND gate, OR gate, NOT gate, the three basic forms. Other logical structure is replaced by the combination of AND gate and OR gate. The relationship types of the logic AND and OR are as follows:

$$Y = A \bullet B \tag{1}$$

$$Y = A + B \tag{2}$$

In the formula (1) and (2), A and B are the input signal; Y is the output signal. From (1), we can know that Y will be produced by A and B; from (2), for A and B, any one occurs, it will generate Y. The logic AND is meaningful when A and B are exist; the logic OR is meaningful when one of A and B is exist at least.

We use the two-domain strand to construct the AND gate and OR gate. As shown in Fig. 2, the input strand Input1and the input strand Input2 react with other strand, and produce the output strand tw. To reduce intermediates product, the double strand Gate b3 is designed to take in the waste product strand tb, vt. These recycling structures can increase the reaction rate and make the reaction more thoroughly.

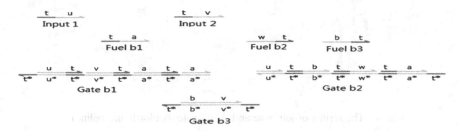

Fig. 2. 2-input 1-output join gate initial state

The principle of join gate reaction is similar to the logical AND, the whole process can meet a logic AND relationship expression. We design a two-domain structure AND gate based on join gate structure, as shown Fig. 3. In the initial state of system, the single strand tx, ty are the input signal single strand, the intermediate consumption strands are ta, zt; each gate structures are different from the gate structure of the join gate. We increase the concentration of intermediate strand at in the Gate b1, which makes the recycling waste strand Gate b3 and Gate b2 reaction at same time in order to increase the reaction rate. The Gate b3 structure can not only recycle strand yt, but also can produce Fuel b1. It slows down the rate of Fuel b1reduction of intermediate reaction in the system. Garbage accumulated will slow down the reaction system. Our design can not only reduce the produce of waste, but also reduce the reaction time and the reaction rate when make the system reach a steady state. To demonstrate this advantage, we make a test using Visual DSD software in the same environment and compare with the results of join gate. The result is shown in Fig. 4 and Table 1 (the horizontal coordinate is the time; the vertical coordinate is the concentration in this paper).

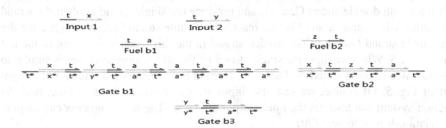

Fig. 3. 2-input 1-output AND gate initial state based on two-domain strand displacement.

In the Fig. 4, the blue line represents the final output strand <t^ z> of AND gate; the purple line represents the final output strand <t^ w> of join gate. The AND gate and the join gate are same initial input in the Fig. 4.

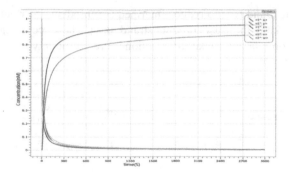

Fig. 4. The results of join gate and AND gate. (Color figure online)

Table 1. The species variation of AND gate and join gate

Species	tx	ty	tz	tu	tv	tw
Initial amount	1	1	0	1	1	0
Final amount	0.0048	0.0054	0.9565	0.0077	0.0083	0.8805
Consumption/generating	0.9952	0.9946	0.9565	0.9923	0.9917	0.8805

In the Table 1, the strand tx, ty tu, tv letter indicate the initial species; the strand tz and tw indicate output species. We set the same initial input amount, $tx = ty = tu = tv = 1$. The generation of the AND gate is 7.6% higher than the join gate. The AND gate reaction is more relatively complete, more product at the same initial concentration of the input strand.

Another important logical relationship is logic OR. According to the formula (2) shows that, in all conditions, as long as there is only one condition which is satisfied, the system will be able to react. For logic OR in the logical circuit, as long as the circuit receives any one signal from all input signal, it will be connected. With this property, we design an OR gate computation model. The initial state of overall system is shown in Fig. 5.

When we input the single strand tx, the strand tx together with single strand Fuel b1 will react with double strand Gate b1, and produce the single strand at; then the strand at together with single strand Fuel b3 react with double strand Gate b3 to generate the final single strand Output tz The needed strand in the this reaction is shown in the left part of Fig. 5. When we input the single strand ty, the all reaction processes is similar to the processes of input strand tx. The needed strand in this reaction is shown in the right part of Fig. 5. Of course, we can also input strand tx and ty at same time, and the reaction system can also get the final Output strand tz. The whole process can display the relationship of logical OR.

We simulate the process by Visual DSD software. Figure 6(a) is shown the OR gate in the case of only input signal strand tx and reaction results; Fig. 6(b) is shown the OR gate in the case of only input signal strand ty and reaction results; Fig. 6(c) is shown in the case of input signal strand tx and ty; tz is final output signal stand.

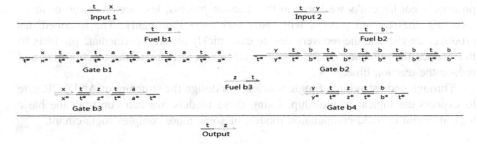

Fig. 5. The initial state of OR gate based on two-domain strand displacement

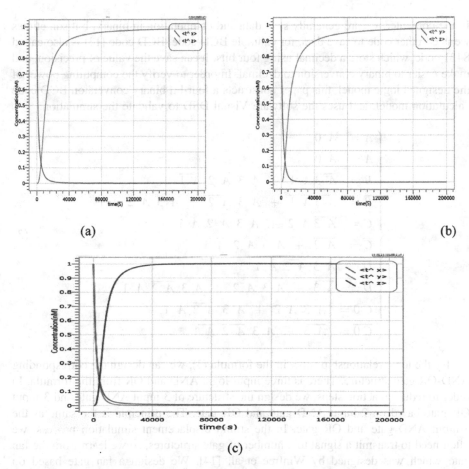

Fig. 6. The result of OR gate reaction. (a) The result when only strand tx exist; (b) the result when only strand ty exist; (c) the result when both strand tx and ty exist.

Through learning the two-domain strand structure, we design AND gate, OR gate for logic operation. The two domain strand is relatively simple and can meet the requirement of Boolean logic operation [15]. The number of strand in the computation process is not large. As we know in the reaction process, less accumulation of inter-mediates, the reaction process will faster and more thoroughly. The intermediates products are less and the recovery device can quickly absorb intermediate products so that reduce the influence on the reaction rate, so as to increase the reaction rate and reduce the reaction time.

Through understanding of logic relation, we design the structure of AND, OR gate to express the logical relationship. Using these models, we can construct the basic logical circuit to build computation models of some more complex logic circuit.

3 Experiments

In the computer, we are generally save data and computation in binary, but sometimes we need other code to save data, for example BCD code. BCD code [13] is also called 8421 code, which store a decimal using four bits. It can save the value of precision, and make easier to binary conversion to decimal. In order to verify the computing power of the designed logic model, this paper constructs a four-bit binary conversion BCD code calculation model, and uses the software Visual DSD to validate the simulation.

$$
\begin{cases}
A = & A\,0 \\
\overline{A} = & \overline{A\,0} \\
B = & \overline{A\,3}\,A\,1 \;+\; A\,3\,A\,2\,\overline{A\,1} \\
\overline{B} = & \overline{A\,3}\,\overline{A\,1} \;+\; A\,3\,\overline{A\,2} \;+\; A\,3\,A\,2\,A\,1 \\
C = & \overline{A\,3}\,A\,2 \;+\; A\,3\,A\,2\,A\,1 \\
\overline{C} = & \overline{A\,2} \;+\; A\,3\,A\,2\,\overline{A\,1} \\
D = & A\,3\,\overline{A\,2}\,\overline{A\,1} \\
\overline{D} = & \overline{A\,3} \;+\; A\,3\,A\,2 \;+\; A\,3\,\overline{A\,2}\,A\,1 \\
C\,0 = & A\,3\,A\,2 \;+\; A\,3\,\overline{A\,2}\,A\,1 \\
\overline{C\,0} = & \overline{A\,3} \;+\; A\,3\,\overline{A\,2}\,\overline{A\,1}
\end{cases}
\tag{3}
$$

By the logic relationship show in the formula (3), we can design the corresponding AND-OR gate structure. There is three input logic AND and OR from the formula. In order to reduce reaction steps, we design the structure of 3-input AND gate and 3-input OR gate based on 2-input AND gate and OR gate. The principle is the same as the 2-input AND gate and OR gate. In the strand displacement simulation process, we often need to transmit a signal to a number of gate structures, so we learn from the fan gate which was designed by Winfree et al. [14]. We design a fan gate based on two-domain strand so that make a signal to split into multiple signal and react with corresponding strand.

We know logical relation could be expressed by the combination of AND and OR. In this work, we construct the logic circuit consists of AND gate and OR gate. The circuit diagram as shown in Fig. 7. The signal A does not need to go through the gate

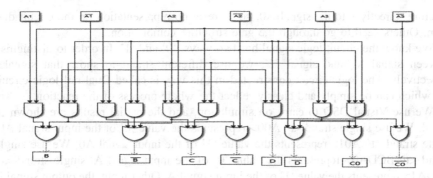

Fig. 7. The circuit diagram of a four-bit binary into BCD code.

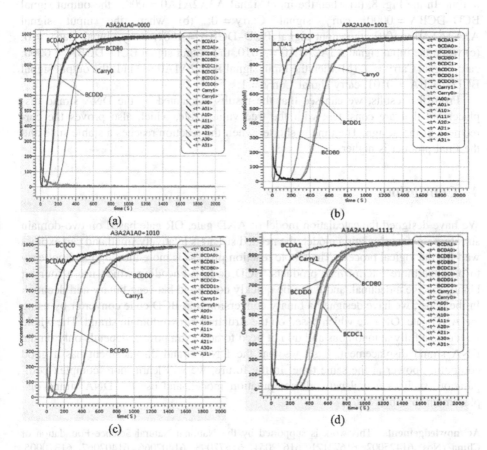

Fig. 8. The simulate result of DSD. (a) The result when the input signal A3A2A1A0 = 0000; (b) the result when the input signal A3A2A1A0 = 1001; (c) the result when the input signal A3A2A1A0 = 1010; (d) the result when the input signal A3A2A1A0 = 1111.

structure, directly into the signal A0, so there is no representation on the circuit diagram. Others need to go through the gate structure combination.

We know the circuit logic signal has two ways, '0' and '1'. In order to distinguish between signal '0' and signal '1', we use different strand to show that signaler respectively. The logic circuit constructed in this way is called Dual-rail logic circuit [2], which can be simply and clearly reflect the whole process of the reaction.

We use Visual DSD to carry on simulation, and the partial results are shown in Fig. 8. We use single strand <t^ A00> represents the value '0' of the input signal A0, single strand <t^ A01> represents the value '1' of the input signal A0; We use single strand <t^ BCDA0> represents the value '0' of the input signal A, single strand <t^ BCDA1> represents the value '1' of the input signal A. Other input, the output signal is similar to the modification. For showing more dynamic process, the stochastic is selected. The horizontal coordinate is the time; the vertical coordinate is the concentration. In the Fig. 8, (a) when the input signal A3A2A1A0 = 0000, the output signal BCD–DCBA = 0000, carry signal Carry = 0; (b) when the input signal A3A2A1A0 = 1001, theoutput signal BCD–DCBA = 1001, carry signal Carry = 0; (c) when the input signal A3A2A1A0 = 1010, the output signal BCD–DCBA = 0000, carry signal Carry = 1; (b) when the input signal A3A2A1A0 = 1111, the output signal BCD–DCBA = 0101, carry signal Carry = 1.

From the results, we can see that the results calculated by the DNA strand displacement that are consistent with the true value table, which also proves the correctness of the design. The time is shorter when the reaction reaches a stable state and the final product is higher.

4 Conclusions

We have designed the calculation model of AND gate, OR gate based on two-domain structure. In the calculation model, the needed strand is relatively small and can express well for the logical process. The whole reaction process is more thorough and not easy to produce more intermediate products. BCD code is encoding which use a four bit binary to save a decimal, and it can be quickly converted between binary and decimal. In the paper, we have designed a binary conversion into BCD code converter by using the AND gate, OR gate which based on the structure of two-domain strand. The whole system is a fully autonomous and do not need to add extra power, depends only on the DNA strand displacement. It shows that the DNA strand displacement has a good computing power at the same time. In the future, we will learn more complex computation, such as neural network computation model, and use the DNA strand displacement to achieve it.

Acknowledgements. This work is supported by the National Natural Science Foundation of China (Nos. 61425002, 61672121, 61672051, 61572093, 61402066, 61402067, 61370005, 31370778), Program for Changjiang Scholars and Innovative Research Team in University (No. IRT_15R07), the Program for Liaoning Innovative Research Team in University (No. LT2015002), the Basic Research Program of the Key Lab in Liaoning Province Educational Department (Nos. LZ2014049, LZ2015004), Scientific Research Fund of Liaoning Provincial

Education (Nos. L2015015, L2014499), and the Program for Liaoning Key Lab of Intelligent Information Processing and Network Technology in University.

References

1. Adleman, L.M.: Molecular computation of solutions to combinatorial problems. J. Sci. **266**, 1021–1024 (1994)
2. Qian, L., Winfree, E., Bruck, J.: Neural network computation with DNA strand displacement cascades. J. Nat. **475**, 368–372 (2011)
3. Zhao, H., Gao, L., Luo, J.F., Zhou, D.R., Lu, Z.H.: Massively parallel display of genomic DNA fragments by rolling-circle amplification and strand displacement amplification on chip. J. Talanta **82**, 477–482 (2010)
4. Seelig, G., Soloveichik, D., Zhang, D.Y., Winfree, E.: Enzyme-free nucleic acid logic circuits. J. Sci. **314**, 1585–1588 (2006)
5. Murieta, I.S., Patón, A.R.: Probabilistic reasoning with a Bayesian DNA device based on strand displacement. J. Nat. Comput. **13**, 549–557 (2014)
6. Condon, A., Kirkpatrick, B., Maňuch, J.: Reachability bounds for chemical reaction networks and strand displacement systems. J. Nat. Comput. **13**, 449–516 (2014)
7. Zhang, D.Y., Seeling, G.: Dynamic DNA nanotechnology using strand displacement reactions. J. Nature Chem. **3**, 103–113 (2011)
8. Zhang, C., Ma, L.N., Dong, Y.F., Yang, J., Xu, J.: Molecular logic computing model based on DNA self-assembly strand branch migration. J. Chin. Sci. Bull. **58**, 32–38 (2013)
9. Ogihara, M., Ray, A.: Simulating boolean circuits on a DNA computer. J. Algorithmic **25**(2–3), 239–250 (1999)
10. Liu, W., Wang, S., Xu, J.: A new DNA computing model for the NAND gate based on induced hairpin formation. J. Bio Syst. **77**(1–3), 87–92 (2004)
11. Cardelli, L.: Two-Domain DNA Strand Displacement. J. Math. Struct. Comput. Sci. **26**(2), 247–271 (2010)
12. Soloveichik, D., Seelig, G., Winfree, E.: DNA as a Universal Substrate for Chemical Kinetics. J. PNAS **107**, 5393–5398 (2012)
13. Babu, H.M.H., Chowdhury, A.R.: Design of a compact reversible binary coded decimal adder circuit. J. Syst. Archit. **52**(5), 272–282 (2006). Elsevier
14. Qian, L., Winfree, E.: A simple DNA gate motif for synthesizing large-scale circuits. In: Goel, A., Simmel, F.C., Sosík, P. (eds.) DNA 2008. LNCS, vol. 5347, pp. 70–89. Springer, Heidelberg (2009). doi:10.1007/978-3-642-03076-5_7
15. Shi, X.L., Wang, Z.Y., Deng, C.Y., Song, T.S., Pan, Q., Chen, Z.H.: A novel bio-sensor based on dna strand displacement. J. PLoS One **9**(10), e108856 (2014)

Several Logic Gates Extended from MAGIC-Memristor-Aided Logic

Lin Chen[1], Zhong He[2($\boxtimes$)], Xiaoping Wang[1], and Zhigang Zeng[1]

[1] Huazhong University of Science and Technology, Road Luoyu, Hubei 1037, China
[2] Chongqing Cigarette Factory, China Tobacco Chongqing Industrial Co., Ltd.,
Road Tushan, 589, Chongqing, China
hzriver@163.com

Abstract. Recently, it has been demonstrated that memristors can be utilized as logic operations and memory elements. In this paper, several logic gates extended from MAGIC–Memristor-Aided Logic, including IMP, XNOR, NAND and OR logic gates, are presented. The extended logic gates (except for the OR logic gate) are not only used as standalone logic but also can be performed within a crossbar array, providing opportunities for novel non-von Neumann computer architectures. Another logic gate (OR gate) is presented to alleviate the issue where the logic state of the output memristor can not fully switch to the desired state in the previous designs.

Keywords: Memristor · Stateful logic · In-memory computing · Nanocrossbar memory

1 Introduction

As it has become increasingly difficult to overcome various physical limits of the traditional CMOS technology [1], alternative elements are desired for higher performance. An element called memristor (short for "memory resistor") is a promising candidate. The concept of a memristor was firstly theoretically postulated by Chua in 1971 [2], and later, Williams's team presented a resistance variable device as a memristor at HP Labs in 2008 [3]. A memristor is a two-terminal device, where the resistance of the device is changed by the electrical current, as shown in Fig. 1. As an emerging nanoscale device, memristor has a lot of advantageous features, such as non-volatility, high-density, low-power, and good-scalability [4].

Recently, researchers have demonstrated memristor's potential applications to programmable analog circuits [5–7], neural networks [8–10]. In addition, memristors have drawn researchers interests in logic circuits [11–13] and logic arrays

This work was supported by the National Natural Science Foundation of China (Grant Nos. 61374150 and 11271146), the State Key Program of the National Natural Science Foundation of China (Grant No. 61134012), and the Doctoral Fund of Ministry of Education of China (Grant No. 20130142130012).

F. Cong et al. (Eds.): ISNN 2017, Part I, LNCS 10261, pp. 170–179, 2017.
DOI: 10.1007/978-3-319-59072-1_21

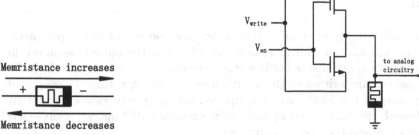

Memristance increases

Memristance decreases

Fig. 1. Memristor symbol.

Fig. 2. Initialization circuit of the extended logic gates.

[14,15]. A memristor-based material implication (IMP) logic gate was proposed in [11] as stateful logic, which can process the data on the memory *in situ* without reading it out or writing back explicitly.

2 Operation Principle of the Logic Gates Extended from MAGIC

Like MAGIC gates in [12], the extended logic gates are also performed by memristive-only logic structure. Here, the states of input and output memristors are presented as memristance, where the high memristance R_{OFF} and low memristance R_{ON} arc considered as state 0 and state 1 respectively. To perform a logic operation, the inputs of the logic gate are the initial logic state of the input memristors $(in_1, in_2, \cdots)$, and the output is the final logic state of the output memristor (out).

Operations for the extended logic gates (except for the XNOR logic gate) consists of two sequential steps. In step-1, we initialize the output memristor to a known logic state. In step-2, a voltage V_0 is applied across the logic gates. For the XNOR logic gate, it needs 3 sequential steps to complete logic function. In this paper, VTEAM model [16] is adopted. For VTEAM model, voltages $V_{T,ON}$ and $V_{T,OFF}$ are both the threshold voltages of a memristor. When the applied positive voltage is larger than $V_{T,OFF}$, the memristor switches to the high resistance state. When the applied negative voltage is less than $V_{T,ON}$, the memristor switches to the low resistance state. For the case where the applied voltage is larger than $V_{T,ON}$ and less than $V_{T,OFF}$ (within threshold voltage), the memristor remains at the initial state.

Like MAGIC, we also choose the circuit in [6] for the initialization of the memristors. In order to make the circuit simple, we make some modifications to it. We use two voltages V_{write} and V_{en}, as shown in Fig. 2. In the next section, the extended logic gates are described.

3   IMP Logic Gate and the Extension to XNOR Logic Gate

Same with MAGIC, the extended IMP logic gate consists of two input memristors ($in1$, $in2$) and an additional memristor (out) as the output, as shown in Fig. 3(a). In step-1, memristor out is written to state 1, and if necessary, memristors $in1$ and $in2$ are written to the input values. In step-2, voltage V_0 is applied at the gateway of the logic gate. The applied voltage produces a current that passes through the circuit and appears at memristor out. Now, we analyse the four input cases and the operation process.

Case 1. $in1$ is logic 0, $in2$ is logic 0. For this case, the voltage of the output memristor is lower than the memristor threshold voltage. Hence, the logic state of the output memristor does not change and remain at logic 1. In order not to change the logic state of $in1$, the voltage across $in1$ should be less than $|V_{T,ON}|$. Assuming $R_{OFF} \gg R_{ON}$, the constraint is

$$V_0 < |V_{T,ON}|. \tag{1}$$

Case 2. $in1$ is logic 0, $in2$ is logic 1. For this case, it is the same with case 1, the logic state of the output memristor remains at logic 1. The voltage across $in1$ should be less than $|V_{T,ON}|$.

Case 3. $in1$ is logic 1, $in2$ is logic 0. For this case, the voltage of the output memristor is greater than the memristor threshold voltage. Hence, the logic state of the output memristor switches to logic 0. Since $in1$ is at logic 1, the positive voltage V_0 does not change the logic state of $in1$. The constraint is

$$2V_{T,OFF} < V_0. \tag{2}$$

Case 4. $in1$ is logic 1, $in2$ is logic 1. For this case, like case 3, since $in1$ is at logic 1, the positive voltage V_0 does not change the logic state of $in1$. The voltage of the output memristor is lower than the memristor threshold voltage. Hence, the logic state of the output memristor does not change and remain at logic 1. The constraint is

$$V_0 < 3V_{T,OFF}. \tag{3}$$

Taking into account the above four cases, we can get the final constraint, the constraint is

$$2V_{T,OFF} < V_0 < \min\left[|V_{T,ON}|, 3V_{T,OFF}\right]. \tag{4}$$

IMP logic gate ($in1$ IMP $in2 \rightarrow out$) can be performed, the corresponding truth table is shown in Fig. 3(b) and the behavior of IMP operation is shown in Fig. 3(c).

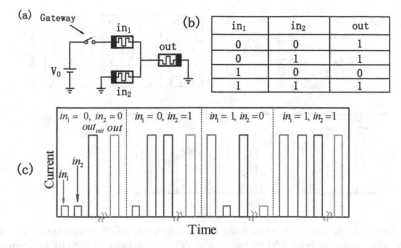

Fig. 3. IMP logic gate extended from MAGIC. (a) Schematic of IMP logic gate. (b) Truth table of IMP logic gate ($in1$ IMP $in2 \rightarrow out$). (c) Simulations of IMP logic gate.

4 XNOR Logic Gate Extended from IMP Logic Gate

In Sect. 3, we have analysed IMP logic gate which can be implemented within 2 steps. Now, we extend IMP logic gate to XNOR logic gate. The schematic of the XNOR logic gate is shown in Fig. 4(a). Similar to IMP logic gate, in step-1, memristor out is initialized to state 1. In step-2, switch S_1 is set to V_0 position, at the same time, switch S_2 is set to Gnd position. Voltage V_0 is constrained to Eq. 4, so IMP operation ($in1$ IMP $in2 \rightarrow out$) is performed successfully. In step-3, switch S_1 is set to Gnd position, and switch S_2 is set to V_0 position at the same time. For the cases where $in1$ and $in2$ are the combinations of $\{(0,0),(0,1),(1,1)\}$, memristor out remains at the original state, namely state 1 after step-2. However, for the case where $in1$ and $in2$ are state 1 and state 0 respectively, out switches from state 1 to state 0 after step-2. So, during step-3, for the cases where $in1$ and $in2$ are the combinations of $\{(0,0),(0,1),(1,1)\}$, IMP operation ($in2$ IMP $in1 \rightarrow out$) is performed. Therefore for the case where $in1$ is state 0, and $in2$ is state 1, out switches from state 1 to state 0, as to other two combinations of $\{(0,0),(1,1)\}$, out still remains at its original state. Now, we discuss the last case where $in1$ is state 1, $in2$ is state 0, and out is state 0 after step-2. When switch S_1 is set to Gnd position and switch S_2 is set to V_0 position, most of voltage V_0 is dropped across memristor $in2$. For the fact that V_0 is constrained to Eq. 4, so $in2$ is still in state 0, and out is not also changed. Based on IMP logic gate, we have extended XNOR logic gate within 3 steps successfully. Same with IMP logic gate, XNOR logic gate is not only used as standalone logic but also can be placed within a crossbar array. A circuit structure is also proposed to implement XNOR logic gate in [7]. In [7], the states of two input variables x and y are represented by voltage level value

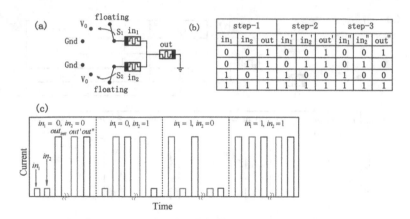

Fig. 4. XNOR logic gate extended from IMP logic gate. (a) Schematic of IMP logic gate. (b) The truth table of XNOR logic gate. (c) Simulations of XNOR logic gate.

V_X and memristance R_Y, and it is not easy to be applied to massive crossbar arrays. Different from [7], we choose the states of memristors (memristance) as the states of input variables, which highlights the nonvolatility of memristors in memory. The comparison of IMP and XNOR logic gates is shown in Table 1.

Table 1. Comparison of IMP and XNOR logic gates.

	IMP in [11]	IMP here	XNOR in [7]	XNOR here
No. of voltages	2 (V_{SET}, V_{COND})	1 (V_0)	3 (V_X, V_P, V_Q)	1 (V_0)
Separate input and output	No	Yes	Yes	Yes
No. of circuit elements	2M1R	3M	3M, 1R5S	3M
No. of steps for logic gates	1	2	3	3
Within crossbar?	Yes	Yes	No	Yes

5   IMP Logic Gate Within a Crossbar Array

In [12], as to perform a two-input MAGIC NOR gate in a crossbar array, $in1$ and $in2$ are, respectively, in columns $j + 1$ and j, and out is in column $j - 1$. While applying V_0 to columns $j + 1$ and j, grounding column $j - 1$, all rows potentially perform NOR operation, it is not just a certain row. This issue can be addressed to isolate unselected rows using half-selected cells. Here, we adopt the structure of a crossbar array shown in Fig. 5. We choose a row of configuration memristors as the output memristors. Different from MAGIC, we can perform IMP, XNOR and NAND operations within a crossbar array. In Sect. 7, we will describe NAND gate in detail.

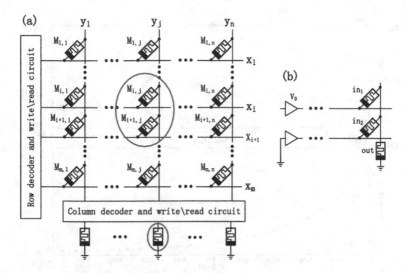

Fig. 5. IMP logic gate within a crossbar array. (a) Schematic of a memristive crossbar structure. IMP logic gate is achieved in column j, where $in1$ and $in2$ are, respectively, in rows i and $i+1$, and out is the load memristor in column j, as marked by an red oval. (b) Schematic of IMP logic gate within a crossbar array. The voltage at the gateway V_0 is the applied voltage at row i, where row $i+1$ and column j are connected ground. (Color figure online)

6   Evaluation and Design Considerations for the IMP Operation Extended from MAGIC

The speed of the IMP logic gate extended from MAGIC is evaluated in Vituoso. Same with MAGIC, VTEAM model [16] is used in the simulations, several important parameters are: $R_{ON} = 1\,\mathrm{k\Omega}$, $R_{OFF} = 300\,\mathrm{k\Omega}$, $V_{T,ON} = -1.5\,\mathrm{V}$, $V_{T,OFF} = 0.3\,\mathrm{V}$. The behavior of IMP logic gate for different values of V_0 is shown in Fig. 6. The case 3 of inputs is chosen to evaluate the delay of the IMP operation. From (4), V_0 can vary from 0.6 to 0.9 V according to the parameters listed above.

7   Additional Logic Gates Extended from MAGIC

In [12], the NAND logic gate can not be applied within a crossbar array. Under the similar operation principle to IMP, a NAND logic gate which can be used in memory is presented. As shown in Fig. 7, out is initialized to logic 1 prior to execution. For correct circuit behavior, assuming $R_{OFF} \gg R_{ON}$, the constraint is

$$\frac{3}{2}V_{T,OFF} < V_0 < 2V_{T,OFF}. \tag{5}$$

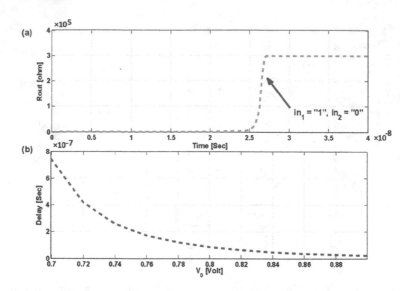

Fig. 6. Simulations of IMP logic gate. (a) Output memristor for the case where $in1$ is logic 1 and $in2$ is logic 0, $V_0 = 0.85$ V. (b) Delay for different values of V_0.

When an input memristor is logic 0, the operation of the NAND logic gate can be destructive. Therefore the maximum applied voltage for a two-input NAND logic gate is

$$V_0 < \min\left[2V_{T,OFF}, |V_{T,ON}|\right]. \tag{6}$$

As shown in Fig. 7(c), for χ input memristors, the design constraint is

$$V_{T,OFF} \cdot \left(1 + \frac{1}{\chi}\right) < V_0 < \frac{V_{T,OFF}}{R_{ON}} \cdot \left[R_{ON} + \left(\frac{R_{ON}}{\chi - 1}\right) \parallel R_{OFF}\right].$$

For nondestructive operation of a χ-input NAND, the maximum applied voltage is

$$V_0 < \min\left\{\frac{V_{T,OFF}}{R_{ON}} \cdot \left[R_{ON} + \left(\frac{R_{ON}}{\chi - 1}\right) \parallel R_{OFF}\right], |V_{T,ON}|\right\}. \tag{7}$$

For the case where one of the input memristors is logic 1, the voltage of out is greater than $|V_{T,ON}|$. As the resistance of out decreases, the voltage of out decreases. When the voltage of out is less than or equal to $|V_{T,ON}|$, the resistance does not decrease. So in this case, out can not switch fully from logic 0 to logic 1. Here, an improved OR logic gate is proposed, shown in Fig. 8(a). out is initialized to logic 0 prior to execution. For correct circuit behavior, for the combination of $\{0, 0\}$, the constraint of the OR gate is

$$|V_{T,ON}| < V_0 < 2|V_{T,ON}|. \tag{8}$$

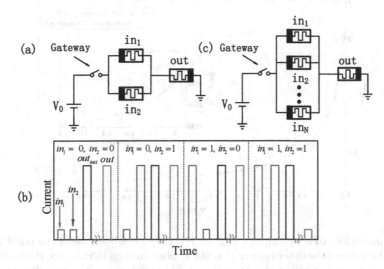

Fig. 7. NAND logic gate extended from MAGIC. (a) Schematic of a two-input NAND gate. (b) Simulation results for a two-input NAND gate. (c) Schematic of an N-input NAND gate.

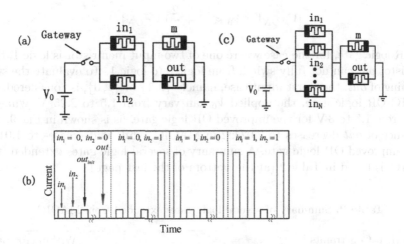

Fig. 8. The improved OR logic gate extended from MAGIC. (a) Schematic of a two-input OR gate. The logic gate consists of two input memristors $in1$ and $in2$ and an additional structure which includes an output memristor out and a memristor m connected in parallel. The resistance of memristor m is equal to R_{OFF}, which keeps unchanged during execution. (b) Simulations of a two-input OR gate. (c) Schematic of a N-input OR gate.

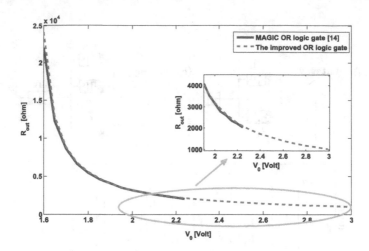

Fig. 9. Simulations of a two-input OR gate. For the case where one of the input memristors is logic 0 and the other is logic 1, as the applied voltage V_0 changes, the resistance of the output memristor *out* also changes. The blue line represents the change of the resistance of *out* in [12], while the red dash line represents the improved OR gate. (Color figure online)

As shown in Fig. 8(c), for χ input memristors, the design constraint is

$$|V_{T,ON}| < V_0 < (1 + \frac{2}{\chi})|V_{T,ON}|. \tag{9}$$

For OR logic gate, for the case where one of two input memristor is logic 1, then memristor *out* can not fully switch from logic 0 to logic 1. To evaluate the state switching of *out*, the worst input case, namely $\{0, 1\}$ or $\{1, 0\}$, is considered. For MAGIC OR logic gate, the applied V_0 can vary from 1.5 to 2.25 V, while V_0 varies from 1.5 to 3 V for the improved OR logic gate, as is shown in Fig. 9. The resistance of *out* decreases to $2.02\,k\Omega$ in MAGIC, while it decreases to $1.01\,k\Omega$ in the improved OR logic gate. A summary of several logic gates extended from MAGIC, is listed in Table 2 (at the bottom of the last page).

Table 2. Summary of several logic gates extended from MAGIC.

Function	Constraints	Within crossbar
IMP	$2V_{T,OFF} < V_0 < \min\left[|V_{T,ON}|, 3V_{T,OFF}\right]$	Yes
XNOR	$2V_{T,OFF} < V_0 < \min\left[|V_{T,ON}|, 3V_{T,OFF}\right]$	Yes
NAND	$V_{T,OFF} \cdot \left(1 + \frac{1}{\chi}\right) < V_0$ $< \min\left\{\frac{V_{T,OFF}}{R_{ON}} \cdot \left[R_{ON} + \left(\frac{R_{ON}}{\chi-1}\right) \parallel R_{OFF}\right], |V_{T,ON}|\right\}$	Yes
OR	$|V_{T,ON}| < V_0 < (1 + \frac{2}{\chi})|V_{T,ON}|$	No

8   Conclusion

In this paper, several logic gates extended from MAGIC are presented. The extended logic gates (except for the OR gate) can be performed within a crossbar array. Another logic gate (OR gate) is presented to alleviate the issue where the logic state of the output memristor can not fully switch to the desired state in the previous designs.

References

1. Kuhn, K.J.: Considerations for ultimate CMOS scaling. IEEE Trans. Electron Devices **59**, 1813–1828 (2012)
2. Chua, L.O.: Memristor - the missing circuit element. IEEE Trans. Circuit Theory **18**, 507–519 (1971)
3. Strukov, D.B., Snider, G.S., Stewart, D.R., Williams, R.S.: The missing memristor found. Nature **453**, 80–83 (2008)
4. Junsangsri, P., Lombardi, F.: Design of a hybrid memory cell using memristance and ambipolarity. IEEE Trans. Nanotechnol. **12**, 71–80 (2013)
5. Shin, S., Kim, K., Kang, S.M.: Memristor applications for programmable analog ICs. IEEE Trans. Nanotechnol. **10**, 266–274 (2011)
6. Pershin, Y.V., Ventra, M.D.: Practical approach to programmable analog circuits with memristors. IEEE Trans. Circuits Syst. I: Reg. Papers **57**, 1857–1864 (2010)
7. Shin, S., Kim, K., Kang, S.M.: Resistive computing: memristors-enabled signal multiplication. IEEE Trans. Circuits Syst. I: Reg. Papers **60**, 1241–1249 (2013)
8. Wu, A., Zhang, J., Zeng, Z.: Dynamic behaviors of a class of memristor-based Hopfield networks. Phys. Lett. A **375**, 1661–1665 (2011)
9. Pershin, Y.V., Di Ventra, M.: Experimental demonstration of associative memory with memristive neural networks. Neural Netw. **23**, 881–886 (2010)
10. Ebong, I.E., Mazumder, P.: CMOS and memristor-based neural network design for position detection. Proc. IEEE **100**, 2050–2060 (2012)
11. Borghetti, J., Snider, G.S., Kuekes, P.J., Yang, J.J., Stewart, D.R., Williams, R.S.: Memristive switches enable stateful logic operations via material implication. Nature **464**, 873–876 (2010)
12. Kvatinsky, S., Belousov, D., Liman, S., Satat, G., Wald, N., Friedman, E.G., Kolodny, A., Weiser, U.C.: Magic-memristor-aided logic. IEEE Trans. Circuits Syst. II: Exp. Briefs **61**, 895–899 (2014)
13. Shin, S., Kim, K., Kang, S.M.: Reconfigurable stateful NOR gate for large-scale logic-array integrations. IEEE Trans. Circuits Syst. II: Exp. Briefs **58**, 442–446 (2011)
14. Kim, K., Shin, S., Kang, S.M.: Field programmable stateful logic array. IEEE Trans. Comput.-Aided Des. Integr. Circuits Syst. **30**, 1800–1813 (2011)
15. Kvatinsky, S., Satat, G., Wald, N., Friedman, E.G., Kolodny, A., Weiser, U.C.: Memristor-based material implication (imply) logic: design principles and methodologies. IEEE Trans. Very Large Scale Integr. (VLSI) Syst. **22**, 2054–2066 (2014)
16. Kvatinsky, S., Ramadan, M., Friedman, E.G., Kolodny, A.: VTEAM: a general model for voltage-controlled memristors. IEEE Trans. Circuits Syst. II: Exp. Briefs **62**, 786–790 (2015)

Static Hand Gesture Recognition Based on RGB-D Image and Arm Removal

Bingyuan Xu$^{(\boxtimes)}$, Zhiheng Zhou, Junchu Huang, and Yu Huang

School of Electronic and Information Engineering,
South China University of Technology, Guangzhou, China
xu.bingyuan@mail.scut.edu.cn

Abstract. A novel hand gesture recognition algorithm is proposed for human-computer interaction, which is based on RGB-D image (RGB image and Depth image) and arm removal. The hand is firstly extracted from the background based on depth data and skin-color features. Then the arm area is removed by using distance transformation operations, and gesture composed of palm and fingers is obtained. Finally Hu moments of the gesture are calculated and entered into Support Vector Machine (SVM) for recognition. Experimental results demonstrate that the proposed algorithm can recognize 8 gestures with an accuracy of 95.83% in the complex background.

Keywords: RGB-D image · Distance transformation · Arm removal · Static hand gesture recognition · SVM

1 Introduction

With the gradual transfer of human-computer interface towards user-orientated, gesture recognition as a natural and intuitive mode, has gradually developed into a research hotspot in the field of human-computer interaction, and has been widely used in somatosensory games, robot control and computer control. Compared with gesture recognition based on data glove, vision-based gesture recognition has the advantages of low requirement for equipment, interactive nature and so on, and becomes the mainstream of gesture recognition.

In recent years, more and more scholars have participated in the study of static gesture recognition based on vision [1–3]. Vision-based gesture recognition uses a single camera or multiple cameras to collect gesture information. The gestures are segmented by computer programs, and identified by specific methods. How to segment gesture accurately from complex background and identify its meaning are the key problems that static hand gesture recognition systems need to solve.

As the first step of gesture recognition, hand segmentation is the most critical step, and its accuracy directly affects the recognition effect. In gesture recognition based on monocular vision, it is difficult to separate the hand from the background. The commonly used methods of hand segmentation are: (1) the use of hand features, such as skin color, hand contour, fingertip, hand size, etc. The hand is usually segmented using skin-color features and geometric features, but this method is susceptible to external light intensity and to skin-like regions in the background. (2) by wearing a marked

© Springer International Publishing AG 2017
F. Cong et al. (Eds.): ISNN 2017, Part I, LNCS 10261, pp. 180–187, 2017.
DOI: 10.1007/978-3-319-59072-1_22

gloves or simplifying complex background, create an environment where hand and background are easily separated. This method can solve the interference of the skin-like regions, but the environmental requirements become more harsh that it cannot meet the needs of human-computer interaction under the realistic conditions. With the popularity of the Kinect camera, many researchers choose to use depth information to solve the problem of hand segmentation in complex backgrounds [4–7]. Kinect includes a RGB camera, an infrared camera and an infrared transmitter, can be used to capture RGB image and Depth image. Using the Depth image alone can segment hand quickly, but the hand detection accuracy is still low in complex background. In [7], it is proposed to combine RGB image and Depth image for hand segmentation, and the depth information and skin color information are used to achieve accurate hand segmentation. A combination of color and depth information can make the system more robust [5, 6].

Static hand gesture refers to the form of the palm and fingers, that is, the gesture information can be expressed only by the palm and fingers. As shown in Fig. 1, the arm existing after hand segmentation is redundant information, whose length will interfere with the gesture recognition results. Since the removal of arm disturbances can improve the recognition rate of the system, some algorithms for arm-area remov-ing have been studied. The arm removal algorithm proposed in [8] is based on the characteristics of narrower wrist, which is not suitable for special situations such as hand tilted. Distance transformation and vector dot product operations are used to determine the cutting line for arm removal [9]. This algorithm is not affected by the rotation of the hand, but it cannot quickly remove the arm area because of the high time complexity. In [1], the method of using structural elements is studied. This me-thod cannot effectively remove the arm, which will affect the recognition results.

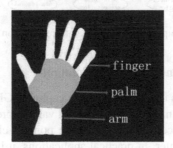

Fig. 1. The composition after hand segmentation

In order to solve the above problems, a static gesture recognition algorithm com-bined with RGB-D image and arm removal is proposed in this paper. The main con-tributions include: firstly, a method of hand segmentation based on depth information and skin-color features is proposed, which can overcome the interference of complicated background and extract the hand accurately; secondly, an arm removal method based on distance transformation is proposed, which can effectively and quickly remove the arm area, and solve the problem that the recognition rate of the system is not high when the arm exists; finally, it is proved that the static gesture recognition system constructed by this algorithm is robust, and has high recognition rate in complex environment.

2 The Proposed Static Gesture Recognition Algorithm

The flow of the static gesture recognition algorithm in this paper is shown in Fig. 2, and includes the following main steps: first, RGB image and Depth image are acquired simultaneously by Kinect camera, and then the depth threshold is used to segment the hand from the background; second, the skin color information is used to further extract the hand, and a binary image is obtained; third, the arm removal algorithm based on distance transformation is used to remove the arm area, and a binary image composed of fingers and palm is obtained; finally, we calculate the Hu moments of the gesture image and use SVM to obtain the recognition result. The above depth threshold segmentation and skin color segmentation constitute the hand segmentation algorithm in this paper.

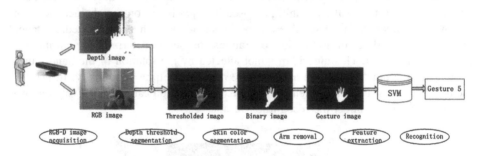

Fig. 2. The proposed static gesture recognition algorithm

2.1 Image Acquisition and Hand Segmentation

As we know, Kinect camera can simultaneously capture RGB image and Depth image. Depth image is a grayscale image, where the pixel value represents the relative distance from the point within the scene to Kinect. If the point is closer to Kinect, the corresponding pixel will be brighter. In practical applications, it is considered that the hand closest to Kinect. Therefore, Depth image can be used to locate the hand. According to the appropriate depth threshold, the corresponding pixels in the RGB image can be extracted and a RGB image containing hand is obtained. The above method avoids the influence of background environment, but the image obtained by depth threshold segmentation may contain other regions except the hand (such as sleeves), which will affect the subsequent operation. Therefore, the skin color information is used to further extract the hand. YCr'Cb' color space [10] has the advantages of strong skin color clustering and little influence by external illumination, and is widely used in skin color segmentation. By converting the image from RGB color space to YCr'Cb' color space and using threshold segmentation method, the skin color region in the image can be accurately detected. After the depth threshold segmentation and skin color segmentation, a binary image is obtained, which successfully separates the hand from background.

2.2   Arm Removal

The distance transformation of binary image is defined as: if the current pixel value is 0, it is still 0 after distance transformation; if the current pixel value is 1, it is the distance from the current pixel to the nearest 0 value pixel after distance transformation. The Euclidean distance between two pixels is calculated as formula (1), where (r_1, c_1), (r_2, c_2) are the coordinate values of pixel 1 and pixel 2 respectively. If the pixel in the foreground target is farther away from the boundary, the pixel value would be larger after distance transformation.

$$d = \sqrt{(r_1 - r_2)^2 + (c_1 - c_2)^2} \tag{1}$$

Figure 3 shows the complete flow of arm removal algorithm based on distance transformation. It includes the following steps:

- Step 1: Take a distance transformation operation on the binary image, which is obtained after hand segmentation. Then the distance from the pixel inside the hand to the hand boundary can be obtained, the maximum of which is generally located at the center of the palm (position P_c in Fig. 3) and is taken as the radius R_0 of the inscribed circle of the palm.
- Step 2: Draw a circle on the image obtained in step 1. In order to keep the palm inside the circle (the blue circle in the Fig. 3), P_c is chosen as the center and $R_1 = 1.35 \times R_0$ is chosen as the radius, and the pixels within the circle are all 0. And then the largest pixel value in the remaining regions is detected, which is taken for P_{max} and corresponding to the pixel P. When the arm area exists, P is usually located at the midline of the arm (position P in Fig. 3), and P_{max} is large; When there is no arm area, P is generally located at the midline of the finger, and P_{max} is relatively small.
- Step 3: Calculate Ratio $= \frac{P_{max}}{R_0}$. If Ratio is greater than T, the image obtained in step 2 is judged to have an arm area, and the following steps are executed to remove the arm area. Otherwise, it is determined that there is no arm area and the hand segmentation image can be used directly for feature extraction. In this paper, the experimental value $T = 0.35$.
- Step 4: Use eight-connected discrimination algorithm to mark the remaining regions. If the label value of a connected region is the same as that of the pixel P, the region is removed and a binary image containing only fingers is obtained.
- Step 5: XOR the result of step 4 with the result of step 3 to obtain a binary image containing only the arm area.
- Step 6: XOR the result of step 5 with the hand segmentation image to obtain the gesture image.

Based on hand segmentation, distance transformation is innovatively used to locate the center of the palm precisely, and palm-based circle combined with eight-connected discrimination algorithm is used to remove the arm interference. Compared with other arm removal algorithms, the proposed algorithm has low computational complexity and can effectively and quickly remove the arm without the influence of the rotated hand.

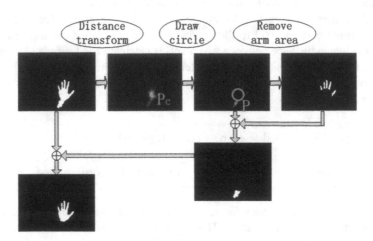

Fig. 3. The proposed arm removal algorithm (Sample image with arm area, ⊕ represents XOR operation)

2.3   Feature Extraction and Recognition

Hu moments are used to capture the shape and spatial information of images, and can solve the problems of scale change, image translation change, coordinate transformation and rotation transformation in the process of gesture feature matching. In addition, the static hand gesture recognition problem in this paper belongs to the small sample and nonlinear classification problem, so SVM with excellent classification performance is selected as the classifier. In this paper, the Hu moments of the gesture image are extracted and selected as input feature vector of SVM. In the training phase, we complete selection of kernelfunction and optimization parameters, and finally obtain a SVM model. In the test phase, the Hu moments are extracted and input into the SVM model to obtain the final recognition result.

3   Experimental Results

In order to verify the effectiveness of this algorithm, the 8 gestures shown in Fig. 4 are used for experiments, which represent number 0 to 7.

All the experiments are performed in MATLAB2013b and Microsoft Visual Studio 2010 on a computer with Intel (R) Core (TM) i5-4440 (3.10 GHz), 8G memory, and Windows 7 operating system. A Microsoft Kinect is used to collect 8 hand gesture

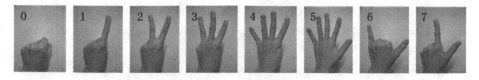

Fig. 4. Eight gestures used for experiments

images of 3 people in complex background of the laboratory, thus 720 test samples (90 images for each gesture) and 2400 training samples (300 images for each gesture) are created.

Figure 5 shows the processing of the 8 hand gesture images using this algorithm. Firstly, the depth threshold is used to extract the corresponding pixels in the RGB image, thus a RGB image containing the hand is obtained. Since the obtained image contains the sleeve region that satisfies the depth threshold, the skin-color features are used to further extract the hand. The experimental results show that, even in complex background, the proposed hand segmentation method can accurately detect hands, which lays a foundation for gesture recognition. The arm removal algorithm proposed in this paper can effectively and quickly remove the arm area and meet the requirements of real-time gesture recognition system. On this basis, we extract the Hu moments of the gesture images, train and test the extracted feature vectors by using SVM. The SVM model implements gesture recognition and classification, and achieves good recognition results. As shown in Table 1, the static hand gesture recognition system constructed in this paper has a recognition rate of 95.83% for 8 gestures.

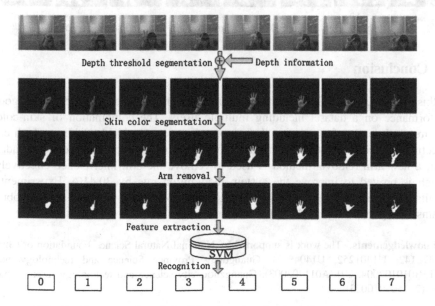

Fig. 5. The processing of the 8 hand gesture images using this algorithm

Table 1. The performance of different algorithms

Recognition rate (%)	This algorithm	This algorithm excluding the arm removal algorithm
The training set	99.58	99.33
The testing set	95.83	65.42

In order to verify the contribution of the proposed arm removal algorithm in improving the system recognition rate, this algorithm (excluding the arm removal algorithm) is used to deal with the same sample set. As shown in Table 1, the recognition rate of the system is only 65.42% without removing the arm interference.

In order to verify the advantages of this algorithm in dealing with complex environments, the same 8 images are processed only using skin color information, and the results are shown in Fig. 6. The binary image obtained by skin color segmentation contains the skin regions and the skin-like regions of the original image, and we need to use other features to further extract the hand. However, this method is not robust enough and is easily influenced by the background, and cannot solve the problem that the hand and skin region (such as face, neck and so on) or skin-like region overlap.

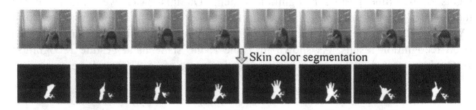

↓ Skin color segmentation

Fig. 6. Only use the skin color information for hand segmentation

4 Conclusion

In this paper, a new hand gesture recognition algorithm is proposed, which shows good performance on a dataset including multiple users. The combination of skin-color features and depth information is shown to realize accurate hand tracking, which can effectively solve the problem of hand segmentation in complex background. In addition, a new arm removal method is used to remove the arm interference effectively, which is proved to improve the system recognition rate by 30.41%. Experimental results show that the gesture recognition system constructed in this paper is robust against complex backgrounds and reaches a recognition rate of 95.83%.

Acknowledgements. The work is supported by National Natural Science Foundation of China (61372142, U1401252, U1404603), Guangdong Province Science and technology plan (2013B010102004, 2013A011403003), Guangzhou city science and technology research projects (201508010023).

References

1. Priyal, S.P., Bora, P.K.: A robust static hand gesture recognition system using geometry based normalizations and Krawtchouk moments. Pattern Recogn. **46**(8), 2202–2219 (2013)
2. Modanwal, G., Sarawadekar, K.: Corrigendum to "towards hand gesture based writing support system for blinds". Pattern Recogn. **57**(C), 50–60 (2016)
3. Lei, J., Han-Fei, Y.I.: A hand gesture recognition method based on SVM. Comput. Aided Drafting Des. Manufact. (English Version) **20**(2), 85–91 (2010)
4. Chan, C., Mirfakhraei, S.S.: Hand gesture recognition using kinect. Bachelor thesis, Boston University, Boston (2013)
5. Pugeault, N., Bowden, R.: Spelling it out: real-time ASL fingerspelling recognition. In: IEEE International Conference on Computer Vision Workshops, vol. 28, pp. 1114–1119 (2011)
6. Dong, L., Wang, H., Hao, Z., Liu, J.: Robust hand posture recognition based on RGBD images. In: Chinese Control and Decision Conference, pp. 2735–2740 (2014)
7. Dominio, F., Donadeo, M., Zanuttigh, P.: Combining multiple depth-based descriptors for hand gesture recognition. Pattern Recogn. Lett. **50**, 101–111 (2014)
8. Gao, J.Y.: Research on gesture recognition technology in the human-computer interaction. M.S. Thesis, Xidian University, Xi'an (2013)
9. Cheng, X.Y.: Number-sign hand posture recognition based on arm-area removing. M.S. Thesis, Southwest University, Chongqing (2014)
10. Cao, J.Q., Wang, H.Q., Lan, Z.L.: Skin color division base on modified YCrCb color space. J. Chongqing Jiaotong Univ. (Nat. Sci.) **29**(3), 488–492 (2010)

Real-Time Classification Through a Spiking Deep Belief Network with Intrinsic Plasticity

Fangzheng Xue[1,2], Xuyang Chen[1,2], and Xiumin Li[1,2(✉)]

[1] Key Laboratory of Dependable Service Computing in Cyber Physical
Society of Ministry of Education, Chongqing University,
Chongqing 400044, China
xuefangzheng@cqu.edu.com
[2] College of Automation, Chongqing University, Chongqing 400044, China
xmli@cqu.edu.cn

Abstract. Deep Belief Networks (DBNs) has made a good effect in machine learning and object classification. However, the current question is how to reduce the computational cost without detrimental to accuracy. To solve this problem, this paper is undertaken to convert the Siegert neuron into LIF neuron in DBNs and analyze the effects of changing the value of parameters for spiking neurons such as thresholds and firing rates. Besides, we also add intrinsic plasticity (IP) into the network to render better adaptive capability. Besides, the most exciting results is the spiking DBN with intrinsic plasticity submits its first correct guess of the output label within an average of 2.5 ms after the onset of the simulated Poisson spike train input with the initial firing rates beyond 200 Hz, and the recognition accuracy is still more than 94 percent.

Keywords: DBNs · Spiking neural network · Intrinsic plasticity · Numeral recognition

1 Introduction

Over the past few years, some science and technology developed by deep learning have a significant impact on all aspects of signal and information processing, which is not only exist with traditional field, also exists in machine learning and artificial intelligence [12,22,23]. Deep belief networks, as a kind of deep neural network, is a part of deep learning. DNBs and the unsupervised layer-by-layer pre-training with Contrastive Divergence(CD) algorithm are bring up in 2006 [9,10], which are widely used in natural language processing [1,10,11] and patterns recognition [3,5]. SNNs are often regarded as the 3rd generation of neural networks, spiking neuron model and synaptic plasticity are the main characteristics of it. Besides, it also taking into account the time of spike firing [17]. However, its far from enough to study the above two neural networks in separately. Combining with the advantages of both, we can get some different results. Besides, other than synaptic plasticity, a single neuron also has the ability

© Springer International Publishing AG 2017
F. Cong et al. (Eds.): ISNN 2017, Part I, LNCS 10261, pp. 188–196, 2017.
DOI: 10.1007/978-3-319-59072-1_23

to change its intrinsic excitability to match its synaptic input. This mechanism is referred to as neuronal intrinsic plasticity (IP) [7,13,14,21]. With this IP mechanism, a single neuron can strengthen the excitability when its input is deprived for a period of time and weaken the excitability when the input is boosted and a spiking DBN has the adaptive ability for the different intensity of the input.

With the background of big data, lots and lots of data are produced on the internet every day. So that how to speed up the computation is a very important problem with computational accuracy changed little. In this paper, we aimed at reducing the time to identify a handwritten number through a spiking deep belief network with self-adaptive capacity.

In this article, we first train the DBN with a time-stepped model and CD algorithm. And then with the network structure remain unchanged and the changed neuron model, the learned parameters are transferred to a functionally equivalent spiking neural network. We add intrinsic plasticity [2] to make it has self-adaptive ability at last. Using DBN, we can regard it as a probability generation model and its efficient to use the unlabeled data to learn. On the other hand, based on dynamic event-driven processing of SNNs, the spiking DBN can process data at a high rate of speed in real time. With using the MNIST dataset [24], we evaluate the spiking DBN by analyzing the parameters such as thresholds and firing rates. Besides, adding intrinsic plasticity is efficient to optimize the network.

This paper is divided into five major sections as follows: Section one of this paper opens with the background of DBNs and spiking neural networks(SNNs) and intrinsic plasticity, it points out the research purpose and research methods of this paper. The basic knowledge and application of DBNs, SNNs and IP are presented in Sect. 2. In this paper, the experimental data and process are introduced in detail under the major heading of experimental setup. In the following, the results are given in Sect. 4.

2 Materials and Methods

2.1 Deep Belief Networks

DBNs can be constructed by stacking Restricted Boltzmann Machines(RBMs) which is a kind of special Markov Random Filed [8]. A RBM consists of two layers, one is called visible layer which is usually used as input layer and the other is called hidden layer. There is fully connected between the random units of the two layers, but no connection within in each single layer (Fig. 1).

For a Bernoulli distribution RBM, we can define the energy function of it as follows:

$$E\left(\mathbf{v},\mathbf{h};\theta\right) = -\sum_{i=1}^{m}\sum_{j=1}^{n} w_{ij}v_i h_j - \sum_{i=1}^{m} b_i v_i - \sum_{j=1}^{n} c_j h_j, \qquad (1)$$

where $\theta = (\mathbf{w},\mathbf{b},\mathbf{c})$. The encoded joint probability can be written as:

$$p\left(\mathbf{v},\mathbf{h};\theta\right) = \frac{\exp\left(-E\left(\mathbf{v},\mathbf{h};\theta\right)\right)}{Z}, \qquad (2)$$

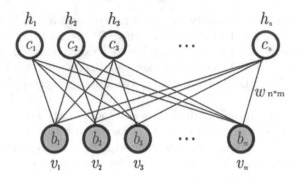

Fig. 1. Restricted Boltzmann machines. Denoting the states of visible units with v_i, the states of hidden units with h_j, the biases of visible units and hidden units with bi and c_j respectively, and w_{ij} represents the weights connecting these units.

where Z is a partition function, $Z = \sum_{\mathbf{v}} \sum_{\mathbf{h}} \exp(-E(\mathbf{v}, \mathbf{h}; \theta))$, so we can get the conditional probability:

$$p(h_j = 1 | \mathbf{v}; \theta) = \sigma \left(\sum_{i=1}^{m} w_{ij} v_i + c_i \right) \tag{3}$$

$$p(v_i = 1 | \mathbf{h}; \theta) = \sigma \left(\sum_{j=1}^{n} w_{ij} h_i + b_i \right) \tag{4}$$

We can get the rule of updating weights of RBM through computing the gradient of log likelihood function that is $\log p(\mathbf{v}; \theta)$:

$$\Delta w_{ij} = E_{data}(v_i h_j) - E_{model}(v_i h_j) \tag{5}$$

In the above equation, $E_{data}(v_i h_j)$ is the expectation of observational data in train dataset. However, its complicated to compute the value of $E_{model}(v_i h_j)$, so we can approximately calculate the value by using CD algorithm (Fig. 2).

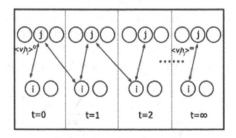

Fig. 2. Process of contrastive divergence. The first step is to initialize v_0 using training data, then to sample $h_0 \sim p(h | v_0)$, the next two steps is to sample v_1 and h_1.

2.2 Neuron Model

In order to smoothly transform a DBN model to a spiking DBN model, we replaced the sigmoid neurons with Siegert neurons [19] which have the mathematically equivalent transfer functions of Leaky Integrate-and-Fire(LIF) neurons [20] with Poisson-process inputs. And Siegert neurons have the similar dynamics with LIF neurons.

A Siegert neuron receiving excitatory and inhibitory rates and corresponding input weights $(\overrightarrow{w}_e, \overrightarrow{w}_i)$. For accurate approximation of LIF rates the input rates have to be independent Poisson spike trains. And the principle of Siegert neuron can be compute as [6]:

$$\mu_Q = \tau_m \sum \left(\overrightarrow{w}_e \overrightarrow{\lambda}_e + \overrightarrow{w}_i \overrightarrow{\lambda}_i \right) \qquad\qquad \Upsilon = V_{rest} + \mu_Q$$

$$\sigma_Q^2 = \tfrac{\tau_m}{2} \sum \left(\overrightarrow{w}_e^2 \overrightarrow{\lambda}_e + \overrightarrow{w}_i^2 \overrightarrow{\lambda}_i \right) \qquad\qquad \Gamma = \sigma_Q$$

$$\lambda_{out} = \Phi(\Upsilon, \Gamma)$$
$$= \left(t_{ref} + \tfrac{\tau_m}{\Gamma} \sqrt{\tfrac{\pi}{2}} \int_{V_{reset}+k\gamma\Gamma}^{V_{th}+k\gamma\Gamma} du \cdot \exp\left[\tfrac{(u-\Upsilon)^2}{2\Gamma^2} \right] \cdot \left[1 + erf\left(\tfrac{(u-\Upsilon)}{\Gamma\sqrt{2}} \right) \right] du \right)^{-1}$$
$$(6)$$

where $k = \sqrt{\tau_{syn}/\tau_m}$ and $\gamma = |\zeta(1/2)|$ for ζ being the Riemann zeta function. The leaky integrate-and-fire neuron is probably the best-known example of a formal spiking neuron model. The standard form is as follows:

$$\tau_m \frac{du}{dt} = -u(t) + RI(t) \qquad\qquad (7)$$

Where $u(t)$ is the membrane potential of the single neuron, R is the membrane resistance and $I(t)$ is the total injected current as the input of the neuron. τ_m is the membrane time constant and it is the product of the membrane resistance and the membrane capacity $\tau_m = RC$. When the membrane reached the threshold, then the neuron makes a spike and the membrane is reset to a new value u_r.

2.3 Intrinsic Plasticity

As we known, in biological visual neural system, we can reduce the size of our pupils when the light is too strong so that the amount of light in our pupils is reduced. On the contrary, when the light is too weak, we can get more light in our eyes with increasing pupil size so that we can see things clearly. Similarly, it has been found that single neurons also have the ability to change their intrinsic excitability over time to match different synaptic input levels. The intrinsic excitability of a neuron will increase when its synaptic input is deprived and decrease when the input is boosted significantly. This long-term learning ability is referred to as neuronal intrinsic plasticity [4,14,18], which makes the spiking DBNs has self-adaptive ability to the different inputs.

The differential equations of the proposed IP learning rule are presented as follows:

$$\tau_{ip}\frac{drC}{dt} = \frac{1}{rC} - yI + \beta(1-y)I$$
$$\tau_{ip}\frac{drR}{dt} = -rR + y - \beta(1-y) \tag{8}$$

where $rC = 1/C$ and $rR = 1/R$, τ_{ip} denotes the relative integration resolution of the IF model and the IP learning rule, and the output y of the neuron in response to the input I is described by a summation of impulsive functions to denote spikes.

$$y = \epsilon \sum_f \delta\left(t - t^{(f)}\right) \tag{9}$$

where $\delta\left(t - t^{(f)}\right)$ is a Dirac delta function representing a spike fired at time $t^{(f)}$ and ϵ indicates the strength of a spike (Fig. 3).

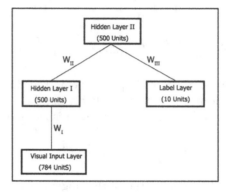

Fig. 3. Architecture of the DBN for handwritten digit recognition.

3   Experimental Setup

We used MNIST dataset of handwritten digits to train and test the network. This set consists of two parts, the training set with 60000 gray-scale images of handwritten digits and the test set with 10000 digits. The spiking BDN includes four layers: the visual input layer with 784 units, the hidden layer I with 500 units, the hidden layer II with 500 units, and the label layer with 10 units. W_I, W_{II}, W_{III} represent the weights of RBMs.

The experiment is divided into recognition mode and generation mode due to the bi-directional weights. In recognition mode the 784-500-500-10 network is used to recognize the digits received by the bottom layer. On the contrary, in generation mode the network architecture is turn to 10-500-500-784, and this mode provide a way to visualize and reconstruct what has been learned in the network. Before the above processing, we have to get the weights through training the DBN with Siegert neurons, which means to use Siegert neurons to calculate the outputs of the hidden neurons with CD algorithm.

As for spiking input, the intensity values of the MNIST images were normalized to values between 0 and 1. And Poisson distributed spike trains were generated for each image pixel with firing rates proportional to the pixels intensity value [15].

4   Results

This section covers the results of accuracy, cost of time and self-adaptive ability of the spiking DBN with IP and without IP in recognition mode and generation mode. Without the condition of adding IP in the spiking DBN, we analyzed the threshold (V_{th}) and τ_m of LIF spiking neuron. Finally, we set threshold to be 1.5 and τ_m to be 9.

4.1   Recognition Mode

In recognition mode ($V_{reset} = 0$, $V_{th} = 1.5$, $\tau_m = 9$), the Visual Input Layer receives the external Poisson spike trains and activity spreads in down-top direction through the network. And there are three RBMs in this kind of network architecture, in the first RBM, we can get the probability model from the energy model and get the maximum likelihood form the probability model. So the last question is to solve the maximum likelihood by Gibbs-sampling procedure (Fig. 4).

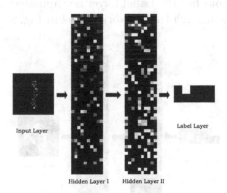

Fig. 4. Recognition mode of the spiking DBN. This is a screen capture of the network while recognizing samples of input activations corresponding to digit 1. With the Poisson spike trains of digit 1 simulating in the Visual Input Layer (left), activity propagated through the whole network and finally the corresponding Label 1 showed in the Label Layer (right).

As shown in Table 1, when the initial firing rate is greater than 700 Hz and the accuracy is trending downward. However, this trend do not happen to the situation of adding IP and the accuracy goes up and down. On the whole, 94

Table 1. Accuracy of recognition mode with IP and without IP

Firing rate (Hz)	100	300	500	700	900	1100	1300	1500	1700	1900
With IP	94.2	94.7	94.4	94.2	94.4	94.4	94.2	94.3	94.2	94.3
Without IP	93.2	93.8	93.8	94.0	93.9	93.8	93.7	93.6	93.6	93.5

Table 2. Elapsed time of recognition mode with IP and without IP

Firing rate (Hz)	100	200	300	400	500	600	700	800	900	1000
With IP (ms)	14.2	3.1	2.7	2.5	2.4	2.5	2.4	2.4	2.3	2.3
Without IP (ms)	28	21.2	12.8	9.0	8.4	7.2	7.0	6.2	5.2	5.3

percent is the demarcation point of the two conditions, so we can get the relation that the accuracy can be improved by adding IP mechanism in the spiking DBN. Besides, Table 2 represents the elapsed time of the two cases. Under the circumstance of adding IP, the network recognizes a single number within an average of 2.5 ms when the firing rate is more than 200 Hz. On the other hand, the network will spend more than 5 ms to recognize a single handwritten digits.

4.2   Generation Mode

In this mode, the network architecture is reversed and the Input Layer does not receive the external input but the Label Layer is stimulated, and activity spreads in top-down direction through the network shows in Fig. 5.

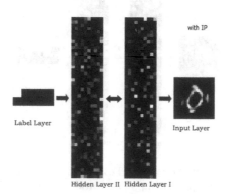

Fig. 5. Generation mode of spiking DBN.

Figure 5 shows that the neuron corresponding to Label 0 was simulated in the Label Layer (left) and finally it was reconstructed to sample 0 by the encoded probability distribution in the weights of the spiking DBN in 10 steps. But it was reconstructed unsuccessfully in 70 steps without IP in the network. Which means that the IP mechanism changed the inner firing rates of the network. And under the same input, the number of spike in Hidden Layer II and Hidden Layer I rises significantly after adding IP so that the network can reconstruct the object in a short period.

5   Conclusion

In this paper, we proposed an optimized spiking DBN via adding IP to the network, which makes the network has higher accuracy and less time cost comparing with the results that the accuracy reaches 94.09 percent and time cost of 5.8 ms for each sample presented by Daniel [16]. For the sake of this IP mechanism, this spiking DBN has adaptive capability. It will increase the inner firing rates of the network when the initial firing rates is too small to make the activity spread precisely and quickly to the Label Layer. On the contrary, when the initial firing rates is too big to accurately recognize the input sample, the IP mechanism will decrease the value of inner firing rates to ensure accuracy.

Acknowledgments. This work was supported by the National Natural Science Foundation of China (Nos. 61304165 and 61473051) and Natural Science Foundation of Chongqing (No. cstc2016jcyjA0015).

References

1. Mohamed, A.R., Yu, D., Deng, L.: Investigation of full-sequence training of deep belief networks for speech recognition. In: Conference of the International Speech Communication Association, Makuhari, Chiba, Japan, September, pp. 2846–2849. INTERSPEECH (2010)
2. Li, C., Li, Y.: A review on synergistic learning. IEEE Access **4**, 119–134 (2016)
3. Ciresan, D.C., Meier, U., Gambardella, L.M., Schmidhuber, J.: Deep, big, simple neural nets for handwritten digit recognition. Neural Comput. **22**(12), 3207–3220 (2010)
4. Erhan, D., Bengio, Y., Courville, A., Manzagol, P.A., Vincent, P., Bengio, S.: Why does unsupervised pre-training help deep learning? J. Mach. Learn. Res. **11**(3), 625–660 (2010)
5. Battenberg, E., Wessel, D.: Analyzing drum patterns using conditional deep belief networks. In: Ismir (2012)
6. Jug, F., Lengler, J., Krautz, C., Steger, A., Lengler, J., Krautz, C.: Spiking networks and their rate-based equivalents: does it make sense to use siegert neurons? Cadmo.ethz.ch
7. Daoudal, G., Debanne, D.: Long-term plasticity of intrinsic excitability: learning rules and mechanisms. Learn. Mem. **10**(6), 456–465 (2003)
8. Hinton, G., Deng, L., Yu, D., Dahl, G.E., Mohamed, A., Jaitly, N., et al.: Deep neural networks for acoustic modeling in speech recognition. IEEE Sig. Process. Mag. **29**(6), 82–97 (2012)
9. Hinton, G.E., Osindero, S., Teh, Y.W.: A fast learning algorithm for deep belief nets. Neural Comput. **18**(7), 1527–1554 (2006)

10. Hinton, G.E., Salakhutdinov, R.R.: Reducing the dimensionality of data with neural networks. Science **313**(5786), 504–507 (2006)
11. Dahl, G.E., Dong, Y., Li, D., Acero, A.: Large vocabulary continuous speech recognition with context-dependent DBN-HMMS. IEEE Int. Conf. Acoust. Speech Sig. Process. **125**, 4688–4691 (2011)
12. Arel, I., Rose, D.C., Karnowski, T.P.: Deep machine learning - a new frontier in artificial intelligence research [research frontier]. IEEE Comput. Intell. Mag. **5**(4), 13–18 (2010)
13. Triesch, J.: A gradient rule for the plasticity of a neurons intrinsic excitability. Int. Conf. Artif. Neural Netw.: Biol. Inspirations **3696**, 65–70 (2005)
14. Desai, N.S., Rutherford, L.C., Turrigiano, G.G.: Plasticity in the intrinsic excitability of cortical pyramidal neurons. Nature Neurosci. **2**(6), 515–520 (1999)
15. Wallisch, P., Lusignan, M.E., Benayoun, M.D., Baker, T.I., Dickey, A.S., Hatsopoulos, N.G.: MATLAB for neuroscientists: an introduction to scientific computing in MATLAB (2014)
16. O'Connor, P., Neil, D., Liu, S.C., Delbruck, T., Pfeiffer, M.: Real-time classification and sensor fusion with a spiking deep belief network. Front. Neurosci. **7**, 178 (2013)
17. Ponulak, F., Kasiński, A.: Supervised learning in spiking neural networks with resume: sequence learning, classification, and spike shifting. Neural Comput. **22**(2), 467–510 (2010)
18. Kourrich, S., Calu, D.J., Bonci, A.: Intrinsic plasticity: an emerging player in addiction. Nat. Rev. Neurosci. **16**(3), 173–184 (2015)
19. Siegert, A.J.F.: On the first passage time probability problem. Phys. Rev. **81**(4), 617–623 (1951)
20. Andrew, A.M.: Spiking neuron models: single neurons, populations, plasticity. Kybernetes **4**(7/8), 277C–280 (2003)
21. Zhang, W., Linden, D.J.: The other side of the engram: experience-driven changes in neuronal intrinsic excitability. Nat. Rev. Neurosci. **4**(11), 885–900 (2003)
22. Bengio, Y.: Learning deep architectures for ai. Found. Trends Mach. Learn. **2**(1), 1–127 (2009)
23. Bengio, Y., Courville, A., Vincent, P.: Representation learning: a review and new perspectives. IEEE Trans. Pattern Anal. Mach. Intell. **35**(8), 1798–1828 (2013)
24. Lecun, Y., Bottou, L., Bengio, Y., Haffner, P.: Gradient-based learning applied to document recognition. Proc. IEEE **86**(11), 2278–2324 (1998)

Hamiltonian-Driven Adaptive Dynamic Programming Based on Extreme Learning Machine

Yongliang Yang[1]($^{(\boxtimes)}$), Donald Wunsch[2], Zhishan Guo[3], and Yixin Yin[1]

[1] School of Automation and Electrical Engineering, University of Science and
Technology Beijing, Beijing 100083, People's Republic of China
y.yang.2016@ieee.org

[2] Department of Electrical and Computer Engineering, Missouri University of
Science and Technology, Rolla, MO 65409-0040, USA

[3] Department of Computer Science, Missouri University of Science and Technology,
Rolla, MO 65409, USA

Abstract. In this paper, a novel frame work of reinforcement learning
for continuous time dynamical system is presented based on the Hamiltonian functional and extreme learning machine. The idea of solution
search in the optimization is introduced to find the optimal control policy in the optimal control problem. The optimal control search consists of
three steps: evaluation, comparison and improvement of arbitrary admissible policy. The Hamiltonian functional plays an important role in the
above framework, under which only one critic is required in the adaptive critic structure. The critic network is implemented by the extreme
learning machine. Finally, simulation study is conducted to verify the
effectiveness of the presented algorithm.

Keywords: Reinforcement learning · Adaptive dynamic programming ·
Extreme learning machine · Hamiltonian functional · Optimization

1 Introduction

The centerpiece of the optimal control problem of dynamical systems falls into
the solution of the Hamilton-Jacobi-Bellman (HJB) equation [3,7]. For the linear
dynamical system, the HJB equation reduces to the Riccati equation, which is
quadratic in the solution [2]. For the nonlinear dynamical system, the HJB equation is a nonlinear partial differential equation, of which the analytical solution
is generally not available. Therefore, it is necessary to develop efficient method
to solve the HJB equation.

Since the exact solution of the HJB equation is difficult to obtain, there are
many works focus on solving the HJB equation approximately. [6] developed an
iterative approach to solve the Riccati equation for the optimal control problem
of linear dynamical systems. Similarly, [11] solved the HJB equation for nonlinear systems in a successive way. Another kind of method to solve the optimal

© Springer International Publishing AG 2017
F. Cong et al. (Eds.): ISNN 2017, Part I, LNCS 10261, pp. 197–205, 2017.
DOI: 10.1007/978-3-319-59072-1_24

control problem of dynamical systems is called approximate/adaptive dynamic programming (ADP) [14] or adaptive critic designs [10], which consists of three networks: the model network, the critic network and the action network. Since then, ADP has attracted many attentions. [13] developed online implementation for ADP. [15] analyzed the finite error property of ADP implemented by neural networks (NNs). [5,9] introduced the idea of off-policy reinforcement learning to ADP. Also, ADP has been extended to discrete-time systems in [8,17], complex valued nonlinear systems in [12].

Inspired by [16], the idea of the solution search method in the optimization problem is introduced to find the optimal control policy for the optimal control problem. Consider the unconstrained optimization problem $\min_x F(x)$, where X is the decision space. Denote x_k as the current estimate of the optimal decision $x^* = \arg\min_x F(x)$. In order to get a better successive estimation x_{k+1} which satisfies $F(x_{k+1}) < F(x_k)$, the condition $\langle x_{k+1} - x_k, \nabla F(x_k) \rangle < 0$ is required. Similarly, the optimal control can be search in the admissible control set as the following frame work solution search problem can be split into the following steps:

– To build a criterion that evaluates an arbitrary admissible control $u_k(\cdot)$, i.e. calculate the corresponding cost $J(u_k(\cdot))$;
– To establish a rule that compares two admissible controls;
– To design a successive control $u_{k+1}(\cdot)$ with a better cost $J(u_{k+1}(\cdot))$, depending on the previous steps and current admissible control $u_k(\cdot)$.

Steps one and three are identical to the policy evaluation and policy improvement in policy iteration. Essentially, the framework above reinterpreted policy iteration from the perspective of optimization. About the critic network training, the Hamiltonian-driven ADP is implemented by extreme learning machine (ELM) [4].

The reminder of the paper is organized as follows. Section 2 gives the problem formulation. The Hamiltonian-driven framework is discussed in Sect. 3. The neural network implementation of the Hamiltonian-driven ADP by extreme learning machine is given in Sect. 4. Simulation studies for linear systems are presented in Sect. 5. Section 6 gives the conclusions.

2   Problem Formulation

This paper considers the stabilizable time-invariant input affine nonlinear dynamic system of the form:

$$\dot{x} = f(x) + g(x)u \tag{1}$$

where $x \in \Omega \subseteq R^n$ is the state vector, $u(\cdot) \in R^m$ is the control policy, $f(\cdot) \in R^n$ and $f(\cdot) \in R^{n \times m}$ are locally Lipschitz functions with $f(0) = 0$. $x(t_0) = x_0$ is the initial state and Ω is a compact set in R^n.

The optimal control problem is to find a policy $u^*(x)$ that minimizes the cost:

$$J\left(u\left(\cdot\right);x_0\right) = \int_{t_0}^{\infty} L\left(x,u\right) dt \qquad (2)$$

where the scalar utility function $L(x,u)$ is differentiable and satisfies $L(x,u) \geq 0$, for $\forall x,u$. In this paper, $L(x,u)$ is chosen as $L(x,u) = Q(x) + \|u\|_R$, with $\|u\|_R = u^T R u$. $Q(x)$ is a positive semidefinite function and R is a symmetric positive definite matrix. $J(u(\cdot);x_0)$ can be viewed as a cost functional of policy $u(\cdot)$ starting from state x_0, which is an evaluation of the given policy.

The following definitions are required for the following discussions.

Definition 1. *The control function is called admissible control if it satisfies:*

- *$u(x)$ is continuous and $u(0) = 0$;*
- *$u(x)$ can stabilize the modified nominal system;*
- *the cost functional (2) is finite for $\forall x_0 \in \Omega$.*

Definition 2. *Define the Hamiltonian functional of a given admissible policy $u(\cdot)$ as*

$$H\left(u;x,V\left(x\right)\right) = L\left(x,u\right) + \left\langle \frac{\partial V}{\partial x}, \dot{x} \right\rangle \qquad (3)$$

where $\langle \cdot, \cdot \rangle$ denotes the inner product between two vectors of the same dimension.

In (3), both state x and the value function $V(x)$ should be viewed as parameters of the Hamiltonian.

Based on the Hamiltonian, a sufficient condition of optimality for optimal control problems centers on the HJB equation [3,7]

$$0 = \min_u H\left(u;x,V\left(x\right)\right) \qquad (4)$$

Assuming that the minimum on the right-hand side of (4) exists and is unique, then the optimal control is

$$u^*\left(x\right) = -\frac{1}{2}R^{-1}g^T\frac{\partial V^*}{\partial x} \qquad (5)$$

Inserting this result into (4), an equivalent formulation of the HJB equation can be found

$$0 = Q\left(x\right) + \left\langle \frac{\partial V^*}{\partial x}, f\left(x\right) \right\rangle - \frac{1}{4}\left[\frac{\partial V^*}{\partial x}\right]^T gR^{-1}g^T\frac{\partial V^*}{\partial x} \qquad (6)$$

Note that the optimal control policy (5) depends on the solution of the HJB equation (4). However, solving the HJB equation is challenging since it is a nonlinear PDE, quadratic in value function gradient, and does not have an analytical solution for general nonlinear systems. In the next section, Hamiltonian-driven ADP with convergence proofs is developed to approximate the solution of HJB equation iteratively.

3   Hamiltonian-Driven ADP

In this section, a novel framework of ADP to find the optimal control policy of the optimal control problem in the previous section will be described.

3.1   Evaluation Step

As shown in Sect. 1, the first step in the Hamiltonian-driven ADP is to set an evaluation for a given admissible policy.

Let $r = u + \frac{1}{2} R^{-1} g^T \frac{\partial V^*}{\partial x}$, then the Hamiltonian functional can be rewritten as

$$H\left(u; x, V\left(x\right)\right) = Q\left(x\right) + \left\langle \frac{\partial V}{\partial x}, f\left(x\right) \right\rangle + r^T R r - \frac{1}{4} \left[\frac{\partial V}{\partial x}\right]^T g R^{-1} g^T \frac{\partial V^*}{\partial x} \quad (7)$$

Therefore, the Hamiltonian functional is quadratic in the control policy $u(\cdot)$. From (2), it can be seen that the cost functional depends on the state trajectory. However, there is no closed-form solution of the general nonlinear dynamic system (1), which makes the calculation of the cost functional difficult. Therefore, an equivalent formulation to the cost functional which does not depend on the state trajectory is introduced as follows.

Definition 3. *The positive definite continuously differentiable scale-valued function $V(x)$ is called the value function of system (1) if it satisfies*

$$\begin{cases} H\left(u; x, V\left(x\right)\right) = 0, \forall x \in \Omega \\ V\left(x\left(\infty\right); u\left(\cdot\right)\right) = 0 \end{cases} \quad (8)$$

The equation in (8) is called generalized HJB (GHJB) equation in [2]. The value function in (8) does not depend on the state trajectory. The equivalence between the value function in (8) and the cost functional in (2) is formulated as the following lemma.

Lemma 1 [2]. *Assume that $x(t)$ is the state trajectory of system (1) when an admissible policy $u(x)$ is applied. Assume further that there exists a positive definite continuously differentiable function $V(x)$ that satisfies (8). Then the value function $V(x)$ is equivalent to the cost functional $J(u(x))$ in (2).*

Lemma 1 shows that the Hamiltonian plays an important role in the calculation of the value function $V(x)$ with respect to an admissible policy $u(x)$, as demonstrated in Fig. 1. The horizontal and vertical axes represent the set of admissible policy and the Hamiltonian respectively. Given an admissible, the value function is the one that make the Hamiltonian identical to 0 for $\forall x \in \Omega$.

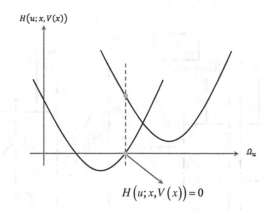

Fig. 1. The system trajectories of $x_1(t)$ (left) and $x_2(t)$ (right).

3.2   Comparison Step

In this subsection, the second step of the Hamiltonian-driven ADP is introduced. In the previous section, the Hamiltonian helps to evaluate a given admissible policy. Here, in order to compare the performance of two different policies, some other aspects about the Hamiltonian functional is studied.

Consider the Hamiltonian in (7), then the minimum of the Hamiltonian functional can be written as

$$
h = \min_u H\left(u; x, V\left(x\right)\right)
$$
$$
= Q\left(x\right) + \left\langle \frac{\partial V}{\partial x}, \dot{x} \right\rangle - \frac{1}{4}\left[\frac{\partial V}{\partial x}\right]^T g R^{-1} g^T \frac{\partial V}{\partial x} \tag{9}
$$

and the control that attains the minimum in (9) is

$$
u = \min_v H\left(v; x, V\left(x\right)\right) = -\frac{1}{2}R^{-1}g^T\frac{\partial V}{\partial x} \tag{10}
$$

Based on (9) and (10), the second step of the Hamiltonian-driven ADP can be described as the following lemma.

Lemma 2 [16]. *Let $u_i(x)$ be two different admissible policies, with corresponding value functions $V_i(x)$. Denote the minimum of the Hamiltonian as $h_i = \min_u H\left(u; x, V_i\left(x\right)\right), i = 1, 2,\ \bar{u}_i = -\frac{1}{2}R^{-1}g^T\frac{\partial V_i}{\partial x}, i = 1, 2$ and $d_i = \left\|\bar{u}_i - u_i\right\|, i = 1, 2$. Then the following conditions hold:*

(1) $h_i \leq 0, i = 1, 2$;
(2) $h_1 \leq h_2 \rightarrow V_1\left(x\right) \geq V_2\left(x\right), \forall x \in \Omega$;
(3) $d_1 \geq d_2 \rightarrow V_1\left(x\right) \geq V_2\left(x\right), \forall x \in \Omega$;

In Lemma 2, the minimum of the Hamiltonian and the control policy that achieves the minimum is crucial to the comparison problem. Similar to Lemma 1, Lemma 2 can be illustrated in Fig. 2, where the Hamiltonian is identical to 0 along the state trajectory, as shown by the intercept of the Hamiltonian curve with the horizontal axis. We then investigate the properties about the minimum of Hamiltonian to compare the performance.

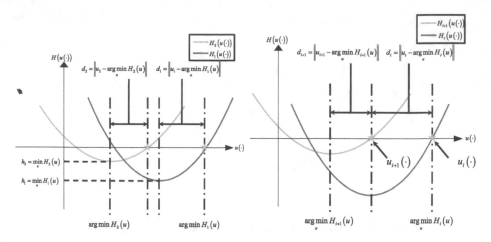

Fig. 2. Admissible policy comparison.   **Fig. 3.** Improvement of an admissible policy.

3.3 Improvement Step

In the previous section, Lemma 2 only describes the comparison between two admissible policies. However, it does not provide the explicit expression of the improved policy when given an admissible policy. In this section, the third step of the Hamiltonian ADP is discussed.

When given an admissible policy, the improved policy can be expressed by the following Lemma.

Lemma 3 [16]. *Suppose the admissible policy sequence $\{u_i(\cdot)\}$ and the corresponding value function sequence $\{V_i(\cdot)\}$ satisfies the GHJB equation in (8). Suppose further that the policy sequence is generated by the following relationship:*

$$u_{i+1} = -\frac{1}{2}R^{-1}g^T\frac{\partial V_i(x)}{\partial x} \tag{11}$$

then: (i) the value function sequence $\{V_i(\cdot)\}$ is non-increasing; (ii) both the value function and the policy sequences $\{u_i(\cdot)\}$ converge to the solution of the HJB equation.

Note that Lemma 3 is a special case of Lemma 2. The improved policy is the policy that attains the minimum of previous Hamiltonian functional, as shown in Fig. 3.

4 Hamiltonian ADP Structure

In this section, the implementation details and network structure of the Hamiltonian-driven ADP is illustrated.

As shown in Fig. 4, there is only one critic in the Hamiltonian-driven ADP. The critic outputs the value function gradient. Since the Hamiltonian of a given

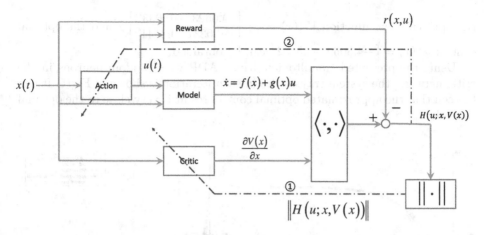

Fig. 4. Hamiltonian-driven ADP structure.

policy is required to be identical to 0, the norm of the Hamiltonian is the objective to be minimized for the critic network training. According to Lemma 3, the optimal control policy can be obtained by successive minimization of the Hamiltonian. Therefore, the Hamiltonian is the target to be minimized for the action network training.

In contrast to the adaptive critic structure for the optimal control of discrete-time dynamical system, there is only one critic network. Therefore, the critic network can be implemented by the ELM due to its fast training speed. In [1], the critic is implemented by the neural network with polynomial basis. However, the order of the value function and the value gradient is usually unknown for general nonlinear systems with admissible policy. Due to the uniform approximation ability of ELM with random neurons has [4], the Hamiltonian-driven ADP with critic network implemented by ELM can approximate the value gradient in the probabilistic meaning. This is the motivation to introduce ELM into the Hamiltonian-driven ADP.

5   Simulation

In this section, the presented Hamiltonian-driven ADP with critic network implemented by the ELM is applied to the linear quadratic regulator problem. Consider the linear system

$$\begin{bmatrix} \dot{x}_1 \\ \dot{x}_2 \end{bmatrix} = \begin{bmatrix} 0.5 & 1.5 \\ 2 & -2 \end{bmatrix} \begin{bmatrix} x_1 \\ x_2 \end{bmatrix} + \begin{bmatrix} 4 \\ 1 \end{bmatrix} u \tag{12}$$

with the utility function $r(x, u) = x_1^2 + 2x_2^2 + 0.05u^2$. The initial state is $x_0 = \begin{bmatrix} -1 & 2 \end{bmatrix}^T$. The initial policy $u = -2x_1 + x_2$ is obtained by the pole placement technique. Since the value function of the linear system is quadratic, the optimum condition results in the Riccati equation [7]. Solving the Riccati equation, the optimal value function $V^*(x) = x^T \begin{bmatrix} 0.0623 & -0.0337 \\ -0.0337 & 0.3141 \end{bmatrix} x$ and the optimal control $u^*(x) = \begin{bmatrix} -4.3069 & -3.5852 \end{bmatrix}$ can be obtained.

Using the presented Hamiltonian-driven ADP with random neurons in the critic network, the system trajectories in each iteration is shown in Fig. 5. It can be seen that the approximated optimal control result is very close to the optimal control.

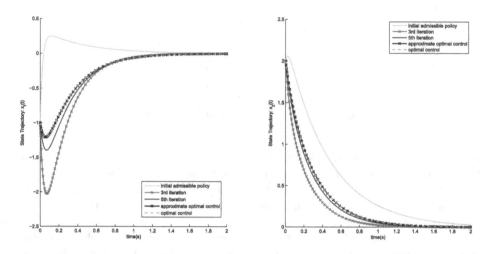

Fig. 5. The system trajectories of $x_1(t)$ (left) and $x_2(t)$ (right).

6   Conclusion

A novel frame work of ADP based on the Hamiltonian for continuous time dynamical system is presented based on the Hamiltonian functional and extreme learning machine. The Hamiltonian based framework can be another interpretation about ADP from the view of optimization. The Hamiltonian-driven ADP consists of three steps: evaluation, comparison and improvement of the admissible policy. The Hamiltonian functional plays an important role in the above framework. With the Hamiltonian functional, only one critic is needed in the adaptive critic structure. The critic network is implemented by the fast and efficient ELM. A simulation is conducted to verify the effectiveness of the presented algorithm at the end.

Acknowledgments. This work was supported in part by the Mary K. Finley Missouri Endowment, the Missouri S&T Intelligent Systems Center, the National Science Foundation, the National Natural Science Foundation of China (NSFC Grant No. 61333002) and the China Scholarship Council (CSC No. 201406460057).

References

1. Abu-Khalaf, M., Lewis, F.L.: Nearly optimal control laws for nonlinear systems with saturating actuators using a neural network HJB approach. Automatica **41**(5), 779–791 (2005)
2. Beard, R.W., Saridis, G.N., Wen, J.T.: Galerkin approximations of the generalized Hamilton-Jacobi-Bellman equation. Automatica **33**(12), 2159–2177 (1997)
3. Bertsekas, D.P.: Dynamic Programming and Optimal Control. Athena Scientific, Belmont (1995)
4. Huang, G.B., Zhu, Q.Y., Siew, C.K.: Extreme learning machine: theory and applications. Neurocomputing **70**(1), 489–501 (2006)
5. Jiang, Y., Jiang, Z.P.: Computational adaptive optimal control for continuous-time linear systems with completely unknown dynamics. Automatica **48**(10), 2699–2704 (2012)
6. Kleinman, D.: On an iterative technique for Riccati equation computations. IEEE Trans. Autom. Control **13**(1), 114–115 (1968)
7. Lewis, F.L., Syrmos, V.L.: Optimal Control. Wiley, New York (1995)
8. Liu, D., Wei, Q.: Policy iteration adaptive dynamic programming algorithm for discrete-time nonlinear systems. IEEE Trans. Neural Netw. Learn. Syst. **25**(3), 621–634 (2014)
9. Modares, H., Lewis, F.L., Jiang, Z.P.: H$_\infty$ tracking control of completely unknown continuous-time systems via off-policy reinforcement learning. IEEE Trans. Neural Netw. Learn. Syst. **26**(10), 2550–2562 (2015)
10. Prokhorov, D.V., Wunsch, D.C.: Adaptive critic designs. IEEE Trans. Neural Netw. **8**(5), 997–1007 (1997)
11. Saridis, G.N., Lee, C.G.: An approximation theory of optimal control for trainable manipulators. IEEE Trans. Syst. Man Cybern. **9**(3), 152–159 (1979)
12. Song, R., Xiao, W., Zhang, H., Sun, C.: Adaptive dynamic programming for a class of complex-valued nonlinear systems. IEEE Trans. Neural Netw. Learn. Syst. **25**(9), 1733–1739 (2014)
13. Vamvoudakis, K.G., Lewis, F.L.: Online actor-critic algorithm to solve the continuous-time infinite horizon optimal control problem. Automatica **46**(5), 878–888 (2010)
14. Wang, F.Y., Zhang, H., Liu, D.: Adaptive dynamic programming: an introduction. IEEE Comput. Intell. Mag. **4**(2), 39–47 (2009)
15. Wei, Q., Wang, F.Y., Liu, D., Yang, X.: Finite-approximation-error-based discrete-time iterative adaptive dynamic programming. IEEE Trans. Cybern. **44**(12), 2820–2833 (2014)
16. Yang, Y., Wunsch, D., Yin, Y.: Hamiltonian-driven adaptive dynamic programming for continuous nonlinear dynamical systems. IEEE Trans. Neural Netw. Learn. Syst. (2017, to be published)
17. Zhang, H., Qin, C., Jiang, B., Luo, Y.: Online adaptive policy learning algorithm for H$_\infty$ state feedback control of unknown affine nonlinear discrete-time systems. IEEE Trans. Cybern. **44**(12), 2706–2718 (2014)

An Enhanced K-Nearest Neighbor Classification Method Based on Maximal Coherence and Validity Ratings

Nian Zhang[1](✉), Jiang Xiong[2], Jing Zhong[2], Lara Thompson[3], and Hong Ying[2]

[1] Department of Electrical and Computer Engineering,
University of the District of Columbia, Washington, D.C. 20008, USA
nzhang@udc.edu
[2] College of Computer Science and Engineering,
Chongqing Three Gorges University, Chongqing 404000, China
{xjcq123,zhongandy}@sohu.com, yinghok@163.com
[3] Department of Mechanical Engineering,
University of the District of Columbia, Washington, D.C. 20008, USA
lara.thompson@udc.edu

Abstract. Traditional k-nearest neighbor methods couldn't be able to correctly classify objects when their k nearest neighbors are dominated by other classes. This paper formulates a two-class classification problem, and applies a modified k-nearest neighbors (KNN) classifier algorithm based on maximal coherence, validity ratings, and k-fold cross validation to classify the test samples. We build a validity score for the pairs of sample and their surroundings according to their labels. The k nearest neighbors (including the unknown test object) of each sample in the training set as well as the unknown test object itself will be determined. The unknown test object will be tentatively assigned to a class membership. Then we use the validity scores to quantify the degree to which a pre-determined group of samples resemble their k nearest neighbors. A classifier is designed which take into account the coherence and validity ratings. A numerical example demonstrates the effectiveness of the algorithm in detail. The enhanced KNN method is compared with the conventional KNN and the modified KNN method on both real world wine data and photo-thermal infrared imaging spectroscopy (PT-IRIS) data for up to 20 different k values. Classification accuracy of KNN method and our method in terms of various combinations of k-value and k-fold cross validation are compared. The experimental results show that the proposed enhanced KNN method outperforms the conventional KNN and the modified KNN method on both real world wine data and PT-IRIS data. In addition, the classification accuracy of both the conventional KNN and our method increase drastically when k = 5. The average classification accuracy of our method on the PT-IRIS data featuring small sample size and high overlap is 97.87%.

Keywords: K-nearest neighbors (KNN) classification · Modified k-nearest neighbors (MKNN) · Supervised classification · K-fold cross validation

© Springer International Publishing AG 2017
F. Cong et al. (Eds.): ISNN 2017, Part I, LNCS 10261, pp. 206–214, 2017.
DOI: 10.1007/978-3-319-59072-1_25

1 Introduction

In security, identification of explosive materials on relevant substrates has become a heightened priority for homeland security and counter-terrorism applications in recent years. In addition, the detection of explosives is also used in many peaceful applica- tions. For example, it can be relevant in environmental areas to monitor the quality of soil, water, and groundwater suspected of being contaminated by explosives and their degradation products, in order to prevent poisoning of populations of humans and animals [1]. However the ability to detect small amounts of analytes across large relevant substrates is complicated by the optical and thermal interactions between analyte and substrate [2]. The primary challenge is to distinguish explosives from the substrates, such as glass or clothing. While glass or clothing is chemically distinct from explosives, they nonetheless have overlapping infrared absorption/emission features with explosives. Further complications are the facts that polymeric materials tend to absorb and emit throughout the IR. Therefore, the development of analytical tools that can identify explosive remains is of tremendous importance in the forensic field for crime-scene reconstruction [3].

Fix and Hodges proposed the k-nearest neighbors algorithm (i.e. KNN) in 1951 where the classification rules do not depend on the underlying distributions [4]. The non-parametric methods meet the needs when the data distributions of are either unknown or unavailable in many real world applications. The inputs contain k closest training data, while the output of the classifier is a class membership. A test sample is assigned to the same class as the majority of the labels of its k nearest neighbors. Various modified KNN have been proposed to improve the accuracy rate of KNN. Wang et al. presented an adaptive distance measure to significantly improve the performance of the KNN rules [5]. Tang and He proposed an extended k-NN method (i.e. ENN), which not only finds those who are the nearest neighbors of the test sample, but also those who identify the test sample to be their nearest neighbor [6]. Parvin et al. generated some kinds of similarities among the training data, and added this additional information to the weight of each neighbor [7]. Taneja et al. proposed a fuzzy logic based KNN algorithm where fuzzy clusters are obtained at pre-processing step while the class membership of the training data is computed in reference with the centroid of the clusters [8].

Although these methods demonstrated that the classification accuracy performs better than the conventional KNN, their performance is unknown when the sample sizes are small or the data are highly overlapped. In addition, to date, there have been insignificant efforts to analyze the photo-thermal infrared (IR) imaging spectroscopy (PT-IRIS) data using computational intelligence and data mining techniques [9]. Therefore, it's important to develop an advanced KNN method to further improve the accuracy and compare with the previous variations. We propose an enhanced KNN variation method by exploring the performance on not only the same datasets (i.e. wine dataset) used by other researchers for the purpose of comparison, but also the PT-IRIS datasets, which has small sample size versus much larger feature size [10]. Unlike the conventional KNN method where only the nearest neighbors are used to determine the class of the test object, we also include the test object to the training dataset to maximize the intra-class coherence and then make a classification decision.

The remaining of the paper is organized as follows. In Sect. 2, a two-class classification problem is formulated. The enhanced k-nearest neighbor method is described. In Sect. 3, we provide a numerical example on the two-class classification problem and derive the output according to our method. In Sect. 4, experimental results are presented. In Sect. 5, the conclusions are provided.

2　The Enhanced K-Nearest Neighbor Method

Unlike the conventional KNN method, the idea of the enhanced KNN method is to find out the k nearest neighbors (including the unknown test object) of each sample in the training dataset, as well as the unknown test object. Then we use a concept of validity rating to quantify the degree to which a pre-determined group of samples resemble their k nearest neighbors. Finally, a classifier will assign the unknown test object to a class membership based on the validity ratings.

In a highly overlapped dataset, it is extremely difficult to classify the data with high accuracy. Part of the reasons is that data belonging to different classes mix with each other or a single data is surrounded by large groups of data with different classes. Figure 1 shows the above situations which would pose several great challenges for classification. First of all, not only X_1 is overlapped with Y_2, but also X_3 is overlapped with Y_3; second, a green unknown test object, P to be classified is closest to X_1 (Class 1), but it is closer to Y_1 and Y_2 (Class 2) on the top left than X_2 and X_3 (Class 1) on the bottom right.

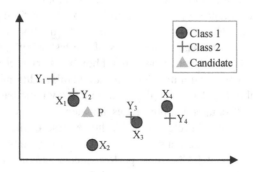

Fig. 1. A two-class classification example. Not only X1 is overlapped with Y2, but also X3 is overlapped with Y3; second, a green unknown test object, P to be classified is closest to X1 (Class 1), but it is closer to Y1 and Y2 (Class 2) on the top left than X2 and X3 (Class 1) on the bottom right. (Color figure online)

Let $T = \{X_1, X_2, \ldots, X_m, Y_1, Y_2, \ldots, Y_n\}$ be the training data, where $\{X_1, X_2, \ldots, X_m\}$ has a given class label, C_1, and $\{Y_1, Y_2, \ldots, Y_n\}$ has a given class label, C_2. P is an unknown test candidate data that we want to classify.

A concept of validity rating is used to measure how similar a training data looks to its k nearest neighbors. The validity of a training data is computed based on the label of the data and the labels of its k nearest neighbors, as defined in (1). The validity for data x, V(x) counts the number of nearest neighbors that have the same labels as x.

$$V(x) = \frac{1}{k}\sum_{i=1}^{k} S(label(x), label(NN_i(x, \mathrm{T})))$$ (1)

Where k is a pre-defined number of nearest neighbors. $label(x)$ is the class membership of data x. $label(NN_i(x, \mathrm{T}))$ is the class membership of the ith nearest neighbor of x. $NN_i(x, T)$ stands for the ith nearest neighbor of x inside T. S is a function representing the similarity between x and its ith nearest neighbor. The function S is defined in (2).

$$S(i,j) = \begin{cases} 1 & i = j \\ 0 & i \neq j \end{cases}$$ (2)

After adding a validity attribute to the training samples, we are able to obtain a score for each training sample based on the labels of the object and its surroundings. Then the k nearest neighbors (including the unknown test object itself) of each sample in the training set as well as the unknown test object will be determined. The unknown test object will be tentatively assigned to a class membership based on some criteria, and form a group. Then we calculate the validity ratings to quantify the degree to which the aforementioned group of samples resembles their k nearest neighbors, as shown in (3).

$$M_i^j(x) = \frac{1}{N+k}\sum_{x \in C}\sum_{i=1}^{k} S(label(x), label(NN_i(x, \mathrm{T})))$$ (3)

Where N is the size of samples determined by criteria, C. k is a pre-defined number of nearest neighbors. The criteria, C are defined in (4):

$$C = \begin{cases} \{\text{Samples in Class}_i\} \cup \{P\}, & \text{when } i = j \\ \{\text{Samples in Class}_i\}, & \text{when } i \neq j \end{cases}$$ (4)

P is the unknown test sample. $label(x)$ is the class membership of data x. $label(NN_i(x, \mathrm{T}))$ is the class membership of the ith nearest neighbor of x. $NN_i(x, T)$ stands for the ith nearest neighbor of x inside T. S is a function representing the similarity between x and its ith nearest neighbor. The function S is defined in (2).

A classifier will take into account the coherence and validity ratings, and assign the class label associated to the maximum coherence to the unknown test sample, as defined in (5).

$$Classifier = \arg\max_{j \in 1,2} \sum_{i=1}^{2} M_i^j(x)$$ (5)

3 Numerical Example

In this section, we provide a numerical example corresponding to the enhanced KNN method described in Sect. 2 and demonstrate the effectiveness of the method.

First we find out the k nearest neighbors (including the unknown test object) of each sample in the training dataset, as well as the unknown test object, as shown in (6). In the training dataset T, $\{X_1, X_2, \ldots, X_m\}$ has a given class label, C_1, and $\{Y_1, Y_2, \ldots, Y_n\}$ has a given class label, C_2. P is an unknown test object that we want to classify.

$$
\begin{aligned}
NN_{1,2,3}(X_1) &= [Y_2, P, Y_1] & NN_{1,2,3}(Y_1) &= [Y_2, X_1, P] \\
NN_{1,2,3}(X_2) &= [P, X_3, Y_3] & NN_{1,2,3}(Y_2) &= [X_1, P, Y_1] \\
NN_{1,2,3}(X_3) &= [Y_3, X_4, Y_4] & NN_{1,2,3}(Y_3) &= [X_3, X_4, Y_4] \\
NN_{1,2,3}(X_4) &= [Y_4, X_3, Y_3] & NN_{1,2,3}(Y_4) &= [X_4, X_3, Y_3] \\
NN_{1,2,3}(P) &= [X_1, Y_2, X_2] & &
\end{aligned} \tag{6}
$$

Then the unknown test object will be tentatively assigned to a class membership based on criteria C, and then get involved in the intra-class correlation computation. First we assume $P \in C_1$, and then solve for M_1^1 and M_2^1, respectively.

Solve for M_1^1: Given $P \in C_1$, i = 1, j = 1, $NN_i(x, \text{T}) = C_1 \cup C_2 \cup \{P\}$, $C = C_1 \cup P = \{X_1, X_2, X_3, X_4, P\}$, $N = 5$, k = 3.

According to (3), $M_i^j(x) = \frac{1}{N+k} \sum\limits_{x \in C} \sum\limits_{i=1}^{k} S(label(x), label(NN_i(x, \text{T})))$. The validity ratings can be calculated using (1), (2), and (6). Thus,

$$
M_1^1 = \frac{1}{5+3} \sum_{x \in (X_1, X_2, X_3, X_4, P)} \sum_{i=1}^{3} S(label(x), label(NN_i(x, \text{T})))
$$

$$
= \frac{1}{8} \times \begin{bmatrix} S_1(X_1, Y_2) + S_2(X_1, P) + S_3(X_1, Y_1) + \\ S_1(X_2, P) + S_2(X_2, X_3) + S_3(X_2, Y_3) + \\ S_1(X_3, Y_3) + S_2(X_3, X_4) + S_3(X_3, Y_4) + \\ S_1(X_4, Y_4) + S_2(X_4, X_3) + S_3(X_4, Y_3) + \\ S_1(P, X_1) + S_2(P, Y_2) + S_3(P, X_2) \end{bmatrix} = 0.875
$$

Similarly, $M_2^1 = 0.571$.

Next we assume $P \in C_2$, and then solve for M_1^2 and M_2^2, respectively.

Solve for M_1^2: Given $P \in C_2$, i = 1, j = 2, $NN_i(x, \text{T}) = C_1 \cup C_2 \cup \{P\}$, $C = C_1 = \{X_1, X_2, X_3, X_4\}$, $N = 4$, k = 3.

$$M_1^2 = \frac{1}{4+3} \sum_{x \in (X_1, X_2, X_3, X_4)} \sum_{i=1}^{3} S(label(x), label(NN_i(x, \text{T})))$$

$$= \frac{1}{7} \begin{bmatrix} S_1(X_1, Y_2) + S_2(X_1, P) + S_3(X_1, Y_1) + \\ S_1(X_2, P) + S_2(X_2, X_3) + S_3(X_2, Y_3) + \\ S_1(X_3, Y_3) + S_2(X_3, X_4) + S_3(X_3, Y_4) + \\ S_1(X_4, Y_4) + S_2(X_4, X_3) + S_3(X_4, Y_3) \end{bmatrix} = 0.429$$

Similarly, $M_2^2 = 0.875$.

$$Classifier = \arg\max_{j \in 1,2} \sum_{i=1}^{2} M_i^j(x)$$
$$= \arg\max\{M_1^1 + M_2^1, M_1^2 + M_2^2\} = \arg\max\{0.875 + 0.571, 0.429 + 0.875\}$$
$$= \{1.446, 1.304\}$$

Therefore, we assign P to Class 1.

4 Experimental Results

The photo-thermal infrared imaging spectroscopy (PT-IRIS) data set [11] is ideal to test the proposed enhanced ENN method, as it features small sample size and much larger feature size (i.e. 428 vs. 728). Each data has 728 features representing the temperature increase for the laser pulse [12]. The labels for all explosive materials are +1, while the non-explosives are set to −1.

First we compare our algorithm to the conventional KNN and the modified KNN [7] on the wine data. We increase the k value from 3 to 7, and observe the classification accuracy of the three algorithms. The result is shown in Fig. 2. The left blue column

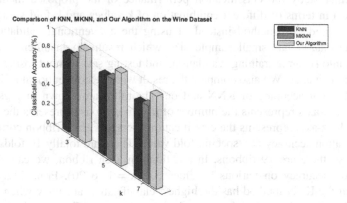

Fig. 2. Comparison of the conventional KNN method, modified KNN method, and our algorithm on the wine data in terms of different k values. (Color figure online)

represents the conventional KNN, the middle green column represents the MKNN, and the right yellow column represents our method. The result shows that our algorithm outperforms both of the conventional KNN and the modified KNN.

We then applied the proposed method on the PT-IRIS data. We compare the classification accuracy of our algorithm to the ENN method. The k value increases from 1 to 20. The result is shown in Fig. 3. The red curve represents the conventional KNN, and the blue curve represents our method. It shows that our method has higher classification accuracy than the conventional KNN except when k = 3. In addition, the classification accuracy of both methods increase drastically when k = 5. At k = 5, the classification accuracy of our method is 97.87%.

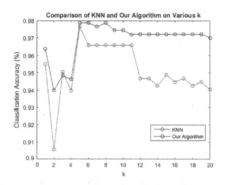

Fig. 3. Comparison of KNN method and our algorithm on various k values from 1 to 20. (Color figure online)

Moreover, when k is greater than 5, our method keeps high classification accuracy with small scale fluctuation. However, the KNN method dropped twice at k = 6 and k = 11, respectively. Furthermore, both methods have shown a declining trend when k = 20.

We further study the classification performance of the proposed method on the PT-IRIS data in terms of different combination of k values and k-fold values. We use k-fold cross-validation method instead of using the conventional validation method because the dataset has small sample size which results in insufficient data to be partitioned into separate training, validation, and testing sets without losing significant modeling competence. We also compare the result with the conventional KNN method. The classification accuracy of KNN and our algorithm are shown in Figs. 4 and 5, respectively. X-axis represents the number of k value, y-axis represents the number of folds, and the z-axis represents the classification accuracy. Each ribbon corresponds to the classification accuracy at a specific fold value. There are totally 19 folds (i.e. 2nd– 20th fold), so there are 19 ribbons. In addition, on each ribbon, we can observe the classification accuracy on various k values (i.e. k = 1 to 20). From Fig. 4, we can observe that the KNN method has the highest classification accuracy when k = 11 for all folds. From Fig. 5, we find that when k = 5 and fold = 17, our method reaches the peak of classification accuracy and remain at peak values when k increases.

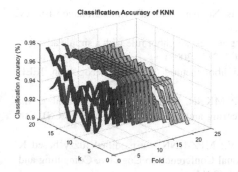
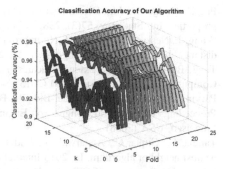

Fig. 4. Classification accuracy of KNN method in terms of various combination of k values and fold values.

Fig. 5. Classification accuracy of our method in terms of various combination of k values and fold values.

5 Conclusions

This paper formulates a two-class classification problem, and proposes an enhanced k-nearest neighbors (KNN) method based on maximal coherence, validity ratings, and k-fold cross validation. Unlike the conventional KNN method, our method finds out the k nearest neighbors (including the unknown test object) of each sample in the training dataset, as well as the unknown test object. Then we use a concept of validity rating to quantify the degree to which a pre-determined group of samples resemble their k nearest neighbors. Finally, a classifier will be designed to assign the unknown test object to a class membership based on the coherence and the validity ratings. We compare the results of our method to the conventional KNN and the modified KNN method on different combination of k values and $fold$ values. The experimental results demonstrate that our method has significantly higher classification accuracy than the conventional KNN and the modified KNN method on both wine dataset and PT-IRIS dataset.

Acknowledgment. This work was supported by the National Science Foundation (NSF) #1505509, #1533479, #1654474, and USGS Grant #2016DC181B.

References

1. Gheyas, I.A., Smith, L.S.: Feature subset selection in large dimensionality domains. Pattern Recogn. **43**(1), 5–13 (2010)
2. Kendziora, C.A., Furstenberg, R., Papantonakis, M.: Infrared photothermal imaging of trace explosives on relevant substrates. In: Proceedings of SPIE, vol. 8709 (2013)
3. Zhang, N., Thompson, L.A.: An intelligent clustering algorithm for high dimensional and highly overlapped photo-thermal infrared imaging data. In: Fall 2016 ASEE Mid-Atlantic Regional Conference, Hempstead, New York, 21–22 October 2016

4. Fix, E., Hodges Jr., J.L.: Discriminatory analysis. Nonparametric discrimination: consistency properties. Int. Stat. Rev. **57**(3), 238–247 (1989)
5. Wang, J., Neskovic, P., Coope, L.N.: Improving nearest neighbor rule with a simple adaptive distance measure. Pattern Recogn. Lett. **28**, 207–213 (2007)
6. Tang, B., He, H.: ENN: extended nearest neighbor method for pattern recognition. IEEE Comput. Intell. Mag. **10**(3), 52–60 (2015)
7. Parvin, H., Alizadeh, H., Minaei-Bidgoli, B.: MKNN: modified K-nearest neighbor. In: Proceedings of the World Congress on Engineering and Computer Science (WCECS), San Francisco, USA (2008)
8. Taneja, S., Gupta, C., Aggarwal, S., Jindal, V.: MFZ-KNN – a modified fuzzy based K nearest neighbor algorithm. In: 2015 International Conference on Cognitive Computing and Information Processing (CCIP), Noida, pp. 1–5 (2015)
9. Ramirez Rochac, J.F., Zhang, N., Behera, P.: Design of adaptive feature extraction algorithm based on fuzzy classifier in hyperspectral imagery classification for big data analysis. In: The 12th World Congress on Intelligent Control and Automation (WCICA 2016), Guilin, China (2016)
10. Zhang, N.: Cost-sensitive spectral clustering for photo-thermal infrared imaging data. In: 2016 Sixth International Conference on Information Science and Technology (ICIST), Dalian, China (2016)
11. Ramirez Rochac, J.F., Zhang, N.: Reference clusters based feature extraction approach for mixed spectral signatures with dimensionality disparity. In: 10th Annual IEEE International Systems Conference (IEEE SysCon 2016), Orlando, Florida (2016)
12. Ramirez Rochac, J.F., Zhang, N.: Feature extraction in hyperspectral imaging using adaptive feature selection approach. In: The Eighth International Conference on Advanced Computational Intelligence (ICACI2016), Chiang Mai, Thailand, pp. 36–40 (2016)

Credit Risk Assessment Based on Flexible Neural Tree Model

Yishen Zhang[1,2], Dong Wang[1,2(✉)], Yuehui Chen[1,2(✉)],
Yaou Zhao[1,2], Peng Shao[3], and Qingfang Meng[1,2]

[1] School of Information Science and Engineering,
University of Jinan, Jinan, People's Republic of China
{ise_wangd, yhchen}@ujn.edu.cn
[2] Shandong Provincial Key Laboratory of Network Based Intelligent
Computing, Jinan 250022, People's Republic of China
[3] School of Mathematics, Dalian University of Technology,
Dalian, People's Republic of China

Abstract. In recent years, as China's credit market continues to expand, a large number of P2P (person-to-person borrow or lend money in Internet Finance) platforms were born and developed. Most of the P2P platforms in China use data mining methods to evaluate the credit risk of loan applicants. Artificial neural network (ANN) is an emerging data mining tool and has good classification ability in many application fields. This paper presents a model of credit risk assessment based on flexible neural tree (FNT), which can reduce the overdue rate and save the analysis time. Overdue and non-overdue sample data are provided by the Jinan Hengxin Micro-Investment Advisory Co., Ltd., and used to build the model. Experiments show that the proposed model is more accurate and has less time cost for the overdue classification of credit risk assessment.

Keywords: Artificial neural network · Credit risk assessment · Flexible neural tree

1 Introduction

Credit loan is an unsecured loan model. In recent years, the credit market has been expanding rapidly in China. On one hand, the rapid development of China's economy has shortened the cycle of capital turnover. On the other hand, because of the improvement of Chinese national consumption capacity, businesses increasingly need high demand for funds, so a large number of P2P Internet inclusive financial platforms came into being. As no complete credit evaluation system like banks in China, P2P platform has small contain ability to non-collateral customers, it obtains better risk prediction results only through the establishment of the corresponding credit risk assessment model. So a large number of platforms are exploring their own methods of credit risk assessment, most of which use data mining approach to try to collect and understand the customer information to better grasp the authenticity and validity of customer information; to evaluation financial situation of customers more reasonable; to predict the business conditions, repayment intention and ability of borrows more accurately.

© Springer International Publishing AG 2017
F. Cong et al. (Eds.): ISNN 2017, Part I, LNCS 10261, pp. 215–222, 2017.
DOI: 10.1007/978-3-319-59072-1_26

The establishment of a good credit risk evaluation model is the biggest challenge to the development of P2P platform and credit market. If the model control is too strict to the customer, the platform will lose some high-quality customers and make it passive in the industry competition. On the contrary, the overdue rate of the platform will continued to rise, which makes financial managers difficult to be responsible and lose credibility. Therefore, it is important to establish the credit risk evaluation model to prevent bad debts happening, to promote the speed of capital flow and to maintain the security and stability of capital. In the field of credit risk assessment, artificial neural networks, genetic programming, genetic algorithms, support vector machines, logistic regression and some hybrid models have achieved gratifying results in terms of performance and precision.

In the past few years, many excellent algorithms and research methods have been tested on the basis of customer information data in the field of credit risk assessment. Khashman used artificial neural network algorithm in Germany customer dataset and achieved the accuracy rate of 83.6% [1]. Bekhet and Eletter applied RBF network algorithm to the Jordanian commercial bank data set, and the test sets had accuracy rate of 86.5% [2]. Wang et al. uses the improved BP neural network algorithm and the accuracy rate is 86% [3]. The traditional Artificial Neural Network has the stationary structure, but Flexible neural tree (FNT) has the special structures which called flexible tree structures, with this characteristic, FNT model can get better property from the learning.

In this paper, a new method based on FNT model was proposed for classification of customer information, and the results in 10-fold cross validation shows our method achieved better performance than the other state-of-arts.

2 Data Collection and Variable Definition

Customer information data can be described from many dimensions. In this paper, we randomly took 300 samples of overdue customers and 300 Negative samples of non-overdue customers all of which were from 2,000 customers of Jinan Hengxin Micro-Investment Advisory Co., Ltd. between 2014 and 2016. In this study, the author chooses 13 dimensions to describe and consider the customer information. The standard of selected dimensions are: (1) do not contain the customer's identity information; (2) exclude the subjective information from the point of view of the actual human audit, such as the use of loans, business models, profits and other objective information which can only be verified by a third party as difficulties to verify and census them.

According to these principles, the selected dimensions can maximize the provided data by customer which objectively and difficulty to forge. The accurate classification based on actual data which can verify and excluding the subjective description. Table 1 shows the variable, values, and definitions of 13 selected dimensions of the study, and the Table 2 shows the examples of datasets.

The 600 samples are based on the statistics in Table 1, and then all the data will processed as "Max_Min standardization" for the next step, and get ready to input to the FNT model, the normalized samples are shown in Table 3. The normalization rule is shown in Eq. (1).

Table 1. Proposed variables for building dataset

Variable	Value	Variable definition
Gender, G	0, 1	0: female 1: male
Degree of education, D	1 to 4	1: graduated from junior high school 2: graduated from high school 3: graduated from junior college 4: get bachelor degree or above
Age, A	Actual value	The number of years the customer has experienced since birth
Marital status, M	0 to 2	0: unmarried 1: married 2: divorce
Account properties, AP	0, 1	0: local registered permanent residence 1: foreign registered permanent residence
The number of years experienced by the company, YC	Actual value	The number of years in which the customer's work unit has been established
Industry categories, IC	0 to 5	Industry category of customer
Job level within the company, JC	0 to 4	0: general staff 1: junior management staff 2: middle management staff 3: senior management staff 4: the founder
Total income, TI	Actual value	Customer's itemized of the savings card which printed by bank, and the difference between total amount and total expenditure in recently six months
Total debt, TD	Actual value	The sum of outstanding loan balance and average usage limit in recently six months shown in the summary of liability information within customer credit report
Housing ownership situation, HS	0 to 2	0: no real estate 1: full purchase 2: mortgage
Vehicle ownership situation, VS	0 to 2	0: no car production 1: full purchase 2: mortgage
Overdue numbers shown in credit reporting, OR	Actual value	The sum of the number of overdue times in the credit transaction details within customer credit report

$$P'_{ij} = \frac{P_{ij} - m_i}{M_i - m_i} \qquad (1)$$

where, P'_{ij} is the normalized customer data. P_{ij} is the original customer data. M_i is the maximum value of the dimension i. m_i is the minimum value of the dimension i.

Table 2. Examples

No.	G	D	A	M	AP	YC	IC	JC	TI	HS	VS	OR	TD
1	1	3	48	1	2	4	0	1	150	2	1	3	137.8
2	1	2	36	1	2	10	1	1	30	0	1	0	41.2
3	2	3	49	1	1	2	1	1	48	1	1	2	91.5
4	1	3	54	1	2	11	3	1	102	2	0	4	16.5
5	1	3	36	1	3	7	1	1	20	0	0	3	69.8

Table 3. Normalized samples

No.	G	D	A	M	AP	YC	IC	JC	TI	HS	VS	OR	TD
0	0.5	0.44	0.33	0.5	0.23	0.17	0	0.05	1	1	0.1	0.13	0
0	0.5	0.35	0.33	0.5	0.14	0.17	0	0.09	1	0	0.1	0.09	0
0	0.5	0.23	1	0	0.05	0	0	0.04	1	1	0.4	0.27	1
0	0.5	0.64	0.33	0	0.08	0	0	0.05	0	1	0.1	0.04	1
0	0.5	0.76	0.33	0	0.20	0.17	0	0.15	0.5	0	0	0.19	1

3   Classification Method

3.1   Flexible Neural Tree

Flexible neural tree (FNT) is a special artificial neural network with flexible tree structures. It is proposed by Chen et al. [4, 5] and relatively easy for this model to reach near-optimal structure by using optimization algorithms. The FNT model consists of tree-structural encoding method and specific instruction set, it is also generated by using function set F and terminal instruction set T, described as follows.

$$S = F \cup T = \{+_2, +_3 \cdots +_N\} \cup \{x_1 \cdots x_n\} \tag{2}$$

where $+_i (i = 1, 2 \cdots N)$ denotes non-leaf nodes with i arguments, the $x_1, x_2 \cdots x_n$ are leaf nodes with none arguments.

Figure 1 shows the output of a non-leaf node which calculated by FNT model. Instruction $+_i$ is also called a flexible neuron operator with i inputs. The output of a flexible neuron +n is calculated as follows and the total excitation of $+_n$ is given by

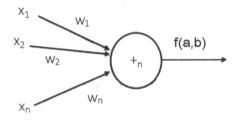

Fig. 1. Non-leaf node of flexible neural tree with a terminal instruction set $T = \{x_1, x_2, \cdots, x_n\}$

$$net_n = \sum_{j=1}^{n} w_j x_j \tag{3}$$

In Eq. (3), $x_j (j = 1, 2, \cdots, n)$ are the input elements to node $+_n$. The output of the node $+_n$ is then calculated by

$$out_n = f(a_n, b_n, net_n) = e^{-(\frac{net_n - a_n}{b_n})^2} \tag{4}$$

A typical FNT model is illustrated in Fig. 2. Its overall output can be computed from left to right by a depth-first method recursively.

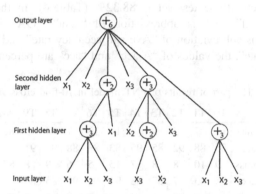

Fig. 2. Typical representation of FNT with function instruction set $\{+_2, +_3, +_4, +_5, +_6\}$ and terminal set $\{x_1, x_2, x_3\}$, which has four layers.

General learning algorithm of FNT

- Step 1. Initialize the values of parameters used in the particle swarm optimization (PSO) algorithms. Set the elitist program as NULL and set the fitness value as the biggest positive real number. Create the initial population.
- Step 2. Construct optimization using PSO algorithm, in which the fitness function is calculated by root mean square error (RMSE).
- Step 3. If the better structure has found, then go to step 4, otherwise go to step 2.
- Step 4. Optimize parameters using PSO algorithm.
- Step 5. If the maximum number of local search is reached, or no better parameter vector is found for a significantly long time (100 steps), then go to step 6; otherwise go to step 4.
- Step 6. If the satisfied solution is found, then stop; otherwise go to step 2.

3.2 Prediction Assessment

In statistical analysis, two methods can be used to check the effectiveness of the classifier in applications, namely, independent dataset tests and 10-fold cross validation tests. For 10-fold cross validation, the full training set will be separated equally into 10 subset. Each subset will regarded as test data set to compute the overall accuracy (OA) of the model trained by the rest of full training data set. In addition, Sensitivity (Sens) and Specificity (Spec) are also used to evaluate the performance of classifier.

4   Discussion and Results

In this study, the FNT model was used to perform a 10-fold cross validation of a data set containing 600 sample data, i.e. 540 training samples and 60 testing samples were used for each experiment and were performed on each data set. The results show that the average accuracy of the test set is 88.32% (Table 4). In the Table 4, "T" is abbreviation of "trail", "D" is abbreviation of "data", "OA" is abbreviation of "Overall", "A-acc" is abbreviation of "Average accuracy rate" and "acc" is abbreviation of "accuracy rate", the values of "A-acc" and "acc" are percentages.

Table 4. The part of results of FNT model in 10-fold cross validation

		T0	T1	T2	T3	T4	T5	T6	T7	T8	T9	A-acc
D0	miss	11	7	5	9	3	10	10	7	4	5	88.17
	acc	83	88	92	85	95	83	83	88	93	91	
D1	miss	9	10	8	8	7	5	5	6	9	8	87.50
	acc	85	83	87	87	88	91	91	90	85	86	
D2	miss	8	7	10	6	9	4	9	8	5	4	88.33
	acc	87	88	83	90	85	93	85	86	91	93	
D3	miss	7	9	6	9	7	9	7	8	6	4	88.00
	acc	88	85	90	85	88	85	88	86	90	93	
D4	miss	9	8	3	3	8	6	7	4	9	5	89.67
	acc	85	87	95	95	87	90	88	93	85	91	
D5	miss	10	10	8	5	6	6	5	7	9	5	88.17
	acc	83	83	87	92	90	90	91	88	85	91	
D6	miss	9	5	9	7	10	8	9	10	5	7	86.83
	acc	85	92	85	88	83	86	85	83	91	88	
D7	miss	8	9	11	5	3	7	7	4	9	6	88.50
	acc	87	85	82	92	95	88	88	93	85	90	
D8	miss	7	10	9	8	6	9	7	5	5	5	88.17
	acc	88	83	85	87	90	85	88	91	91	91	
D9	miss	9	4	7	6	4	3	8	6	5	9	89.83
	acc	85	93	88	90	93	95	86	90	91	85	
OA												88.32

We compared the average accuracy, sensitivity and specificity between our model and other methods. The results are shown in Table 5, we can see that our method has higher accuracy compared to other method, and the specificity is slightly better than the others. Another point to make is this: the sensitivity value of Improved BP Neutral

Network method is 91.6%, and this value was calculated by once experiment result form with 14 positive simples and 6 negative samples, totally 20 simples. The proportion of positive samples is much higher, so the sensitivity value also high, besides the sensitivity index is mentioned there only and no mention of any other place, so this value is included in Table 5 for reference.

Table 5. The comparison of our method and other methods

Algorithm	Accuracy (%)	Sens (%)	Spec (%)
Improved BP neutral network	86	91.6	62.5
Radial basis function scoring model	86.5	84.2	87.9
Artificial neural networks	83.6	Null	Null
This method (average)	88.32	85.67	92.79

5 Conclusion

In this study, we proposed a redesigned and redefined customer information feature dimension and FNT model for the field of credit risk assessment. Compared with other methods, the method proposed in this study has different degrees of improvement in various evaluation indexes, while the validity of the FNT model is proved. In the future, we will continue to improve the algorithm method and search for more effective classifiers in order to obtain better classification accuracy in this field.

Acknowledgements. This research was supported by the National Natural Science Foundation of China (Grant No. 61302128, 61573166, 61572230, 61671220, 61640218), the Youth Science and Technology Star Program of Jinan City (201406003), the Shandong Distinguished Middle-aged and Young Scientist Encourage and Reward Foundation, China (Grant No. ZR2016FB14), the Project of Shandong Province Higher Educational Science and Technology Program, China (Grant No. J16LN07), the Shandong Province Key Research and Development Program, China (Grant No. 2016GGX101022).

References

1. Khashman, A.: Neural network for credit risk evaluation: investigation of different neural models and learning schemes. Exp. Syst. Appl. **37**(9), 6233–6239 (2010)
2. Bekhet, H., Eletter, S.: Credit risk assessment model for Jordanian commercial banks: neural scoring approach. Rev. Dev. Financ. **4**(1), 20–28 (2014)
3. Wang, L., Chen, Y., Zhao, Y., Meng, Q., Zhang, Y.: Credit management based on improved BP neural network. IHMSC **1**, 497–500 (2016)
4. Chen, Y., Yang, B., Dong, J., Abraham, A.: Time-series forecasting using flexible neural tree model. Inf. Sci. **174**, 219–235 (2005)
5. Yang, B., Chen, Y., Jiang, M.: Reverse engineering of gene regulatory networks using flexible neural tree models. Neurocomputing **99**, 458–466 (2013)
6. Abdou, H., Pointon, J., El-Masry, A.: On the applicability of credit scoring models in Egyptian banks. Banks Bank Syst. **2**(1), 4–19 (2007)

7. Bensic, M., Sarlija, N., Zekic-Susac, M.: Modeling small-business credit scoring by using logistic regression, neural networks and decision trees. Intell. Syst. Account. Financ. Manag. **13**(3), 133–150 (2005)
8. Blanco, A., Mejias, R., Lara, J., Rayo, S.: Credit scoring models for the microfinance industry using neural networks: evidence from Peru. Exp. Syst. Appl. **40**(1), 356–364 (2013)
9. Heiat, A.: Comparing performance of data mining models for computer credit scoring. J. Int. Financ. Econ. **12**(1), 78–83 (2012)
10. Koh, H., Tan, W., Goh, C.: A two-step method to construct credit scoring models with data mining techniques. Int. J. Bus. Inf. **1**(1), 96–118 (2006)
11. Jagric, V., Kracun, D., Jagric, T.: Does non-linearity matter in retail credit risk modeling? Financ. uver-Czech J. Econ. Financ. **61**(4), 384–402 (2011)
12. Wu, C., Guo, Y., Zhang, X., Xia, H.: Study of personal credit risk assessment based on support vector machine ensemble. Int. J. Innov. **6**(5), 2353–2360 (2010)
13. Xie, T., Yu, H., Wilamowski, B.: Comparison between traditional neural networks and radial basis function networks. In: Proceedings of 2011 IEEE International Symposium on Industrial Electronics, pp. 1194–1199 (2011)
14. Yap, P., Ong, S., Husain, N.: Using data mining to improve assessment of credit worthiness via credit scoring models. Exp. Syst. Appl. **38**(10), 1374–1383 (2011)
15. Memarian, H., Balasundram, S.: Comparison between multi-layer perceptron and radial basis function networks for sediment load estimation in a tropical watershed. J. Water Resour. Prot. **4**, 870–876 (2012)

A Portable Prognostic System for Bearing Monitoring

Bulent Ayhan[1], Chiman Kwan[1(✉)], and Steven Liang[2]

[1] Signal Processing, Inc., Rockville, MD 20850, USA
{bulent.ayhan, chiman.kwan}@signalpro.net
[2] Georgia Institute of Technology, Atlanta, GA 30332, USA
steven.liang@me.gatech.edu

Abstract. This paper summarizes the development of a practical and high performance bearing prognostic system, which contains a portable hardware data acquisition system with flexible and modular prognostic tools. The data acquisition system has a multi-sensor analog to digital (A/D) card with USB connection, a laptop, and a modular software based on Labview. The low-cost A/D card from National Instruments can simultaneously acquire multiple sensor data (such as accelerometer, tachometer and load cell) at high sampling rates (48 KS/s). The Labview based software can run in any laptops and PCs. The basic functions of the software include: (1) data acquisition control (sampling rate, sensor selection, etc.); (2) application configuration manager (each configuration addresses one application); (3) feature selection (spectrum-based features or time-domain based features) (4) prognostic tool library (the library will be expandable); (5) visualization of data acquisition, feature trend plots and prognostic results; (6) data management (raw data and log data storage, retrieval, etc.). Simulation experiments using actual bearing test data demonstrated the functionalities of the system.

Keywords: Prognostic system · Bearing · Remaining life prediction

1 Introduction

The ability to accurately predict the early stages of failures of critical electro-mechanical components such as bearing [1–3, 5–8], pumps [11], actuators [12, 14–16], motors [10, 13, 17], communication equipment [4, 9], and gear [19, 20] is critical for affordable system operation and can also be used to enhance system safety. Early detection of potential failures not only saves costs, but also increases reliability and availability, and even save lives. Usually the failure goes through a series of transitions from normal, to minor degradation, and then progresses to a significantly degraded state, and finally to a complete failure. Each stage of degradation generally has unique characteristics that can be identified from sensor outputs. Hence, for high quality prognosis system, it is essential to have a library of advanced, robust, and reliable algorithms that can perform accurate diagnostics and prognostics.

In this research, we focus on bearing prognostics. The goal is to develop a portable system with fast data acquisition hardware, and flexible, expandable, and high

© Springer International Publishing AG 2017
F. Cong et al. (Eds.): ISNN 2017, Part I, LNCS 10261, pp. 223–233, 2017.
DOI: 10.1007/978-3-319-59072-1_27

performance prognostics algorithms. We have successfully developed such a system. The data acquisition card can simultaneously collect data from up to 8 sensors. The prognostic software contains 6 feature extraction modules and 1 prognostic module. The software modules can be expanded. Two sets of actual bearing data were used to demonstrate the functionalities of the proposed system.

The paper is organized as follows. In Sect. 2, we will focus on the system description. Details of the software implementation will be described in a journal paper. Section 3 summarizes the experimental results. Finally, some concluding remarks and ideas for future research will be mentioned in Sect. 4.

2   Bearing Prognostic System

2.1   Portable Prognostic System

We developed a bearing prognostic system as shown in Fig. 1. This is a portable system with data acquisition hardware and flexible and modular prognostic software tools. The data acquisition system has a multi-sensor A/D card (USB) and a laptop. The function generator in the figure is only used for system test. In actual applications, the data acquisition system can be connected to accelerometers, load cells, etc. The low-cost USB card from National Instruments can simultaneously acquire multiple sensor data (accelerometer, tachometer and load cell) at high sampling rates such as 48 KS/s. The laptop and USB data acquisition card can be easily carried to anywhere for periodic equipment inspection. The hardware cost is low.

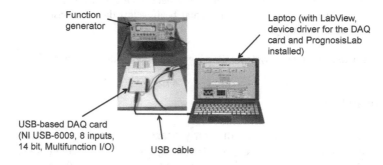

Fig. 1. Portable prognostic system.

2.2   Prognostic Software

The laptop contains a prognostic tool known as PrognosisLab, which is modular and based on Labview platform. The Labview based software can run in laptops and/or PCs. The basic functions of the software include: (1) data acquisition control (sampling rate, sensor selection, etc.); (2) application configuration creation/selection tool; (3) feature selection (spectrum-based features or time-domain based features) (4) prognostics tool library (currently there is one prognostic algorithm which is called

Parallel-DCA-RLS); (5) visualization of data acquisition, feature trend plots and prognostic results; (6) data management (storage, etc.).

Key features of "PrognosisLab" are:

- Flexible and a modular software structure

  The software is applicable for prognosis of different bearing types with the use of a configuration creation/selection tool. Application configurations for different bearing types and with different operational and feature settings can be created with the configuration creation/selection tool.

- Feature extraction library

  Feature extraction phase is a critical phase for prognostics. The extracted features are going to be used as the degradation signatures by the prognostics algorithm for remaining lifetime predictions. There are currently 6 feature extraction ways in PrognosisLab.

- Prognostics Library

  Currently, there is a single technique in the library (Parallel-DCA-RLS), which is an adaptive, data driven prognostics technique that is based on the physics of a failure without the need of any off-line training. Extending this library by incorporating other prognostic tools such as hidden Markov model [1] or neural networks is one of our future objectives.

- Operation condition check

  PrognosisLab is designed to check the operating condition by loadcell and tachometer sensors and enables the prognostics approach to function only when the system operates within the identified operating conditions (speed and load) by the user.

LabView (32 bit) is currently the programming platform. The Parallel-DCA-RLS technique (which is used in PrognosisLab as the prognostics technique) is a data driven adaptive prognostics method which is based on the physics of failure model. This technique requires that the extracted feature from the collected accelerometer data has a gradually increasing pattern in time (as the degradation in bearing worsens, the feature amplitude increases). This trend has generally been observed in the bearing lifetime tests even though there are some cases which the feature amplitude increase is not monotonic at some stages of the bearing degradation, but rather follows a fluctuating trend at these stages due to the varying nature of the damage; this phenomenon is known as "healing" [6]. The spectrum amplitudes extracted from the accelerometer data at bearing fault specific frequencies and/or user specified frequencies and the root mean squared (RMS) time domain feature of the collected vibration data are the current features used in PrognosisLab. PrognosisLab software may not provide reliable results if the extracted features from the collected accelerometer data do not gradually increase in amplitude. Fluctuating feature values during the lifetime can be compensated for up to a degree in DCA-RLS; however, if the extracted feature decreases in amplitude and and/or shows a highly fluctuating pattern (due to extreme healing effects), since this will conflict with the failure model of the DCA-RLS technique, PrognosisLab may not provide meaningful remaining useful lifetime predictions for this type of cases.

PrognosisLab software requires the following hardware units:

1. Laptop or PC with LabView - 2012 software (32 bits) being installed in the laptop and the device driver for the DAQ card needs to be installed; the device driver for the USB-based DAQ card can be found in the NIDAQ mx 9.5.5 CD library that comes with the NI-DAQ card.
2. The USB-based DAQ card (NI USB-6009, 8 inputs, 14 bit, Multifunction I/O) should be connected to the laptop via a USB cord. This will power the USB card automatically and the card should be detected by the laptop (we purchased this USB-based DAQ from National Instruments).
3. Accelerometer sensor needs to be connected to the one set of the analog inputs of the DAQ card; the voltage reading from the sensor should not exceed ±5 V at any time (currently this sensor is not available, a function generator is used to imitate the accelerometer sensor data in the form of a sinusoidal signal and this data is collected by the DAQ card and a data file which consists of scaling coefficients are used to scale the collected data from the function generator so that the extracted feature time-series data behaves as the physics of failure model that is observed in degrading bearings)
4. Tachometer sensor needs to be connected to the one set of the analog inputs of the DAQ card; the voltage reading from the sensor should not exceed ±5 V at any time (currently this sensor is not available, the sensor is simulated by directly feeding the tachometer value in the application configuration file as if it is collected from a sensor)
5. Load-cell sensor needs to be connected to the one set of the analog inputs of the DAQ card; the voltage reading from the sensor should not exceed ±5 V at any time (currently this sensor not available, the sensor is simulated by directly feeding the tachometer value in the application configuration file as if it is collected from a sensor)

During the PrognosisLab implementation and in testing the modules' operational functionality, the actual data collection testing/debugging was done with the use of a function generator instead of a test rig with an actual bearing. A sinusoidal waveform has been generated in the function generator and this has been fed to the analog channels of the NI-DAQ unit. This collected data is then scaled inside the software with some pre-prepared coefficients to imitate a vibration feature acquired from a degrading bearing. The tool requires the operating conditions to be within the range identified by the user in the selected/created configuration file. The operating condition values in the application configuration file are currently fed to the operating condition check module in the tool as if they were collected from the tachometer and load cell sensors. The reason the delivered software codes retrieve the operating condition values directly from the application configuration file whereas these should come from the sensors is that an actual experiment test rig with the above sensors (accelerometer, load cell and tachometer) was not available during the software development. For a realistic test, the users need to integrate these sensors and make the corresponding revisions/customizations in the delivered Main Layer source code so that PrognosisLab can be used in real testing environments with these three sensors. It should however be noted that the actual bearing lifetime data sets have been used to test the feasibility of the adaptive prognostics algorithm embedded in PrognosisLab; these actual data sets have been collected when the bearing operating conditions were fixed (IMS and Georgia Tech data sets, same speed and load conditions); however since a real lifetime situation would involve varying operating conditions, it is considered that the PrognosisLab software should have the functionality of an operation condition check for both speed and load; if the readings for the operating conditions are not met, PrognosisLab will warn the user about this via its status indicator in the Main Layer and the prognostics technique is not applied to this data which is out of range of the accepted operating condition.

PrognosisLab consists of a Main Layer from which user can access to three other layers. The Main Layer has also a display section that the user is informed both about the lifetime status and system status. Configuration Layer is the layer which user can create a new application configuration file, and/or edit an existing application config-uration file before a test is started. The Trend Plots Layer allows the user visualizing the trend plots of the extracted features and the RUL estimations during real-time prog-nostics. Configuration Visualization Layer is another layer which the user can examine the application configuration settings during a real-time prognostics test. The layers in PrognosisLab can be seen in Fig. 2.

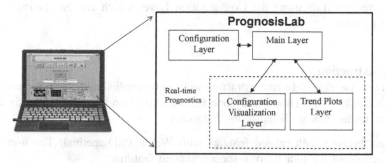

Fig. 2. The layers in PrognosisLab. Here, we briefly describe the Main Layer. Other layers will be described in a journal version of this paper.

After running the "PrognosisLab" which is a LabView file, the Main Layer comes to screen by default. A screenshot of the user interface of the Main Layer can be seen in Fig. 3. There is a total of 6 pushbuttons (Start, Stop, Config, Trend Plots, Exit and Brief Results), a "Normal/Warning/Alert" remaining lifetime status indicator with green/yellow/red colors, estimated remaining lifetime estimation box in which the estimation is displayed and another indicator that provides information about the system status. Additionally, the configuration file that is currently in use is displayed in the Main Layer at the bottom of the screen. In the same folder that the PrognosisLab LabView file resides, there are two other data files. These files are:

1. "LastUsedConfig.dat": This is the text data file which stores the application con-figuration file that was last used in PrognosisLab. This file is created and/or updated whenever the user stops an ongoing test with the "STOP" button of the Main Layer.

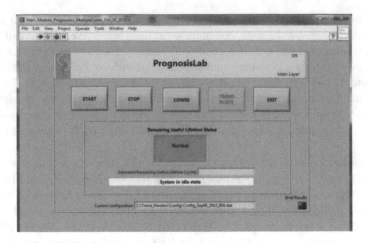

Fig. 3. Main Layer

2. "Default_Config.dat": This is the default application configuration file that comes with PrognosisLab, the user can create his/her own application configuration files with PrognosisLab using the Configuration Layer which can be opened with the Config button.

Feature Extraction

The software provides 6 feature extraction ways to be applied to the accelerometer data. The user is allowed to choose only one feature extraction technique. The six feature extraction techniques are listed in the following.

1. Extract bearing fault related features with Welch PSD method: The user has to choose one of the four bearing specific related features,
2. Extract Bearing Fault Related Feature with HFRT (High Frequency Resonance Technique),
3. Extract Feature at a customized frequency with Welch PSD method (the user needs to provide a customized frequency value in the application configuration file),
4. Extract Feature at a customized frequency with HFRT (the user needs to provide a customized frequency value in the application configuration file),
5. Extract time domain features on raw data (RMS),
6. Extract time-domain features on the envelope data (RMS).

In two of the feature extraction ways, the Welch PSD spectrum method is used. This method is used to extract the PSD amplitudes at the four identified fault frequencies and user-customized frequency. The four bearing fault specific frequencies are ball pass outer raceway frequency, ball pass inner raceway frequency, ball rotational frequency and fundamental train frequency. The computations of these frequencies are shown in (1)–(4) where F_S is the rotational shaft frequency, N_B is the number of rollers, D_b is the mean roller diameter, D_c is the pitch diameter, θ is the roller contact angle.

$$F_{BPFO} = \frac{N_B}{2} F_S \left(1 - \frac{D_b \cos \theta}{D_c} \right) \quad : \text{ ball pass outer raceway frequency} \qquad (1)$$

$$F_{BPFI} = \frac{N_B}{2} F_S \left(1 + \frac{D_b \cos \theta}{D_c} \right) \quad : \text{ ball pass inner raceway frequency} \qquad (2)$$

$$F_{BRF} = \frac{D_c}{D_b} F_S \left(1 - \frac{D_b^2 \cos^2 \theta}{D_c^2} \right) \quad : \text{ ball rotational frequency} \qquad (3)$$

$$F_{FTF} = \frac{1}{2} F_S \left(1 - \frac{D_b \cos \theta}{D_c} \right) \quad : \text{ fundamental train frequency} \qquad (4)$$

HFRT (High Frequency Resonance Technique) is used in two other feature extraction ways. This technique is used to extract the HFRT features at the four identified fault frequencies and also the user-customized frequency. Other than the spectrum-based features, the RMS (root mean square) value of the collected accelerometer data, which is a time-domain based feature, is another feature extraction way. The RMS of the envelope of the bandpass-filtered accelerometer data is the last feature extraction way provided in this tab. Among the 6 sets of potential features (Welch spectrum at a bearing fault specific frequency, Welch spectrum at user defined frequency, HFRT at a bearing fault specific frequency, HFRT at user defined frequency, RMS for raw data, RMS for envelope data) to be used in the prognostics; only one feature extraction technique is allowed to be selected by the user in the application configuration settings.

Robust Prognostic Algorithm: Parallel-DCA-RLS

Damage Curve Approach-Recursive Least Square (DCA-RLS) [5] is an adaptive technique; however, in order to prevent any instability issues that might result from poor choice of initial parameters, a parallel DCA-RLS framework as shown in Fig. 4 is developed; this framework has multiple DCA-RLS units that run in parallel and each being initialized with a different set of initial parameter values.

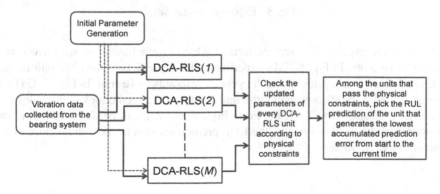

Fig. 4. Robust prognostic algorithm: Parallel-DCA-RLS.

3 Demonstration Experiments

We used two sets of actual bearing data for demonstrations. One data set was generated by the NSF I/UCR Center for Intelligent Maintenance Systems (IMS–www.imscenter. net) with support from Rexnord Corp. in Milwaukee, WI. We became aware of this bearing lifetime dataset through the Prognostics Data Repository website of NASA [18].

Another data set came from Georgia Tech. Three bearing lifetime experiments have been conducted in a laboratory at Georgia Tech. The experimental scheme has three sub-systems: a test housing system, an oil circulation system, and a data acquisition and processing system, which is shown in Fig. 5. The radial load is provided by a Power Team P59 hydraulic hand-pump that supplies pressure to the load cylinder on the housing. The shaft is driven by a vector drive motor with a speed controller. The oil circulation system regulates the flow and temperature of the lubricant. The operating system is lubricated by ISO VG 32 mineral oil.

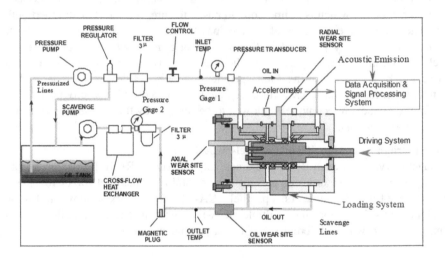

Fig. 5. Experiment setup [8]

Quite a few experiments were performed. Due to page limitation, we include two representative plots. In Fig. 6 (IMS experiments RUL prediction plot), the unit for the RUL predictions is in cycles and each cycle corresponds to 10 min. In Fig. 7 (GaTech experiments RUL prediction plots), the unit for the RUL predictions is in cycles and each cycle corresponds to a minute. From Figs. 6 and 7, it can be observed that in the last portion of the bearing lifetime, the RUL predictions start to come quite close to the groundtruth RUL values.

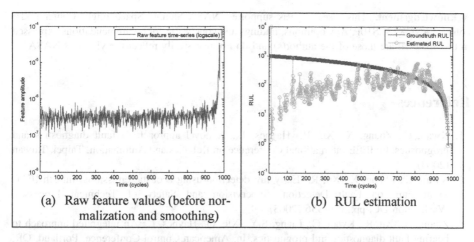

Fig. 6. RUL estimations for Bearing#1 in IMS-Experiment#2 using Welch-PSD features at BPFI (StatDur = 11)

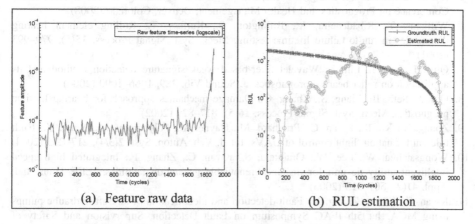

Fig. 7. GaTech data set Experiment#2 Channel 1 (BPFO) (StatDur = 3)

4 Conclusions

We presented a brief summary of the development of a portable prognostic system for bearings. The data acquisition hardware can simultaneously measure up to 8 sensors. The software is flexible, modular, and expandable. Preliminary demonstrations using actual bearing data demonstrated the efficacy of the proposed system. One future research direction is to expand the tools in the prognostic library. For example, some data driven algorithms based on neural network will be considered for inclusion into the system.

Acknowledgement. This research was supported NASA Stennis Space Flight Center under contract #NNX13CS10P. Any opinions, findings and conclusions or recommendations expressed in this material are those of the author(s) and do not necessarily reflect the views of NASA.

References

1. Kwan, C., Zhang, X., Xu, R., Haynes, L.: A novel approach to fault diagnostics and prognostics. In: IEEE International Conference on Robotics and Automation, Taipei, Taiwan (2003)
2. Kwan, C., Zhang, X., Xu, R.: Early fire detection using acoustic emissions. In: 5th IFAC Symposium on Fault Detection, Supervision and Safety of Technical Processes, Washington DC, pp. 351–356 (2003)
3. Zhang, X., Xu, R., Kwan, C., Liang, S.Y., Xie, Q., Haynes, L.: An integrated approach to bearing fault diagnostics and prognostics. In: American Control Conference, Portland, OR, pp. 2750–2755 (2005)
4. Zhang, G., Kwan, C., Xu, R., Vichare, N., Pecht, M.: An enhanced prognostic model for intermittent failures in digital electronics. In: IEEE Aerospace Conference (2007)
5. Ayhan, B., Kwan, C., Liang, S.: Adaptive prognostics of bearing. In: IEEE International Conference on Prognostics and Health Management, Denver, Colorado (2008)
6. Williams, T., Ribadeneira, X., Billington, S., Kurfess, T.: Rolling element bearing diagnostics in run-to-failure lifetime testing. Mech. Syst. Signal Process. **15**(5), 979–993 (2001)
7. Qiu, H., Lee, J., Lin, J.: Wavelet filter-based weak signature detection method and its application on roller bearing prognostics. J. Sound Vib. **289**, 1066–1090 (2006)
8. Qiu, J., Seth, B., Liang, S., Zhang, C.: Damage mechanics approach for bearing lifetime prognostics. Mech. Syst. Signal Process. **16**(5), 817–829 (2002)
9. Zhang, X., Xu, R., Kwan, C., Pritchard, M., Haynes, L., Polycarpou, M., Yang, Y.: Fault tolerant formation flight control of UAVs. Int. J. Veh. Auton. Syst. **2**(3/4), 217–235 (2004)
10. Wongsaichua, W., Lee, W., Oraintara, S., Kwan, C., Zhang, F.: Integrated high speed intelligent utility tie unit for disbursed/renewable generation facilities. IEEE Trans. Ind. Appl. **41**(2), 507–513 (2005)
11. Kwan, C., Xu, R., Zhang, X.: Fault detection and identification of aircraft hydraulic pumps using MCA. In: 5th IFAC Symposium on Fault Detection, Supervision and Safety of Technical Processes, Washington DC, pp. 1137–1142 (2003)
12. Polycarpou, M., Zhang, X., Xu, R., Yang, Y., Kwan, C.: A neural network based approach to adaptive fault tolerant flight control. In: IEEE International Symposium on Intelligent Control, pp. 61–66 (2004)
13. Zhang, H., Lee, W.-J., Kwan, C., Ren, Z., Chen, H., Sheeley, J.: Artificial neural network based on-line partial discharge monitoring system for motors. In: IEEE-IAS, I&CPS Annual Conference, Saratoga Spring, NY (2005)
14. Zhang, X., Polycarpou, M., Xu, R., Kwan, C.: Actuator fault diagnosis and accommodation for improved flight safety. In: Joint IEEE International Symposium on Intelligent Control and Mediterranean Conference on Control and Automation Conference, pp. 640–645 (2005)
15. Zhang, X., Miller, D., Xu, R., Kwan, C., Chen, H.: A maximum entropy based approach to fault diagnosis using discrete and continuous variables. In: 6th IFAC Symposium on Fault Detection, Supervision and Safety of Technical Processes, Beijing (2006)

16. Xu, R., Zhang, G., Zhang, X., Haynes, L., Kwan, C., Semega, K.: Sensor validation using nonlinear minor component analysis. In: Wang, J., Yi, Z., Zurada, J.M., Lu, B.-L., Yin, H. (eds.) ISNN 2006. LNCS, vol. 3973, pp. 352–357. Springer, Heidelberg (2006). doi:10.1007/11760191_52
17. Kwan, C., Qian, T., Ren, Z., Chen, H., Xu, R., Lee, W., Zhang, H., Sheeley, J.: A novel approach to corona monitoring. In: Wang, J., Liao, X.-F., Yi, Z. (eds.) ISNN 2005. LNCS, vol. 3498, pp. 494–500. Springer, Heidelberg (2005). doi:10.1007/11427469_80
18. http://ti.arc.nasa.gov/tech/dash/pcoe/prognostic-data-repository/
19. Kwan, C., Davydov, A., Xu, R.: Helicopter gearbox fault isolation. In: Proceedings of SPIE on Component and Systems Diagnostics, Prognosis, and Health Management, vol. 4389 (2001)
20. Kwan, C., Xu, R., Haynes, L.: Gearbox failure prediction using infrared cameras. In: Proceedings of SPIE on Thermosense XXIII, vol. 4360 (2001)

Parameter Estimation of Linear Systems with Quantized Innovations

Changchang Hu$^{(\boxtimes)}$

Department of Automation, Tsinghua University,
Beijing 100084, China
hcz@tsinghua.edu.cn

Abstract. A recursive identification algorithm for linear discrete time-invariant systems with quantized observations is developed based on the minimum mean square error (MMSE) criterion. It is demonstrated that a persistently exciting input is rich enough to identify the parameter vector of the system with only quantized observations. Under the Gaussian assumption on the estimate of the parameter vector, an identification algorithm is proposed based on a quantized Recursive Least-Squares Estimation (RLSE) scheme. It is interesting that the effect of the quantization can be approximately characterized by a scalar coefficient of a modified Riccati difference equation which is dependent on the number of quantization levels.

Keywords: Linear systems · Parameter estimation · Quantized innovations · MMSE

1 Introduction

Research on networked systems has attracted a lot of interests recently due to their many advantages including flexibility and low implementation cost as compared to traditional point-to-point wired systems and also due to new research opportunities arising from constrains in communication resources and lack of reliability of networks. In the past few years, control, estimation and system identification with quantized observations have attracted significant interests. In particular, the studies on system identification with quantized observations are of importance in understanding modeling capability under limited observations and the trade-off between communication resources and identification complexity.

Traditional system identification using linear sensors is a relatively mature research area that bears a vast body of literature. There are numerous textbooks and monographs on the subject in a stochastic or worst-case framework, see e.g. [10,11,13]. Many significant results have been obtained for identification and adaptive control involving random disturbances in the past decades [1,2,8–11]. Note that set-valued observations are often encountered in digital systems and have been studied in many branches of signal processing problems. Gradient algorithms

© Springer International Publishing AG 2017
F. Cong et al. (Eds.): ISNN 2017, Part I, LNCS 10261, pp. 234–241, 2017.
DOI: 10.1007/978-3-319-59072-1_28

for adaptive filtering using quantized data were studied in [16]. One class of adaptive filtering problems that has recently drawn considerable attention uses "hard limiters" to reduce computational complexity. The idea employs the sign operator in the error and/or the regressor, leading to a variety of sign-error, sign-regressor, and sign-sign algorithms. Some recent work in this direction can be found in [3–5].

We apply the quantized MMSE to estimate system parameters. Noting that the estimation performance is measured by the estimation error covariance, quantizer and estimator are jointly designed to minimize the error covariance. It turns out that the quantizer applied on the innovation of measurement is optimal under the assumption that the estimate is Gaussian. The striking feature of the estimator is that it has a similar form as the classical Recursive Least-Squares Estimation(RLSE) [11] algorithm and the quantization effects on the estimation performance can be characterized by a scalar coefficient in a modified Riccati equation of error covariance.

2   Problem Formulation

Consider the parameter identification problem for the discrete linear time-invariant system:

$$y_k = \phi_k^T \theta_k + d_k, \tag{1}$$

where $\theta_k = [a_0, \ldots, a_{n-1}]^T \in \mathcal{R}^n$ is a parameter vector, $\phi_k = [u_k, \ldots, u_{k-n+1}]^T \in \mathcal{R}^n$ is the vector of inputs with $u_k = 0$, $k < 1 - n$, $d_k \in \mathcal{R}$ is an independent and identically distributed Gaussian noise with zero mean and covariance σ_k^2: $0 < \underline{\lambda}_{\sigma^2} \leq \overline{\lambda}_{\sigma^2} < \infty$. The support of the prior distribution for the system parameter vector θ is assumed in a given area Ω.

Assume the estimator (identifier) can only access the quantized measurements transmitted over a network. The network configuration to be considered is characterized in Fig. 1, where the raw measurement y_k is quantized as s_k and then transmitted to a remote estimator for identification of the parameter vector through a noiseless digital channel. At time k, given a partition $P_k^N = \{R_k^i; \ i = 1, \cdots, N\}$ of the measurement space $\mathcal{R}^m$, y_k is quantized into s_k by the quantizer q_k which is embedded in the sensor node. The explicit form of the quantizer is given by:

$$s_k := q_k(y_k) = E[y_k | R_k^i] = \frac{\int_{R_k^i} y_k p(y_k) dy_k}{\int_{R_k^i} p(y_k) dy_k} := l_k^i \text{ if } y_k \in R_k^i. \tag{2}$$

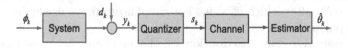

Fig. 1. Network configuration

Following the definition of the quantizer $q_k(\cdot)$ in turn reads

$$E[y_k|s_k = l_k^i] = \int_{\mathcal{R}} y_k p(y_k|s_k = l_k^i) dy_k = \frac{\int_{R_k^i} y_k p(y_k) dy_k}{\int_{R_k^i} p(y_k) dy_k} = l_k^i,$$

which implies that

$$s_k = E[y_k|s_k] \text{ and } E[s_k] = E[y_k]. \tag{3}$$

Actually, s_k is chosen to minimize the quadratic distortion within each cell R_k^i, i.e.,

$$\frac{\partial \int_{R_k^i} (y_k - s_k)^2 dy_k}{\partial s_k} = -2 \int_{R_k^i} (y_k - s_k) dy_k = 0 \Rightarrow s_k = l_k^i.$$

Remark 1. It should be noted that without the exact distribution information of y_k, the quantizer doesn't work by Eq. (2). However, once we obtain the distribution of y_k, the quantizer is implementable.

Our purpose is to estimate the system parameter vector θ under the MMSE criterion with persistently exciting inputs ϕ_k and the information of the quantized observations $s_{1:k} = \{s_1, \cdots, s_k\}$ available up to time k.

3   Identification Algorithms

In this section, we will first derive the recursive quantized MMSE estimation and then proceed to design the optimal quantizer for application. Some stability results on the solution of a modified Riccati difference equation will be reviewed as well.

Define the optimal current parameter estimate $\hat{\theta}_k$ in the sense of MMSE and

$$P_k := E[(\hat{\theta}_k - \theta)(\hat{\theta}_k - \theta)^T]$$

with the initial values $\hat{\theta}_0 \in \Omega$ and finite positive definite matrix P_0.

Consider that the prior information of the parameter vector is $\theta \in \Omega$, the estimate is modified to be

$$\hat{\theta}_k^r := \hat{\theta}_k|\Omega = \hat{\theta}_k I_{\{\hat{\theta}_k \in \Omega\}} + \hat{\theta}_0 I_{\{\hat{\theta}_k \notin \Omega\}}$$

and the variance $P_k^r := P_k I_{\{\hat{\theta}_k \in \Omega\}} + P_0 I_{\{\hat{\theta}_k \notin \Omega\}}$, correspondingly.

Actually, the prior information of the parameter vector ensures that the estimate is bounded. However, for an asymptotically convergent algorithm, there exists a time k^r, after which $\hat{\theta}_k^r = \hat{\theta}_k$. For simplification, we use $\hat{\theta}_k$ in the discussion of the paper.

Assumption A1: The distribution of the estimate $\hat{\theta}_k$ is assumed to be Gaussian with mean θ and variance P_k.

The recursive quantized identification algorithm is given in the following theorem.

Theorem 1. *Consider the stochastic system described by Eq. (1) with the quantized measurement (3). Under Assumption A1, the quantized MMSE estimation algorithm can be recursively computed as follows:*

$$\hat{\theta}_k = \hat{\theta}_{k-1} + K_k(s_k - \phi_k^T \hat{\theta}_{k-1}) \tag{4}$$

$$P_k = P_{k-1}^* + K_k E(y_k - s_k)^2 K_k^T \tag{5}$$

where K_k is defined by $K_k = P_{k-1}\phi_k S_k^{-1}$ and $P_k^* = P_{k-1} - P_{k-1}\phi_k S_k^{-1}\phi_k^T P_{k-1}$, where $S_k = \phi_k^T P_{k-1}\phi_k + \sigma_k^2$.

Proof. Using the quantized observation $s_{1:k}$ available at time k, our task is to find the estimate of θ, denote by $\hat{\theta}_k = \hat{\theta}(s_{1:k})$, to minimize $E[(\theta - \hat{\theta}_k)(\theta - \hat{\theta}_k)^T]$. To this end, we introduce an intermediate estimate $\hat{\theta}_k^* = \hat{\theta}^*(s_{1:k-1}, y_k)$ to minimize $E[(\theta - \hat{\theta}_k^*)(\theta - \hat{\theta}_k^*)^T]$. Together with assumption A1, the joint probability density of $\hat{\theta}_{k-1}$ and y_k is given by

$$p(\hat{\theta}_{k-1}, y_k) = p(\hat{\theta}_{k-1})p(y_k) \tag{6}$$

$$= Ce^{-\frac{1}{2}[\theta^T(P_{k-1}^{-1} + \phi_k \phi_k^T/\sigma_k^2)\theta - 2\theta^T(P_{k-1}^{-1}\hat{\theta}_{k-1} + \phi_k y_k)]}, \tag{7}$$

where C is independent of θ. Thus, the joint distribution of $\hat{\theta}_{k-1}$ and y_k is Gaussian, which implies that $\hat{\theta}_k^*$ is also the maximum likelihood estimate of θ. Then, completing the square of (7) yields that

$$\hat{\theta}_{MLE} = \hat{\theta}_k^* = (P_{k-1}^{-1} + \phi_k \phi_k^T/\sigma_k^2)^{-1}(P_{k-1}^{-1}\hat{\theta}_{k-1} + \phi_k y_k).$$

Using the matrix inversion lemma [17] on $(P_{k-1}^{-1} + \phi_k \phi_k^T/\sigma_k^2)^{-1}$, we obtain

$$\hat{\theta}_k^* = \hat{\theta}_{k-1} + K_k(y_k - \phi_k^T \hat{\theta}_{k-1}), \tag{8}$$

$$P_k^* = E[(\theta - \hat{\theta}_k^*)(\theta - \hat{\theta}_k^*)^T]. \tag{9}$$

It can readily verified that

$$P_k = E[(\hat{\theta}_k^* - \theta)(\hat{\theta}_k - \hat{\theta}_k^*)^T] + E[(\hat{\theta}_k - \hat{\theta}_k^*)(\hat{\theta}_k^* - \theta)^T]$$
$$+ P_k^* + E[(\hat{\theta}_k - \hat{\theta}_k^*)(\hat{\theta}_k - \hat{\theta}_k^*)^T]. \tag{10}$$

However, based on the property of the conditional expectation,

$$E[(\hat{\theta}_k^* - \theta)(\hat{\theta}_k - \hat{\theta}_k^*)^T] = E[E[(\hat{\theta}_k^* - \theta)(\hat{\theta}_k - \hat{\theta}_k^*)^T | s_{1:k-1}, y_k]]$$
$$= E[E[\hat{\theta}_k^* - \theta | s_{1:k-1}, y_k](\hat{\theta}_k - \hat{\theta}_k^*)^T] = 0 \text{ by (8).}$$

Similarly, $E[(\hat{\theta}_k - \hat{\theta}_k^*)(\hat{\theta}_k^* - \theta)^T] = 0$. Thus, to minimize P_k, it is equivalent to minimize $E[(\hat{\theta}_k - \hat{\theta}_k^*)(\hat{\theta}_k - \hat{\theta}_k^*)^T]$ by (10). Consequently,

$$\hat{\theta}_k = E[\hat{\theta}_k^* | s_{1:k}] = \hat{\theta}_{k-1} + K_k(E[y_k | s_{1:k}] - \phi_k^T \hat{\theta}_{k-1})$$
$$= \hat{\theta}_{k-1} + K_k(s_k - \phi_k^T \hat{\theta}_{k-1})$$

and $P_k = P_{k-1}^* + K_k E(y_k - s_k)^2 K_k^T$.

Remark 2. The resulting estimation error variance can be decomposed into two parts: one is P_k^* which cannot be reduced as it is the exact estimation error variance derived from the RLSE; the other nonnegative part is $K_k E[(y_k - s_k)^2 | s_{1:k}] K_k^T$ induced by quantization.

From Eq. (5), it is clear that a higher resolution quantizer will result in a smaller estimation error variance. In the rest of this subsection, we will concentrate on the design of optimal quantizer of y_k to minimize P_k and refer to it as the *optimal* quantized MMSE estimation.

Note that the performance of the estimation can be measured by the quantity of $\mathrm{tr}(P_k)$. Next, the focus will be on investigating the optimal quantizer to minimize $\mathrm{tr}(P_k)$, which is equivalent to minimizing the term

$$\mathrm{tr}(E(y_k - s_k)^2) = E[\mathrm{tr}(E[(y_k - s_k)^2 | s_{1:k-1}])] \tag{11}$$

by the variance update formula in Eq. (5). Using the monotone property of conditional expectation, the minimization is converted to the following optimal quantizer design problem:

$$q_k^* = \mathrm{argmin}_{s_k = q_k(y_k)} \mathrm{tr}(E[(y_k - s_k)^2 | s_{1:k-1}])$$
$$= \mathrm{argmin}_{s_k = q_k(y_k)} E[\|y_k - s_k\|_2^2 | s_{1:k-1}] \tag{12}$$

In the sequel, we shall elaborate how to get the optimal quantizer q_k^*.

Under Assumption A1, we have $y_k - \phi_k^T \hat{\theta}_{k-1}$ belongs to the Gaussian family with the mean 0 and variance $S_k = \sigma_k^2 + \phi_k^T P_{k-1} \phi_k$. Notice that

$$E[\|y_k - s_k\|_2^2 | s_{1:k-1}] = S_k \times E[\|\frac{y_k - \phi_k^T \hat{\theta}_{k-1}}{\sqrt{S_k}} - \frac{s_k - \phi_k^T \hat{\theta}_{k-1}}{\sqrt{S_k}}\|_2^2 | s_{1:k-1}],$$

consequently, we have proved the following facts.

Proposition 1. *Under the Gaussian assumption A1, the optimal quantizer in Eq. (12) is the same as the optimal quantizer to minimize the average quadratic distortion of a Gaussian random variable with mean $\phi_k^T \hat{\theta}_{k-1}$ and variance S_k. Thus, denote $q_G^*(\cdot)$ as the optimal scalar quantizer to minimize the quadratic distortion for a standard Gaussian random variable with quantization level N, the optimal quantizer $q_k^*(\cdot)$ in Eq. (12) is given by*

$$q_k^*(y_k) = \sqrt{S_k} \times q_G^*(\frac{y_k - \phi_k^T \hat{\theta}_{k-1}}{\sqrt{S_k}}) + \phi_k^T \hat{\theta}_{k-1}.$$

Denote the associated minimum quadratic distortion as $\underline{D}_N^G$, which is given by

$$\underline{D}_N^G = E[(x - q_G^*(x))^2], \tag{13}$$

where x is assumed to be a standard Gaussian random variable. For example, when $N = 2$, the optimal scalar quantizer for the standard Gaussian random variable x would be quantized as follows [12]:

$$q_G^*(x) = \begin{cases} \sqrt{\frac{2}{\pi}}, & \text{if } x \geq 0; \\ -\sqrt{\frac{2}{\pi}}, & \text{otherwise,} \end{cases} \tag{14}$$

and the corresponding minimum quadratic distortion is computed as $\underline{D}_2^G = 1 - 2/\pi$.

However, to our best knowledge, there are no known close-form solutions to the problem of optimal quantization for an arbitrary random variable. Most of solutions are iterative algorithms, i.e., Lloyd's method I [14] or stochastic gradient method for the Gaussian signal [15]. Fortunately, the optimal quantizer of a standard Gaussian random variable is readily available, e.g., the optimal quantizer parameters and the associated minimum quadratic distortion are explicitly computed and tabulated in [12], which makes it possible to efficiently use the optimal quantizer for the measurement at each time step. We shall elaborate in Theorem 2 below that the optimal quantized estimation has a simple recursive structure and the estimation error variance is given by a simple modified Riccati difference equation (MRDE), which is close to the RLSE without quantization in terms of complexity.

Theorem 2. *Consider system (1) with the quantized measurement (3). Under Assumption A1, the optimal quantized MMSE identification algorithm is given by the following update formulae:*

$$\hat{\theta}_k = \hat{\theta}_{k-1} + K_k' q_G^* \left(\frac{y_k - \phi_k^T \hat{\theta}_{k-1}}{\sqrt{S_k}} \right) \tag{15}$$

$$P_k = P_{k-1} - (1 - \underline{D}_N^G) S_k^{-1} P_{k-1} \phi_k \phi_k^T P_{k-1} \tag{16}$$

where $S_k = \phi_k^T P_{k-1} \phi_k + \sigma_k^2$ and $K_k' = P_{k-1} \phi_k S_k^{-1/2}$.

Proof. Based on the Proposition 1, the optimal quantizer of y_k is solved as

$$s_k^* = \sqrt{S_k} \times q_k^* \left(\frac{y_k - \phi_k^T \hat{\theta}_{k-1}}{\sqrt{S_k}} \right) + \phi_k^T \hat{\theta}_{k-1} = \sqrt{S_k} \times q_G^* \left(\frac{y_k - \phi_k^T \hat{\theta}_{k-1}}{\sqrt{S_k}} \right) + \phi_k^T \hat{\theta}_{k-1}$$

Substituting the above into Eq. (4) follows Eq. (15). Moreover,

$$E[(y_k - s_k^*)^2] = E[E[(y_k - s_k^*)^2 | s_{1:k-1}]] = \underline{D}_N^G S_k.$$

In view of Eqs. (5), (16) can be readily obtained.

The final result of this subsection investigates the asymptotic properties of the estimation.

Corollary 1. *As the quantization level N goes to infinity, the quantized estimation algorithms in Theorems 1–2 weakly converge to the classical estimation with measurements of y_k.*

Proof. It follows from [7] that $\underline{D}_N^G \to 0$ and $q^*(y) \to y$ as $N \to \infty$. The rest is very straightforward from Theorems 1–2.

Remark 3. From Theorems 1–2, we can see that the identification algorithm works for any number of given quantization levels, even for the binary case. The effect of quantization is characterized by D_N^G which monotonically decreases to 0 as $N \to \infty$. Note that $D_2^G = 1 - \frac{2}{\pi}$.

4   Illustrative Examples

Consider a gain system: $y_k = \phi_k^T \theta + d_k$, where the output is quantized with two levels. The optimal quantizer is defined in (14). Suppose the true parameter is $\theta = 2$ and the prior information of the parameter is its belonging to $[1, 7]$, $\{d_k\}$ is a sequence of i.i.d. Gaussian variables with mean zero and variance σ_d^2. The input ϕ_k is a sequence bounded in $[1, 5]$, which satisfies the persistent exciting condition given in Assumption A1.

The convergence of the estimate $\hat{\theta}_k$ within 1000 steps can be illustrated in Fig. 2. We can see that at the beginning the estimation error could be large. Note that at the begining, $\hat{\theta}_k$ may not be Gaussian. However, $\hat{\theta}_k$ is asymptotically Gaussian, which leads to an asymptotically optimal identification.

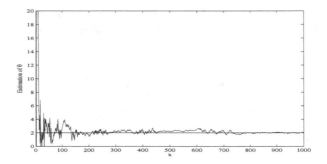

Fig. 2. Convergence of the estimation $\hat{\theta}_k$

The curve of kP_k is plotted in Fig. 3 with 1000 steps. We can see clearly that kP_k has its upper and lower bounds, which shows that P_k converges to 0 with order 1. However, under persistently exciting conditions, the limit kP_k may not exist.

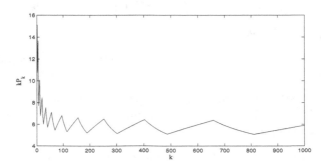

Fig. 3. Curve of kP_k

5   Conclusions

This paper studies system identification under quantized observations based on the minimum mean square error (MMSE) criterion. Under the assumption that the estimate of parameter is Gaussian, the estimation algorithm is constructed and the optimal quantization is designed.

References

1. Caines, P.: Linear Stochastic Systems. Wiley, Hoboken (1988)
2. Chen, H.F., Guo, L.: Identification and Stochastic Adaptive Control. Birkhäuser, Boston (1991)
3. Chen, H.F., Yin, G.: Asymptotic properties of sign algorithms for adaptive filtering. IEEE Trans. Autom. Control 48(9), 1545–1556 (2003)
4. Elvitch, C.R., Sethares, W.A., Rey, G.J., Johnson, C.R.: Quiver diagrams and signed adaptive fiters. IEEE Trans. Acoust. Speech Sig. Process. 37(2), 227–236 (1989)
5. Eweda, E.: Convergence analysis of an adaptive filter equipped with the sign-sign algorithm. IEEE Trans. Autom. Control 40(10), 1807–1811 (1995)
6. Gersho, A., Gray, R.: Vector Quantization and Signal Compression. Kluwer Academic Publishers, Norwell (1991)
7. Graf, S., Luschgy, H.: Foundations of Quantization for Probability Distributions. Springer, Heidelberg (2000)
8. Hakvoort, R.G., Van den Hof, P.M.J.: Consistent parameter bounding identification for linearly parameterized model sets. Automatica 31(7), 957–969 (1995)
9. Kumar, P.R.: Convergence of adaptive control schemes using least-squares parameter estimates. IEEE Trans. Autom. Control 35(4), 416–424 (1990)
10. Kushner, H.J.K., Yin, G.: Stochastic Approximation and Recursive Algorithms and Applications, 2nd edn. Springer, New York (2003)
11. Ljung, L.: System Identification: Theory for the User, I edn. Prentice-Hall, Englewood Cliffs (1987). II edn. (1999)
12. Max, J.: Quantizing for minimum distortion. IEEE Trans. Inf. Theory 6(1), 7–12 (1960)
13. Milanese, M., Vicino, A.: Optimal estimation theory for dynamic systems with set membership uncertainty: an overview. Automatica 27(6), 997–1009 (1991)
14. Kieffer, J.: Exponential rate of convergence for Lloyds method I. IEEE Trans. Inf. Theory 28(2), 205–210 (1982)
15. Pagés, G., Printems, J.: Optimal quadratic quantization for numerics: the Gaussian case. Monte Carlo Methods Appl. 9(2), 135–165 (2003)
16. Wigren, T.: Adaptive filtering using quantized output measurements. IEEE Trans. Sig. Process. 46(12), 3423–3426 (1998)
17. Horn, R., Johnson, C.: Matrix Analysis. Cambridge University Press, Cambridge (1985)

LSTM with Matrix Factorization for Road Speed Prediction

Jian Hu[1(✉)], Xin Xin[1], and Ping Guo[1,2]

[1] School of Computer Science and Technology,
Beijing Institute of Technology, Beijing 100081, China
{jhu,xxin}@bit.edu.cn, pguo@ieee.org
[2] Laboratory of Graphics and Pattern Recognition,
Beijing Normal University, Beijing 100875, China

Abstract. Road speed prediction is a key point of Intelligent Transport System. Plenty of work have proved the effectiveness and efficiency of neural network in forecasting freeway velocity. However, the missing values are obstacles when applying the widely used trajectory data to neural network. In trajectory data, most roads may not be covered by enough trajectories in a short time. Due to highly sparsity, it will bring extra cost if we first fill missing data then perform training. To solve this issue, we propose a collaborative model that combines LSTM neural network with matrix factorization to reduce sparsity and make prediction simultaneously. We conduct experiments with a sufficient amount of trajectories and the results show that our model outperforms cascaded methods in both MAE and RMSE.

Keywords: Speed prediction · Sparse trajectories · Neural network · Matrix factorization

1 Introduction

With the rapid development of cities, traffic congestion becomes more and more serious. There is a contradiction between accelerating pace of life and long commute. Hence it is necessary to plan travel route and keep away from busy roads.

Predicting road speed is a crucial part of intelligent transportation. Most traditional work, including time series analysis [1] and neural network [2], are based on the data from static sensors. It has been proved that the feedforward neural network is effective for forecasting road velocity [3] and recurrent neural network is even more accurate [2]. However, only the input data with no missing values are suitable for neural network. Hence these methods are designed for speed estimation of freeways where the loop detectors are embedded.

Trajectory data of floating cars can provide a better coverage of the city road network so we are able to deduce traffic condition in different districts. Unfortunately, data sparsity and data irrelevance are difficulties to forecast road speed utilizing the trajectory data.

F. Cong et al. (Eds.): ISNN 2017, Part I, LNCS 10261, pp. 242–249, 2017.
DOI: 10.1007/978-3-319-59072-1_29

To solve the above issues, in this paper we focus on the sparsity and model training at the same time rather than isolate filling missing values from learning pattern from data. We construct a recurrent neural network to dig out the latent relation of road speeds between different time slots. Meanwhile we take advantage of matrix factorization which helps to learn the similarity of different roads and alleviate sparsity. Finally, we combine the two models and train them together in order to benefit each other from different aspects of learning. The experimental results demonstrate that the proposed method outperforms traditional work by 7% in MAE and RMSE.

2   Related Work

There are a lot of researches in the past few decades that elaborate on estimation of speed using complete data. Common methods contain autoregressive integrated moving average model [1,4], hidden Markov model [5,6], conditional random fields [7] and neural network [2,3,12]. These methods are designed for the loop detector data which are dense and complete and their performances decrease when utilizing trajectory data. Hence we come up with a collaborative model to improve the accuracy.

Trajectory data provide a large coverage of road network which are suitable for traffic research [9]. Because of the low cost and easy access, trajectory data are widely used in Intelligent Transport System [10,11]. There are some work to do before utilizing these data such as route inference [13] and travel time allocation [14]. In this paper, we focus on forecasting road speed with sparse trajectories.

3   Problem Definition

The time slot is set to 10 min. From the trajectory data, we are able to calculate the rough time that a taxi spends on each road and we record the corresponding time slot of the trajectory. If a road is covered by more than five floating cars in same time slot, the average travel speed is considered to be reliable. As shown in Fig. 1, V_t is a column vector that indicates all the road speeds in time slot

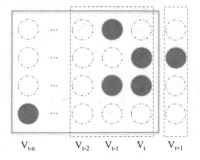

Fig. 1. Problem definition

t and each element in V_t corresponds to the velocity of a certain road. The reliable observations are represented by solid circles while the missing values are represented by dotted circles. The task is to accurately estimate V_{t+1} in the future according to $V_t, V_{t-1}, V_{t-2}, ..., V_{t-n}$.

4 Long Short Term Memory with Matrix Factorization

4.1 The Model

Parameter Definitions. Figure 2 gives a general impression of LSTMMF. The detail elements of the model are defined as follows.

Definition 1: Velocity matrix. The velocity matrix is represented by V and each element $V_{a,b}$ is the speed of ath road in bth time slot. There are M rows and N columns in V and $V \in \mathbb{R}^{M \times N}$. V_t is the column vector of V which is used to indicate all the travel speeds in time slot t. Zero-one matrix F denotes whether the corresponding element in V is a reliable value. 0 indicates missing and 1 indicates non-missing.

Definition 2: Context matrix. To perform the matrix factorization, we introduce road context matrix R and time slot context matrix T. According to low rank assumption, we set the rank of R and T to $P(P \ll M$ and $P \ll N)$. Hence matrix $R \in \mathbb{R}^{M \times P}$ and matrix $T \in \mathbb{R}^{P \times N}$. R_a is the row vector in R which denotes the contexts of ath road and T_b is the column vector in T which denotes the contexts of bth time slot. By filling the missing values in V with the product of R and T, we get a complete matrix called V'.

Definition 3: LSTM layer. The recurrent hidden layer in our model is traditional LSTM architecture [16,18–20]. The following equations give the complete algorithm of our hidden layer [15]. At time slot t, $\mathbf{x}^t$ is the input to LSTM layer and $\mathbf{h}^t$ is the output of hidden layer. Input gate, forget gate and output gate are denoted as $\mathbf{i}^t, \mathbf{f}^t$ and $\mathbf{o}^t$. Input node and cell state are represented by $\mathbf{g}^t$ and $\mathbf{c}^t$ [17].

$$\mathbf{g}^t = tanh(W^{gx}\mathbf{x}^t + W^{gh}\mathbf{h}^{t-1} + \mathbf{b}_g) \tag{1}$$

$$\mathbf{i}^t = sigm(W^{ix}\mathbf{x}^t + W^{ih}\mathbf{h}^{t-1} + \mathbf{b}_i) \tag{2}$$

$$\mathbf{f}^t = sigm(W^{fx}\mathbf{x}^t + W^{fh}\mathbf{h}^{t-1} + \mathbf{b}_f) \tag{3}$$

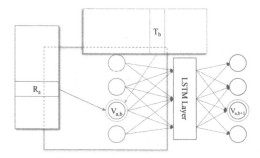

Fig. 2. Proposed model

$$\mathbf{o}^t = sigm(W^{ox}\mathbf{x}^t + W^{oh}\mathbf{h}^{t-1} + \mathbf{b}_o) \tag{4}$$

$$\mathbf{c}^t = \mathbf{g}^t \odot \mathbf{i}^t + \mathbf{c}^{t-1} \odot \mathbf{f}^t \tag{5}$$

$$\mathbf{h}^t = tanh(\mathbf{c}^t) \odot \mathbf{o}^t \tag{6}$$

Inference. Based on the definitions above, we expect the product of matrix R and matrix T is closed to velocity matrix V at all non-missing values [8] and at the same time, if we take k successive column vectors in V' $(V'_{i-k+1}, V'_{i-k+2}, ..., V'_{i-1}, V'_i)$ as an input sequence to neural network, the error between last output and V'_{i+1} will also be minimum.

4.2   Training Algorithm

The complete work flow is demonstrated as follows.

1. Perform standardizing on original data matrix V.
2. Initialize R with standard normal distribution, and initialize T by minimizing the error between V and product of R and T at non-missing values. Initialize V' by replacing missing data with product of R and T. Initialize all parameters inside the neural network.
3. Generate input vector sequences and target vectors from V'. Calculate output vectors and form them into matrix V'' so V'' represents the output from neural network. The square error between V' and V'' is recorded as e_1. The square error between V'' and product of R and T is recorded as e_2.
4. Update the neural network parameters based on backpropagation through time to reduce e_1.
5. Modify R and T based on gradient descent to reduce e_2. Generate new V'.
6. Iterate steps 3 to 5 until convergence.

There are two errors governing the process of training. They are defined by the following equations. λ_1 is a parameter that indicates the degree of influence of matrix factorization on neural network. On the contrary λ_2 represents the degree of influence of neural network on matrix factorization.

$$e_1 = \sum_{i=1}^{M} \sum_{j=1}^{N} \left[F_{i,j}(V'_{i,j} - V''_{i,j})^2 + \lambda_1(1 - F_{i,j})(V'_{i,j} - V''_{i,j})^2 \right] \tag{7}$$

$$e_2 = \sum_{i=1}^{M} \sum_{j=1}^{N} \left[F_{i,j}(V'_{i,j} - R_i \cdot T_j)^2 + \lambda_2(1 - F_{i,j})(V''_{i,j} - R_i \cdot T_j)^2 \right] \tag{8}$$

4.3   Complexity Analysis

For our network, the computational complexity is dominated by the feedforward and feedback operations. Because LSTM algorithm is very efficient, with an excellent update complexity of $O(W)$ where W is the number of weights [16], the computational complexity of our model is $O(Wk)$ where k denotes the number of time steps in recurrent layer. Hence the proposed model is capable to handle large scale data.

5   Experiments

5.1   Data Set

The trajectory data are generated by 10,176 taxis during July to November in 2014. We select 5,014 roads inside the 4th ring in Beijing with a total length of 2,359 km and the whole road network covers an area of 256 km^2 shown in Fig. 3. There are 8,830,789 complete travel trajectories with 401,232,582 GPS points in our data set and the total travel distance is 37,517,780 km. The average frequency of GPS sampling is 38 s/point. We use the data of July to September for training, October for evaluation and November for test. The metrics to evaluate results are mean absolute error(MAE) and root mean square error(RMSE). The definitions are as follows. H is the set of index i where y_i is non-missing.

$$\text{MAE} = \frac{\sum_{i\in H} |y_i - \hat{y}_i|}{Card(H)} \tag{9}$$

$$\text{RMSE} = \sqrt{\frac{\sum_{i\in H} (y_i - \hat{y}_i)^2}{Card(H)}} \tag{10}$$

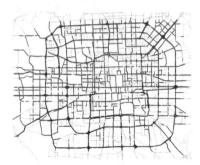

Fig. 3. The road network used in our experiments which covers the majority of roads inside the 4th ring in Beijing

5.2   Baseline Methods

We compare LSTMMF with the following baseline methods and TensorFlow [21,22] is utilized to accomplish our experiments.

1. *Historical Travel Speed (HTS).* The prediction of each road speed is determined by its own average speed in historical data.
2. *Most Recent Travel Speed (MRTS).* MRTS estimates the future travel speed based on the most recent observation of each road.
3. *Traditional LSTM (LSTM).* First we fill the missing values with the average speed of each road, then utilize Long Short Term Memory network to estimate road speed in the future.

4. *Matrix Factorization and LSTM (MF+LSTM).* MF+LSTM performs non-negative matrix factorization in the first place to reduce sparsity and then utilizes Long Short Term Memory network.
5. *RNN with Matrix Factorization (RNNMF).* This method is a simplified version of LSTMMF. It replaces Long Short Term Memory cell with traditional recurrent neural network architecture.

5.3   Overall Performances

Table 1 presents the overall performances of our model and other methods. The percentages are improvements from our best method. It is obvious that LSTMMF outperforms all the baselines in terms of MAE and RMSE. The first two methods are inaccurate and the largest MAE is 2.88. Comparing our model with LSTM and MF+LSTM in the middle, it is observed that the proposed method outperforms MF+LSTM by 7.4% in MAE and 7.1% in RMSE. The previous methods, which make the estimation in a cascaded way, suffer from the propagating errors. There is a slight difference between RNNMF and LSTMMF which is able to prove the effectiveness of Long Short Term Memory in memorizing long term dependency.

Figure 4 shows how MAE and RMSE change during a day. It is observed that the errors are increasing at rush hour. Figure 5 demonstrates the relation between neural network architecture and the results. The best result is achieved when using 20 hidden units.

Table 1. Overall performances of different methods

Methods	MAE (m/s)	RMSE
HTS	2.88 (34.3%)	3.91 (29.4%)
MRTS	2.50 (24.4%)	3.47 (20.5%)
LSTM	2.14 (11.7%)	3.12 (11.5%)
MF+LSTM	2.04 (7.4%)	2.97 (7.1%)
RNNMF	1.95 (3.1%)	2.85 (3.2%)
LSTMMF	1.89	2.76

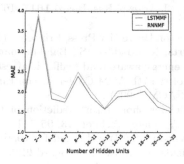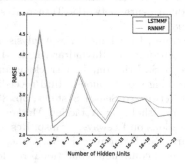

Fig. 4. Performances differ over time of day

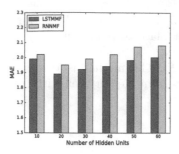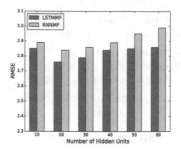

Fig. 5. Results on different architecture

6   Conclusion

In this paper, we propose a novel model applied for the previous problem of road speed prediction. To train neural network and matrix factorization together, LSTMMF can handle sparse trajectory data well. Experiments verify that the traditional cascaded ways of utilizing sparse data to neural network suffer from error propagation and our model is capable of alleviating the disadvantage.

Acknowledgement. The work described in this paper was mainly supported by the National Nature Science Foundation of China (Nos. 61672100, 61375045), the Ph.D Programs Foundation of Ministry of Education of China (No. 20131101120035), the Joint Research Fund in Astronomy under cooperative agreement between the National Natural Science Foundation of China and Chinese Academy of Sciences (No. U1531242), Beijing Natural Science Foundation (Nos. 4162054, 4162027), and the Excellent young scholars research fund of Beijing Institute of Technology.

References

1. Min, W.L., Wynter, L.: Real-time road traffic prediction with spatio-temporal correlations. Transp. Res. Part C: Emerg. Technol. **19**, 606–616 (2011)
2. Lint, J.W.C.V., Hoogendoorn, S.P., Zuylen, H.J.V.: Accurate freeway travel time prediction with state-space neural networks under missing data. Transp. Res. Part C: Emerg. Technol. **13**, 347–369 (2005)
3. Park, D., Rilett, L.R.: Forecasting multiple-period freeway link travel times using modular neural networks. Transp. Res. Rec.: J. Transp. Res. Board **1617**, 163–170 (1998)
4. Hamilton, J.D.: Time Series Analysis. Princeton University Press, Princeton (1994)
5. Thiagarajan, A., Ravindranath, L., LaCurts, K., Madden, S., Balakrishnan, H., Toledo, S., Eriksson, J.: VTrack: accurate, energy-aware road traffic delay estimation using mobile phones. In: Proceedings of the 7th ACM Conference on Embedded Networked Sensor Systems, pp. 85–98. ACM (2009)
6. Qi, Y., Ishak, S.: A hidden Markov model for short term prediction of traffic conditions on freeways. Transp. Res. Part C: Emerg. Technol. **43**, 95–111 (2014)
7. Djuric, N., Radosavljevic, V., Coric, V., Vucetic, S.: Travel speed forecasting by means of continuous conditional random fields. J. Transp. Res. Board **2263**, 131–139 (2011)

8. Funk., S.: Netflix update: Try this at home. http://sifter.org/~simon/journal/20061211.html
9. Pang, L.X., Chawla, S., Liu, W., Zheng, Y.: On detection of emerging anomalous traffic patterns using GPS data. Data Knowl. Eng. **87**, 357–373 (2013)
10. Wang, Y., Zheng, Y., Xue, Y.: Travel time estimation of a path using sparse trajectories. In: ACM SIGKDD International Conference on Knowledge Discovery and Data Mining, pp. 25–34. ACM (2014)
11. Yuan, J., Zheng, Y., Xie, X., Sun, G.: T-drive: enhancing driving directions with taxi drivers' intelligence. IEEE Trans. Knowl. Data Eng. **25**, 220–232 (2013)
12. Kumar, K., Parida, M., Katiyar, V.: Short term traffic flow prediction for a non urban highway using artificial neural network. Procedia-Soc. Behav. Sci. **104**, 755–764 (2013)
13. Deka, L., Quddus, M.: Trip-based weighted trajectory matching algorithm for sparse GPS data. In: Transportation Research Board Annual Meeting (2015)
14. Jenelius, E., Koutsopoulos, H.N.: Travel time estimation for urban road networks using low frequency probe vehicle data. Transp. Res. Part B Methodol. **53**, 64–81 (2013)
15. Lipton, Z.C., Berkowitz, J., Elkan, C.: A critical review of recurrent neural networks for sequence learning. Computer Science (2015)
16. Hochreiter, S., Schmidhuber, J.: Long short-term memory. Neural Comput. **9**, 1735–1780 (1997)
17. Zaremba, W., Sutskever, I., Vinyals, O.: Recurrent neural network regularization. eprint arxiv (2014)
18. Sutskever, I., Vinyals, O., Le, Q.V.: Sequence to sequence learning with neural networks. Adv. Neural Inf. Process. Syst. **4**, 3104–3112 (2014)
19. Werbos, P.J.: Backpropagation through time: what it does and how to do it. Proc. IEEE **78**, 1550–1560 (1990)
20. Graves, A.: Supervised Sequence Labelling with Recurrent Neural Networks. Springer, Heidelberg (2012)
21. Abadi, M., Agarwal, A., Barham, P., Brevdo, E., Chen, Z., Citro, C., et al.: Tensorflow: large-scale machine learning on heterogeneous distributed systems. arXiv preprint arXiv:1603.04467 (2016)
22. Abadi, M., Barham, P., Chen, J., Chen, Z., Davis, A., Dean, J., et al.: Tensorflow: a system for large-scale machine learning. In: Proceedings of the 12th USENIX Symposium on Operating Systems Design and Implementation (OSDI) (2016)

Cognition Computation and Neural Networks

Adaptive Control Strategy for Projective Synchronization of Neural Networks

Abdujelil Abdurahman$^{(\boxtimes)}$, Cheng Hu, Ahmadjan Muhammadhaji, and Haijun Jiang

College of Mathematics and System Sciences, Xinjiang University, Urumqi, Xinjiang Uyghur Autonomous Region, People's Republic of China
abjil1@163.com

Abstract. In this paper, we studied the projective synchronization of a type of chaotic neural networks (NNs) by introducing a novel adaptive control strategy. We obtained some useful sufficient criteria for the projective synchronization of considered networks via designing novel adaptive controller and introducing a suitable Lyapunov function. In addition, we gave a numerical example to validate the feasibility of the obtained results. It is worth to mention that the projective synchronization is a very general and it includes chaos stabilization, anti-synchronization and complete synchronization as its special cases.

Keywords: Neural network · Projective synchronization · Adaptive control

1 Introduction and Preliminaries

Since Pecora and Carroll [1] first realized the synchronization between the master-slave systems, the synchronization study of chaotic systems, including NNs, becomes a hot research topic over the past two decades due to their amazing applications in number of areas, ranging from pattern recognition to image processing [2–4]. Meanwhile, many useful synchronization approaches have been proposed for the chaotic systems such as complete synchronization [5], finite-time synchronization [3,6,7], lag synchronization [8], generalized synchronization [9], anti-synchronization [10], and projective synchronization [11,12], etc.

Projective synchronization can be understood that the master and slave systems realized the synchronization up to a scaling factor p_i, i.e., $y_i(t) \to p_i x_i(t)$, $t \to \infty$. As compared with complete synchronization, there are a lot of advantages in projective synchronization since the irregularity of the scaling constant can greatly improve the security of communication [12,13]. As a result, projective synchronization of chaotic systems received interest of many scholars in different fields [14–17]. In [16], the projective synchronization of class of NN with time varying delay was studied via Krasovskii-Lyapunov approach. [17] investigates the weak projective synchronization of NNs with mixed time-varying delays and parameter mismatch.

© Springer International Publishing AG 2017
F. Cong et al. (Eds.): ISNN 2017, Part I, LNCS 10261, pp. 253–260, 2017.
DOI: 10.1007/978-3-319-59072-1_30

It is worth the mention that all of the above mentioned works realized the projective synchronization by designing very complex controller that mainly consisted of linear term $k_i e_i(t)$ and activation functions $f_j(e_j(t))$. However, in some special cases, for example when the solutions of master system are bounded we can optimize the controller by removing the term relevant to $f_j(e_j(t))$. In this paper, we studied the projective synchronization for a class of chaotic NN via designing a novel adaptive control strategy. By using some inequality techniques and constructing a suitable Lyapunov function, we derived some useful sufficient criteria for the projective synchronization of considered networks. Finally, an example is given to illustrate the effectiveness of the obtained results.

Considered the projective synchronization of following NNs model

$$\dot{x}_i(t) = -c_i x_i(t) + \sum_{j=1}^{n} a_{ij} f_j(x_j(t)) + I_i, \tag{1}$$

where $i \in \mathscr{I} \triangleq \{1, 2, \cdots, n\}$, $c_i > 0$ denotes to the self connection of ith neuron, $f_j(x_j(t))$ is the activation function, a_{ij} are connection weights, and I_i corresponds to the external input of the ith neuron.

Throughout the paper, we assume that the following hypotheses are satisfied for the model (1):

$\mathbf{H_1}$: For any $i \in \mathscr{I}$, the any solution $x_i(t, x_i^0)$ of system (1) with initial value $x_i^0 \in R$ is bounded. That is there exists a positive constant M_i such that

$$|x_i(t, x_i^0)| \leq M_i \quad \text{for any } i \in \mathscr{I} \text{ and } t \geq 0.$$

$\mathbf{H_2}$: For any $j \in \mathscr{I}$, the activation functions $f_j(v)$ is continuous and differentiable with bounded derivatives. That is there exist constants N_j such that,

$$\dot{f}_j(v) \leq N_j, \ \forall \ v \in R.$$

Remark 1. When the activation functions satisfy the hypothesis $\mathbf{H_1}$, then from Lagrange mean value theorem, it is not difficult to check that $f_j(v)$ satisfies the globally Lipschitz condition. That is,

$$|f_j(v_1) - f_j(v_2)| \leq L_j |v_1 - v_2|, \ \forall \ v_1, v_2 \in R,$$

where $L_j = N_j$.

In the paper, we consider system (1) as the master system, its slave system is given as follows

$$\dot{y}_i(t) = -c_i y_i(t) + \sum_{j=1}^{n} a_{ij} f_j(y_j(t)) + I_i + u_i(t), \tag{2}$$

where $u_i(t)$ is a controller which will be designed.

We define the error term as

$$e_i(t) = y_i(t) - p_i x_i(t), \quad i \in \mathscr{I}, \tag{3}$$

where p_i is a scaling constant.

The main aim of this paper is to design a suitable control input $u_i(t)$ such that the master-slave networks (1) and (2) can achieve projective synchronization. To do this, first we give a following definition.

Definition 1. The master-slave networks (1) and (2) are said to achieve projective synchronization if there exist $\delta \geq 1$ and $\varepsilon > 0$ such that

$$\|e(t)\| = \|y(t) - Px(t)\| \leq \delta\|y^0 - Px^0\|e^{-\varepsilon t}, \quad \text{for all } t \geq 0,$$

where $P = diag(p_1, p_2, \cdots, p_n)$, $x(t)$ and $y(t)$ are the state solutions of master-slave systems (1) and (2) with initial values $x^0 = (x_1^0, x_2^0, \cdots, x_n^0)^T$ and $y^0 = (y_1^0, y_2^0, \cdots, y_n^0)^T$, respectively.

Remark 2. If the scaling constant $p_i = -1$ or $p_i = 1$ for all $i \in \mathscr{I}$, then the synchronization problem between the master-slave systems (1) and (2) will degenerate to the anti-synchronization or complete synchronization. If the scaling constant $p_i = 0$, then the synchronization problem will reduced to a chaos control problem.

2   Main Results

Now, we consider the projective synchronization between the master-slave systems (1) and (2). First, from the definition of $e_i(t) - y_i(t) - p_i x_i(t)$, the error system can be expressed as

$$\dot{e}_i(t) = -c_i e_i(t) + \sum_{j=1}^{n} a_{ij}\varpi_{ij}(e_j(t)) + (1 - p_i)I_i + u_i(t), \tag{4}$$

where $\varpi_{ij}(e_j(t)) = f_j(y_j(t)) - p_i f_j(x_j(t))$.

Now design the control input $u_i(t)$ in slave system (2) as the following form

$$\begin{cases} u_i(t) = -\beta_i(t)sign(e_i(t)) - \xi_i(t)e_i(t), \\ \dot{\beta}_i(t) = b_i|e_i(t)|, \\ \dot{\xi}_i(t) = d_i e_i^2(t), \end{cases} \tag{5}$$

where $i \in \mathscr{I}$, and b_i and d_i are arbitrary positive constants determined in later. Then we have a following result on the projective synchronization between master-slave systems (1) and (2).

Theorem 1. *Assume that the hypotheses* $\mathbf{H_1}$ *and* $\mathbf{H_2}$ *are true. If the slave system (2) is controlled with the adaptive controller (5), then the master-slave systems (1) and (2) are projective synchronized.*

Proof. First, noting the definition of $\varpi_{ij}(e_j(t))$, one has

$$
\begin{aligned}
\varpi_{ij}(e_j(t)) &= (f_j(y_j(t)) - p_i f_j(x_j(t))) \\
&= (f_j(y_j(t)) - f_j(p_j x_j(t)) + f_j(p_j x_j(t)) - p_i f_j(x_j(t))).
\end{aligned}
\tag{6}
$$

From the Lagrange mean value theorem, we get

$$
\begin{aligned}
f_j(p_j x_j(t)) - p_i f_j(x_j(t)) &= f_j(p_j x_j(t)) - f_j(0) - p_i(f_j(x_j(t)) - f_j(0)) + (1 - p_i)f_j(0)) \\
&= \dot{f}_j(\eta_j^1)p_j x_j(t) - p_i \dot{f}_j(\eta_j^2)x_j(t) + (1 - p_i)f_j(0)),
\end{aligned}
\tag{7}
$$

where $\eta_j^1 \in (\min\{0, p_j x_j(t)\}, \max\{0, p_j x_j(t)\})$ and $\eta_j^2 \in (\min\{0, x_j(t)\}, \max\{0, x_j(t)\})$

Using the assumptions $\mathbf{H_1}$ and $\mathbf{H_2}$, we get

$$
\begin{aligned}
\dot{f}_j(\eta_j^1)p_j x_j(t) &\le L_j M_j |p_j|, \\
p_i \dot{f}_j(\eta_j^2)x_j(t) &\le L_j M_j |p_i|.
\end{aligned}
\tag{8}
$$

In view of the inequalities (6), (7) and (8), $\varpi_{ij}(e_j(t))$ can be estimated as

$$
|\varpi_{ij}(e_j(t))| = |(f_j(y_j(t)) - p_i f_j(x_j(t)))| \le L_j |e_j(t)| + r_{ij}.
\tag{9}
$$

where $r_{ij} \triangleq L_j M_j(|p_j| + |p_i|) + |(1 - p_i)||f_j(0)|$.

Now construct the following Lyapunov function

$$
V(t) = \sum_{i=1}^{n} \left\{ \frac{1}{2}e_i^2(t) + \frac{1}{2b_i}(\beta_i(t) - \beta_i)^2 + \frac{1}{2d_i}(\xi_i(t) - \xi_i)^2 \right\},
$$

where β_i and ξ_i are positive constant determined in later. Calculating the derivative of $V(t)$, we get

$$
\begin{aligned}
\frac{dV(t)}{dt} &= \sum_{i=1}^{n} \left\{ e_i(t)\left[-c_i e_i(t) + \sum_{j=1}^{n} a_{ij}\varpi_{ij}(e_j(t)) + (1 - p_i)I_i + u_i(t) \right] \right. \\
&\quad \left. + (\xi_i(t) - \xi_i)e_i^2(t) + (\beta_i(t) - \beta_i)|e_i(t)| \right\} \\
&\le \sum_{i=1}^{n} \left\{ -c_i e_i(t)^2 + \sum_{j=1}^{n} |a_{ij}||e_i(t)|(L_j|e_j(t)| + r_{ij}) - \beta_i(t)|e_i(t)| - \xi_i(t)e_i^2(t) \right. \\
&\quad \left. + (\xi_i(t) - \xi_i)e_i^2(t) + [(\beta_i(t) - \beta_i) + (1 - p_i)I_i]|e_i(t)| \right\} \\
&\le \sum_{i=1}^{n} \left\{ -c_i e_i(t)^2 + \frac{1}{2}\sum_{j=1}^{n} |a_{ij}|L_j(e_i^2(t) + e_j^2(t)) + \sum_{j=1}^{n} r_{ij}|a_{ij}||e_i(t)| \right. \\
&\quad \left. - \xi_i e_i^2(t) - [\beta_i + (1 - p_i)I_i]|e_i(t)| \right\} \\
&= \sum_{i=1}^{n} \left[-c_i - \xi_i + \frac{1}{2}\sum_{j=1}^{n} |a_{ij}|L_j + \frac{1}{2}\sum_{j=1}^{n} |a_{ji}|L_i \right] e_i(t)^2 \\
&\quad \sum_{i=1}^{n} \left[-\beta_i + \sum_{j=1}^{n} r_{ij}|a_{ij}| + (1 - p_i)I_i \right]|e_i(t)|,
\end{aligned}
$$

where we used the inequality $2ab \leq a^2 + b^2$ for any $a, b \in R$.

Choosing β_i and ξ_i large enough such that

$$\xi_i \geq -c_i + \frac{1}{2}\sum_{j=1}^{n}|a_{ij}|L_j + \frac{1}{2}\sum_{j=1}^{n}|a_{ji}|L_i + \varepsilon_i,$$

$$\beta_i \geq \sum_{j=1}^{n}r_{ij}|a_{ij}| + (1-p_i)I_i,$$

where $\varepsilon_i > 0$ for $i \in \mathscr{I}$ is arbitrarily chosen constant.

Let $\varepsilon = \min_{i \in \mathscr{I}}\{\varepsilon_i\} > 0$, then we get

$$\frac{dV(t)}{dt} \leq -\sum_{i=1}^{n}\varepsilon_i e_i^2(t) \leq -\varepsilon e^T(t)e(t).$$

Therefore,

$$e^T(t)e(t) \leq 2V(t) = 2V(0) + 2\int_0^t \dot{V}(s)ds$$

$$\leq 2V(0) - 2\varepsilon\int_0^t e^T(s)e(s)ds.$$

From the Grownwall inequality, we get

$$e^T(t)e(t) \leq 2V(0)e^{-2\varepsilon t},$$

this implies that

$$\|e(t)\| \leq \Lambda^*\|y^0 - x^0\|e^{-\varepsilon t}, \tag{10}$$

where positive constant Λ^* satisfies the following inequality

$$2V(0) \leq \|y^0 - x^0\|^2 + \sum_{i=1}^{n}\left\{\frac{1}{b_i}(\beta_i(0) - \beta_i)^2 + \frac{1}{d_i}(\xi_i(0) - \xi_i)^2\right\}$$

$$\leq \Lambda^{*2}\|\|y^0 - x^0\|^2. \tag{11}$$

Thus, from Definition 1, the system (1) and system (2) are projective synchronized under the adaptive controller (5). The proof is completed.   □

Remark 3. In the adaptive controller (5), we have to take smaller adaptive gains b_i and d_i to reduce the control inputs, but this may causes a slower synchronization speed. Therefore, when the adaptive controller used to realize the projective synchronization, the adaptive gains b_i and d_i should be chosen in accordance with the synchronization speed to be quick and the controller input $u_i(t)$ not to be very large, considering the designer requirements.

3   Numerical Simulations

For $n = 3$, consider the following NNs model

$$\dot{x}_i(t) = -c_i x_i(t) + \sum_{j=1}^{3} a_{ij} f_j(x_j(t)) + I_i, \tag{12}$$

where $i = 1, 2, 3$, $f_1(u) = f_2(u) = \tanh(u)$, $c_1 = c_2 = 0.94$, $a_{11} = 1.650$, $a_{12} = a_{13} = a_{21} = -4.2240$, $a_{22} = 1.452$, $a_{23} = -5.808$, $a_{31} = -4.2240$, $a_{32} = 5.808$, $a_{33} = 1.320$ and $I_1 = I_2 = 0$.

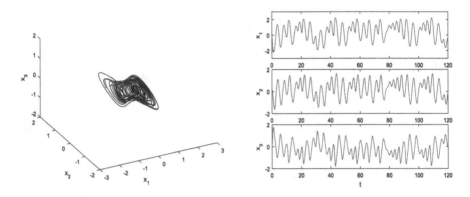

Fig. 1. The transient behavior of system (12).

The Matlab simulation of (12) under the initial conditions $x_1^0 = -0.1$, $x_2^0 = 0.2$ and $x_3^0 = 0.1$ is given in Fig. 1, from Fig. 1 we can see that the master system (12) has a chaotic attractor.

In the below, we will consider the projective synchronization of master system (12) and its slave system described by

$$\dot{y}_i(t) = -c_i y_i(t) + \sum_{j=1}^{3} a_{ij} f_j(y_j(t)) + I_i + u_i(t), \tag{13}$$

where c_i, a_{ij}, f_j and I_j are the same as defined in system (12) and $u_i(t)$ is given by (5).

It is not difficult to check that the hypothesis $\mathbf{H_2}$ is satisfied $N_1 = N_2 = 1$. Also from the simulation of system (12) in Fig. 1, we can see that the solutions of systems (12) are bounded. Thus the hypothesis $\mathbf{H_1}$ is also satisfied. Therefore, feom Theorem 1, the systems (12) and (13) can realize projective synchronization. Taking $y^k(0) = [-0.12 + 0.35k, 0.212 + 0.45k, 0.13 + 0.35k]$, $\beta^k(0) = [0.12 + 0.02k, 0.13 + 0.035k, 0.17 + 0.025k]$, $\xi^k(0) = [0.22 + 0.068k, 0.23 + 0.028k, 0.21 + 0.048k]$ for $k \in \{-3, -2, \cdots 3\}$, $b_i = 0.02$ and $d_i = 0.08$ for $i \in \{1, 2, 3\}$, then the time evolution of the synchronization errors (left) and curves (right) between

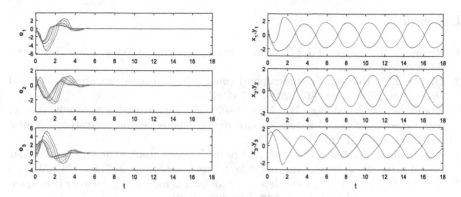

Fig. 2. The evaluation of synchronization errors and curves for $p_i = -1$.

master-slave systems (12) and (13) with scaling constant $p_i = -1$ are shown in Fig. 2. The adaptive feedback gains $\beta^k(t)$ (left) and $\xi^k(t)$ (right) are given in Fig. 3, respectively.

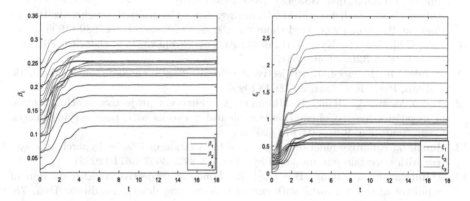

Fig. 3. Time evolution of the adaptive feedback gains β_i and ξ_i for $p_i = -1$.

4   Conclusion

In the paper, we studied the projective synchronization of class of NNs. By using the analysis technique and designing a novel adaptive controller, some simple but useful synchronization criteria have been derived. In addition, an example is provided to validate the feasibility of the introduced synchronization scheme.

Acknowledgments. This work was founded by the National Natural Science Foundation of P.R. China (Grant Nos. 11601464 and 61164004).

References

1. Pecora, L.M., Carroll, T.L.: Synchronization in chaotic systems. Phys. Rev. Lett. **64**, 821–824 (1990)
2. Ojalvo, J.G., Roy, R.: Spatiotemporal communication with synchronized optical chaos. Phys. Rev. Lett. **86**, 5204–5207 (2001)
3. Abdurahman, A., Jiang, H.: New results on exponential synchronization of memristor-based neural networks with discontinuous neuron activations. Neural Netw. **84**, 161–171 (2016)
4. Abdurahman, A., Jiang, H., Teng, Z.: Finite-time synchronization for fuzzy cellular neural networks with time-varying delays. Fuzzy Sets Syst. **297**, 96–111 (2016)
5. Yao, C., Zhao, Q., Yu, J.: Complete synchronization induced by disorder in coupled chaotic lattices. Phys. Lett. A **377**, 370–377 (2013)
6. Abdurahman, A., Jiang, H., Teng, Z.: Finite-time synchronization for memristor-based neural networks with time-varying delays. Neural Netw. **69**, 20–28 (2015)
7. Abdurahman, A., Jiang, H., Hu, C., Teng, Z.: Parameter identification based on finite-time synchronization for Cohen-Grossberg neural networks with time-varying delays. Nonlinear Anal. Modell. Control **20**(3), 348–366 (2015)
8. Abdurahman, A., Jiang, H., Teng, Z.: Exponential lag synchronization for memristor-based neural networks with mixed time delays via hybrid switching control. J. Frankl. Inst. **353**(13), 2859–2880 (2016)
9. Rulkov, N.F., Sushchik, M.M., Tsimring, L.S.: Generalized synchronization of chaos in directionally coupled chaotic systems. Phys. Rev. E **51**, 980 (1995)
10. Kim, C.M., Rim, S., Kye, W.H., et al.: Anti-synchronization of chaotic oscillators. Phys. Lett. A **320**, 39–46 (2003)
11. Mainieri, R., Rehacek, J.: Projective synchronization in three-dimensional chaotic systems. Phys. Rev. Lett. **82**, 304 (1999)
12. Abdurahman, A., Jiang, H., Rahman, K.: Function projective synchronization of memristor-based Cohen-Grossberg neural networks with time-varying delays. Cogn. Neurodyn. **9**(6), 603–613 (2015)
13. Runzi, L.: Adaptive function project synchronization of Rössler hyperchaotic system with uncertain parameters. Phys. Lett. A **372**, 3667–3671 (2008)
14. Abdurahman, A., Jiang, H., Teng, Z.: Function projective synchronization of impulsive neural networks with mixed time-varying delays. Nonlinear Dyn. **78**, 2627–2638 (2014)
15. Bao, H., Cao, J.: Projective synchronization of fractional-order memristor-based neural networks. Neural Netw. **63**, 1–9 (2015)
16. Ghosh, D., Banerjee, S.: Projective synchronization of time-varying delayed neural network with adaptive scaling factors. Chaos Solitons Fractals **53**, 1–9 (2013)
17. Huang, J., Li, C., Zhang, W., Wei, P.: Weak projective lag synchronization of neural networks with parameter mismatch. Neural Comput. Appl. **24**, 155–160 (2014)

Real-Time Decoding of Arm Kinematics During Grasping Based on F5 Neural Spike Data

Narges Ashena[1(✉)], Vassilis Papadourakis[2], Vassilis Raos[2], and Erhan Oztop[1]

[1] Ozyegin University, Istanbul, Turkey
narges.ashena@ozu.edu.tr
[2] University of Crete, Iraklion, Greece

Abstract. Several studies have shown that the information related to grip type, object identity and kinematics of monkey grasping actions is available in macaque cortical areas of F5, MI, and AIP. In particular, these studies show that the neural discharge patterns of the neuron populations from the aforementioned areas can be used for accurate decoding of action parameters. In this study, we focus on single neuron decoding capacity of neurons in a given region, F5, considering their functional classification, i.e. as to whether they show the mirror property or not. To this end, we recorded neural spike data and arm kinematics from a monkey that performed grasping actions. The spikes were then used as a regressor to predict the kinematic parameters. Results show that single neuron real-time decoding of the kinematics is not perfect, but reasonable performance can be achieved with selected neurons from both populations. Considering the neurons that we have studied (N:32), non-mirror neurons seem to act as better single-neuron decoders. Although it is clear that population-level activity is needed for robust decoding, single-neuron decoding capacity may be used as a quantitative means to classify neurons in a given region.

Keywords: Neural decoding · Grasping · Arm kinematics · Ventral premotor cortex (F5) · Image processing

1 Introduction

In this study, we aim to assess the real-time kinematic decoding capacity of ventral premotor cortex (area F5) neurons. Area F5, representing hand action [1], includes neurons called mirror neurons [2–4]. Mirror neurons show significant activity when the monkey performs an action as well as when it observes a similar action being done by another monkey or experimenter. In the literature, there have been efforts to apply the decoding paradigm for generating data to help neural prostheses development [5,6]. Grip type classification has been the target for several decoding studies based on ventral premotor cortex [6–8,11], dorsal premotor cortex [9], as well as parietal cortex [7] neural activity. In [7], Bayesian classifier approach is used to classify power and precision grips with

© Springer International Publishing AG 2017
F. Cong et al. (Eds.): ISNN 2017, Part I, LNCS 10261, pp. 261–268, 2017.
DOI: 10.1007/978-3-319-59072-1_31

5 differently oriented targets. A Support Vector Machines classifier is used as another possible approach for decoding grip types in [6,9]. In [11], it has been shown that grip type prediction based on single neuron activity can be made for some F5 neurons. A recent study shows that, in addition to grip type decoding, the details of hand shaping and reaching phase can be accurately decoded by using population level neural activity [10]. However, to our knowledge the specific decoding power of mirror vs. non-mirror neurons in F5 has not been addressed in the literature. To this end, with this study we aim to investigate the single neuron kinematics decoding power of F5 mirror and non-mirror neurons comparatively. For extracting monkey hand kinematics, grasping actions of the monkey are captured by a video camera together with the neural spike data. Image processing techniques are used to extract kinematics data related to arm movement. These kinematics data are the angle of lower arm with respect to the vertical axis and the approximate distance between monkey wrist and the object center. The results of this study show that single neuron decoding of the kinematics data is possible and non-mirror neurons seem to be better decoders than mirror neurons.

2   Materials and Methods

2.1   Data Definition and Experimental Setup

The monkey was seated in front of the behavioral apparatus which was a rotating turntable on which 3D geometrical solids were accommodated [12]. The objects were presented one at a time, always in the same central position. The following objects and grips were used: large sphere, whole hand prehension with all the fingers wrapped around the object and the palm in contact with the object; cylinder, finger prehension, using all fingers but the thumb; ring, hook grip with the index finger inserted into a ring; cube in vertical groove, advanced precision grip using the pulpar surface of the distal phalanx of the index finger opposed to the pulpar surface of the distal phalanx of the thumb. At the beginning of each trial, a LED above the selected object turned on and the monkey had to fixate it and press a key. Following a fixation period, a dimming of the LED signaled the onset of the reach-to-grasp movement. The monkey had to reach for, grasp, pull and hold the object while fixating it until the extinction of the LED cuing its release. The period from the start of the monkey's movement to the time of object pulling is defined as the movement epoch. Monkey's movement was recorded at 120 frames per second by a camera that viewed the experimental setup from a constant distance. A second LED, hidden from the monkey's view, was used to align the video frames with neural activity. This LED changed its state when the monkey's hand left the home position to reach for the object and when the object was grasped. For alignment purposes, it was crucial to robustly detect the state of this LED. We recorded the activity of neurons from area F5 of the ventral premotor cortex from the monkey's left hemisphere (contralateral to the moving arm). Neurons that responded during the execution of grasping movements were also tested for the mirror property. The monkey observed while the experimenter grasped the same objects using the same grips. Neurons responding to the observation task were further classified as mirror neurons whereas the rest were classified as non-mirror neurons.

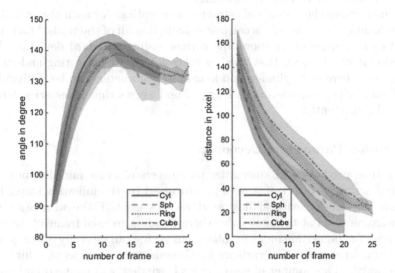

Fig. 1. Average of extracted kinematics of all the trials and the neurons ± standard deviation for each object

2.2 Kinematics Data Extraction

The first step for extracting kinematics data from videos is to find the frames corresponding to the movement epoch. As mentioned earlier, there is an LED hidden from the monkey's view and it is used for aligning the video frames with the neural activity. Measuring the average intensity of pixels in the rectangular area which contains this LED followed by a mixed semi-autonomous thresholding is used to find the time that LED changes its state. Having frames simultaneous with the movement epoch extracted, motion parameters can be obtained from each frame. Because the camera location is fixed in the experiment setup, the physical measures of the setup are as analogous to pixel scale in the frames. Moreover, fixed location of the camera causes the initial position of the monkey wrist and the center of the object to be constant in all the frames of all the trials. Knowing the initial location of the monkey's wrist and the location of the target, the sub-area of the frames where the action happened is obtained. Using foreground detection, moving pixels from the sub-area are extracted and regions of connected foreground pixels (components) are defined. In order to reduce noise and foreground pixels related to sudden jerks in the setup while the monkey is performing the task, a time profile is assigned to each extracted component. Components with short time profile are considered as noise and jerks and eliminated from frames. Finally, only arm pixels remain in the frame which can be processed further by image processing techniques in order to obtain the motion parameters which are the angle of lower arm with respect to the vertical axis parallel to monkey body and the approximate distance between monkey wrist and the object center. For the sake of simplicity, the former is named as angle and the latter is named as distance.

Figure 1 shows the average of extracted kinematics. For each object and neuron combination, the average is calculated including all of the trials. Shaded area around each average curve shows the corresponding standard deviation. From the figure, it can be seen that monkey takes longer to grasp ring and cube as compare to sphere and cylinder, and kinematic parameters can be distinguished for different grips. Considering each grip, Fig. 1 shows that kinematics data are extracted consistently.

2.3   Motion Parameters Decoding

For real-time prediction of kinematics parameters of angle and distance based on neural activity, spike trains were preprocessed by the following steps. First, Gaussian convolution with sliding windows was used. These windows (50 ms wide) were centered at the time index corresponding to each frame of the movement epoch (a frame time index in spike train vector can be easily obtained using alignment information). The variance for Gaussian was taken as one-third of the window width. The number of spikes in each window was counted and applied as a coefficient of the Gaussian for the same window. Because the windows were overlapping, the final signal was the accumulation of the results of Gaussian convolution applied on each window. As a result $s(t)$ is obtained to represent the spike train in continuous manner (t refers to the time index of each frame).

As the signals during the movement may be envisioned to drive the monkey arm during grasping, which is a plant with non negligible dynamics, it is reasonable to assume that effect of a single spike is not instantaneous but should affect the future. Therefore as the final step of pre-processing each spike related activity is expanded in time in a decaying fashion as shown in Eq. 2 where $S(t)$ is the aggregated neural activity for time index t which is the summation of current and weighted previous activities. δ is the rate of decay and it depends on the duration of movement. Intuitively, $S(t)$ is simply the weighted sum of activities until time t weighted with an exponential factor to discount for distance in time.

$$S(t) = \sum_{tidx=1}^{t} s(tidx) * e^{\frac{-(t-tidx)^2}{\delta}} \tag{1}$$

To decide the best model for predicting angle and distance, we used polynomial family as model space. As the result of cross validation based model selection, 2nd degree polynomials were found to give best generalization for the data. Subsequently, the prediction models are defined with 3 parameters as shown in Eq. 2 where $W_{1...3}^{D}$ and $W_{1...3}^{A}$ refer to model parameters corresponding to distance and angle respectively. S represents the aggregated neural activity that is described above. The model parameters can be found easily by pseudo-inverse once the $S(t)$, $Distance(t)$ and $Angles(t)$ are available.

$$Distance(t) = W_0^D + W_1^D S(t) + W_2^D S(t)^2$$
$$Angle(t) = W_0^A + W_1^A S(t) + W_2^A S(t)^2$$

(2)

For each neuron and object combination, there are up to 10 trials to be included in regression. In order to assess the prediction ability of neurons leave one out (LOO) cross validation is used. As a result, for each neuron and object combination, we obtain an average training and test (cross validation) error, which is used as a criterion to compare neurons' performance in decoding kinematics data. The final model is obtained as the average of the models obtained in each cross validation iteration and used to report decoding performance of single neuron (see Figs. 2 and 3).

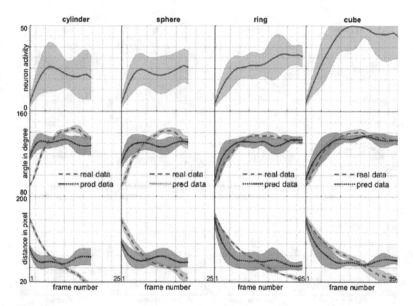

Fig. 2. Non-mirror neuron 449 performance as average ± standard deviation over all the trials for each object: the first row is the average of the aggregated neural activity, second and third rows represent angle and distance respectively: blue curve is the average of extracted kinematics and the red curve is the average of the corresponding predicted kinematics. (Color figure online)

3 Results and Discussion

In this study, arm kinematics decoding capacity of 32 neurons (14 mirror and 18 non-mirror) is investigated. Polynomials with degree 2 are found to be suitable for decoding kinematics. For each neuron, regression problem is solved 8 times considering all 4 objects and 2 types of kinematics data. The evaluation and comparison of the neurons' performance for each kinematics decoding are represented by the regression output as test error and fitting curve. If a neuron is

not a good decoder for a specific kinematics data, the regression performs poorly
and it fails in providing acceptable fit by typically generating an approximately
constant line which is equal to the average of real data. On the other hand, if
a neuron is a good decoder, the regression performance improves considerably
and it generates an output curve that approximates the real data. Figures 2 and
3 show the decoding results related to two non-mirror and mirror neurons with
good decoding performance. The first row of the figures shows neural activity,
the second row shows the result of decoding the angle and last row shows the
result of decoding the distance. All the subplots are the average of performed
trials for a specific object manipulation. The bold curve in the subplots depicts
the average of related data in the respective subplot over the trials and for each
bold curve standard deviation can be seen as shaded area around it.

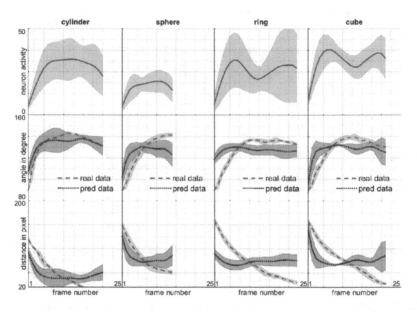

Fig. 3. Mirror neuron 453 performance as average ± standard deviation over all the
trials for each object: the first row is the average of the aggregated neural activity,
second and third rows represent angle and distance respectively: blue curve is the
average of extracted kinematics and red curve is the average of the corresponding
predicted kinematics. (Color figure online)

In order to compare mirror and non-mirror neurons contribution in kinematics, their decoding performance can be considered as a quantitative comparison
criteria. Each neuron performance is observed for angle and distance separately.
For decoding each kinematics data, each neuron's performance is defined as
all-object error which is the average value of decoding errors for all objects.
Figure 4 summarizes decoding results for mirror and non-mirror neurons considering both kinematics. Vertical axis shows mean of all-object decoding error

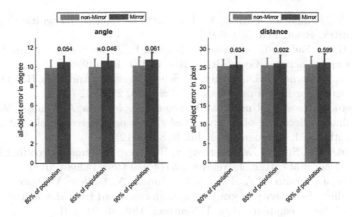

Fig. 4. Comparing mirror and non-mirror neurons decoding performance based on mean all-object prediction error for each group

for each group. In order to eliminate outliers and noisy neurons, a few worst units are left out from both groups: Horizontal axis of subplots represents the percentage of included neurons from each group (corresponding to the exclusion of 2–3 neurons from each group). For each percentage, mirror and non-mirror neurons performance are shown as blue and red bars respectively. T-test with 0.05 significance level is applied to determine the differences and the corresponding p-value of each comparison is reported above the bar graphs. The significant differences between the decoding performance of non-mirror and mirror neurons are marked with * above the bars.

Considering all-object error as neuron performance criteria, non-mirror neurons seem to be a better decoder than mirror neurons. The difference becomes significant by focusing on angle kinematics and involving 85% of each population. From this result, it can be suggested that non-mirror neurons are better general decoders while mirror neurons are possibly more object specific decoders. Experiments with more neurons should be carried out to back up the current results.

Acknowledgments. This work was supported by the grant OBSERVENEMO within the framework of the bilateral S&T Cooperation Program between the Republic of Turkey and the Hellenic Republic. Grant No. 113S391 funded by TUBITAK and grant 14UR OBSERVENEMO co- Financed by the European Union and the Greek State, MCERA/GSRT.

References

1. Rizzolatti, G., Camarda, R., Fogassi, L., Gentilucci, M., Luppino, G., Matelli, M.: Functional organization of inferior area 6 in the macaque monkey. II. area F5 and the control of distal movements. Exp. Brain Res. **71**, 491–507 (1988)
2. Di Pellegrino, G., Fadiga, L., Fogassi, L., Gallese, V., Rizzolatti, G.: Understanding motor events: a neurophysiological study. Exp. Brain Res. **91**, 176–180 (1992)

3. Gallese, V., Fadiga, L., Fogassi, L., Rizzolatti, G.: Action recognition in the pre-motor cortex. Brain **119**, 593–609 (1996)
4. Rizzolatti, G., Fadiga, L., Gallese, V., Fogassi, L.: Premotor cortex and the recognition of motor actions. Cogn. Brain. Res. **3**, 131–141 (1996)
5. Musallam, S., Corneil, B.D., Greger, B., Scherberger, H., Andersen, R.A.: Cognitive control signals for neural prosthetics. J. Sci. **305**, 258–262 (2004)
6. Carpaneto, J., Raos, V., Umilta, M.A., Fogassi, L., Murata, A., Gallese, V., Micera, S.: Continuous decoding of grasping tasks for a prospective implantable cortical neuroprosthesis. J. Neuroeng. Rehabil. **9**, 84 (2012)
7. Townsend, B.R., Subasi, E., Scherberger, H.: Grasp movement decoding from pre-motor and parietal cortex. J. Neurosci. **31**(40), 14386–14398 (2011)
8. Carpaneto, J., Umilta, M.A., Fogassi, L., Murata, A., Gallese, V., Micera, S., Raos, V.: Decoding the activity of grasping neurons recorded from the ventral premotor area F5 of the macaque monkey. J. Neurosci. **188**, 80–94 (2011)
9. Hao, Y., Zhang, Q., Controzzi, M., Cipriani, C., Li, Y., Li, J., Zhang, S., Wang, Y., Chen, W., Carrozza, M.C., Zheng, X.: Distinct neural patterns enable grasp types decoding in monkey dorsal premotor cortex. J. Neural Eng. **11**, 066011 (2014)
10. Menz, V.K., Schaffelhofer, S., Scherberger, H.: Representation of continuous hand and arm movements in macaque areas M1, F5, and AIP: a comparative decoding study. J. Eng. **12**(5), 056016 (2015)
11. Kirtay, M., Papadourakis, V., Raos, V., Oztop, E.: Neural representation in F5: cross-decoding from observation to execution. J. BMC Neurosci. **16**, 190 (2015)
12. Papadourakis, V., Raos, V.: Cue-dependent action-observation elicited responses in the ventral premotor cortex (area F5) of the macaque monkey. Society for Neuroscience Abstracts, Program No. 263.08 (2013)

Application of Deep Belief Network to Land Cover Classification Using Hyperspectral Images

Bulent Ayhan and Chiman Kwan[(✉)]

Signal Processing, Inc., Rockville, MD 20850, USA
{bulent.ayhan, chiman.kwan}@signalpro.net

Abstract. This paper summarizes some preliminary results of applying deep belief network (DBN) to land classification using hyperspectral images. The performance of DBN is then compared with several conventional classification approaches. A fusion approach is also proposed to combine spatial and spectral information in the classification process. Actual hyperspectral image data were used in our investigations. Based on the particular data and experiments, it was found that DBN has slightly better classification performance if only spectral information is used and has slightly inferior performance than a conventional method if both spatial and spectral information are used.

Keywords: Deep learning · DBN · SVM · SAM · Hyperspectral image · Land classification

1 Introduction

According to the book by Liang et al. [1], land cover classification is very important for quantifying the location, extent, and variability of change; the causes and processes of change; and the responses to and consequences of change. Existing data products related to land cover classification use raw data from MODIS and SPOT, which are multispectral imagers. Hyperspectral images has great potential in discriminating different targets due to the availability of high spectral resolution. Multispectral and hyperspectral images have been used in anomaly detection, classification, and change detection [2–8, 15, 17–19] in recent years. Actually, NASA is planning a HyspIRI mission that incorporates a hyperspectral imager with more than 200 spectral bands, 30 m resolution, and global coverage [9].

Deep learning has gaining popularity in recent years and has been applied to many applications, including target recognition, speech recognition, and many others [10]. With more layers in the neural network, it is believed that deep neural network will be able to capture the complicated relationships between various features. The goal of our study is to investigate the application of deep learning in land cover classification. First, we would like to see how well deep belief network (DBN), which is one type of deep neural network, can achieve in pixel classification for land cover type using hyperspectral images. Second, we would like to see how DBN performs as compared to some of theconventional algorithms based on spectral information only. Third, we

© Springer International Publishing AG 2017
F. Cong et al. (Eds.): ISNN 2017, Part I, LNCS 10261, pp. 269–276, 2017.
DOI: 10.1007/978-3-319-59072-1_32

would like to explore whether the fusion of spatial and spectral information can further improve the land cover classification performance.

Our paper is organized as follows. In Sect. 2, we will briefly describe several land cover classification algorithms, including DBN, Support Vector Machine (SVM), spectral angle mapper (SAM), and a variant of SAM called M-SAM. Section 3 will summarize a comparative study of the various algorithms to a notable hyperspectral image data. Finally, a few concluding remarks and ideas for future research will be summarized in Sect. 4.

2 Technical Approach

In this research, we have applied 4 algorithms in our land classification investigations. In the following sections, we briefly describe the key algorithms applied in our studies.

2.1 SAM [13] and M-SAM [4]

SAM (Spectral Angle Measure) [13] is a well-known and simple technique for pixel signature classification. Given a known signature vector and a test pixel signature, SAM computes the angle between the two and use the angle for discrimination. The formula for SAM can be seen in (1). We used an extended version of SAM in such a way that multiple radiance profiles of the same target are used for detection. When constructing the similarity score image, we formed the detection scores by picking the highest similarity of the test data pixel to these extracted target signatures; we called this M-SAM (Multiple-SAM) [4]. Suppose $\mathbf{s}_i = (s_{i1}, s_{i2}, \ldots, s_{iL})^T$ is pixel in the hyperspectral image cube with L bands and $\mathbf{r}_j = (r_{j1}, r_{j2}, \ldots, r_{jL})^T$ is the jth target signature variant, where $j = 1, \ldots, K$ and K is the total number of signature variants for the target of interest.

The similarity measure, SAM, between the two radiance profiles, $\mathbf{s}_i$ and $\mathbf{r}_j$, SAM $(\mathbf{s}_i, \mathbf{r}_j)$ is then computed as:

$$SAM(\mathbf{s}_i, \mathbf{r}_j) = \cos^{-1}\left(\frac{\langle \mathbf{s}_i, \mathbf{r}_j \rangle}{\|\mathbf{s}_i\| \|\mathbf{r}_j\|}\right) \tag{1}$$

where $\langle \mathbf{s}_i, \mathbf{r}_j \rangle = \sum_{l=1}^{L} s_{iL} r_{jL}$ and $\|\mathbf{s}_i\| = \left(\sum_{l=1}^{L} s_{iL}^2\right)^{1/2}$ and $\|\mathbf{r}_j\| = \left(\sum_{l=1}^{L} r_{jL}^2\right)^{1/2}$. M-SAM computes the similarity score between the test pixel, $\mathbf{s}_i$, and each of the target radiance profile variant, $\mathbf{r}_j$, and assigns the highest similarity value as the final similarity value.

2.2 Support Vector Machine (SVM) [11, 12, 16]

In the past two decades, theoretical advances and experimental results have drawn considerable attention to the use of kernel functions in data clustering and classification. Among them stands out the SVM, which is a general architecture that can be applied to pattern recognition and classification, regression, estimation and other problems such as speech and target recognition. Notable advantages of SVM are: (1) there is no over training problem; (2) the training data set can be small; (3) convergence is guaranteed. Since SVM is well known, we omit the details and would like to mention that we used the SVM toolbox [12] in our experiments.

2.3   Deep Neural Network (DNN)

In the past few years, there have been intense research in using DNN for various applications. The idea of DNN is not new. It is a multilayer perceptron network with many layers. The learning algorithm is based on the well-known back-propagation algorithm. However, recent progresses in combining back-propagation with effective pretraining, in incorporating different kinds of nonlinearities, in regularizing net weight estimation, and in GPU based parallel processing have demonstrated tremendous potentials of deep learning in many machine learning tasks. According to [10], DNN has a number of advantages such as better capturing of hierarchical feature representations, learning more complex behavior and using distributed representations to learn the interactions of many different factors on different levels.

There are two popular DNNs: DBN (Deep Belief Network) and CNN (convolutional neural network). In this paper, we selected DBN and the toolbox developed in [14] was used in our studies.

3   Comparative Studies

3.1   About the Test Hyperspectral Image Data

The hyperspectral image data used in this example is called "NASA-KSC" image [15]. The image shown in Fig. 1 corresponds to the mixed vegetation site over Kennedy Space Center (KSC), Florida. The image data was acquired by the National Aeronautics and Space Administration (NASA) Airborne Visible/Infrared Imaging Spectrometer instrument, on March 23, 1996. AVIRIS acquires data in a range of 224 bands with wavelengths ranging from 0.4 μm to 2.5 μm. The KSC data have a spatial resolution of 18 m. Excluding water absorption and low signal-to noise ratio (SNR) bands, there are 176 spectral bands for classification. In the NASA-KSC image, there are 13 different land-cover classes available. It should be noted that only a small portion of the image has been tagged with the ground truth information and these pixels with the tagged ground truth information have been used in the classification study.

3.2   DBN Structure Used in This Study

DBNs can be viewed as a composition of simple, unsupervised networks such as restricted Boltzmann machines (RBMs). We used a 3-Level DBN architecture:

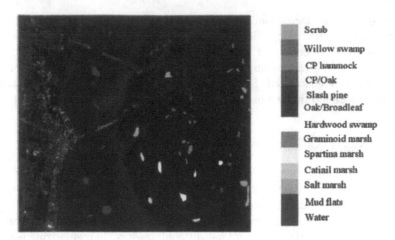

Scrub
Willow swamp
CP hammock
CP/Oak
Slash pine
Oak/Broadleaf
Hardwood swamp
Graminoid marsh
Spartina marsh
Catiail marsh
Salt marsh
Mud flats
Water

Fig. 1. NASA-KSC image and tagged pixels with ground truth information.

(a) Level-1 (RBM with 200 hidden units)
(b) Level-2 (RBM with 200 × 200 hidden units)
(c) Level-3 (connection to y (output) with neural network)

3.3   Results and Comparison with Benchmark Techniques SAM and SVM

In the image, tagged parts of the image are separated into three sets, i.e., training, validation, and testing data, with a separation ratio of 6:2:2. 60% of the tagged samples are randomly chosen as the training set, and 20% and 20% of the tagged samples are randomly selected for the validation and testing sets. The validation set is to pick the training parameters and the network architecture for DBN such that the resultant trained model on the training set provides the best performance on the validation set. Once the best trained model is identified using the validation set, the model is then applied on the test data set to produce the final classification results. In addition to test data set, the classification results are generated on the validation and training sets as well.

For the benchmark techniques to compare with DBN, SVM (Support Vector Machine), SAM (Spectral Angle Mapper) and an extended version of SAM, Multiple-SAM, techniques are used. In SAM, signatures for each land-cover class are randomly picked (one signature for each class); this is the SAM version without utilizing the validation data set. For SAM with utilizing the validation set, the class signatures are picked from the training set such that these picked signatures would provide the best similarity scores on the corresponding class samples in the validation data set.

In Multiple-SAM (M-SAM), for each class, a total of 20 and 50 representative signatures are picked from the training set such that these signatures provide the best similarity scores on the corresponding class samples in the validation set. These picked signatures are then used on the testing set. The similarity score between each of the test data sample and each of the picked signatures is computed. The averaged similarity scores for each class are generated for each test data sample and the class label is decided based on the highest similarity. In SVM, LIBSVM toolbox is used with a kernel type of Radial Basis Function and automatically regularized support vector classification SVM method type (nu-SVC).

For using spatial information, the bands 12 (Blue), 20 (Green) and 29 (Red) in the hyperspectral image data are picked and a spatial window with a size of 7×7 is formed for each pixel such that the pixel of interest is in the center of the local window size. The local RGB image cube with a size of $7 \times 7 \times 3$ is then transformed into a vector form and added to the end of the spectral information of the corresponding pixel. It should be noted that the number of spectral bands for a pixel in this hyperspectral image is 176; after adding the vector form of the pixel spatial information which is 147; the size of the vector for an input pixel becomes 323.

The correct classification rates are shown in Tables 1, 2, 3, 4 and 5. In particular, only spectral information was used in Tables 1, 2 and 3. It can be seen that DBN results are slightly better than SVM. SAM and M-SAM do not work well perhaps due to the presence of mixed pixels in the image data. When spatial and spectral information is fused, both SVM and DBN have been improved by more than 3 percentage points. Now, SVM performed slightly better.

Table 1. Preliminary results with applying SAM to hyperspectral data- *Spectral information only* (random selection of training/testing/validation sets)

	Utilizing validation set	Number of signatures for each class	Train set (correct classification rate)	Validation set (correct classification rate)	Test set (correct classification rate)
SAM	no	1	0.6511	0.6676	0.6479
SAM	yes	1	0.7819	0.7801	0.7847
M-SAM	yes	20	0.7848	0.7820	0.7818
M-SAM	yes	50	0.7887	0.7957	0.7876

Table 2. Preliminary results with applying DBN to hyperspectral data - *Spectral information only* (random selection of training/testing/validation sets)

	Architecture	Epochs	Train set (correct classification rate)	Validation set (correct classification rate)	Test set (correct classification rate)
DBN	200 and (200 × 200)	1,500	0.7493	0.7566	0.7488
DBN	200 and (200 × 200)	5,000	0.8940	0.8895	0.8826
DBN	200 and (200 × 200)	20,000	0.9299	0.9247	0.9234
DBN	200 and (200 × 200)	30,000	0.9436	0.9374	0.9389

Table 3. Preliminary results with applying SVM to hyperspectral data - *Spectral* information only

	LIBSVM options	Train set (correct classification rate)	Validation set (correct classification rate)	Test set (correct classification rate)
SVM	nu-SVC with radial basis function kernel type(LIBSVM parameters: '-s 1 -t 2 -d 3 -r 0 -c 1 -n 0.1 -p 0.1 -m 100 -e 0.000001 -h 1 -b 0 -wi 1)	0.9658	0.9374	0.9340

Table 4. Preliminary results with applying DBN to hyperspectral data – *Spectral + Spatial information* (random selection of training/testing/validation sets)

	Architecture	Epochs	Train set (correct classification rate)	Validation set (correct classification rate)	Test set (correct classification rate)
DBN	200 and (200 × 200)	1,500	0.8024	0.7947	0.7924
DBN	200 and (200 × 200)	5,000	0.9504	0.9169	0.9389
DBN	200 and (200 × 200)	20,000	0.9863	0.9589	0.9602
DBN	200 and (200 × 200)	30,000	0.9925	0.9648	0.9631

Table 5. Preliminary results with applying SVM to hyperspectral data – *Spectral + Spatial* information only

	LIBSVM options	Train set (correct classification rate)	Validation set (correct classification rate)	Test set (correct classification rate)
SVM	nu-SVC with radial basis function kernel type(LIBSVM parameters: '-s 1 -t 2 -d 3 -r 0 -c 1 -n 0.1 -p 0.1 -m 100 -e 0.000001 -h 1 -b 0 -wi 1)	0.9863	0.9795	0.9709

One reason that DBN did not outperform other conventional algorithms is perhaps due to the fact of limited training data. Usually, the more the training data, the better the performance of DBN is. Another reason could be due to the selection of the structure in the applied DBN and that a more proper DBN structure might further improve DBN's performance.

4   Conclusions

We presented some preliminary results of applying DBN to land cover classification using hyperspectral image data. DBN is also compared with several conventional algorithms. It was observed that DBN is comparable to SVM in this study. Since the experiments were carried out by using a limited data set, it is premature to claim that DBN is better or worse than conventional methods. Extensive studies which investigate different DBN structures and other deep learning approaches are needed in the future.

References

1. Liang, S., Li, X., Wang, J.: Advanced Remote Sensing: Terrestrial Information Extraction and Applications. Academic Press, Cambridge (2012)
2. Zhou, J., Kwan, C., Ayhan, B., Eismann, M.: A novel cluster kernel RX algorithm for anomaly and change detection using hyperspectral images. IEEE Trans. Geosci. Remote Sens. **54**(11), 6497–6504 (2016)
3. Zhou, J., Kwan, C., Budavari, B.: Hyperspectral image super-resolution: a hybrid color mapping approach. SPIE J. Appl. Remote Sens. **10**, 035024 (2016)
4. Ayhan, B., Kwan, C.: On the use of radiance domain for burn scar detection under varying atmospheric illumination conditions and viewing geometry. J. Sig. Image Video Process. **11**, 605–612 (2016)
5. Kwan, C., Choi, J.H., Chan, S., Zhou, J., Budavari, B.: Resolution enhancement for hyperspectral images: a super-resolution and fusion approach. In: IEEE International Conference on Acoustics, Speech, and Signal Processing, New Orleans (2017)
6. Nguyen, D., Tran, T., Kwan, C., Ayhan, B.: Endmember extraction in hyperspectral images using l-1 minimization and linear complementary programming. In: SPIE, vol. 7695 (2010)
7. Li, S., Wang, W., Qi, H., Ayhan, B., Kwan, C., Vance, S.: Low-rank tensor decomposition based anomaly detection for hyperspectral imagery. In: IEEE International Conference on Image Processing (ICIP), Quebec City, Canada (2015)
8. Qu, Y., Guo, R., Wang, W., Qi, H., Ayhan, B., Kwan, C., Vance, S.: Anomaly detection in hyperspectral images through spectral unmixing and low rank decomposition. In: International Geoscience and Remote Sensing Symposium (IGARSS), Beijing (2016)
9. Lee, C.M., Cable, M.L., Hook, S.J., Green, R.O., Ustin, S.L., Mandl, D.J., Middleton, E.M.: An introduction to the NASA hyperspectral infrared imager (HyspIRI) mission and preparatory activities. Remote Sens. Environ. **167**, 6–19 (2015)
10. Lecun, Y., Ranzato, M.: Deep learning tutorial. In: 30th International Conference on Machine Learning, Atlanta (2013)
11. Qian, T., Li, X., Ayhan, B., Xu, R., Kwan, C., Griffin, T.: Application of support vector machines to vapor detection and classification for environmental monitoring of spacecraft. In: Wang, J., Yi, Z., Zurada, J.M., Lu, B.-L., Yin, H. (eds.) ISNN 2006. LNCS, vol. 3973, pp. 1216–1222. Springer, Heidelberg (2006). doi:10.1007/11760191_177
12. Chang, C.-C., Lin, C.-J.: LIBSVM: a library for support vector machines. ACM Trans. Intell. Syst. Technol. **2**, 27:1–27:27 (2011)
13. Kwan, C., Ayhan, B., Chen, G., Chang, C., Wang, J., Ji, B.: A novel approach for spectral unmixing, classification, and concentration estimation of chemical and biological agents. IEEE Trans. Geosci. Remote Sens. **44**, 409–419 (2006)

14. Palm, R.B.: Prediction as a candidate for learning deep hierarchical models of data. Technical University of Denmark (2012)
15. Chen, Y., Lin, Z., Zhao, X., Wang, G., Gu, Y.: Deep learning-based classification of hyperspectral data. IEEE J. Sel. Top. Appl. Earth Observations Remote Sens. **7**(6), 2094–2107 (2014)
16. Qian, T., Xu, R., Kwan, C., Linnell, B., Young, R.: Toxic vapor classification and concentration estimation for space shuttle and international space station. In: Yin, F.-L., Wang, J., Guo, C. (eds.) ISNN 2004. LNCS, vol. 3173, pp. 543–551. Springer, Heidelberg (2004). doi:10.1007/978-3-540-28647-9_90
17. Ayhan, B., Kwan, C., Li, X., Trang, A.: Airborne detection of land mines using mid-wave infrared (MWIR) and laser-illuminated-near infrared images with the RXD hyperspectral anomaly detection method. In: Fourth International Workshop on Pattern Recognition in Remote Sensing, Hong Kong (2006)
18. Ayhan, B., Kwan, C., Vance, S.: On the use of a linear spectral unmixing technique for concentration estimation of APXS spectrum. J. Multi. Eng. Sci. Technol. **2**(9), 2469–2474 (2015)
19. Zhou, J., Kwan, C.: Fast anomaly detection algorithms for hyperspectral images. J. Mult. Eng. Sci. Technol. **2**(9), 2521–2525 (2015)

Reservoir Computing with a Small-World Network for Discriminating Two Sequential Stimuli

Ke Bai[1], Fangzhou Liao[2], and Xiaolin Hu[3,4(✉)]

[1] Department of Physics, Tsinghua University, Beijing, China
[2] Department of Biomedical Engineering, Tsinghua University, Beijing, China
[3] Department of Computer Science and Technology, Tsinghua National Laboratory for Information Science and Technology, Beijing, China
xlhu@tsinghua.edu.cn
[4] Brain-Inspired Computing Research Center, Tsinghua University, Beijing, China

Abstract. Recently, reservoir network was used for simulating the sequential stimuli discrimination process of monkeys. To deal with the inefficient memory problem of a randomly connected network, a winner-take-all subnetwork was used. In this study, we show that a network with the small-world property makes the WTA subnetwork unnecessary. Using the reinforcement learning in the output layer only, the proposed network successfully learns to accomplish the same discrimination task. In addition, the model neurons exhibit heterogeneous firing properties, which is consistent with the physiological data.

Keywords: Reservoir computing · Small-world network · Decision making · Reinforcement learning

1 Introduction

Working memory is a mechanism that maintains and processes information for several seconds during cognitive tasks [1]. One behavior experiment based on the delayed discrimination task [12] is usually used as working memory paradigm. A monkey received two vibrating tactile signals with different frequencies (f_1, f_2) consecutively. Each of them lasts for 0.5 s and they are separated with a 3 s time interval. The monkey's task is to discriminate which one has the higher frequency.

Several classical models have been proposed to model this task [9,11]. Reservoir computing [5,8] is attractive for simulating this task because it avoids the artificially designed learning rule except for the last readout layer, and its randomly initialized weights lead to heterogeneous neuron property. Barak et al. [2] used it to achieve a good result. Cheng et al. [3] used reinforcement learning algorithm to train the readout layer, but it still needs a biologically implausible winner take all layer.

© Springer International Publishing AG 2017
F. Cong et al. (Eds.): ISNN 2017, Part I, LNCS 10261, pp. 277–284, 2017.
DOI: 10.1007/978-3-319-59072-1_33

From the perspective of biological plausibility and the idea of the complex system, we constructed a constrained E-I balance reservoir computing model with a small-world connection constraint. This model meets the three standards mentioned above when simulating this task.

2   Methods

2.1   Experiment Paradigm

The simulation paradigm is consistent with the biology experiment. The input is a time series with two square waves standing for different frequencies (f_1, f_2), each lasts for 500 ms. A 500 ms initial time is at the beginning with no inputs. The delay between the two stimuli is 3000 ms. Before the final decision-making, 200 ms are left (Fig. 1a).

2.2   Input

Biology experiments revealed that the firing rate of the input signal to PFC is proportional to the stimulus frequency [13], so in our network, the input signal f_1 and f_2 are represented by two random number sampled from 0.1 to 1. The input neuron is randomly connected to 3% neurons in the reservoir network and the synapse weights are randomly initialized by sampling from standard log-normal distribution. The weights are fixed after the initialization.

2.3   Reservoir

The reservoir network is the core part of this model. It contains a total of $N = 1200$ neurons. To make the network consistent with biology, we divided all the neurons into two categories: excitatory and inhibitory (960 excitatory, 240 inhibitory), the output weights for these two kinds of neurons are constrained to be positive and negative, respectively. The connection probability of excitatory-excitatory (EE), excitatory-inhibitory (EI), inhibitory-excitatory (IE), inhibitory-inhibitory (II) is 0.05, 0.05, 0.2, 0.2, respectively. The absolute connection weights of these excitatory and inhibitory neurons are sampled randomly from standard log-normal distribution. Others are set to be 0. The connection matrix can be seen in Fig. 1b. The weights are fixed after the initialization.

All neurons share the same activation function:

$$r_i = \begin{cases} r_0 + r_0 \tanh(x/r_0), & \text{if } x \leq 0, \\ r_0 + (r_{\max} - r_0) \tanh(x/(r_{\max} - r_0)), & \text{if } x > 0, \end{cases} \tag{1}$$

where r_0 is 0.1, and $r_{\max}$ is 1.5.

The dynamic equation:

$$\frac{dx_i}{dt} = -x_i + g \sum_{i=1}^{N} w_{ij} r_j + \sigma, \tag{2}$$

where x_i and r_i denote the membrane potential and the firing rate of i-th neuron, w_{ij} denotes the connection weight from j-th neuron to i-th neuron, $\tau = 0.3\,\mathrm{s}$ is the time constant, σ is an Gaussian noise with standard deviation of 0.001 independent of i and t. g denotes the gain factor, which is selected purposely, the detail will be discussed in Sect. 2.5.

The differential equation is discretized with Euler method. The time step is 5 ms. After iteration for 4.7 s, the firing rates are read out by the output neurons.

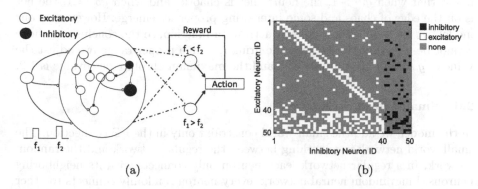

Fig. 1. The model architecture. (a) The network is composed of three parts: the input neuron, reservoir network, output neurons. The dashed and solid lines represent trainable and fixed weights, respectively. (b) The connection matrix of the reservoir network. Only a subset of the network (50 of 1200) is shown here. $p_{rc} = 0.2$ in this case. Notice the small-world structure in the EE part.

2.4 Output

There are two output neurons in the network, which correspond to the choice of f_1 and f_2 respectively. 50% neurons in the reservoir network are connected to them and the synapses weights are also randomly initialized by sampling from standard log-normal distribution. The plasticity weights are updated using a reward-based reinforcement learning rule [7].

$$v_l = \sum_{k=1}^{N} w_{kl} r_k, \tag{3}$$

$$w_{kl} := w_{kl} + \eta(R - \mathrm{E}[R])(r_k v_l) - \alpha w_{kl}, \tag{4}$$

where v_l is the activation of l-th output neuron, R is the reward, it is 1 or 0 determined by whether the network get the correct choice, when $f_1 = f_2$, the reward is given randomly. $\mathrm{E}[R]$ is the expected reward, approximated by

$$\mathrm{E}[R] = \frac{1}{1 - \exp(v_{correct} - v_{wrong})}. \tag{5}$$

2.5 Gain Factor

Gain factor is the g mentioned in Eq. (2). This quantity scales the transfer matrix W of the reservoir network. It is a very important factor. For example, in linear models, if the maximum absolute eigenvalue of gW is larger than one, the firing rates will diverge to infinity, if it is smaller than one, the firing rates will astringent to zero. In the nonlinear recurrent model with weights $w_{ij} \sim N(0, J^2/N)$, and activation function $\tanh(gx)$, Sompolinsky et al. [14] found that when $gJ > 1$, the neural net is chaotic, and when $gJ = 1$, the net is on the edge of chaos and some interesting properties emerge. However, in our model, we use a different nonlinear activation function, so the conclusions above cannot be used directly. To keep the firing rate in a proper range, we adjust the value of g so that the product of g and the maximum absolute eigenvalue is 1.2.

2.6 Small-World Structure

In the model, we add the small-world constraint only in the EE connection. The small-world network is something between the regular network and the random network. In a regular network, each neuron only connects with its neighboring neurons. In a random neural network, every neuron randomly connects to other neurons. The small-world network can be viewed as a regular network with some randomly connected synapses. The proportion of the randomly connected synapses is denoted by p_{rc}. For small-world network, $p_{rc} \in (0,1)$, for regular network, $p_{rc} = 0$, and for random network, $p_{rc} = 1$. Notice that the small-world property only changes the connection topology, and does not change the overall connection probability and connection strength sampling. To investigate the optimal p_{rc}, we evaluated $p_{rc} = 0, 0.1, 0.2, 0.3 \cdots 0.9, 1$.

2.7 Training and Testing Procedure

We run 80 experiments for each p_{rc}. In every experiment, weights are initialized independently. f_1 is evenly sampled 13 times between 0.1 and 1, so does f_2, and the number of training set is the number of (f_1, f_2) combinations: 169. The number of epochs is 2. The learning rate is exponentially decayed by $1/e$ every 100 iterations. In the testing stage, f_1 and f_2 are evenly sampled 10 times between 0.1 and 1, so that testing set has 100 examples. During the test phase, each testing example is tested 22 times with independent noise.

3 Result

3.1 Classification Accuracy

The classification result is not only different under different p_{rc}, but also varies a lot in parallel experiments. Here we show a typical good result, which is an example under $p_{rc} = 0.1$. The accuracy mainly depends on $|f_1 - f_2|$. The closer frequency, the lower right rate. The result of the experiment data and the simulation data is shown in Fig. 2a. Also, notice the asymmetry: the accuracy at $f_1 = 0, f_2 = 0.1$ is lower than that of $f_1 = 0.1, f_2 = 0$, and the accuracy at $f_1 = 1.0, f_2 = 0.9$ is lower than that of $f_1 = 0.9, f_2 = 1.0$. It indicate that when f_1 is close to the extreme value (0 and 1) the network can not remember the exact value, so its decision relies more on the f_2, when f_2 is large, it choose f_2, when f_2 is small, it choose f_1. It is very similar with the biological experiment data [9].

3.2   The Importance of the Small-World Network

The experiment result is very random, so that we have to run many parallel experiments and use statistics to evaluate whether the small-world network improves the probability of getting a good result. The answer is yes. We can see it from Fig. 2b. There are two baselines which are worth paying attention to. First, this is a two-classification task, so the baseline is 0.5. Because the output reinforcement layer is a weak learner and we only have 338 training examples, the model may be trapped in the local minimal point, which leads the result of some experiments lower than 50%. Second, if the network has poor memory ability, which is to say, it can only remember the information of f_2 and forget f_1. Under this situation, the optimal policy is choosing f_2 if $f_2 > 0.5$ and choosing f_1 is $f_2 < 0.5$, so that the accuracy should be 75%. Considering the noise level, we consider accuracy higher than 80% as good performance.

Among the eleven reconnection probability, $p_{rc} = 0.2$ gets the highest median right rate, 81%. And this value is significantly larger than most other p_{rc}. The Wilcoxon rank-sum test results between $p_{rc} = 0.2$ and other groups show that the differences between $p_{rc} = 0.2$ and groups $p_{rc} = 0, 0.3, 0.4, 0.5...1$ are significant (FDR $= 0.05$).

3.3   Heterogeneous Firing Rate

In paper of Machens et al. [10], the activity of neurons often varies strongly in time. The author illustrated this by smoothed peristimulus time histogram of nine cells in prefrontal cortex during the f_1 stimulus and the delay interval. In our model, we found several corresponding neurons for each pattern and the result can be seen in Fig. 3. The display sequence of these nine neurons is exactly the same as the biological data shown in Fig. 1 in [10].

3.4   Regression Analysis

We use the least-square method to do the linear regression:

$$r(t) = b_0(t) + b_1(t)f_1 + b_2(t)f_2, \tag{6}$$

where $r(t)$ is the firing rate of a neuron at time t, b_0, b_1, b_2 are coefficients. We applied it at three time points: the last point before the input of f_2 ($t = 4.0\,\mathrm{s}$),

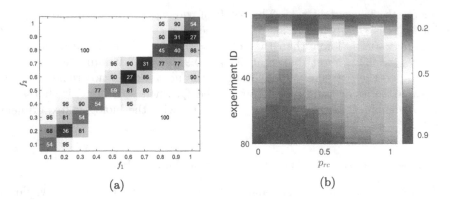

(a)   (b)

Fig. 2. The results on the test set. (a) The classification results at $p_{rc} = 0.1$. The accuracy at each grid is averaged from 22 parallel runs with independent noise. (b) The classification results at different p_{rc} level. The experiment id is sorted by accuracy.

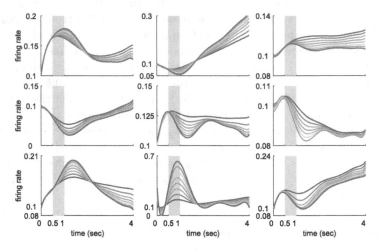

Fig. 3. The smoothed peristimulus time histograms from nine neurons in the reservoir network. The curves are colored according to f_1. The red color represents the largest value. As the color become closer to blue the value gets lower. (Color figure online)

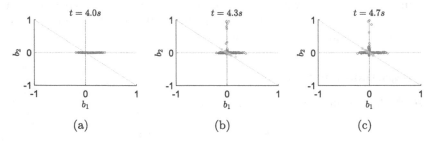

(a)   (b)   (c)

Fig. 4. The regression results. The three subplots show regression results at three time points. Notice the emergence of f_2 tuning cells in (b) and $f_1 - f_2$ tuning cells in (c).

the point during the input of f_2 ($t = 4.3$s) and the last point which is used as the input of the output layer ($t = 4.7$s). We conducted a test [4] to find the neurons tuned to $f_1, f_2, f_1 - f_2$, which corresponds to the neurons with $b_1 \neq 0, b_2 \neq 0, b_1 - b_2 \neq 0$ respectively. Their coefficients distribution is shown in Fig. 4. Notice that there gradually appears neurons tuning to f_1, f_2 and $f_1 - f_2$ as time goes by.

There is actually another kind of $f_1 - f_2$ tuning neurons in our model: the two output neurons can also be considered as $f_1 - f_2$ tuning cells. Theoretically, the output neurons can take input solely from the $f_1 - f_2$ tuning neurons in the reservoir, or it can take input from f_1 tuning cells and f_2 tuning cells and conduct the minus operation on synapses. But we did not further inspect which mechanism it use. Biological results showed that there does exist many neurons tuning to $f_1 - f_2$ [6]. But we can not identify which kind of neuron they are according to our categorization.

4   Conclusion and Discussion

In this work, we adopted the small-world architecture in reservoir network and found that when p_{rc} is at the proper range, it can improve the performance relative to both the regular network and the random network. And it also shows similar firing rate pattern with the biological result. We believe that the reason is the reservoir network with small-world constraint achieved both good information storage ability (like the regular network) and neuron function variety (like the reservoir network). It is possible to generalize our model to more reservoir computing tasks, or be used as a strategy for recurrent network initialization method.

Acknowledgments. This work was supported in part by the National Basic Research Program (973 Program) of China under Grant 2013CB329403, in part by the National Natural Science Foundation of China under Grant 91420201, Grant 61332007, and Grant 61621136008, and in part by the German Research Foundation (DFG) under Grant TRR-169.

References

1. Baddeley, A.D.: Working Memory. Oxfoid Psychology Series, vol. 11. Oxford University Press, New York (1986)
2. Barak, O., Sussillo, D., Romo, R., Tsodyks, M., Abbott, L.: From fixed points to chaos: three models of delayed discrimination. Prog. Neurobiol. **103**, 214–222 (2013)
3. Cheng, Z., Deng, Z., Hu, X., Zhang, B., Yang, T.: Efficient reinforcement learning of a reservoir network model of parametric working memory achieved with a cluster population winner-take-all readout mechanism. J. Neurophysiol. **114**(6), 3296–3305 (2015)
4. Draper, N.R., Smith, H.: Applied Regression Analysis. Wiley, Hoboken (2014)

5. Jaeger, H.: The "echo state" approach to analysing and training recurrent neural networks-with an erratum note. Bonn Ger.: Ger. Natl. Res. Center Inf. Technol. GMD Tech. Rep. **148**, 34 (2001)

6. Jun, J.K., Miller, P., Hernández, A., Zainos, A., Lemus, L., Brody, C.D., Romo, R.: Heterogenous population coding of a short-term memory and decision task. J. Neurosci. **30**(3), 916–929 (2010)

7. Loewenstein, Y., Seung, H.S.: Operant matching is a generic outcome of synaptic plasticity based on the covariance between reward and neural activity. Proc. Natl. Acad. Sci. **103**(41), 15224–15229 (2006)

8. Maass, W., Natschläger, T., Markram, H.: Real-time computing without stable states: a new framework for neural computation based on perturbations. Neural Comput. **14**(11), 2531–2560 (2002)

9. Machens, C.K., Romo, R., Brody, C.D.: Flexible control of mutual inhibition: a neural model of two-interval discrimination. Science **307**(5712), 1121–1124 (2005)

10. Machens, C.K., Romo, R., Brody, C.D.: Functional, but not anatomical, separation of "what" and "when" in prefrontal cortex. J. Neurosci. **30**(1), 350–360 (2010)

11. Miller, P., Wang, X.J.: Inhibitory control by an integral feedback signal in prefrontal cortex: a model of discrimination between sequential stimuli. Proc. Natl. Acad. Sci. Unit. States Am. **103**(1), 201–206 (2006)

12. Romo, R., Hernández, A., Zainos, A., Lemus, L., Brody, C.D.: Neuronal correlates of decision-making in secondary somatosensory cortex. Nat. Neurosci. **5**(11), 1217–1225 (2002)

13. Romo, R., Salinas, E.: Flutter discrimination: neural codes, perception, memory and decision making. Nat. Rev. Neurosci. **4**(3), 203–218 (2003)

14. Sompolinsky, H., Crisanti, A., Sommers, H.J.: Chaos in random neural networks. Phys. Rev. Lett. **61**(3), 259–262 (1988)

Single Channel Speech Separation Using Deep Neural Network

Linlin Chen[1,2], Xiaohong Ma[1(✉)], and Shuxue Ding[2]

[1] School of Information and Communication Engineering,
Dalian University of Technology, Dalian, China
`maxh@dlut.edu.cn`
[2] School of Computer Science and Engineering,
University of Aizu Fukushima, Aizuwakamatsu, Japan
`sding@u-aizu.ac.jp`

Abstract. Single channel speech separation (SCSS) is an important and challenging research problem and has received considerable interests in recent years. A supervised single channel speech separation method based on deep neural network (DNN) is proposed in this paper. We explore a new training strategy based on curriculum learning to enhance the robustness of DNN. In the training processing, the training samples firstly are sorted by the separation difficulties and then gradually introduced into DNN for training from easy to complex cases, which is similar to the learning principle of human brain. In addition, a strong discriminative objective function for reducing the source interference is designed by adding in the correlation coefficients and negentropy. The efficiency of the proposed method is substantiated by a monaural speech separation task using TIMIT corpus.

Keywords: Single channel speech separation · Deep neural network · Discriminative object function

1 Introduction

Single channel speech separation (SCSS) is to extract speech from one signal that is a mixture of multiple sources. It is a vital issue of speech separation and may play an important role in many applications. Researchers have devoted to solving SCSS problems from various perspectives, which can be categorized into computational auditory scene analysis (CASA) based and model based.

Approaches based on CASA have proven to be effective to attack the SCSS problem in an unsupervised mode. In [15], Wang et al. proposed to utilize temporal continuity and cross-channel correlation for speech segregation. The method can segregate speech from interfering sounds but it does not perform well for high frequency part of speech. Hu et al. improve it by segregating resolved and unresolved harmonics differently [5]. A common problem occurred in many CASA-based methods is that the recovered speech usually miss some parts. In order

© Springer International Publishing AG 2017
F. Cong et al. (Eds.): ISNN 2017, Part I, LNCS 10261, pp. 285–292, 2017.
DOI: 10.1007/978-3-319-59072-1_34

to solve this problem, shape analysis techniques in image processing such as labeling and distance function are applied to speech separation in [10].

In model based approaches, non-negative matrix factorization (NMF) is one of the most popular techniques for SCSS in recent years. Conventionally, the basis of each source is trained separately first and then the magnitude spectra of mixed signal is decomposed into a linear combination of the trained basics. Finally, the separated signals can be obtained from the corresponding parts of decomposed mixture [4,12]. However, the separation become difficult when sources are overlap in subspaces. Various attempts have been made to solve this problem [13,16].

Deep neural networks (DNN) has achieved state-of-art results in many applications such as object detection [3], speech recognition [17] owing to its strong mapping ability. Kang et al. use DNN to learn the mapping between the mixture and the corresponding encoding vectors of NMF [8]. In [6], a simple discriminative training criterion which takes into account the squared error of prediction and other sources is proposed.

In this paper, we focus on training strategy and objective function. This paper considers DNN as a kind of system learning rules from a mess of training samples. These training samples are sorted by a ranking function and fed to DNN in ascending order of learning difficulty. Furthermore, we use correlation coefficients and negentropy, rather than the criterion in [6], to model the similarity of recovered signals, which aim at reducing the interference of other sources. Experimental results demonstrated that the proposed method outperformed NMF and approach in [6].

The organization of this paper is as follows: Sect. 2 introduces the proposed methods, including the learning strategy and discriminative objective function, Sect. 3 presents the experimental setting and results based on the TIMIT corpus and conclusion is given in Sect. 4.

2   Proposed Method

2.1   Problem Formulation

In this paper, we assume the observed signal is a mixture of source signals of two speakers. Ignored the attenuations of the path, the problem can be formulated as

$$\mathbf{x}(t) = \mathbf{s}^T(t) + \mathbf{s}^I(t) \tag{1}$$

where $\mathbf{s}^T(t)$ and $\mathbf{s}^I(t)$ represent the target speech and interfering speech respectively. Denoting $\mathbf{x}(t, f)$ as the short time Fourier transform (STFT) of $\mathbf{x}(t)$, the formula in the STFT domain can be represented as

$$\mathbf{x}(t, f) = \mathbf{s}^T(t, f) + \mathbf{s}^I(t, f) \tag{2}$$

Phase recovery is ignored in this paper since the human is not so sensitive to phase distortion. The magnitude spectrum in Eq. (2) can be written in matrix form as follows

$$\mathbf{X} \approx \mathbf{S}^T + \mathbf{S}^I \tag{3}$$

where $\mathbf{S}^T$ and $\mathbf{S}^I$ are the unknown magnitude spectrums need to be estimated by DNN.

2.2 System Framework

The overall framework of the proposed method is showed in Fig. 1. Firstly, pairs of mixed signal and the sources are transformed to time-frequency domain by STFT and frames of magnitude spectrum can be obtained. Then, these training samples are sorted by a ranking function from "easy" to "hard" and fed to the DNN gradually. After the model is mature enough, it is applied to process the magnitude spectrum of test data and predict the spectrums of sources. Finally, an overlap add method is used to synthesize the waveform of the estimated signals [7].

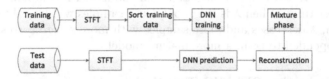

Fig. 1. System framework

2.3 DNN Learning Strategy

For all we know, one starts with small and easy curriculums, and then gradually increases the difficulty level. Inspired by this principle, Bengio et al. proposed a new learning paradigm in machine learning framework called curriculum learning (CL) [1]. The main idea is sorting the training samples by a difficulty measurement and then introducing them from easy to complex to the model. This strategy is proved to be effective to alleviate the bad local optimum problem in non-convex optimization and improve generalization ability [9,11]. The key of CL methodology is to find a ranking function that assigns learning importance to training samples. So the key to us is to find an appropriate ranking function for source separation problem.

In source separation, the target speech is corrupted by interfering speech to varying degrees over time. Empirically, we define the ranking function as follows:

$$f(\mathbf{x}_i, \mathbf{s}_i^T) = 10 log \frac{P_{\mathbf{s}_i^T}}{P_{\mathbf{x}_i}} \tag{4}$$

where $\mathbf{x}_i$ denotes ith frame of mixed magnitude spectrum which is fed to DNN, $\mathbf{s}_i^T$ represents ith frame of target speech, $P_{\mathbf{s}_i^T}$ and $P_{\mathbf{x}_i}$ are the energy of $\mathbf{s}_i^T$ and $\mathbf{x}_i$ respectively. It is easy to see that the bigger the value of f is the "easier" a sample is, since it means a larger proportion of energy the source accounts for. According to the function, the training samples can be sorted and the system will learn from easy to hard.

Formally, let $J(g(\mathbf{x}_i, \mathbf{w}), \mathbf{y}_i)$ denote the objective function of neural network which calculates the cost between the target output $\mathbf{y} = [\mathbf{s}_i^T; \mathbf{s}_i^I]$ and the estimated magnitude spectrums $g(\mathbf{x}_i, \mathbf{w})$. Here $\mathbf{w}$ represents the model parameters inside DNN. Then DNN can be optimized by minimizing:

$$\min_{\mathbf{w},\mathbf{v}} \sum_{i=1}^{m} v_i J(g(\mathbf{x}_i, \mathbf{w}), \mathbf{y}_i) \tag{5}$$

where v_i is determined by:

$$v_i = \begin{cases} 1 & f(\mathbf{x}_i, \mathbf{s}_i^T) > \lambda \ \text{or} \ f(\mathbf{x}_i, \mathbf{s}_i^I) > \lambda \\ 0 & otherwise \end{cases} \tag{6}$$

Equation (6) indicates that a sample is considered as easy one as long as either one of two sources occupies most energy of mixture, since it implies this sample could be separated more easily and should be introduced to the model earlier. The parameter λ controls the pace at which the model learns new samples. It decreases over time. When λ is large, only easy samples will be fed to DNN. As time goes on, λ decreases and more samples with more severe corrosion will be gradually appended to train a more mature model.

2.4   Discriminative Objective Function

In the training stage, the magnitude spectrums from pairs of mixed signal and the sources are utilized to train the DNN. Given $\mathbf{x}_i$ as input of DNN, the output $g(\mathbf{x}_i, \mathbf{w}) = [\tilde{\mathbf{s}}_i^T; \tilde{\mathbf{s}}_i^I]$ are expected to have small error with the target output $\mathbf{y}$, where $\tilde{\mathbf{s}}_i^T$ and $\tilde{\mathbf{s}}_i^I$ represent DNN estimates of target output $\mathbf{s}_i^T$ and $\mathbf{s}_i^I$, so conventionally one can optimize the neural network parameters by minimizing the squared error:

$$J_{mse}(g(\mathbf{x}_i, \mathbf{w}), \mathbf{y}_i) = \frac{1}{2} \sum_{i=1}^{m} (||\mathbf{s}_i^T - \tilde{\mathbf{s}}_i^T||^2 + ||\mathbf{s}_i^I - \tilde{\mathbf{s}}_i^I||^2) \tag{7}$$

Equation (7) enables DNN to separate two sources after training a set of samples. In order to further improve the separation quality, here we propose a different criteria to enhance the discrimination of the two predicted sources. An important fact is that Eq. (7) does not take into account source interference. If the two sources are similar, DNN may be confused and mistakes the target speech for the interfering speech. Correlation coefficient is a metric that measures the correlation between two signals and we expect to minimize the correlation coefficients of the sources to reduce the interference. Moreover, starting from an information theoretic viewpoint, the discrimination problem can be formulated as reducing the mutual information. Mutual information is a natural measure of the dependence between random variables. The mutual information can be approximated by the negentropy. Minimizing the mutual information is roughly equivalent to finding directions in which the negentropy is maximized. Taking into account these two measures, we add the following two parts to the original objective function in Eq. (7):

$$\begin{aligned} J_{cor}(g(\mathbf{x}_i, \mathbf{w}), \mathbf{y}_i) &= \sum_{i=1}^{m} \text{corr}(\tilde{\mathbf{s}}_i^T, \mathbf{s}_i^I) + \text{corr}(\tilde{\mathbf{s}}_i^I, \mathbf{s}_i^T) \\ &= \sum_{i=1}^{m} \frac{\text{cov}(\tilde{\mathbf{s}}_i^T, \mathbf{s}_i^I)}{\sqrt{D(\tilde{\mathbf{s}}_i^T)}\sqrt{D(\mathbf{s}_i^I)}} + \frac{\text{cov}(\tilde{\mathbf{s}}_i^I, \mathbf{s}_i^T)}{\sqrt{D(\tilde{\mathbf{s}}_i^I)}\sqrt{D(\mathbf{s}_i^T)}} \end{aligned} \tag{8}$$

$$J_{neg}(g(\mathbf{x}_i, \mathbf{w}), \mathbf{y}_i) = \sum_{i=1}^{m} (\mathrm{H}_G(\tilde{\mathbf{s}}_i^T) + \mathrm{H}_G(\tilde{\mathbf{s}}_i^I))$$

$$= \sum_{i=1}^{m} (\int p(\tilde{\mathbf{s}}_i^T) \log \frac{p(\tilde{\mathbf{s}}_i^T)}{p_G(\tilde{\mathbf{s}}_i^T)} d\tilde{\mathbf{s}}_i^T + \int p(\tilde{\mathbf{s}}_i^I) \log \frac{p(\tilde{\mathbf{s}}_i^I)}{p_G(\tilde{\mathbf{s}}_i^I)} d\tilde{\mathbf{s}}_i^I) \tag{9}$$

where Eqs. (8) and (9) represent the correlation coefficients and negentropy of the sources, respectively. In Eq. (8), $\mathrm{cov}(\cdot)$ denotes covariance and $D(\cdot)$ denotes variance. In Eq. (9), $p_G(\theta)$ is the density of a Gaussian random variable with the same covariance matrix as θ. To simplify the calculations, here we use nonlinear correlation coefficients [2] and negentropy approximate formula [7] instead,

$$J_{cor}(g(\mathbf{x}_i, \mathbf{w}), \mathbf{y}_i) = \sum_{i=1}^{m} (\sum_{j=1}^{n} \tilde{s}_{i,j}^T g(s_{i,j}^I) + \sum_{k=1}^{n} \tilde{s}_{i,k}^I g(s_{i,k}^T)) \tag{10}$$

$$J_{neg}(g(\mathbf{x}_i, \mathbf{w}), \mathbf{y}_i) = -\frac{1}{a} \sum_{i=1}^{m} E(e^{-\frac{a(\tilde{s}_i^T)^2}{2}} + e^{-\frac{a(\tilde{s}_i^I)^2}{2}}) \tag{11}$$

In Eq. (10), $r(\cdot)$ denotes nonlinear function and $2n$ denotes the dimension of DNN output. $E(\cdot)$ in Eq. (11) represents the statistical expectation of variables and the parameter a is usually chosen as 1. We expect to minimize J_{cor} and maximize J_{neg} for enhancing the discrimination. In order to estimate the unknowns in the model, we solve the following problem:

$$argmin \quad J(g(\mathbf{x}_i, \mathbf{w}), \mathbf{y}_i) \tag{12}$$

where

$$J(g(\mathbf{x}_i, \mathbf{w}), \mathbf{y}_i) = J_{mse} + \eta_1 J_{cor} - \eta_2 J_{neg} \tag{13}$$

is the joint discriminative function which we seek to minimize. η_1 and η_2 are regularization parameters which are chosen experimentally.

3 Experiments

3.1 Experimental Setup

In order to evaluate the performance of the proposed method, we conduct experiments on speech separation with TIMIT corpus. Two speakers, one male and one female, are chosen from database. To each speaker, 80% of the sentences are for training and 20% for testing. Mixed speech utterances were generated by mixing the sentences randomly from the two speakers at 0 dB signal-to-noise ratio (SNR). For increasing the number of training samples, we circularly shift points of the signal from male speaker and mix it with the female source. The time frequency representations are computed by the 512 point short time Fourier transform using a 32 ms window with a step size of 16 ms. Then 257-dimensional magnitude spectrums are used as input features to train DNN.

The separation result obtained from the proposed algorithm is compared with that of the standard NMF and DNN-based separation method [6]. In the NMF experiment, The number of basis vectors is set to 40 for each source. As for DNN, the architecture which jointly optimizes time-frequency masking functions as a layer with DNN in [6] is applied here. The neural network has 2 hidden layers with 160 nodes each, 2 hidden layers with 300 nodes each and 3 hidden layers with 160 nodes each. Pre-training is not adopted here benefits from the activation function Rectified Linear Unit (ReLU), which can reach the same performance without requiring any unsupervised pre-training on purely supervised tasks with large labeled datasets [6]. Empirically, the nonlinear function in Eq. (8) is chosen as $tanh(\cdot)$ and the value of parameters η_1 and η_2 is in the range of 0.1 and 0.3.

The separation performance is evaluated in terms of three metrics, signal to distortion ratio (SDR), signal to interference ratio (SIR), and signal to artifacts ratio (SAR) [14].

3.2 Experimental Results

The separation results with 2 hidden layers and 160 nodes are reported in Table 1. The DNNori in Table 1 means the basic DNN model with the objective function in Eq. (7). It is obvious from the results that all DNN-based methods outperform NMF, which confirms that neural network has better generalization and separation capacity. Compared the results between DNNori and DNNori-cl, which denotes the DNNori using the learning strategy proposed in Sect. 2, the SDR, SIR, and SAR all have been improved. It confirms the need for curriculum learning. Sorting the training samples by the ranking function and making the neural network learn like human brain from easy to complex does help the system to be more robust. As for DNNori-dis, we use the discriminative cost function in (13) instead of the function in Eq. (7). It is interesting to find that the improvement in SDR is so slight that can be ignored, but the SIR achieves around 1.2 dB gain compared to DNNori. The results match our expectation when we design the objective function which aims at enhancing the discrimination and reducing the interfering of the other source. The disadvantage is that some artifacts are introduced into the separation and the SAR is lower than DNNori.

Table 1. Speech separation results of various separation algorithms.

Method	Measurement (dB)		
	SDR	SIR	SAR
NMF	6.008	8.722	7.624
DNN[13]	7.70	11.53	8.07
DNNori	7.58	10.81	8.31
DNNori-cl	7.73	10.88	8.78
DNNori-dis	7.62	12.06	7.86
DNNori-cl-dis	7.87	12.12	8.15

Finally, in DNNori-cl-dis, the curriculum learning strategy and the discriminative objective function are both added in. We can see from the results that the two techniques both play an effective role in single channel source separation. To strongly demonstrate the jointly function of the curriculum learning strategy and the discriminative objective function proposed, we do some experiments on different layers and different notes respectively. And the results are shown in Table 2. According to Table 2, we can see that the case having 3 hidden layers and 160 nodes for each layer and the case having 2 hidden layers and 200 nodes for each layer have achieved different level improvement, which is fit for the conclusion we make above. The model achieves best performance in SDR and SIR.

Table 2. Speech separation results of various network structures.

Method	Measurement (dB)		
	SDR	SIR	SAR
DNNori-cl-dis (3*160)	7.883	11.693	8.522
DNN[13] (3*160)	7.486	10.962	8.257
DNNori-cl-dis (2*300)	7.79	11.477	8.459
DNN[13] (2*300)	7.763	11.343	8.442

4   Conclusions

In this paper, the DNN is used to model each source signal and trained to separate the mixed signal. Two novel improvements have been proposed to further enhance the separation performance: a learning strategy based on curriculum learning and a discriminative objective function that reduces the interference from the other source. We have proved that the proposed algorithm achieves better results through a series of experiments on speech separation. The future work will focus on improving the proposed method by combining the phase separation with DNN training.

References

1. Bengio, Y., Louradour, J., Collobert, R., Weston, J.: Curriculum learning. In: Proceedings of the 26th Annual International Conference on Machine Learning, pp. 41–48. ACM (2009)
2. Cichocki, A., Unbehauen, R., Rummert, E.: Robust learning algorithm for blind separation of signals. Electron. Lett. **30**(17), 1386–1387 (1994)
3. Erhan, D., Szegedy, C., Toshev, A., Anguelov, D.: Scalable object detection using deep neural networks. In: Proceedings of the IEEE Conference on Computer Vision and Pattern Recognition, pp. 2147–2154 (2014)

4. Grais, E.M., Erdogan, H.: Single channel speech music separation using nonnegative matrix factorization and spectral masks. In: 2011 17th International Conference on Digital Signal Processing (DSP), pp. 1–6. IEEE (2011)

5. Hu, G., Wang, D.: Monaural speech segregation based on pitch tracking and amplitude modulation. IEEE Trans. Neural Netw. **15**(5), 1135–1150 (2004)

6. Huang, P.S., Kim, M., Hasegawa-Johnson, M., Smaragdis, P.: Deep learning for monaural speech separation. In: 2014 IEEE International Conference on Acoustics, Speech and Signal Processing (ICASSP), pp. 1562–1566. IEEE (2014)

7. Hyvarinen, A.: Fast and robust fixed-point algorithms for independent component analysis. IEEE Trans. Neural Netw. **10**(3), 626–634 (1999)

8. Kang, T.G., Kwon, K., Shin, J.W., Kim, N.S.: Nmf-based target source separation using deep neural network. IEEE Sig. Process. Lett. **22**(2), 229–233 (2015)

9. Khan, F., Mutlu, B., Zhu, X.: How do humans teach: on curriculum learning and teaching dimension. In: Advances in Neural Information Processing Systems, pp. 1449–1457 (2011)

10. Lee, Y.K., Kwon, O.W.: Application of shape analysis techniques for improved casa-based speech separation. IEEE Trans. Consum. Electron. **55**(1), 146–149 (2009)

11. Ni, E.A., Ling, C.X.: Supervised learning with minimal effort. In: Zaki, M.J., Yu, J.X., Ravindran, B., Pudi, V. (eds.) PAKDD 2010. LNCS, vol. 6119, pp. 476–487. Springer, Heidelberg (2010). doi:10.1007/978-3-642-13672-6_45

12. Smaragdis, P., Raj, B., Shashanka, M.: Supervised and semi-supervised separation of sounds from single-channel mixtures. In: Davies, M.E., James, C.J., Abdallah, S.A., Plumbley, M.D. (eds.) ICA 2007. LNCS, vol. 4666, pp. 414–421. Springer, Heidelberg (2007). doi:10.1007/978-3-540-74494-8_52

13. Sun, D.L., Mysore, G.J.: Universal speech models for speaker independent single channel source separation. In: 2013 IEEE International Conference on Acoustics, Speech and Signal Processing, pp. 141–145. IEEE (2013)

14. Vincent, E., Gribonval, R., Févotte, C.: Performance measurement in blind audio source separation. IEEE Trans. Audio Speech Lang. Process. **14**(4), 1462–1469 (2006)

15. Wang, D.L., Brown, G.J.: Separation of speech from interfering sounds based on oscillatory correlation. IEEE Trans. Neural Netw. **10**(3), 684–697 (1999)

16. Wang, Z., Sha, F.: Discriminative non-negative matrix factorization for single-channel speech separation. In: 2014 IEEE International Conference on Acoustics, Speech and Signal Processing (ICASSP), pp. 3749–3753. IEEE (2014)

17. Xue, S., Abdel-Hamid, O., Jiang, H., Dai, L., Liu, Q.: Fast adaptation of deep neural network based on discriminant codes for speech recognition. IEEE/ACM Trans. Audio Speech Lang. Process. **22**(12), 1713–1725 (2014)

Sparse Direct Convolutional Neural Network

Vijay Daultani[✉], Yoshiyuki Ohno, and Kazuhisa Ishizaka

System Platform Research Laboratories, NEC Corporation, Tokyo, Japan
v-daultani@ax.jp.nec.com, y-ohno@ji.jp.nec.com, k-ishizaka@ay.jp.nec.com
http://www.nec.com

Abstract. We propose a new computation and memory efficient algorithm to speed up Convolutional Neural Networks (CNNs). Equipped with several millions of parameters, leveraging large datasets, CNNs have achieved state-of-the-art recognition accuracy. Recently utilizing sparsity of parameters, several acceleration techniques for CNNs have been introduced, causing a paradigm shift in type of computation from dense to sparse, leading to opportunity for designing a new convolution algorithm suiting high bandwidth performance architecture like SX-ACE.

In this paper we propose a new computation and memory efficient convolution algorithm for inference phase, Sparse Direct Convolution (SDC) and a new representation for sparse filters, Compressed Sparse Offset (CSO). We evaluate our implementation of SDC together with CSO on high bandwidth SX-ACE architecture and show inference time of single convolution layer can reduce with up to 95% of lcnet, 65% of alexnet and 69% of VGG-16 without drop in accuracy.

Keywords: Deep learning · Convolutional Neural Networks · Accelerated convolution algorithm · Vector architecture

1   Introduction

Increasing recognition accuracy of successive state-of-the-art CNNs can be correlated to depth i.e. increasing number of convolutional layers, for example CNNs Alexnet [2] 2012, VGG-16 [3] 2014 and Resnet [4] 2015, had 5, 13, and 152 convolutional layers respectively. Convolution operation takes almost 90% of total execution time for inference phase [5] and given above trend for increasing number of convolutional layers it is evident that convolutional operation is and will remain one of the dominant component of CNN.

Inevitable end of "Moores Law" predicted around 2025–2030 [6] will cease computation or system FLOPS to improve, but bandwidth or BYTES will continue to improve at least for a few decades beyond.

Convolution operation can be realized using multiple algorithms like Direct Convolution (DC), Matrix Multiplication (MM) and Fast Fourier Transform (FFT). Most of the implementations of these algorithms are computation limited (i.e. large FLOPS/BYTES) on general purpose hardware architectures (GPHA). Since these algorithm's arithmetic intensity resides on right of ridge point in roofline [7] analysis on most of the hardware architectures, hence usually are

© Springer International Publishing AG 2017
F. Cong et al. (Eds.): ISNN 2017, Part I, LNCS 10261, pp. 293–303, 2017.
DOI: 10.1007/978-3-319-59072-1_35

computation limited. This can be verified by calculating arithmetic intensity and performing roofline [7] analysis of the implementation.

Recently researchers have found that CNNs are often over-parametrized in [13, 15, 16] and have shown same recogintion accuracy can be achieved with much less number of parameters by pruning redundant parameters. Motivated by the idea of realizing CNNs on hardware constrained devices researchers have proposed techniques to reduce the number of computations by targeting over-parametrization. Such techniques eliminate redundancy between parameters and introduce sparsity, which leads to transition of convolution operation from dense/regular memory access/computation limited to sparse/irregular (stride) memory access/bandwidth limited operation.

Techniques to utilize sparsity of fully connected layers on customized hardware architecture (CHA) like EIE [15] have shown significant speed ups, compared to speed ups achieved on GPHA. One reason for comparatively less speedup on GPHA is absence of hardware level optimizations specific to CNNs, since they have to serve wide range of applications.

These three reasons, first: system FLOPS will cease to improve early on contrary to bandwidth, second: change of computation type of convolution from dense to sparse, and third: absence of efficient algorithm for utilizing sparsity for convolution operation on GPHA lead to requirement of a convolution algorithm that is both computation and bandwidth efficient.

In this paper we present Sparse Direct Convolution (SDC) algorithm for convolution and Compressed Sparse Offset (CSO) for encoding non-zero values of sparse filters. We show together both of the algorithms have universal effect of reducing the inference time of state-of-the-art CNNs. The main contributions of this paper are:

- First, we present a simple Sparse Direct Convolution (SDC) algorithm, which is both computation and memory efficient algorithm for convolution operation which works without transforming the input feature maps before each convolution layer, as required by MM or FFT.
- Second, we present Compressed Sparse Offset (CSO) technique, both computation and memory efficient way to store the indices of the non-zero values of filters 4D tensor.
- Finally, we evaluate our algorithms on a high memory bandwidth architecture like SX-ACE, and demonstrate for various different convolution layers our method SDC plus CSO outperforms the state-of-the-art and most commonly used MM for convolution operation with a difference of magnitude for total execution time of inference phase.

2   Background

2.1   Convolutional Neural Networks

Figure 1 shows how different components are stacked over each other to form a CNN. Conventional CNN is made up of the following components.

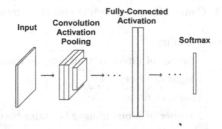

Fig. 1. A typical CNN architecture showing the layers discussed in the Sect. 2.1

Input Layer holds the input image as input to pass it on to the next layer in CNN.

Convolutional Layer extracts features from input feature maps as a result of spatial convolution between filter's 3D tensor and a patch of input feature maps 3D tensor as shown in Eqs. 2 and 3. Since each convolutional layer usually consists of multiple filters they are represented by 4D tensor i.e. filter number, filter height, filter width, and filter depth. Also inorder to maximally utilize the available hardware, multiple images are processed simultaneously in a batch and are also represented by 4D tensor i.e. image number, image height, image width, image depth. Input feature maps for multiple input images in a single batch (4D tensor) $I \in \mathbb{R}^{NCHW}$, when convolved with multiple fitlers (4D tensor) $F \in \mathbb{R}^{KCRS}$ results in output feature maps corresponding to each input image in batch (4D tensor) $O \in \mathbb{R}^{NKPQ}$. Table 1 explains parameters and their meaning for convolutional layer. Height and width of output feature maps i.e. P and Q are calculated using Eq. 1.

$$f(H, pad\_h, R, u) = \left\lceil \frac{H + 2 * pad\_h + 1 - R}{u} \right\rceil \tag{1}$$

$$m(p, u, pad\_h, R, r) = p * u - pad\_h - R - r - 1 \tag{2}$$

$$O[n][k][p][q] = \sum_{c=0}^{C-1} \sum_{r=0}^{R-1} \sum_{s=0}^{S-1} I[n][c][m(p, u, pad\_h, R, r)][m(q, v, pad\_w, S, s)] \cdot F[k][c][r][s] \tag{3}$$

Activation Layer. Each output feature map from convolutional layer is passed through non-linearity i.e. activation layer, which aids in visual recognition task. Many previous state-of-the-art CNNs used sigmoid and tanh non-lineartiy, but ReLU has gained popularity recently and is most commonly used, for its simplicity and effectiveness. **Pooling Layer:** It is common to insert pooling layer in between successive convolution layers. Pooling layer reduces the spatial size of the representation to reduce the number of parameters and hence computation in the network. Pooling layer provides the translation invariance property to the image recognition task and helps to avoid over-fitting. Most common filter size for pooling is 2×2 with a stride of 2. However other filter size for pooling can be chosen. **Fully Connected Layer:** A series of group of Convolution, Activation,

Table 1. Convolution parameters and their meaning

Parameter	Meaning
K	Number of filters/output feature maps
R	Height of filter
S	Width of filter
N	Number of input images i.e. mini-batch size
C	Number of input feature maps
H	Height of input feature map
W	Width of input feature map
u	Stride in vertical direction
v	Stride in horizontal direction
pad_h	Zero-padding in horizontal direction
pad_w	Zero-padding in vertical direction
P	Height of output feature map
Q	Width of output feature map

Pooling layers are usually followed by a series of fully connected layers. Fully connected layers are realized by first transforming the 4D arrays of input feature maps and filters for neurons in layer to 2D matrix and then matrix multiplication is used to generate the output of fully connected layer. **Softmax Layer:** Softmax is the final layer of CNN and is used to calculate the probability distribution over different output classes.

2.2   Different Implementations of Convolution

Some convolution algorithm used in state-of-the-art deep learning frameworks are as follows.

Direct Convolution (DC). Cuda-Convnet [8] one of the first highly optimized CNNs implementation for GPUs realizes convolution operation using DC. Being one of the straight forward implementation of convolution operation, it uses CHWN memory layout for 4D tensors. Advantage of this algorithm is it does not require any data transformation from 4D tensor to 2D tensor. Disadvantage of this algorithm is it requires non regular memory access.

Matrix Multiplication (MM). Caffe [9] one of the famous machine learning framework realizes convolution operation using MM. This algorithm requires first to transform, both input feature maps for each image and filters to 2D tensor i.e. matrix followed by MM. Advantage of this algorithm is that one can use highly optimized MM implementation available in well tuned BLAS libraries for GPHA. Disadvantage of this algorithm is that it involves overhead of data transformation of input feature maps to 2D tensor.

Fast Fourier Transform (FFT). FFT [11] and its variation in fbfft [12] and cuDNN [10] have recently shown FFT can be used to realize convolution operation efficiently. This algorithm first transforms input feature maps and filters from time domain to frequency domain, followed by performing dot product in frequency domain, finally followed by transform of output feature maps from frequency domain to time domain. Advantage of this algorithm is that the number of computations are reduced when convolution performed in frequency domain. Disadvantage of this algorithm is it incurs overhead of data transformation from time domain to frequency domain and vice versa and is effective for large kernels [10,12].

3   Related Work

Memory Efficiency for Deep Convolutional Neural Networks on GPUs: This work [14] proposed heuristics to estimate the best memory layout (CHWN or NCHW) for 4D tensors, and implemented a low overhead routine to transform the tensors from one layout to another if required to assist convolution algorithm for acceleration. On contrary we propose a new accelerated convolution algorithm itself. This work is independent of nature of values in tensors (i.e. dense or sparse) on contrary our work focus on sparsity in filters tensor. This work improves memory efficiency of GPUs on contrary our work is based on improving both computation and memory efficiency of SX-ACE architecture. **Efficient Inference Engine:** This work [15] similar to ours is based on technique of deep compression [13] This work proposed CHA i.e. EIE on contrary we propose a new convolution algorithm irrespective of hardware architecture. This work uses variant of CSC (compressed sparse column) storing value, row index, and column pointer on contrary we propose a new sparse filter representation CSO (compressed sparse offset) storing only value and offset. This work focus on FC layers (bandwidth limited) on contrary our work focuses on convolution layer (computation limited). **Sparse Convolutional Neural Netowrks:** This work [16] introduced sparsity using inter-channel and intra-channel decomposition, on contrary we use pruning and retraining [13]. This work incurs the overhead of transforming the input feature maps 3D tensor to matrix on contrary our method does not require such transformation. This work focuses on sparse-dense matrix (2D tensor) multiplication, on contrary our work focuses on list-tensor (4D tensor) multiplication.

4   Proposed Technique

Several recent work [13,16] have shown the redundancy in parameters of CNNs exists and it can be eliminated to significant level. Deep compression [13] technique proposed using pruning and retraining redundancy can be removed. This procedure results in sparse parameters, while achieving original CNN accuracy. We extend this work by first proposing a new convolution algorithm SDC and second proposing a new representation for sparse filters CSO and sparse filters preprocessing algorithm.

Previous techniques have focused on compressing parameters to reduce memory footprints of CNNs therefore more beneficial for FC layer, whereas we in our work have shown with reasonable sparsity and using our algorithm we can reduce execution time of convolutional layer which is more computation intensive.

4.1 Preprocessing Filters

Once sparse filters 4D tensor are learned in training phase [13], Algorithm 1 scans sparse filters (NCHW memory layout). It first initialize list all_filters, whose each entry is of form single_filter. It then for each filter initializes a list i.e. single_filter whose entry is of form (input_offset, filter_value). For loops in line 4,5,6 represents three different dimensions of each filter i.e. depth, height, width. If a filter value is found non zero in line 9, corresponding 1D index is saved in input_offset and an entry (input_offset, filter_value) is inserted in single_filter in line 16. Unique to our algorithm input_offset is the 1D index in CHWN memory layout of pixel in first patch of the input feature map of first image in batch corresponding to the non zero filter_value. Figure 2 explains CSO format using simple example, Input feature map of size 6×6 (H/W = 6) is convolved with filter of size 4×4 (R/S = 4) with stride 1 (u/v = 1). Color cells of Kernel represents non zero value, which are stored in Filter Values list (filter_value in Algorithm 1) and corresponding offsets for the non zero values of filters in input feature maps are stored in Input Offset (input_offset in Algorithm 1). Filter Values and Input Offset remains same for all the patches of the input feature maps. With CSO each non zero filter's value in 3D tensor is uniquely identified by two items i.e. filter_value and input_offset on contrary to three items i.e. filter_value, row_index and column_pointer used in CSC format used for sparse matrix representation in original deep compression [13] technique. Reducing the number of items required to represent also reduces the number of scalar computations to generate the index of input feature map in Algorithm 2 and hence contribute to reduce the inference time further.

4.2 Sparse Direct Convolution Algorithm

Figure 3 explain how CSO and SDC work together. Once sparse filters are preprocessed, corresponding CSO format representation is used by Algorithm 2. SDC algorithm then for each convolution layer, performs the convolution between input feature maps i.e. I and filters i.e. all_filters. The output is generated in O. For each p, q pixel in output feature map f_nz, we first calculate starting row and column for the input feature map patch using formula i = p * u and j = q * v in line 5 and 7 respectively. Then 1D starting index of patch is calculated in line 8 and stored in patch_index. Pixel value for O for the batch size N, is first copied onto the vector register in line 10. Multiplication between non-zero value filter_value and corresponding pixel of input feature map I for all images is then performed in line 13. After all non-zero entries i.e. input_offset and filter_value in set single_filter have been used to determine the value of output

Algorithm 1. Preprocessing Filters

```
1: initialize all_filter
2: for f ∈ {0, ..., F − 1} do
3:    initialize single_filter
4:    for c ∈ {0, ..., C − 1} do
5:       for r ∈ {0, ..., R − 1} do
6:          for s ∈ {0, ..., S − 1} do
7:             index = NCHW_INDEX(f, c, r, s, F, C, R, S)
8:             filter_value = filters[index]
9:             if filter_value != 0 then
10:                input_offset = CHWN_INDEX(0, c, r, s, N, C, H, W)
11:                insert entry (input_offset, filter_value) in single_filter
12:            end if
13:         end for
14:      end for
15:   end for
16:   insert entry (f, single_filter) in all_filters
17: end for
18: return all_filters
```

pixel value at height p and width q, output is written back from vector register to memory in line 15.

$$NCHW\_index(n, c, h, w, N, C, H, W) = (((n * C + c) * H + h) * W + w)$$
$$CHWN\_index(n, c, h, w, N, C, H, W) = (((c * H + h) * W + w) * N + n) \quad (4)$$

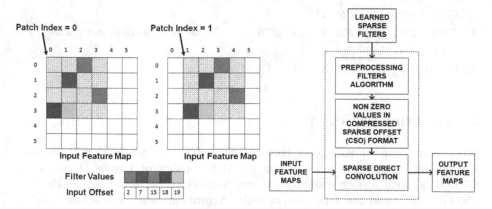

Fig. 2. Compressed sparse format

Fig. 3. Preprocessing filters and sparse direct convolution algorithm

---

Algorithm 2. Sparse Direct Convolution

1: **all_filters** = Preprocessing Filters
2: **for** $(filter\_no,$ **single_filter**$) \in$ **all_filters do**
3:    $f\_nz = filter\_no$
4:    **for** $p \in \{0, \ldots, P - 1\}$ **do**
5:       $i = p * u$
6:       **for** $q \in \{0, \ldots, Q - 1\}$ **do**
7:          $j = q * v$
8:          $patch\_index = \text{CHWN\_INDEX}(0, 0, i, j, N, C, H, W)$
9:          $output\_index = \text{CHWN\_INDEX}(0, f\_nz, p, q, N, K, P, Q)$
10:          $output\_reg[0 : 0 + N] = O[output\_index : output\_index + N]$
11:          **for** $(input\_offset, filter\_value) \in$ **single_filter do**
12:             $input\_index = patch\_index + input\_offset$
13:             $output\_reg[0 : 0+N]$ += $I[input\_index : input\_index+N] * filter\_value$
14:          **end for**
15:          $O[output\_index : output\_index + N] = output\_reg[0 : 0 + N]$
16:       **end for**
17:    **end for**
18: **end for**

Table 2. CNNs and their layers used for the experiments.

Layer	C	H/W	F	R/S	u/v	Sparsity	Description
CONV1	1	28	20	5	1	34	Lenet [1] on MNIST dataset
CONV2	20	12	50	5	1	88	
CONV3	3	227	96	4	4	16	Alexnet [2] on Imagenet dataset
CONV4	48	27	128	5	1	62	
CONV5	256	13	384	3	1	65	
CONV6	192	13	192	3	1	63	
CONV7	192	13	128	3	1	63	
CONV8	256	56	256	3	1	76	VGG-16 [3] on Imagenet dataset
CONV9	256	56	512	3	1	68	
CONV10	512	28	512	3	1	66	

5 Experiment Set up

We conduct experiments on 1 core of Single CPU of SX-ACE with Super-UX operating system. Each CPU on SX-ACE consists of 4 cores (1 core - 64 GFLOP/s) with a peak computation and bandwidth performance of $64 \times 4 = 256$ GFLOP/s and 256 GB/s respectively. Algorithm used for baseline is MM, which first uses im2col operation (tuned for SX-ACE) to transform 3D tensor to 2D tensor, and followed by dense MM from Mathkeisan [18]. In Table 2 columns C, H, F, R, u have meanings as defined in Table 1. Column sparsity specifies the percentage of values in filters which are zero. Sparsity% is same as $100 - \text{Weight}\%$ where Weight% been shown to retain original accuracy of deep compression [13]

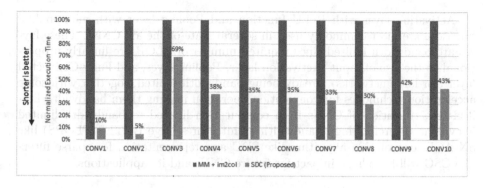

Fig. 4. Percentage reduction in inference time of convolution layers from Table 2

technique. A dry run to warm up cache was followed by 10 iterations of each algorithm's implementation.

6   Results & Analysis

Figure 4 shows percentage reduction in inference time on average for each convolution layer from Table 2. Reduction of 90%, 95%, 31%, 62%, 65%, 65%, 67%, 70%, 58%, 57% were observed for convolution layers of Table 2 respectively. One reason for high percentage reduction for CONV1 and CONV2 (Lenet being inefficiency of MM for small matrix size. Utilizing CSO format SDC algorithm works without any data transformation at inference time and helps to achieve good equilibrium between computation and memory access. Results shows SDC algorithm is both computation efficient: runs near peak performance like MM and memory efficient: loads only necessary input feature maps pixel values. Also CSO format is both computation efficient: storing only 1D input offset saving scalar operations at inference time and memory efficient: encoding with only 1D input offset saves memory access.

7   Conclusion

We proposed fast SDC convolution algorithm and CSO format for encoding sparse filters. SDC possess cherry-pick properties like, no data transformation, computation and memory efficient not available together in other convolution algorithms. With reasonable sparsity and single memory layout (CHWN) SDC when used with CSO can outperform known state-of-the-art implementation of convolution like MM in terms of execution time of convolution operation in inference phase by substantial factor. The experiments demonstrate effectiveness of our method and their universal effect on convolution layers of different state-of-the-art CNNs.

Activation layer with ReLU function and Pooling layer with max-pooling function, are most commonly used in several state-of-the-art CNNs. These layers are bandwidth limited since required number of FLOP are linearly proportional to the size of input. Activation layer, Pooling layer, and Fully Connected layer (for large networks) already being bandwidth limited, together with some acceleration techniques for convolution operation making them bandwidth limited too, transition of CNNs from computation limited to bandwidth limited is evident. Therefore high bandwidth architectures(large BYTES/FLOPS) like SX-ACE, convolution algorithm like SDC and representation for sparse filters like CSO will have huge impact for future CNNs and its applications.

References

1. LeCun, Y., Bottou, L., Bengio, Y., Haffner, P.: Gradient-based learning applied to document recognition. In: Proceedings of the IEEE (1998)
2. Krizhevsky, A., Sutskever, I., Hinton, G.: ImageNet classification with deep convolutional neural networks. In: NIPS (2012)
3. Simonyan, K., Zisserman, A.: Very deep convolutional networks for large-scale image recognition. In: ICLR (2015)
4. He, K., Zhang, X., Ren, S., Sun, J.: Deep residual learning for image recognition. In: arXiv (2015)
5. Jia, Y.: Learning semantic image representations at a large scale. Ph.D. dissertation: UC Berkeley (2014)
6. Matsuoka, S., Amano, H., Nakajima, K., Inoue, K., Kudoh, T., Maruyama, N., Taura, K., Iwashita, T., Katagiri, T., Hanawa, T., Endo, T.: From FLOPS to BYTES: disruptive change in high-performance computing towards the post-moore era. In: CF (2016)
7. Williams, S.W., Waterman, A., Patterson, A.: Roofline: an insightful visual performance model for floating-point program and multicore architecture. Technical report No. UCB/EECS-2008-134. https://www2.eecs.berkeley.edu/Pubs/TechRpts/2008/EECS-2008-134.pdf
8. cuda-convnet. https://code.google.com/p/cuda-convnet/
9. Jia, Y., Shelhamer, E., Donahue, J., Karayev, S., Long, J., Girshick, R., Guadarrama, S., Darrell, T.: Caffe: Convolutional Architecture for Fast Feature Embedding. https://arxiv.org/abs/1408.5093
10. Chetlur, S., Woolley, C., Vandermersch, P., Cohen, J., Tran, J., Catanzaro, B., Shelhamer, E.: cuDNN: Efficient Primitives for Deep Learning. https://arxiv.org/abs/1410.0759
11. Mathieu, M., Henaff, M., LeCun, Y.: Fast training of convolutional networks through FFTs. In: arxiv (2013)
12. Vasilache, N., Johnson, J., Mathieu, M., Chintala, S., Piantino, S., LeCun, Y.: Fast convolutional nets with fbfft: a GPU performance evaluation. In: ICLR (2015)
13. Han, S., Mao, H., Dally, W.J.A.: Deep compression: compressing deep neural networks with pruning, trained quantization and Huffman coding. In: ICLR (2016)
14. Li, C., Yang, Y., Feng, M., Chakradhar, S., Han, H.Z.: Optimizing memory efficiency for deep convolutional neural networks on GPUs. In: SC (2016)
15. Han, S., Liu, X., Mao, H., Pu, J., Pedram, A., Horowitz, M., Dally, J.: EIE: efficient inference engine on compressed deep neural network. In: ISCA (2016)

16. Liu, B., Wang, M., Foroosh, H., Tappen, M., Penksy, M.: Sparse convolutional neural networks. In: CVPR (2015)
17. Vuduc, R.W.: Automatic performance tuning of sparse matrix kernels. Ph.D. dissertation: UC Berkeley (2003)
18. MathKeisan. http://www.mathkeisan.com/

Fuzzy Modeling from Black-Box Data with Deep Learning Techniques

Erick de la Rosa[1], Wen Yu[1]([✉]), and Humberto Sossa[2]

[1] Departamento de Control Automático, CINVESTAV-IPN, Mexico City, Mexico
yuw@ctrl.cinvestav.mx
[2] Instituto Politécnico Nacional, Centro de Investigación en Computación,
Mexico City, Mexico

Abstract. Deep learning techniques have been successfully used for pattern classification. These advantage methods are still not applied in fuzzy modeling. In this paper, a novel data-driven fuzzy modeling approach is proposed. The deep learning methods is applied to learn the probability properties of input and output pairs. We propose special unsupervised learning methods for these two deep learning models with input data. The fuzzy rules are extracted from these properties. These deep learning based fuzzy modeling algorithms are validated with three benchmark examples.

Keywords: Fuzzy system · Black-box modeling · Deep learning

1 Introduction

Human always use "IF-THEN" rules in their thinking. These linguistic propositions are applied in fuzzy system [21]. Compared with the other modeling techniques, the fuzzy modeling of nonlinear systems using fuzzy system is easier to explain, it can directly use information in different forms, it is simple and can be trained easily [16]. There are the following two basic methods to construct the fuzzy rules [11]: (1) extract rules from human knowledge, (2) analyze data to derive rules. The first method need complex work of human [13]. When only the data of the unknown nonlinear systems are available, it is data-based fuzzy modeling. The fuzzy modeling using data includes two stages: (1) structure identification: extract fuzzy rules from input/output data; (2) parameters identification: updated the membership functions of the fuzzy rules with observation data.

Extracting rules from data always uses data partition [7,10,12]. Kernel method [19], neural networks [10], genetic method [13], SVD based algorithm [6] and SVM [5] are also popular. In [18] SVM is extended into on-line version, such that the fuzzy modeling based on SVM can be carried out on-line. [1,15] use adaptive resonance to construct fuzzy sets from input data. In [17] we use online clustering method to partition both input (precondition) and the output (consequent) at the same time.

© Springer International Publishing AG 2017
F. Cong et al. (Eds.): ISNN 2017, Part I, LNCS 10261, pp. 304–312, 2017.
DOI: 10.1007/978-3-319-59072-1_36

In this paper, we will use a complete novel method to extract fuzzy rules. The fuzzy modeling of this paper uses deep learning as the structure identification (rule extraction). We first use deep neural model to analyze the data. Then we use decision tree to extract fuzzy rules from the deep neural model.

Deep learning techniques have achieved impressive good results in many difficult tasks in recent years [2]. Deep learning models are neural networks, which have been widely used for nonlinear systems identification. Multilayer perceptron is a common model [3]. The main difficult of deep learning for nonlinear modeling is to find the structure of the deep neural model [8]. In fact, no any optimal method could be found [4]. Some popular methods are the grid search and random search [4]. In this paper, we will use data-based fuzzy modeling to help the extraction of deep neural model.

In the sense of probability theory, the objective of system identification is to find the best conditional probability distribution $P(y|x)$ between the input x and the output y. Recent results show that deep learning techniques can be applied for nonlinear system modeling [3]. The unsupervised learning is used to obtain the input features and sent to hidden layers. The modeling accuracy can be improved.

Almost all classification tasks use binary values [9]. However, the input/ output values of the nonlinear systems cannot be binary values. In [2], the structure of the deep models are changed, such that the probability can be calculated.

In this paper, we use the restricted Boltzmann machines to design the input unsupervised learning. Then we use the popular decision tree [14] to construct the fuzzy rules from the deep neural networks. The data-driven fuzzy modeling using deep learning is shown in Fig. 1. The membership functions will be updated automatically. The gas furnace data are applied to test our deep learning methods.

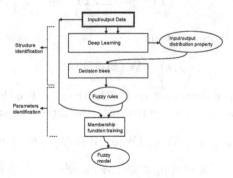

Fig. 1. Data-driven fuzzy modeling using deep learning

2   Fuzzy Modeling Based on Input-Output Data

A unknown nonlinear system is described by the difference equation

$$\mathbf{y}(k) = \Phi\left[\mathbf{x}(k), k\right] \tag{1}$$

where $\Phi(\cdot)$ is an unknown nonlinear function,

$$\mathbf{x}(k) = [\mathbf{y}(k-1), \cdots \mathbf{y}(k-n_y), u(k), \cdots u(k-n_u)]^T \tag{2}$$

$u(k)$ and $\mathbf{y}(k)$ are the control and the output, n_y and n_u correspond to the system order, $\mathbf{x}(k) \in \Re^n$ can be regarded as a new input to the nonlinear function $f(\cdot), n = n_y + n_u + 1$. It is the well known NARMAX model.

For the unknown nonlinear system (1), only $x(k)$ and $y(k)$ are available. We use the following Mamdani fuzzy model

$$\mathrm{R}^j : \text{IF } x_1(k) \text{ is } A_1^j \text{ and } x_2(k) \text{ is } A_2^j \text{ and} \cdots x_n(k) \text{ is } A_n^j \text{ THEN } y(k) \text{ is } B^j$$

here $j = 1 \cdots J$, $A_1^j, \cdots A_n^j$ and B^j are fuzzy sets. The membership function is

$$\mu_i(x_j) = \exp\left(-\frac{(x_j - c_{ji})^2}{\sigma_{ji}^2}\right) \tag{3}$$

where i is the condition in the part "IF", $i = 1 \cdots n$, j is rule number, $j = 1 \cdots J$. From [16], the output of the fuzzy system is

$$\hat{y} = \left(\sum_{j=1}^J w_j \left[\prod_{i=1}^n \mu_{A_i^j}\right]\right) / \left(\sum_{j=1}^J \left[\prod_{i=1}^n \mu_{A_i^j}\right]\right) \tag{4}$$

where $\mu_{A_i^j}$ is for A_i^j, w_j is for $\mu_{B_j} = 1$. Define

$$\phi_i = \prod_{j=1}^n \mu_{A_{ji}} / \sum_{i=1}^{sv_j} \prod_{j=1}^n \mu_{A_{ji}}$$

(4) can be expressed in matrix form

$$\hat{y}(k) = \mathbf{W}(k)\,\Phi\,[\mathbf{x}(k)] \tag{5}$$

In order to obtain a decision tree for the data $[x(k, y(k))]$, we use the following deep neural networks to approximate the input-output mapping

$$\widehat{y}(k) = \phi_p(W_p \phi_{p-1} \dots W_3 \phi_2 \{W_2 \phi_1 [W_1 x(k) + b_1] + b_2\} \dots + b_p) \tag{6}$$

where $\widehat{y}(k) \in \Re^m$, $W_1 \in \Re^{l_1 \times n}$, $b_1 \in \Re^{l_1}$, $W_2 \in \Re^{l_2 \times l_1}$, $b_2 \in \Re^{l_2}$, $W_p \in \Re^{m \times l_{p-1}}$, $b_p \in \Re^m$.

The deep model has p layers, each layer has l_i ($i = 1, \cdots, p-1$) nodes. In this paper, we let $p \geq 3$

$\phi_i \in \Re^{l_i}$ ($i = 1 \cdots p$) are Sigmoid function

$$\phi_i(w_j) = \alpha_i / \left(1 + e^{-\beta_i^T w_j}\right) - \gamma_i$$

In the output layer, ϕ_p is linear, $\Re^m \to \Re^m$, i.e., $\phi_p = [\sum_1 \cdots \sum_m]$.

The deep neural model increased the layer number p, not as the normal neural model to increase the node number l_i.

We first use the random search to find p and l_i. Besides the structure identification, which needs the layer number p and node numbers l_i, parameter identification is to update the weights $(W_1 \cdots W_p)$ to minimize the modeling error with respect to an index.

$$e(k) = \widehat{y}(k) - y(k) \tag{7}$$

3   Modelling Input-Output Property with Deep Learning

The restricted Boltzmann machines (RBMs) are generative energy based models. They learn from the input data through the usage of latent or hidden variables. The latent variables capture features of the data, which help RBM to obtain better representation of the empirical distribution.

RBM is a very successful method in feature extractions from image and text data. It is also an excellent pre-training tool to set the initial parameters for the discriminative models.

The input features extraction is to learn the probability distribution of its input set. The input data to the RBM is $\mathbf{x}(k) = [x_1 \cdots x_n] \in R^n$. In the sense of the coding phase, the output of each RBM is $\bar{\mathbf{h}} = [\bar{h}_1 \cdots \bar{h}_s] \in R^s$. For the $i-th$ hidden node and the $j-th$ visible node, the conditional probabilities are

$$\begin{aligned}
p\left(\bar{h}_i = 1 \mid \mathbf{x}\right) &= \phi\left[W\mathbf{x} + b\right] \\
p\left(x_j = 1 \mid \bar{\mathbf{h}}\right) &= \phi\left[W^T\mathbf{h} + c\right] \\
\bar{h}_i &= \begin{cases} 1 & a < p\left(\bar{h}_i = 1 \mid \mathbf{x}\right) \\ 0 & a \geq p\left(\bar{h}_i = 1 \mid \mathbf{x}\right) \end{cases}
\end{aligned} \tag{8}$$

where ϕ is the sigmoid function, W is a weight matrix, a is a number drawn from a uniform distribution over $[0, 1]$, b and c are visible and hidden biases respectively, $i = 1, \ldots, s$, $j = 1, \ldots, n$.

We define a probability vector $\mathbf{h}$ as

$$\mathbf{h} = \left[p\left(\bar{h}_1 = 1 \mid \mathbf{x}\right) \cdots p\left(\bar{h}_s = 1 \mid \mathbf{x}\right)\right] = [h_1 \cdots h_s]$$

The probability distribution $p(\mathbf{x})$ is an energy-based model

$$p(\mathbf{x}) = \sum_h p(\mathbf{x}, h) = \sum_h \frac{e^{-E(\mathbf{x}, h)}}{Z} \tag{9}$$

where $Z = \sum_h \sum_{\mathbf{x}} e^{-E(\mathbf{x}, h)}$ denotes the sums over all possible values of $\mathbf{h}$ and $\mathbf{x}$, the energy function $E(x, \mathbf{h})$ is

$$E(\mathbf{x}, h) = -c^T\mathbf{x} - b^T\mathbf{h} - \mathbf{h}^T W\mathbf{x} \tag{10}$$

The loss function for the training is

$$L(\theta) = \log \prod_{\mathbf{x}} p(\mathbf{x}) = \log\left[\sum_{\mathbf{x}} e^{-E(\mathbf{x}, h)}\right] - \log\left[\sum_{\mathbf{x}, h} e^{-E(\mathbf{x}, h)}\right]$$

Define the free energy $F(\mathbf{x})$ as

$$F(\mathbf{x}) = \sum_{\mathbf{x}} \log p(\mathbf{x}) = -c^T\mathbf{x} - \sum_{p=1}^{l_i} \log \sum_{h_p} e^{h_p(b_p + W_p\mathbf{x})}$$

If $\mathbf{x}$ and h_p are binary values

$$\begin{aligned} p(h_p = 1|\mathbf{x})_{p=1\cdots l_i} &= \phi[W_p\mathbf{x} + b_p] \\ p(x_t = 1|\bar{\mathbf{h}})_{t=1\cdots l_{i-1}} &= \phi[W_t^T h + c_t] \end{aligned} \tag{11}$$

The weights and biases of each RBM are updated as

$$\theta(k+1) = \theta(k) - \eta_1 \frac{\partial[-\log p(\mathbf{x})]}{\partial\theta(k)} \tag{12}$$

where

$$\frac{\partial \log p(\mathbf{x})}{\partial\theta(k)} = \sum_x p(\mathbf{x})\frac{\partial F(\mathbf{x})}{\partial\theta(k)} - \frac{\partial F(\mathbf{x})}{\partial\theta(k)}$$

We estimate $\sum_z p(\mathbf{x})\frac{\partial F(\mathbf{x})}{\partial\Lambda(k)}$ with the Monte Carlo sampling method

$$\sum_z p(\mathbf{x})\frac{\partial F(\mathbf{x})}{\partial\Lambda(k)} \approx \frac{1}{s}\sum_{z\in S}\frac{\partial F(\mathbf{x})}{\partial\Lambda(k)} \tag{13}$$

A sample z from the training data $\mathbf{x}$ needs the following Monte Carlo algorithm: (1) Calculate $p(\mathbf{h}|\mathbf{x})$ using the current W and b; (2) Sample $\mathbf{h}$ using the conditional distribution $p(\mathbf{h}|\mathbf{x})$. (3) Calculate $p(\mathbf{x}|\mathbf{h})$ using the current W and c. (4) Sample z using the conditional distribution $p(\mathbf{x}|\mathbf{h})$. (5) Repeat steps (1)–(4) k times using the new sample z obtained in step (4), and the new $\mathbf{x}$ in step (1). After k times, we get a sample z for the set S.

Figure 2 gives one layer of the RBM model. The transformation (11) is repeated s times generating s samples needed for the learning process, see Fig. 2.

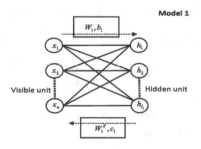

Fig. 2. Markov sampling in a restricted Boltzmann machine

The modified deep learning algorithms, the autoencoders and the restricted Boltzmann machine, proposed in this paper can form a deep neural model (6) from the distribution properties of the input and output.

The next step of the structure identification is to extract fuzzy rules from the neural networks (6). We use the standard decision tree method [22]. We use C4.5 to generate the decision tree from the weights of (6). The fuzzy rules are obtained from the decision tree.

Here for the split point j, this point value is the weight obtained from the deep learning. We use this weight as the center value of the Gaussian membership function (3). The parameter training is to determine the parameters of membership functions of B^j, as well as the membership functions $A_1^j, \cdots, A_n^j$, such that $\hat{y} \to y$. Since the identification error $e(k)$ is

$$e(k) = \hat{y}(k) - y(k) \tag{14}$$

The nonlinear process (1) can be represented by the fuzzy model

$$y(k) = \mathbf{W}^* \Phi[\mathbf{x}(k)] - \mu(k) \tag{15}$$

the modeling error is $\mu(k)$.

The backpropagation like algorithm can train the membership functions

$$
\begin{aligned}
\mathbf{W}_{k+1} &= \mathbf{W}_k - \eta_k e(k) \mathbf{Z}(k)^T \\
c_{ji}(k+1) &= c_{ji}(k) - 2\eta_k z_i \frac{W_i - \hat{y}}{b} \frac{x_j - c_{ji}}{\sigma_{ji}^2} (\hat{y} - y) \\
\sigma_{ji}(k+1) &= \sigma_{ji}(k) - 2\eta_k z_i \frac{W_i - \hat{y}}{b} \frac{(x_j - c_{ji})^2}{\sigma_{ji}^3} (\hat{y} - y)
\end{aligned} \tag{16}
$$

where $\mathbf{Z}(k) = [z_1/b \cdots z_l/b]^T$, $z_i = \prod_{j=1}^{n} \exp\left(-\frac{(x_j - c_{ji})^2}{\sigma_{ji}^2}\right)$, $a = \sum_{i=1}^{sv_j} W_i z_i$, $b = \sum_{i=1}^{l} z_i$, $\hat{y} = \frac{a}{b}$. The normalized identification error $e_N(k) = \frac{e(k)}{1+\max_k(\|\Phi[\mathbf{x}(k)]\|^2)}$ satisfies the following average performance $\limsup_{T \to \infty} \frac{1}{T} \sum_{k=1}^{T} \|e_N(k)\|^2 \leq \bar{\mu}$ where $\bar{\mu} = \max_k \left[\|\mu(k)\|^2\right]$. The proofs of the above algorithms can be found in [20].

4   Simulations

One of the most utilized benchmark examples in system identification is in Box-Jenkins textbook. In this example, a mixed of air and methane is set in order to create a mixture of gases which contained carbon dioxide among other gases. The input of the system is methane gas which is represented with $u(k) = 0.6 - 0.4x(k)$ while the output is the CO_2 concentration $y(k)$. The dataset is composed of 296 successive pairs of readings $[x(k), y(k)]$ that are sampled from the continuous records 9-s intervals (Fig. 3). The model is:

$$y(k) = f[y(k-1), \ldots, y(k-n_y), u(k), \ldots, u(k-n_u)] \tag{17}$$

In our experiments, the delays n_y and n_u are given the values 8 and 5 using a random search method in the interval $[5, 10]$, the training data are 200 examples

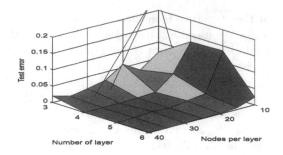

Fig. 3. Random search for the structure parameters.

while the rest are used for validation, early stopping criteria is not used to train the DBMs. Following the approach presented in this work, the regression task is done following the next steps: normalization, coding, training, decoding. $n_y = 8$ and $n_u = 5$.

Our neural model (6) has 4 hidden layers and 1 linear output layer, each hidden layer has 50 nodes. The parameters are the training rate is

$$\eta_3 = 0.1$$
$$\eta_4 = 0.15$$

The average error is defined as

$$\frac{1}{N} \sum_{k=1}^{N} \|\hat{y}(k) - y(k)\|^2$$

The testing errors of the proposed models are shown in Fig. 4.

The errors are
$$AC : 6.862 \times 10^{-5}$$
$$RBM : 4.239 \times 10^{-5}$$

In the pre-training stage, AC obtains better initial weights. Because RBMs need more examples, while this dataset only have 200. The improvement of RBM is not so good.

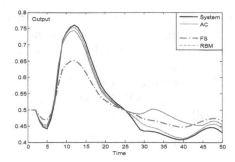

Fig. 4. The testing errors.

5 Conclusions

The restricted Boltzmann machines is extended for nonlinear dynamic system modeling. As an alternative model for fuzzy modelling and an efficient approach for extracting fuzzy rules, this method is more effective than the normal neural modelling and fuzzy modelling.

Acknowledgments. This paper is supported by the National Council of Science and Technology of Mexico (CONACYT), under the project Frontiers of Science (Grant No. 65).

References

1. Angelov, P.: An approach for fuzzy rule-base adaptation using online clustering. Int. J. Approx. Reason. **35**(3), 275–289 (2004)
2. Bengio, Y., Lamblin, P., Popovici, D., Larochelle, H.: Greedy layer-wise training of deep networks. In: Advances in Neural Information Processing Systems (NIPS 2006), pp. 153–160. MIT Press (2007)
3. Bengio, Y., Delalleau, O.: Justifying and generalizing contrastive divergence. Neural Comput. **21**(6), 1601–1621 (2009)
4. Bergstra, J., Bengio, Y.: Algorithms for hyper-parameter optimization. J. Mach. Learn. Res. **13**, 281–305 (2012)
5. Cristianini, N., Shawe-Taylor, J.: An Introduction to Support Vector Machines. Cambridge University Press, Cambridge (2000)
6. Chiang, J.H., Hao, P.Y.: Suuport vector learning mechanism for fuzzy rule-based modelling: a new approach. IEEE Trans. Fuzzy Syst. **12**(1), 1–12 (2004)
7. Chiu, S.L.: Fuzzy model identification based on cluster estimation. J. Intell. Fuzzy Syst. **2**(3), 267–278 (1994)
8. Erhan, D., Manzagol, P.A., Bengio, Y., Bengio, S., Vincent, P.: The difficulty of training deep architectures and the effect of unsupervised pretraining. In: 12th International Conference on Artificial Intelligence and Statistics (AISTATS 2009), pp. 153–160. AISTATS Press (2009)
9. Hinton, G.E., Sejnowski, T.J.: Learning and relearning in Boltzmann machines. In: Parallel Distributed Processing: Explorations in the Microstructure of Cognition, Volume 1: Foundations, pp. 282–317. MIT Press, Cambridge (1986)
10. Jang, J.S.: ANFIS: adaptive-network-based fuzzy inference system. IEEE Trans. Syst. Man Cybern. **23**, 665–685 (1993)
11. Leski, J.M.: TSK-fuzzy modelling based on ε-insensitive learning. IEEE Trans. Fuzzy Syst. **13**(2), 81–193 (2005)
12. Mitra, S., Hayashi, Y.: Neuro-fuzzy rule generation: survey in soft computing framework. IEEE Trans. Neural Netw. **11**(3), 748–769 (2000)
13. Rivals, I., Personnaz, L.: Neural-network construction and selection in nonlinear modelling. IEEE Trans. Neural Netw. **14**(4), 804–820 (2003)
14. Saito, K., Nakano, R.: Extracting regression rules from neural networks. Neural Netw. **15**(10), 1279–1288 (2002)
15. Tzafestas, S.G., Zikidis, K.C.: NeuroFAST: online neuro-fuzzy ART-based structure and parameter learning TSK model. IEEE Trans. Syst. Man Cybern. Part B **31**(5), 797–803 (2001)

16. Wang, L.X.: Adaptive Fuzzy Systems and Control. Prentice-Hall, Englewood Cliffs (1994)
17. Yu, W.: A novel fuzzy neural networks modelling approach to crude oil blending. IEEE Trans. Control Syst. Technol. **17**(6), 1424–1431 (2009)
18. Yu, W.: Fuzzy modelling via on-line support vector machines. Int. J. Syst. Sci. **41**(11), 1325–1335 (2010)
19. de la Rosa, E., Yu, W.: Randomized algorithms for nonlinear system identification with deep learning modification. Inf. Sci. **364**, 197–212 (2016)
20. Yu, W., Li, X.: Fuzzy identification using fuzzy neural networks with stable learning algorithms. IEEE Trans. Fuzzy Syst. **12**(3), 411–420 (2004)
21. Zadeh, L.A.: Fuzzy sets. Inf. Control **8**, 338–353 (1998)
22. Zilke, J.R.: Extracting rules from deep neural networks. TU Darmstadt, Master thesis (2015)

Matrix Neural Networks

Junbin Gao[1], Yi Guo[2(✉)], and Zhiyong Wang[1]

[1] University of Sydney, Sydney, Australia
{junbin.gao,zhiyong.wang}@sydney.edu.au
[2] Western Sydney University, Parramatta, Australia
y.guo@westernsydney.edu.au

Abstract. Traditional neural networks assume vectorial inputs as the network is arranged as layers of single line of computing units called neurons. This special structure requires the non-vectorial inputs such as matrices to be converted into vectors. This process can be problematic for loss of spatial information and huge solution space. To address these issues, we propose matrix neural networks (MatNet), which takes matrices directly as inputs. Each layer summarises and passes information through bilinear mapping. Under this structure, back prorogation and gradient descent combination can be utilised to obtain network parameters efficiently. Furthermore, it can be conveniently extended for multimodal inputs. We apply MatNet to MNIST handwritten digits classification and image super resolution tasks to show its effectiveness. Without too much tweaking MatNet achieves comparable performance as the state-of-the-art methods in both tasks with considerably reduced complexity.

1 Introduction

Neural networks especially deep networks [4,6] have attracted a lot of attention recently due to their superior performance in several machine learning tasks. The basic structure of the most widely used neural networks remains almost the same, i.e. hierarchical layers of computing units (called neurons) with feed forward information flow from previous layer to the next layer [2]. Although there is no restriction on how the neurons should be arranged spatially, traditionally they all line in a row or a column just like elements in a vector. This special structure requires the non-vectorial inputs especially matrices (e.g. images) to be converted into vectors. Unfortunately this process can be problematic. Firstly, the spatial information among elements of the data may be lost during vectorisation. Secondly, the solution space becomes very large which demands very special treatments to the network parameters. These bring in many adverse effects such as great difficulties in training.

To address these issues, we propose matrix neural networks or MatNet for short, which takes matrices directly as inputs. Therefore the input layer neurons form a matrix, for example, each neuron corresponds to a pixel in a grey scale image. The upper layers are also but not limited to matrices. Different from the

© Springer International Publishing AG 2017
F. Cong et al. (Eds.): ISNN 2017, Part I, LNCS 10261, pp. 313–320, 2017.
DOI: 10.1007/978-3-319-59072-1_37

convolutional neural network (CNN) [7] and alike (e.g. neocognitron [3]) where input layers are feature extraction layers and its core is still the traditional vector based neural network, in MatNet matrices are passing through each layer without vectorisation at all. MatNet passes information through bilinear mapping from layer to layer. The neurons in each layer activate according to some activation function e.g. sigmoid, tanh, and rectified linear unit (reLU) [8] to generate its output for the next layer. In order not to disturb the flow, we simple discuss the model in later sections and leave the mathematical details to supplement materials, where interested readers can find the details.

To demonstrate the usefulness of the proposed MatNet, we will test it in two image processing tasks, the well-known MNIST handwritten digits classification and image super resolution. For image super resolution, we show in Sect. 3 the multimodal inputs in MatNet. As shown in Sect. 4, MatNet can achieve comparable classification rate as those sophisticated deep learning neural networks. Due to early research nature of this work, MatNet is not yet optimised for this task and the choices of the key network parameters such as the number of layers and neurons are somewhat arbitrary. For super resolution task, MatNet is very competitive in terms of peak signal to noise ratio (PSNR) compared to the state-of-the-art methods such as the sparse representation (SR) [10]. Once again, this result can be further optimised and we will discuss some further developments that will be carried out in near future in Sect. 5.

2   Matrix Neural Network Model

The basic model of a layer of MatNet is the following bilinear mapping

$$Y = \sigma(UXV^T + B) + E, \tag{2.1}$$

where X and Y are input and output matrices of that layer, U, V, B and E are matrices with *compatible dimensions*, determined by input X and model flexibility. U and V are connection weights, B is the offset of current layer, $\sigma(\cdot)$ is the activation function acting on each element of matrix and E is the error.

2.1   Network Structure

The MatNet consists of multiple layers of neurons in the form of (2.1). Let $X^{(l)} \in \mathbb{R}^{I_l \times J_l}$ be the matrix variable at layer l where $l = 1, 2, \ldots, L, L+1$. Layer 1 is the input layer that takes matrices input directly and Layer $L + 1$ is the output layer. All the other layers are hidden layers. Layer l is connected to Layer $l + 1$ by

$$X^{(l+1)} = \sigma(U^{(l)} X^{(l)} V^{(l)T} + B^{(l)}), \tag{2.2}$$

where $B^{(l)} \in \mathbb{R}^{I_{l+1} \times J_{l+1}}$, $U^{(l)} \in \mathbb{R}^{I_{l+1} \times I_l}$ and $V^{(l)} \in \mathbb{R}^{J_{l+1} \times J_l}$, for $l = 1, 2, ..., L\text{-}1$. For the convenience of explanation, we define

$$N^{(l)} = U^{(l)} X^{(l)} V^{(l)T} + B^{(l)} \tag{2.3}$$

for $l = 1, 2, ..., L$. Hence

$$X^{(l+1)} = \sigma(N^{(l)}).$$

The shape of the output layer is determined by the functionality of the network, i.e. regression or classification, which in turn determines the connections from Layer L. We discuss in the following three cases.

- Case 1: Normal regression network. The output layer is actually a matrix variable as $O = X^{(L+1)}$. The connection between layer L and the output layer is defined as (2.2) with $l = L$.
- Case 2: Classification network. The output layer is a multiple label (0–1) vector $\mathbf{o} = (o_1, ..., o_K)$ where K is the number of classes. In $\mathbf{o}$, all elements are 0 but one 1. The final connection is then defined by

$$o_k = \frac{\exp(\mathbf{u}_k X^{(L)} \mathbf{v}_k^T + tb_k)}{\sum_{k'=1}^{K} \exp(\mathbf{u}_{k'} X^{(L)} \mathbf{v}_{k'}^T + tb_{k'})}, \tag{2.4}$$

where $k = 1, 2, ..., K$, $\overline{U} = [\mathbf{u}_1^T, ..., \mathbf{u}_K^T]^T \in \mathbb{R}^{K \times I_L}$ and $\overline{V} = [\mathbf{v}_1^T, ..., \mathbf{v}_K^T]^T \in \mathbb{R}^{K \times J_L}$. That is both $\mathbf{u}_k$ and $\mathbf{v}_k$ are rows of matrices $\overline{U}$ and $\overline{V}$, respectively. Similar to (2.3), we denote

$$n_k = \mathbf{u}_k X^{(L)} \mathbf{v}_k^T + tb_k. \tag{2.5}$$

(2.4) is the softmax that is frequently used in logistic regression [5]. Note that in (2.4), the matrix form is maintained. However, one can flatten the matrix for the output layer.

Assume that we are given a training dataset $\mathcal{D} = \{(X_n, Y_n)\}_{n=1}^{N}$ for regression or $\mathcal{D} = \{(X_n, \mathbf{t}_n)\}_{n=1}^{N}$ for classification problems respectively. Then we define the following loss functions

$$L = \frac{1}{N} \sum_{n=1}^{N} \frac{1}{2} \|Y_n - X_n^{(L+1)}\|_F^2 \tag{2.6}$$

for regression and

$$L = -\frac{1}{N} \sum_{n=1}^{N} \sum_{k=1}^{K} t_{nk} \log(o_{nk}) \tag{2.7}$$

for classification. MatNet is open to any other cost functions as long as the gradient with respect to unknown variables can be easily obtained.

MatNet reduces the solution space heavily. For Layer l, MatNet has $I_{l+1}I_l + J_{l+1}J_l$ parameters while $I_{l+1}I_lJ_{l+1}J_l$ parameters in traditional vector neural networks. The resultant effects and advantages include less costly training process, less local minima, easier to handle and most of all, direct and intuitive interpretation.

2.2   Regularisation

Regularisation can be considered in ManNet. For example clamping mapping weights

$$\lambda \sum_l (\|U^{(l)}\|_F^2 + \|V^{(l)}\|_F^2),$$

where λ is a nonnegative regularisation parameter and the summation of Frobenius norms includes the output layer as well.

Sparsity is also possible. Here we discuss one method that is straightforward to build into MatNet to eliminate excessive neurons. let $\overline{\rho}^{(l)} = \frac{1}{N} \sum_{n=1}^N X_n^{(l)}$ be the average activations of hidden layer l (averaged over the training set). By enforcing the following

$$\overline{\rho}_{ij}^{(l)} = \rho,$$

one can achieve sparsity in reducing the number of neurons [9]. Therefore, ρ is called a sparsity parameter, typically a small value close to zero, e.g. $\rho = 0.05$. The above equality constraint is implemented by

$$R_l = \text{sum}\left(\rho \log \frac{\rho}{\overline{\rho}^{(l)}} + (1-\rho) \log \frac{1-\rho}{1-\overline{\rho}^{(l)}}\right) \tag{2.8}$$

where $\text{sum}(M)$ summing over all the elements in matrix M; log and division are applied to matrix elementwise. To screen out neurons that are not necessary, we add the following extra term in the cost function of MatNet

$$\beta \sum_{l=2}^L R_l.$$

The optimisation of this is detailed in the supplement materials.

3   Multimodal Matrix Neural Networks

We demonstrate a three layer multimodal MatNet autoencoder here for super resolution application. It is not difficult to extend to other type of multimodal MatNet with multiple hidden layers using the same methodology.

Assume D modalities as matrices in consideration denoted by $X^j \in \mathbb{R}^{K_{j1} \times K_{j2}}$ ($j = 1, 2, ..., D$). Similarly there are D output matrix variables of the same sizes. Let $\mathcal{X} = (X^1, ..., X^D)$. In the hidden layer, we only have one matrix variable $H \in \mathbb{R}^{K_1 \times K_2}$. The transformation from input layer to hidden layer is

$$H = \sigma(\sum_{j=1}^D U_j X^j V_j^T + B), \tag{3.1}$$

and from hidden layer to output layer is

$$\widehat{X}^j = \sigma(R_j H S_j^T + C_j), \quad j = 1, 2, ..., D. \tag{3.2}$$

The objective function to be minimised is defined by

$$L = \frac{1}{2N} \sum_{i=1}^{N} \sum_{j=1}^{D} \|\widehat{X}_i^j - X_i^j\|_F^2. \tag{3.3}$$

L is a function of all the parameters $W = \{U_j, V_j, R_j, S_j, C_j, B\}_{j=1}^{D}$.

We leave the derivation of the optimisation schemes in the supplementary materials.

4   Experimental Evaluation

In this section, we apply MatNet to MNIST handwritten digits classification and image super resolution. The network settings are somewhat arbitrary. For handwritten digits recognition, MatNet was configured as a classification network, i.e. the output layer was a vector of softmax functions. For illustration purpose, we selected a simple MatNet. It contained 2 hidden layers, each with 20×20 and 16×16 neurons. As the numbers of layers and neurons were very conservative, we turned off sparsity constraint as well as weights decay. For super resolution task, the only hidden layer was of size 10×10. The activation function in both networks was sigmoid.

4.1   MNIST Handwritten Digits Classification

The MNIST handwritten digits database is available at http://yann.lecun.com/exdb/mnist/. The entire database contains 60,000 training samples and 10,000 testing samples, and each digit is a 28×28 gray scale image. We use all training samples for modeling and test on all testing samples. The final test accuracy is 97.3%, i.e. error rate of 2.7%, which is inferior to the best MNIST performance by DropConnect with error rate 0.21%.

However, as we stated earlier, MatNet has much less computational complexity. To see this clearly, we carried out a comparison between MatNet and CNN. The CNN consisted of two convolutional layers of size $20 \times 1 \times 5 \times 5$ and $50 \times 20 \times 5 \times 5$ one of which is followed by a 2×2 max pooling, and then a hidden layer of 500 and output layer of 10, fully connected. This is the structure used in Theano [1] demo. The total number of parameters to optimise is 430500, while the total number of parameters in MatNet is 5536. The server runs a 6-core i7 3.3 GHz CPU with 64 GB memory and a NVIDIA Tesla K40 GPU card with 12 GB memory. We used Theano for CNN which fully utilises GPU. On contrast, MatNet is implemented with Matlab without any parallelisation. The difference of training time is astounding. It costed the server more than 20 h for CNN with final test accuracy of 99.07%, whereas less than 2 h for MatNet with test accuracy of 97.3%, i.e. 1.77% worse. In order to see if MatNet can approach this CNN's performance in terms of accuracy, we varied the structure of MatNet in both number of neurons in each layer and number of layers (depth). However, we limited the depth to the maximum of 6 as we did not consider deep structure for the time being. Due to the randomness of the stochastic gradient descent employed in MatNet, we ran through one structure multiple times and collected the test accuracy. Figure 1 shows the performance of different MatNet compared against CNN. The model complexity is rendered as the number of parameters in the model, which is the horizontal axis in the plot. So when MatNet gets more complex, it approaches CNN steadily.

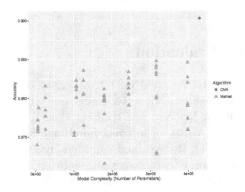

Fig. 1. Test accuracy of MatNet vs CNN

4.2   Image Super Resolution

For image super resolution, we use the multimodal MatNet detailed in Sect. 3. For the training and detailed procedures of obtaining super resolution images please see supplementary materials.

We applied MatNet to the data set used in SR [10]. There are 69 images for training. The patch size was 15×15. We randomly sampled 10,000 patches altogether from all images for training. Some additional parameters for MatNet are $\lambda = 0.001$, $\rho = 0.05$ and $\beta = 1$. So we turned on weight decay and sparsity constraints.

Figure 2 shows the results on two testing images. Multimodal MatNet has comparable performance as SR, the state-of-the-art super resolution method, evaluated by PSNR: for Lena image, multimodal MatNet, SR and bicubic interpolation achieved PSNR 33.966 dB, 35.037 dB and 32.795 dB respectively; for kingfisher image, they had PSNR 36.056 dB, 36.541 dB and 34.518 dB respectively. We applied to a number of images of similar size (256×256) and we observed similar scenario. Figure 3(a) shows the all the test images, including the two in Fig. 2, and PSNR's obtained by different methods is shown in Fig. 3(b). MatNet is very close to SR in terms of PSNR, especially for image 5 and 8.

(a) Lena image (128 × 128)

(b) Kingfisher image (256 × 256)

Fig. 2. Super resolution on 2 sets of testing images. From left to right: input small size image, true high resolution image, up-scaled images (2 times) produced by multimodal MatNet, SR and bicubic interpolation respectively.

(a) All 12 test images   (b) PSNR results

Fig. 3. Super resolution results comparison. The images are indexed from left to right, from top to bottom.

5 Discussion

We proposed a matrix neural network (MatNet) in this paper, which takes matrices input directly without vectorisation. The most prominent advantage of MatNet over the traditional vector based neural works is that it reduces the complexity of the optimisation problem drastically, while manages to obtain comparable performance as the state-of-the-art methods. This has been demonstrated in applications of MNIST handwritten digits classification and image super resolution.

As we mentioned several times in the text, MatNet was not specially optimised for the tasks we showed in experiment section. There is a lot of potentials for further improvement. Many techniques used for deep networks can be readily applied to MatNet with appropriate adaptation, e.g. reLU activation function, max-pooling, etc., which certainly become our future research.

Supplementary Materials

Refer to http://arxiv.org/abs/1601.03805 for all details about the optimisation (including back propagation) and experiments.

References

1. Al-Rfou et al.: Theano: a Python framework for fast computation of mathematical expressions. arXiv e-prints abs/1605.02688, May 2016
2. Bishop, C.: Neural Networks for Pattern Recognition. Clarendon Press, Oxford (1995)
3. Fukushima, K.: Artificial vision by multi-layered neural networks: Neocognitron and its advances. Neural Netw. **37**, 103–119 (2013)
4. Hinton, G.E., Osindero, S., Teh, Y.W.: A fast learning algorithm for deep belief nets. Neural Comput. **18**(7), 1527–1554 (2006)
5. Hosmer Jr., D.W., Lemeshow, S., Sturdivant, R.X.: Applied Logistic Regression, vol. 398. Wiley, Hoboken (2013)
6. LeCun, Y., Bengio, Y., Hinton, G.: Deep learning. Nature **521**, 436–444 (2015)
7. LeCun, Y., Bottou, L., Bengio, Y., Haffner, P.: Gradient-based learning applied to document recognition. Proc. IEEE **86**(11), 2278–2324 (1998)
8. Nair, V., Hinton, G.E.: Rectified linear units improve restricted Boltzmann machines. In: Proceedings of the 27th International Conference on Machine Learning (ICML 2010), pp. 807–814 (2010)
9. Shu, M., Fyshe, A.: Sparse autoencoders for word decoding from magnetoencephalography. In: Proceedings of the Third NIPS Workshop on Machine Learning and Interpretation in NeuroImaging (MLINI) (2013)
10. Yang, J., Wright, J., Huang, T., Ma, Y.: Image super-resolution via sparse representation. IEEE Trans. Image Process. **19**(11), 2861–2873 (2010)

Simplified Particle Swarm Optimization Algorithm Based on Improved Learning Factors

Wei Gao[1], Chuyi Song[1], Jingqing Jiang[2,3(✉)],
and Chenggang Zhang[2]

[1] College of Mathematics, Inner Mongolia University for Nationalities,
Tongliao 028043, China
[2] College of Computer Science and Technology,
Inner Mongolia University for Nationalities, Tongliao 028043, China
jiangjingqing@aliyun.com
[3] Key Laboratory of Symbolic Computation and Knowledge
Engineering of Ministry of Education, Jilin University,
Changchun 130012, People's Republic of China

Abstract. To overcome the shortcomings of the traditional particle swarm optimization algorithm, which are easy to fall into local extreme, a new algorithm based on the simplified particle swarm optimization algorithm is proposed. Firstly, the proposed algorithm removes the speed term, so that it makes the algorithm simple. And then it improves the displacement term. Finally the nonlinearity of the trigonometric function is utilized in the algorithm to improve learning factors. It balances the global search and local search. The six basic test functions are used to compare the standard particle swarm optimization algorithm, the simplified particle swarm optimization algorithm and the improved algorithm proposed in this paper. Experimental results show that the performance of the improved particle swarm optimization is better than the other two algorithms.

Keywords: Particle swarm optimization · Learning factors · Classical functions

1 Introduction

Particle swarm optimization (PSO) is a swarm intelligent global optimization algorithm, originated in the simulation of simplified social model. Particle swarm optimization studied on the birds, fish and human social systems. It confirmed that the information shared between individuals in the community benefit to the global evolution. PSO was proposed by Kennedy and Eberhart in 1995 [1]. After PSO was put forward, many people pay more and more attention to it for the simple process and principle, easy understanding, and fast convergence speed. Therefore it is applied to many fields, such as image processing [2], hardware accelerators [3], industry [4, 5], biology [6], agriculture [7] and so on.

© Springer International Publishing AG 2017
F. Cong et al. (Eds.): ISNN 2017, Part I, LNCS 10261, pp. 321–328, 2017.
DOI: 10.1007/978-3-319-59072-1_38

In particle swarm optimization algorithm, the learning factor is a very important parameter. It can be used to control the global search and local search ability of PSO. So the influences that the particle's experience and the group's experience impact on the particle movement trajectory can be determined by adjust the value of the learning factors. While the self learning factor is bigger, particle flight trajectory mainly refers to the history information of the particles themselves. While the social learning factor is bigger, particle flight trajectory mainly refers to the social information of particles movement. So the appropriate change of the learning factor value is a good way to improve the performance of the algorithm.

But with further research and the emergence of the various problems in reality, the traditional PSO algorithm exposes some defects, such as falling into local optimum easily, the lower convergence speed in the later and the lower search precision. In order to improve the performance of the PSO algorithm, people improved it in many aspects. This paper simply introduces some improved particle swarm optimization algorithms. In the literature [8], Ratnaweera and Halgamuge put forward the linear adjustment of learning factor. The self learning factor changes from bigger to smaller. The social learning factor changes from smaller to bigger. In the literature [9], Chen put forward an improved PSO algorithm which improved the learning factor using concave function and cosine function strategy. In the literature [10], Zhang put forward a hybrid particle swarm optimization algorithm which improved learning factor and constraints factor. The literature [11] put forward the improved PSO algorithm which does not include the speed term parameters. It utilizes the individual optimal location information of the particles to improve the performance of the algorithm. Based on the literature [11], this paper proposed an improved PSO algorithm which removes the speed term and dynamically adjusts the learning factors according to the iteration number. It improved the convergence speed and overcome the shortcoming of falling into local optimum easily.

2   The Standard Particle Swarm Optimization Algorithm

PSO algorithm creates an initial group of particles randomly. Each particle does not have volume quality. Each particle has the position term and the velocity term. The particle is regarded as a feasible solution. But the best solution is determined by the fitness function. Usually each particle searches the solution space by following the best particles of the current population. And it keeps on updating its velocity and position. In PSO, the optimal solution is obtained by searching in each generation.

The PSO algorithm searches on a D dimension space which is composed of m particles. In the process of evolution, each particle maintains two vectors. They are the velocity vector $V_i = (v_{i1}, v_{i2}, \cdots, v_{iD})$ and the position vector $X_i = (x_{i1}, x_{i2}, \cdots, x_{iD})$. The optimal position of the individual obtained in the current searching is $P_i^t = (p_{i1}^t, p_{i2}^t, \cdots, p_{iD}^t)$. $i = 1, 2, \cdots, m$. The global optimal position is $P_g^t = \left(p_{g1}^t, p_{g2}^t, \cdots, p_{gD}^t \right)$.

The velocity and position of the d dimension particle i is updated by the following formula,

$$v_{id}^{t+1} = wv_{id}^{t} + c_1 r_1 \left(p_{id} - x_{id}^{t}\right) + c_2 r_2 \left(p_{gd} - x_{id}^{t}\right) \tag{1}$$

$$x_{id}^{t+1} = x_{id}^{t} + v_{id}^{t+1} \tag{2}$$

Where t is the current iteration number, w is the inertia weight, c_1 and c_2 are learning factor, r_1 and r_2 are random numbers among [0, 1].

In formula (1) there are three parts. The first part is the inheritance of the last generation. And it expresses the particle's recognition of the current state. The second part is the recognition of itself. And it represents learning from its past experience. The third part is the cognition of the particle population. It expresses the information sharing and cooperation between each particle.

If the velocity or the position is illegal we must revise. The usual method is to reset the velocity or the location randomly or set them to the boundary. Usually the termination condition of the PSO algorithm is to reach the maximum number of the iterations or achieve the required accuracy.

3 The Improved Particle Swarm Optimization Algorithm

3.1 The Simplified PSO

The standard PSO algorithm has some defects. It is easy to fall into local optimum. It has slow convergence speed in the later evolution and reaches low accuracy. The literature [10] simplifies the updating formula of the standard PSO. It removes the velocity term. So the search process is controlled by the position. Therefore the PSO algorithm is simplified. In the process of standard PSO algorithm, the particles change its position based on the velocity. But it does not consider the influence between particles. The simplified PSO removes the particle's velocity term. Moreover, the current optimal position of the individual is changed by the average of the current optimal position of all individuals. Therefore, the particles swarm optimization algorithm is simplified.

The updated formulas are as follows:

$$x_{id}^{t+1} = wx_{id}^{t} + c_1(t)r_1 \left(p_{ad} - x_{id}^{t}\right) + c_2(t)r_2 \left(p_{gd} - x_{id}^{t}\right) \tag{3}$$

$$p_{ad} = \sum_{i=1}^{m} \frac{p_{id}}{m} \tag{4}$$

In (4), p_{ad} is the average of the current optimal positions of all individuals. p_{id} is the current optimal position of the particle i. p_{gd} is the current global optimal position of the particle population.

3.2   Improve the Learning Factor

The learning factor adopts a fixed value in the above simplified particle swarm algorithm. But the learning factor plays an important role in the PSO. The learning factor can effectively improve the performance of particle swarm optimization algorithm. So the improvement of the learning factor can effectively improve the search performance of PSO algorithm. There are three common forms of learning factor. They are fixed value [12] (typically set to 2.5), in the form of nonlinear [13] and linear [14] respectively.

The standard PSO algorithm generally adopts linear decreasing learning factor [14]. As the searching scope is gradually decreasing, it makes the objective function falling into local extreme points within the neighborhood, and easily convergences to the local extreme value. In order to overcome the shortcomings of the above PSO algorithm, the learning factor is further improved on the basis of literature [11]. The improvement of the learning factor is as follows:

$$c_1(t) = c_1(t) + \sin(t) \tag{5}$$

$$c_2(t) = c_2(t) + \cos(t) \tag{6}$$

$$1 \le c_1(t) \le 2.5 \tag{7}$$

$$1 \le c_2(t) \le 2.5 \tag{8}$$

Where t is the number of the iteration; $c_1(t)$ is the self- learning factor; $c_2(t)$ is the social learning factor; $\sin(t)$ is a sine function of the number of iterations. $\cos(t)$ is a cosine function of the number of iterations. The formula (7) and formula (8) constrain the scope of learning factor.

In the formula (5), the self-learning factor is improved by using the sine function. At the beginning, the $c_1(t)$ is larger. It benefits the local search. With the increase of the number of iterations, the self-learning factor changes up and down as a cycle. It prevented the PSO algorithm trap into the local optimum. In the formula (6) social learning factor is improved by using the cosine function of the number of iterations. At the beginning, the $c_2(t)$ is small. It benefits the algorithm searching in the solution space. With the increase of the number of iterations, the social learning factor changes down and up as a cycle. While the number of iterations is set properly, the self-learning factor declines and the social learning factor increases finally. At last the smaller self-learning factor and the bigger social factor benefit the global searching.

The formula (7) and (8) is introduced by the literature [12]. The searching ability of the PSO algorithm is good for the learning factor lies between [1, 2.5]. The first part of the PSO algorithm is not only decreasing monotonically, but also related to the objective function closely. At the same time, through the adjustment of the second part, the learning factor can't change dramatically. This makes the improved particle swarm optimization algorithm not only converges fast but also not likely to fall into the local optimum.

4   The Steps of the Improved PSO Algorithm

The steps of improved particle swarm optimization algorithm are as follows:

Step 1.   The population initialization. The position (X_i^0) is created randomly in the search space. Each particle in the population has a position vector x_{id}. m is the number of particles in the population. $d \in [1, D]$ and D is the dimension of a particle.

Step 2.   Let X_{max} is the maximum value of the position. p_i is the current optimal position of the particle i. p_g is the optimal position of the population.

Step 3.   Calculating learning factors according to the formula (5) and (6), and then calculate p_{id}, p_{gd}.

Step 4.   Calculate the fitness of particle i, and f(x) is the fitness function. $i \in [1, m]$.

Step 5.   If the fitness value $f(x_i)$ of the particle i is superior to the fitness value $f(p_i)$ of the current optimal position p_i of particle i, p_i is replaced by x_i. If $f(x_i)$ is better than the fitness of global optimal position $f\left(p_g\right)$, p_g is replaced by x_i.

Step 6.   Updating position according to the formula (3).

Step 7.   If the termination condition is reached by the iteration, output the global optimal position. Otherwise returns to step 3, and continue to search.

5   Simulation Experience

Six basic test functions are used to test the efficiency of the improved PSO. And the simulation results are compared with the standard particle swarm optimization algorithm [1] and the simplified particle swarm optimization algorithm. The improved PSO sets the parameters as follows: Population size of particle swarm is $m = 30$. Dimension is $D = 20$. The maximum iteration number is 50; the inertia weight is w = 0.927; learning factors are $c_1(1) = c_2(1) = 1$. Six basic test functions [15] are as follows:

(a) Sphere function

$$f(x) = \sum_{d=1}^{D} x_d^2$$

Global optimum is $f(X^*) = 0$, search range is $-100 \leq x_d \leq 100$.

(b) Shaffer's f7 function

$$f(x) = \left(x_1^2 + x_2^2\right)^{0.25} \left[sin^2 \left(50 \left(x_1^2 + x_2^2\right)^{0.1} \right) + 1.0 \right]$$

Global optimum is $f(X^*) = 0$, search range is $-100 \leq x_d \leq 100$.

(c) Griewank function

$$f(x) = \frac{1}{4000} \sum_{i=1}^{D} x_i^2 - \prod_{i=1}^{D} \cos \frac{x_i}{\sqrt{i}} + 1$$

Global optimum is $f(X^*) = 0$, search range is $-100 \le x_d \le 100$.

(d) Rosenbrock function

$$f(x) = \sum_{i=1}^{D-1} \left[100 \left(x_{i+1} - x_i^2 \right)^2 + (x_i - 1)^2 \right]$$

Global optimum is $f(X^*) = 0$, search range is $-30 \le x_d \le 30$.

(e) Rastrigrin function

$$f(x) = \sum_{i=1}^{D} \left[x_i^2 - 10\cos(2\pi x_i) + 10 \right]$$

Global optimum is $f(X^*) = 0$, search range is $-10 \le x_d \le 10$.

(f) Shaffer's f6 function

$$f(x) = 0.5 + \frac{\left(\sin \sqrt{x_1^2 + x_2^2} \right)^2 - 0.5}{\left[1.0 + 0.001 \left(x_1^2 + x_2^2 \right) \right]^2}$$

Global optimum is $f(X^*) = 0$, search range is $-100 \le x_d \le 100$.

Table 1 shows the results of three algorithms for six test functions. For each algorithm, the first column is the best fitness value and the second column is the number of iteration that needed to reach the best value. It can be seen from Table 1 that the improved PSO can reach the global optimum for five of six functions and reach better solution for the other functions. The iterations that needed are less than the other two algorithms.

Figures 1 and 2 show the performance compared the Simplified PSO and the improved PSO for the Rosenbrock and Shaffer's f6 function. From the two diagrams it can be seen clearly that the performance of the improved PSO algorithm is superior to the simplified PSO algorithm. It makes the change of position faster in the first half part

Table 1. The best fitness value of the test functions

Test functions	Standard PSO		Simplified PSO		Improved PSO	
	Best fitness	Iterations	Best fitness	Iterations	Best fitness	Iterations
Sphere	16.1610	50	0.2624	50	0	10
Shaffer's f7	7.2028	50	0.0152	50	0	19
Griewank	9.0726	50	0	11	0	4
Rosenbrock	12.7292	50	0.0266	50	0	19
Rastrigrin	6.7522	50	0.2286	50	0	10
Shaffer's f6	1.6979	50	0.219	50	0.0097	8

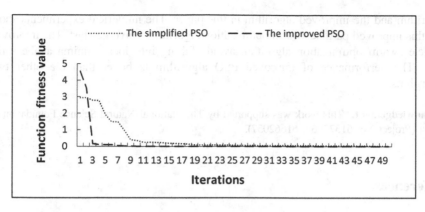

Fig. 1. Rosenbrock function fitness value change in the two algorithms

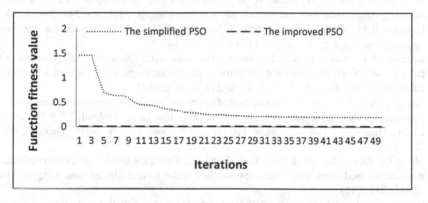

Fig. 2. Shaffer's f6 functions fitness value change in the two algorithms

of searching and slower in the last half part. The improved PSO algorithm can search the optimal solution of the objective function efficiently and it voids tofall into the local optimum.

6   Conclusion

To overcome the shortcomings of the standard particle swarm optimization algorithm which have lower convergence speed and easy to fall into local optimum, an improved algorithm is put forward in this paper which based on simplified particle swarm optimization algorithm to improve the learning factors. Firstly, this algorithm removes the velocity term of the standard PSO algorithm. And it simplified the PSO. Secondly, it uses the relationship between the particle and the number of iterations to improve the position. Finally, it improves the learning factors. The change of learning factors in this way benefits the local search. Six basic test functions are used to test the standard particle swarm optimization algorithm, the simplified particle swarm optimization algorithm and the improved algorithm in this paper. The numerical experiments show that this improved particle swarm optimization algorithm converges fast. The improved particle swarm optimization algorithm avoids falling into local minima at the same time. The performance of improved PSO algorithm is better than the other two algorithms.

Acknowledgement. This work was supported by The National Natural Science Foundation of China (Project No. 61373067, 61662057).

References

1. Kennedy, J., Eberhart, R.C.: Particle swarm optimization. Proc. IEEE Int. Conf. Neural Netw. **4**, 1942–1948 (1995)
2. Setayesh, M., Zhang, M., Johnston, M.: A novel particle swarm optimization approach to detecting continuous, thin and smooth edges in noisy images. Inf. Sci. **246**, 28–51 (2013)
3. Calazan, R.M., Nedjah, N., Mourelle, L.M.: A hardware accelerator for particle swarm optimization. Appl. Soft Comput. **14**, 347–356 (2014)
4. Echevarr, L.C., Santiago, O.L., Fajardo, J.A.H., Neto, A.J.S., Sanchez, D.J.: A variant of the particle swarm optimization for the improvement of fault diagnosis in industrial systems via faults estimation. Eng. Appl. Artif. Intell. **212**, 1–16 (2014)
5. Sun, Z., Zhao, J., Wang, W.: Application of improved particle swarm optimization based on gaussian search to grinding predictive control. J. Dalian Univ. Technol. **1**, 89–96 (2015)
6. Li, N., Huang, Z.: Improved niche particle swarm optimization. Softw. Guide **2**, 45–47 (2015)
7. Meng, Q., Zhang, M., Yang, G., Qiu, R., Ming, X.: Guidance line recognition of agricultural machinery based on particle swarm optimization under natural illumination. J. Agric. Mach. **47**, 11–20 (2016)
8. Ratnaweera, A., Halgamuge, S.: Self-organizing hierarchical particle swarm optimizer with time-varying acceleration coefficients. Evol. Comput. **8**, 240–255 (2004)
9. Zhao, Z., Huang, S., Wang, W.: Simplified particle swarm optimization algorithm based on stochastic inertia weight. Comput. Appl. Res. **22**, 361–363 (2014)
10. Chen, S., Cai, G., Guo, W., Chen, G.: Study on nonlinear strategy of acceleration coefficient in particle swarm optimization (PSO) algorithm. J. Yangtze Univ. (Nat. Sci. Ed.) **4**, 1–4 (2007)
11. Zhang, S., Zhong, W.: Hybrid particle swarm optimization algorithm of new learning factors and constraint factor. Appl. Res. Comput. **32**, 3626–3628 (2015)
12. Clerc, M.: The swarm and the queen: towards a deterministic and adaptive particle swarm optimization. In: Proceeding of IEEE International Conference on Evolutionary Computation, pp. 1951–1957 (1999)
13. Zhu, X., Li, Y., Li, N., Fan, B.: Improved PSO algorithm based on swarm prematurely degree and nonlinear periodic oscillating strategy. J. Commun. **35**, 182–189 (2014)
14. Zhang, L., Wu, Y., Wei, X.: An adaptive particle swarm optimization algorithm based on linear dynamic parameter. Res. Dev. **3**, 15–18 (2011)
15. Wang, Q., Wang, Z., Wang, S.: A modified particle swarm optimizer using dynamic inertia weight. Chin. Mech. Eng. **16**, 945–948 (2005)

Synchronization Analysis for Complex Networks with Interval Coupling Delay

Dawei Gong[1]($\boxtimes$), Xiaolin Dai[1], Jinliang Song[2], and Bonan Huang[3]

[1] School of Mechatronics Engineering, University of Electronic Science and Technology of China, Chengdu, China
pzhzhx@126.com, www_dxl@126.com
[2] The National Network of Liaoning Electric Power Research Institute, Shenyang, China
sjl1224@163.com
[3] College of Information Science and Engineering, Northeastern University, Shenyang, China
78778322@qq.com

Abstract. This paper concerns synchronization problems for complex networks with interval delays. By using an inequality that is introduced from Newton-Leibniz formula and combining the Finsler's Lemma with homogenous matrix, convergent LMI relaxations for synchronization analysis are proposed with matrix-valued coefficients. Finally, a numerical example is provided to illustrate the effectiveness of the proposed methods.

Keywords: Synchronization · Complex networks · Hybrid coupling · Finsler's lemma

1 Introduction

Complex networks, which have large size and non-trivial complex topological features, have been intensively studied by many researchers in recent years. Such networks have connections which are neither purely regular nor purely random. These networks are used to understand and predict the behaviour of many structures, e.g. internet, medicine, society and biology. It has been found that lots of phenomena in real world can be modeled by complex networks [1,2]. Amongst all the topics which are studied by complex networks, synchronization phenomena plays an important role due to their real world potential applications. There are many interesting synchronization phenomena in nature world. Lots of efforts have been putting into the development of the synchronization problems in complex networks [3–5].

However, it should be noticed that for the complex dynamical networks in all of the aforementioned contributions and most of the existing studies, time delays occur commonly in complex networks because of the network traffic congestions as well as the finite speed of signal transmission over the links. And the synchronization problem for various types of networks with delayed coupling has been

© Springer International Publishing AG 2017
F. Cong et al. (Eds.): ISNN 2017, Part I, LNCS 10261, pp. 329–336, 2017.
DOI: 10.1007/978-3-319-59072-1_39

extensively studied [6, 7]. However, in those papers, much computation complexities, and the number of variables are usually very huge. So, how to improve system performance while removing the redundant variables and reducing computation complexities still remains largely unsolved and challenging. Thus, how to utilize the information in time-varying delay in order to further obtain less conservative synchronization for complex networks with time-varying delays still remains largely unsolved and challenging.

Therefore, we study synchronization problem for a general complex system with time-varying delays by using a useful inequality and introducing Finsler's lemma. Combining with Finsler's Lemma, convergent LMI relaxations for synchronization analysis are proposed. The obtained results can be expanded to many existing research papers. Finally, our main results are illustrated through some numerical simulation examples.

Notations: R^n is the n-dimensional Euclidean space; $R^{m \times n}$ denotes the set of $m \times n$ real matrices. I_n represents the n-dimensional identity matrix. The notation $X \geq 0$ (respectively, $X > 0$) means that X is positive semidefinite (respectively, positive definite); $diag(\cdots)$ denotes a block-diagonal matrix; $\begin{bmatrix} X & Y \\ * & Z \end{bmatrix}$ stands for $\begin{bmatrix} X & Y \\ Y^T & Z \end{bmatrix}$. Matrix dimensions, if not explicitly stated, are assumed to be compatible for algebraic operations.

2   Preliminaries

Consider a delayed complex dynamical network consisting of N coupled nodes, in which each node is an n-dimensional dynamical subsystem

$$\dot{x}_i(t) = f(x_i(t)) + c \sum_{j=1}^{N} g_{ij} A x_j(t - \tau(t)), (i = 1, 2, \ldots, N) \tag{1}$$

where $x_i(t) = (x_{i1}(t), x_{i2}(t), \cdots, x_{in}(t))^T \in R^n$ is the state vector of the ith node. $f(x_i(t)) \in R^n$ is a continuously differentiable vector function. The constant $c > 0$ represents the coupling strength. $A = (a_{ij})_{n \times n}$ is a inner-coupling matrix, $G = (g_{ij})_{N \times N}$ represent the outer-coupling connections. The time delays $\tau(t)$ is time-varying differentiable functions which satisfy

$$0 \leq \tau(t) \leq \tau, \dot{\tau}(t) \leq \mu < 1$$

In the following, some elementary situations are introduced, which play an important role in the proof of the main result.

Throughout this paper, the following assumptions are needed.

Assumption 1. The outer-coupling configuration matrices of the neural networks satisfy

$$\begin{cases} g_{ij} = g_{ji} \geq 0, i \neq j \\ g_{ii} = - \sum_{j=1, j \neq i}^{N} g_{ij}, i, j = 1, 2, \cdots, N \end{cases}$$

Next, we give some useful definitions and lemmas.

Definition 1. The delayed dynamical networks (1) are said to achieve asymptotic synchronization if

$$x_1(t) = x_2(t) = \ldots = x_N(t) = s(t), t \to \infty \tag{2}$$

where $s(t)$ is a solution of an isolate node, satisfying $\dot{s}(t) = f(s(t))$.

Lemma 1. [8] Consider the delayed dynamical network (1), the eigenvalues of outer coupling matrix G are denoted by

$$0 = \lambda_1 > \lambda_2 \geq \lambda_3 \geq \ldots \geq \lambda_N,$$

if the following $N-1$ of n-dimensional time-varying delayed differential equations are asymptotically stable about their zero solution

$$\dot{w}_k(t) = J(t)w_k(t) + c\lambda_k Aw_k(t - \tau(t)) \tag{3}$$

where $J(t)$ is the Jacobian of $f(x(t))$ at $s(t)$, then the synchronized states are asymptotically stable.

Lemma 2. (Jensen's inequality) For constant matrix $\Upsilon \in R^{n \times n}, \Upsilon^T = \Upsilon > 0$, scalar $\rho > 0$ and vector function $\varpi : [0, \rho] \to R^n$, we have:

$$\rho \int_0^\rho \varpi^T(s)\Upsilon\varpi(s)ds \geq (\int_0^\rho \varpi(s)ds)^T \Upsilon(\int_0^\rho \varpi(s)ds)$$

Lemma 3. (Finsler's Lemma) Let $\xi \in R^n, \phi = \phi^T \in R^{n \times n}$, and $\mathbb{B} \in R^{m \times n}$, such that $rank(\mathbb{B}) < n$. Then the following statements are equivalent:

(1) $\xi^T \phi \xi < 0, \mathbb{B}\xi = 0, \xi \neq 0$
(2) $(\mathbb{B}^\perp)^T \phi \mathbb{B}^\perp < 0$,
(3) $\exists L \in R^{n \times m}, \Phi + LB + B^T L^T < 0$,

where $\mathbb{B}^\perp$ is a right orthogonal complement of $\mathbb{B}$

Lemma 4. [9] Let $x(t) \in R^n$ be a vector-valued function with first-order continuous-derivative entries. Then, the following integral inequality holds for any matrices $M_1, M_2 \in R^{n \times n}$, $X = X^T > 0 \in R^{n \times n}$, and $Z \in R^{2n \times 2n}$, and a scalar function $\tau \geq 0$:

$$-\int_{t-\tau}^t \dot{x}^T(s)X\dot{x}(s)ds \leq \xi^T(t)\Upsilon\xi(t) + \tau\xi^T(t)Z\xi(t),$$

where

$$\Upsilon := \begin{bmatrix} M_1^T + M_1 & -M_1^T + M_2 \\ * & -M_2^T - M_2 \end{bmatrix}, \xi(t) := \begin{bmatrix} x(t) \\ x(t-\tau) \end{bmatrix}, Z = \begin{bmatrix} M_1^T \\ M_2^T \end{bmatrix} X^{-1} [M_1 \ M_2].$$

Remark 1. This Lemma was proposed by He Yong in 2005, and it has been applied widely in many literatures. However, it is rarely used in the complex networks. It was deduced from Newton-Leibniz formula, so it can use fewer free-weighting matrices to acquire better results than some existing references in which a number of free-weighting matrices were introduced by Newton-Leibniz formula.

3   Main Results

In this section, we present our main results by introducing Finsler's lemma. Combining with the Finsler's Lemma, convergent LMI relaxations for synchronization analysis are proposed.

Theorem 1. From Lemma 3, dynamical system (1) is asymptotically synchronized if there exist positive definite symmetric matrices P_k, Q_k, S_k, Y_k, and any real matrices M_{ik}, $(i = 1, 2)$, such that the following LMIs hold for all $2 \leq k \leq N$:

$$(\mathbb{B}^{\perp})^T \Xi_{2k} \mathbb{B}^{\perp} < 0, \tag{4}$$

where

$$\mathbb{B} = [J(t), c\lambda_k A, -I_N],$$

$$\Xi_{2k} = \begin{bmatrix} H_{2k} & \tau \Gamma_1^T Y_k^T & \tau(1-\mu)\Gamma_2^T \\ * & -Y_k - Y_k^T + \tau S_k & 0 \\ * & * & -\tau(1-\mu)S_k \end{bmatrix} < 0, \tag{5}$$

$$\Gamma_1 = [J(t), c\lambda_k A], \Gamma_2 = [M_{1k}, M_{2k}],$$

$$H_{2k} = \begin{bmatrix} \Omega_{11} & \Omega_{12} \\ * & \Omega_{22} \end{bmatrix}.$$

$$\Omega_{11} = P_k J(t) + J(t)^T P_k + Q_k + (1-\mu)(M_{1k}^T + M_{1k})$$
$$\Omega_{12} = P_k c\lambda_k A - (1-\mu)(M_{1k}^T - M_{2k})$$
$$\Omega_{22} = -(1-\mu)Q_k - (1-\mu)(M_{2k}^T + M_{2k}).$$

Proof: Choose a Lyapunov-Krasovskii functional candidate as

$$V(t) = w_k^T(t)P_k w_k(t) + \int_{t-\tau(t)}^{t} w_k^T(s)Q_k w_k(s)ds + \int_{t-\tau(t)}^{t}\int_{\theta}^{t} \dot{w}_k^T(s)S_k \dot{w}_k(s)dsd\theta,$$

$$\tag{6}$$

Then the time derivative of $V(t)$ along the trajectory will satisfy

$$\dot{V}(t) = 2w_k^T(t)P_k\left[J(t)w_k(t) + c\lambda_k Aw_k(t-\tau(t))\right]$$
$$+ w_k^T(t)Q_k w_k(t) - (1-\dot{\tau}(t))w_k^T(t-\tau(t))Q_k w_k(t-\tau(t))$$
$$+ \tau(t)\dot{w}_k^T(t)S_k\dot{w}_k(t) - (1-\dot{\tau}(t))\int_{t-\tau(t)}^{t}\dot{w}_k^T(s)S_k\dot{w}_k(s)ds$$
$$\leq 2w_k^T(t)P_k\left[J(t)w_k(t) + c\lambda_k Aw_k(t-\tau(t))\right]$$
$$+ w_k^T(t)Q_k w_k(t) - (1-\mu)w_k^T(t-\tau(t))Q_k w_k(t-\tau(t))$$
$$+ \tau\dot{w}_k^T(t)S_k\dot{w}_k(t) - (1-\mu)\int_{t-\tau(t)}^{t}\dot{w}_k^T(s)S_k\dot{w}_k(s)ds \tag{7}$$

From Lemma 4, for any M_{1k}, M_{2k} with an appropriate dimension yields the following integral inequality

$$-\int_{t-\tau(t)}^{t}\dot{w}_k^T(s)S_k\dot{w}_k(s)ds$$
$$\leq \eta_k^T(t)\begin{bmatrix} M_{1k}^T + M_{1k} & -M_{1k}^T + M_{2k} \\ * & -M_{2k}^T - M_{2k} \end{bmatrix}\eta_k(t) + \tau\eta_k^T(t)\begin{bmatrix} M_{1k}^T \\ M_{2k}^T \end{bmatrix}S_k^{-1}\left[M_{1k}\ M_{2k}\right]\eta_k(t), \tag{8}$$

where

$$\eta_k^T(t) = [w_k^T(t), w_k^T(t-\tau(t))].$$

On the other hand, it is easy to see from the system (3) of that the following equation also holds for any matrices $Y_k, 2 \leq k \leq N$,

$$0 = 2\dot{w}_k^T(t)Y_k[-\dot{w}_k(t) + J(t)w_k(t) + c\lambda_k Aw_k(t-\tau(t))] \tag{9}$$

Combing (7), (8), and (9), we can obtain

$$\dot{V}_k(t) \leq \zeta_k^T(t)\varXi_{2k}\zeta_k(t) \tag{10}$$

where

$$\zeta_k^T(t) = [w_k^T(t), w_k^T(t-\tau(t)), \dot{w}_k^T(t)].$$

Note $\mathbb{B}\zeta_k(t) = 0$, it follows from Lemma 3 that $(\mathbb{B}^\perp)^T\varXi_{2k}\mathbb{B}^\perp < 0$ is equivalent to $\zeta_k^T(t)\varXi_{2k}\zeta_k(t) < 0$. This completes the proof.

Remark 2. Convergent LMI relaxations are introduced by Finsler's Lemma with homogenous matrix. Then, more matrix-valued coefficients can be introduced and the aim of reducing conservatism can also be achieved. Therefore, our method significantly improvements the performance of the synchronization results for complex networks, and can be applied to most of the existing synchronization results, such as [10–12].

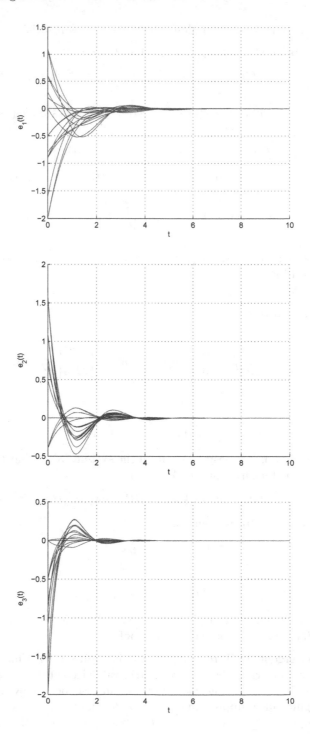

Fig. 1. Synchronization errors for networks with time-varying delays: $e_j(t), j = 1, 2, 3$

4   Numerical Examples

In this section, some numerical examples will be given to illustrate the usefulness of the obtained results.

Example 1: Consider a complex network model with 20 nodes, where each node is a simple three-demensional stable linear system described by

$$\begin{bmatrix} \dot{x}_1(t) \\ \dot{x}_2(t) \\ \dot{x}_3(t) \end{bmatrix} = \begin{bmatrix} -x_1 \\ -2x_2 \\ -3x_3 \end{bmatrix} \tag{11}$$

which is asymptotically stable at $s(t) = 0$, and it's Jacobian is $J = diag(-1, -2, -3)$. We assume that the inner-coupling matrix is $A = 0.5 diag(1, 1, 1)$, and the outer-coupling matrix is defined as:

$$G = \begin{bmatrix} -2 & 1 & 0 & 0 & 1 \\ 1 & -3 & 1 & 1 & 0 \\ 0 & 1 & -2 & 1 & 0 \\ 0 & 1 & 1 & -3 & 1 \\ 1 & 0 & 0 & 1 & -2 \end{bmatrix} \tag{12}$$

By using the Matlab LMI Toolbox, according to the three theorems, the complex networks (1) with time-varying delays can achieve global synchronization. To further illustrate the efficiency of our method, we plot the synchronization errors between the states of nodes in Fig. 1. In this simulation, $e_j(t) = x_{1j}(t) - x_{ij}$ for $i = 2, \cdots, 20; j = 1, 2, 3$, with $c = 0.4, \tau = 1$, and $\mu = 0.5$. The initial values are randomly chosen.

5   Conclusion

We have derived delay-dependent asymptotically stability criteria in terms of LMIs for network synchronization that are very easy to verify. These synchronization conditions are applicable to networks with different topologies and different sizes. We have also shown a numerical example to verify the theoretical results. This paper is only a first step toward network model with coupling delays, which can describe more realistic complex networks. There are also some limitations in our models. For example, in our models all the delays are the same, and the coupling strengths are all constants. In addition, how to extend the existing results to these complex systems is still a challenging problem.

Acknowledgment. This work was supported by the National Natural Science Foundation of China (51305066, 61603076).

References

1. Tan, S.L., Lü, J.H., Chen, G.R.: When structure meets function in evolutionary dynamics on complex networks. IEEE Trans. Circ. Syst. Mag. **14**, 36–50 (2014)

2. Zou, X.J., Gong, D.W., Wang, L.P., Chen, Z.Y.: A novel method to solve inverse variational inequality problems based on neural networks. Neurocomputing **173**, 1163–1168 (2016)

3. Gao, H.J., Lam, J., Chen, G.R.: New criteria for synchronization stability of general complex dynamical networks with coupling delays. Phys. Lett. A **360**, 263–273 (2006)

4. Shen, H., Park, J.H., Wu, Z.G., Zhang, Z.Q.: Finite-time H_∞ synchronization for complex networks with semi-Markov jump topology. Commun. Nonlinear Sci. Numer. Simul. **24**, 40–51 (2015)

5. Qin, J.H., Gao, H.J., Zheng, W.X.: Exponential synchronization of complex networks of linear systems and nonlinear oscillators: a unified analysis. IEEE Trans. Neural Netw. Learn. Syst. **26**, 510–521 (2015)

6. Gong, D.W., Lewis, F.L., Wang, L.P., Xu, K.: Synchronization for an array of neural networks with hybrid coupling by a novel pinning control strategy. Neural Netw. **77**, 41–50 (2016)

7. Gong, D.W., Zhang, H.G., Huang, B.N., Ren, Z.Y.: Synchronization criteria and pinning control for complex networks with multiple delays. Neural Comput. Appl. **22**, 151–159 (2013)

8. Li, C.G., Chen, G.R.: Synchronization in general complex dynamical networks with coupling delays. Phys. A **343**, 263–278 (2004)

9. Zhang, X.M., Wu, M., She, J.H., He, Y.: Delay-dependent stabilization of linear systems with time-varying state and input delays. Automatica **41**, 1405–1412 (2005)

10. Dai, X.L., Gong, D.W., Huang, B.N., Li, J.J.: Synchronisation analysis for coupled networks with multiple delays. Int. J. Syst. **46**, 2439–2447 (2015)

11. Gong, D.W., Liu, J.H., Zhao, S.Y.: Chaotic synchronisation for coupled neural networks based on T-S fuzzy theory. Int. J. Syst. Sci. **46**, 681–689 (2015)

12. Huang, B.N., Zhang, H.G., Gong, D.W., Wang, J.Y.: Synchronization analysis for static neural networks with hybrid couplings and time delays. Neurocomputing **148**, 288–293 (2015)

FPGA Implementation of the L Smallest k-Subsets Sum Problem Based on the Finite-Time Convergent Recurrent Neural Network

Shenshen Gu$^{(\boxtimes)}$ and Xiaowen Wang

School of Mechatronic Engineering and Automation,
Shanghai University, Shanghai 200072, China
gushenshen@i.shu.edu.cn

Abstract. For a given set S of n real numbers and a positive integer $k < n$, there are totally $\binom{n}{k}$ subsets of S with k elements. Among these subsets, to find L subsets whose summation of their elements are the L smallest is so called the L smallest k-subsets sum problem. It is widely applied in the real operations and computational research. However it is very computationally challenging to process large scale L smallest k-subsets sum problem. To solve this problem, this paper presents a FPGA implementation of a finite-time convergent recurrent neural network of L smallest k-subsets sum problem. And the neural network model is tested on a Digilent Nexys 4 DDR board with Xilinx Artix 7 FPGA. Experimental results show that the proposed hardware implementation method has a high degree of parallelism and fast performance.

Keywords: Neural networks · FPGA · k-Subset

1 Introduction

The k-subset is defined as a subset of S containing k different elements, for a given set S with n real numbers, where $k < n$ [1]. The binomial coefficient $\binom{n}{k}$ indicates the number of k-subsets on n elements. For example, there are $\binom{4}{2} = 6$ 2-subsets of $\{3, 5, 9, 10\}$. The L smallest k-subsets sum problem is an operation that defines L k-subsets whose summation of subset elements are the L smallest among all possible combinations. It is obvious that the 2 smallest 2-subsets of $\{3, 5, 9, 10\}$ is $\{3, 5\}$ and $\{3, 9\}$.

There are numerous applications of the subset sum problem in the operations research. In computer science, it is widely implemented in the optimal memory management in multiple programming [2]. The paper [3] presents its usage in supporting multiple scalable video sequences by wireless resources. And in [4] the

The work described in the paper was supported by the National Science Foundation of China under Grant 61503233.

© Springer International Publishing AG 2017
F. Cong et al. (Eds.): ISNN 2017, Part I, LNCS 10261, pp. 337–345, 2017.
DOI: 10.1007/978-3-319-59072-1_40

subset sum problem was researched as a specific case of the Knapsack problem in optimization. Chang [5] presented a quantum algorithms for solving an instance of the subset-sum problem and test it on a quantum computer. For the hardware implementation, a GPU implementation is proposed in paper [6] to solve the subset sum problem.

In the past time, its difficult to compile digital signal processing by field-programmable-gate-array (FPGA), because of the complexity of describing digital signal processing through hardware description language (HDL). However, the appearance of a high level modeling tool Xilinx System Generation make it easier to process mathematical model by constructing digital signal processing in FPGA. More and more signal processing with high performance are compiled with FPGA.

In this paper the hardware implementation of the k-subset neural network model is studied by using a field-programmable-gate-array (FPGA) chip. To solve the L smallest k-subsets sum problem, a efficient algorithm was proposed in the paper [7] by combining the L shortest paths algorithm and the finite-time convergent recurrent neural network. Based on this algorithm a neural network model was constructed. After verification of the neural network model based on Xilinx Blockset in Simulink, the Verilog hardware description language (HDL) is used to describe the neural network in Xilinx company's comprehensive FPGA development software Vivado.

This paper is organized as follows. Section 2 presents the problem formulation and illustrates the description of the problem with a network. Section 3 introduces the neural network design procedure, architecture and properties. And the FPGA implementation is also described in this section. Next, in Sect. 4, experimental results are given to verify the efficiency of the FPGA implementation. Finally, a conclusion was drawn in the end to summarize this work.

2   Problem Formulation and Neural Network Model

A proper mathematic model is very indispensable to solve the L smallest k-subsets problem effectively. Mathematically, the lth $(0 < l \leq L)$ smallest k-subset problem can be formulated as a function

$$x_i = \begin{cases} 1, & \text{if } v_i \in \{\text{the } lth \text{ smallest } k\text{-subset}\}; \\ 0, & \text{otherwise}; \end{cases} \tag{1}$$

for $i = 1, \ldots, n$; where $v \in R^n$ and $k \in \{1, \ldots, n-1\}$. Figure 1 shows the operation graphically.

Considering the L smallest k-subsets sum problem can be regarded as the further application of the L shortest paths problem, many algorithms have been proposed for finding the shortest path. To find the shortest path several neural networks were proposed in [8], the shortest path problem can be converted to the linear program as follows:

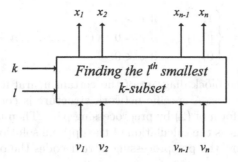

Fig. 1. Diagram of finding the lth $(0 < l \leq L)$ smallest k-subset operation.

$$\text{minimize} \quad \sum_{i=1}^{n} \sum_{j=1}^{n} c_{ij} x_{ij}$$

$$\text{subject to} \quad \sum_{k=1}^{n} x_{ik} + \sum_{l=1}^{n} x_{li} = \delta_{il} - \delta_{in}$$

$$x_{ij} \geq 0, \quad i, j = 1, 2, \ldots, n$$

where c_{ij} denotes the weight between node i and node j, and δ_{ij} is referred to as Kronecker delta function which is indicated by $\delta_{ij} = 1(i = j)$ and $\delta_{ij} = 0(i \neq j)$. The dual problem of the previous problem is

$$\text{maximize} \quad y_n - y_1$$

$$\text{subject to} \quad y_j - y_i \leq c_{ij}, \quad i, j = 1, 2, \ldots, n$$

where y_i denotes the dual decision variable according to node i. By defining $z_i = y_i - y_1$ for $i = 1, 2, \ldots, n$ the dual problem can be further expressed by following program.

$$\text{minimize} \ z_n, \qquad\qquad\qquad\qquad\qquad (2)$$
$$\text{subject to} \ z_j - z_i \leq c_{ij}, \quad i \neq j, \ i, j = 1, 2, \ldots, n$$

where $z_1 = 0$. The effectiveness and high efficiency of the finite-time convergent recurrent neural network has been proved in [9], because of its flexible structure and the capability of selecting parameters for global convergence. By combining (2) and the finite-time convergent recurrent neural network, the neural network model can be modulated as follows:

$$\epsilon \frac{dz}{dt} = -\sigma A^T g(Az - b) - c, \qquad\qquad\qquad (3)$$

where ϵ denotes a positive scaling constant, $\sigma > 1$ is a gain parameter, $z = (z_2, z_3, \ldots, z_n)^T$, $b = (c_{12}, \ldots, c_{1n}, c_{21}, c_{23}, \ldots, c_{2n}, \ldots, c_{n1}, \ldots, c_{n,n-1})^T \in \mathbb{R}^{n(n-1)}$, A is an $n(n-1) \times n$ matrix of 0, 1 and -1 to construct $n(n-1)$ inequality constraints, $g : R^n \rightarrow [0, 1]^n$ is a piecewise linear function which is defined by $g(v) = (g(v_1), \ldots, g(v_{n-1}))$ and

$$g(v_i) = \begin{cases} 1, & \text{if } v > 0, \\ [0,1], & \text{if } v = 0, \quad (i = 1, 2, \ldots, n-1) \\ 0, & \text{if } v < 0. \end{cases} \tag{4}$$

Figure 2 shows the block diagram of the certain neural network which consists of three main parts. A specific network structure is converted to a linear programming in the form of (2) by preprocessing part. Then the neural network processing part works as the calculation of the optimal solution. At last, according to the connections, the postprocessing part decodes the optimal solution to the solution.

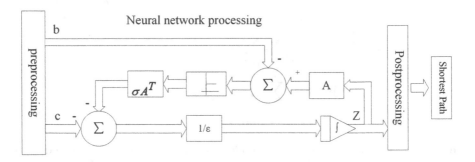

Fig. 2. Block diagram of the neural network for finding the shortest path.

3   FPGA Implementation

Based on the neural network model, a circuit implementation model was built. Figure 3 shows the circuit model, which consists of four kinds of modules: adder modules, function modules, multiplier modules and integrator modules. According to the circuit model, a hardware implementation architecture is constructed with Xilinx Blockset in Simulink. The function module is modeled by comparator, i.e., if the input is positive then the output equals one, when the input is less the or equal to zero, the output equals zero. As to the integrator, considering that the integration process equals to small accumulation process, the numerical value of the integrand is divided into many small ones which are added to the accumulator step by step then stored in the register. Figure 4 shows the hardware model of the integrator. Since the valve of the step is very significant, if the step is too small, the output will converge very slowly. And if the step is too big, the output will turn out divergent. After many experimental tries, the hardware model performs well with the step of 0.02.

Figure 5 shows the hardware implementation model of the neural networks. The neural networks are unsupervised which consist of three layers. The first layer is composed of adder modules and function modules. The middle layer consist of multiplier modules and adder modules. The multiplier modules work as connection weights (i.e. the gain parameter σ in (3)) with the valve of 2. The

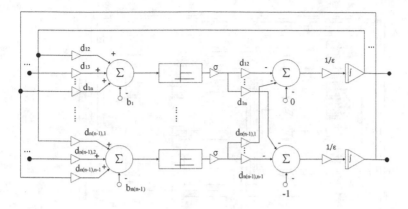

Fig. 3. Circuit implementation of the neural network for finding the shortest path.

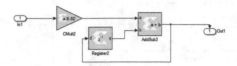

Fig. 4. The model for an integrator.

last layer is composed of integrator modules, it is also the output layer of the neural networks, the output will feedback to the first layer.

System Generator must work in Simulink of Matlab based on the method of model design. User can construct the mathematical model by Xilinx Blockset in Simulink and get the simulation results. Then System Generator can generate Verilog HDL code and test bench which can be compiled in Vivado an integrated development environment (IDE) software of Xilinx. The System Generator has a high compatibility of floating-point calculation which makes the iteration process more accurate and guarantees the convergence of the system. In this paper, the generation environment is based on Xilinx Artix-7 xc7a100T.

4   Simulation Results

To show the effectiveness and efficiency of the neural network hardware implementation, the following simulations are performed. In the first case where $L = 4$, $k = 2$, $S = \{2, 4, 6, 9\}$, the promised optimal value should be six. The type of the data in this model is fixed binary floating point with size of 20 bits, the sample step is 1s, the initial value of the accumulator is 15. The simulation results of the model in Simulink are shown in Fig. 6. In the second case where $L = 4$, $k = 2$, $S = \{3, 5, 9, 10\}$, the promised optimal value should be eight. Figure 7 shows the simulation results of the model. The two cases above show that the hardware model of the neural network constructed in simulink can work correctly and efficiently, the simulation result is convergent and accurate.

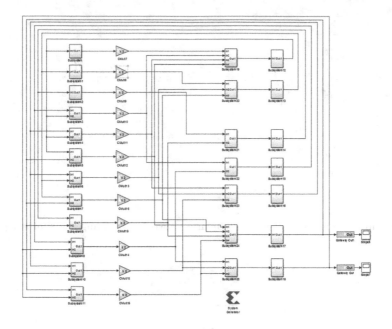

Fig. 5. The hardware implementation model of the neural networks.

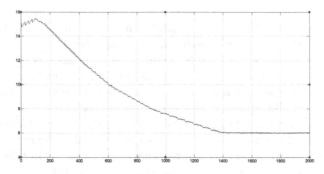

Fig. 6. The simulation results of the model in Simulink with $S = \{2, 4, 6, 9\}$.

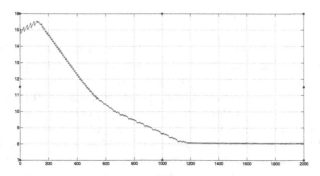

Fig. 7. The simulation results of the model in Simulink with $S = \{3, 5, 9, 10\}$.

In term of the two cases, Verilog HDL code of the neural network can be generated by Xilinx System Generation, after running behavioral simulation in Vivado the wave simulation results are shown in Fig. 8 which illustrates that the output of the first case converges to 6.25 in 13.49us, its very close to the promised result. From Fig. 9, the output of the second case converges in 9.94us and the result is 8.8125.

Fig. 8. The result of wave simulation in Vivado with $S = \{2, 4, 6, 9\}$.

Fig. 9. The result of wave simulation in Vivado with $S = \{3, 5, 9, 10\}$.

In order to display the output on the FPGA chip, the Convert module is employed to convert the output signal into a 4 bit binary integer. The input data of the first case are kept unvaried, the converted result is shown in Fig. 10. Figure 11 illustrates the result of the output displayed by LEDs on the FPGA chip, the type of the FPGA chip implemented in this experiment is Digilent Nexys 4 DDR board with Xilinx Artix 7 FPGA. The power estimation available in the SDK, the total on-chip power under the default environment setting is 0.148W.

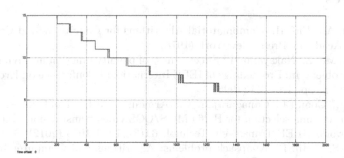

Fig. 10. The simulation result of the converted output with $S = \{2, 4, 6, 9\}$.

Fig. 11. The simulation result on Digilent Nexys 4 DDR board with Xilinx Artix 7 FPGA.

5   Conclusion

In this paper, based on a the finite-time convergent recurrent neural network model combined with FPGA chips, a hardware implementation method was proposed for the the the L smallest k-subsets problem. Xilinx System Generation makes it easier to build the DSP system of neural network. The experimental results indicated that this hardware implementation demonstrated a high degree of parallelism and capable of pipelining of neural networks. The FPGA implementation performed efficient and effective as well. This implementation can be considered for further study by researchers. Considering the inputs data is kept unvaried, the results may be regarded as a benchmark for conceptual and preliminary studies. More neural networks based on various algorithm can be implemented by FPGA.

References

1. Nijenhuis, A., Wilf, H.: Combinatorial Algorithms for Computers and Calculators, 2nd edn. Academic Press, New York (1978)
2. Wang, H., Ma, Z., Nakayama, I.: Effectiveness of penalty function in solving the subset sum problem. In: Proceedings of IEEE International Conference on Evolutionary Computation, pp. 422–425 (1996)
3. Bocus, M., Coon, J., Canagarajah, C., Armour, S.M.D., Doufexi, A.M.J.: Per-subcarrier antenna selection for H.264 MGS/CGS video transmission over cognitive radio networks. IEEE Trans. Veh. Technol. **61**(3), 1060–1073 (2012)
4. Martello, S., Toth, P.: Knapsack problems: Algorithms and Computer Interpretations. Wiley-Interscience, Hoboken (1990). pp. 105–136
5. Chang, W.L.: Quantum algorithms of the subset-sum problem on a quantum computer. In: WASE International Conference on Information Engineering, ICIE 2009, pp. 54–57 (2009)

6. Wan, L., Li, K.L.J.: GPU implementation of a parallel two-list algorithm for the subset-sum problem. Concurr. Comput. **27**(1), 119–145 (2015)
7. Gu, S., Cui, R.: An efficient algorithm for the subset sum problem based on finite-time convergent recurrent neural network. Neuro Comput. **149**, 13–21 (2015)
8. Wang, J.: Primal and dual neural networks for shortest-path routing. IEEE Trans. Syst. Man Cybern. Part A: Syst. Hum. **28**(6), 864–869 (1998)
9. Liu, Q., Wang, J.: Finite-time convergent recurrent neural network with a hard-limiting activation function for constrained optimization with piecewise-linear objective functions. IEEE Trans. Neural Netw. **22**(4), 601–613 (2011)

Accelerating Stochastic Variance Reduced Gradient Using Mini-Batch Samples on Estimation of Average Gradient

Junchu Huang[✉], Zhiheng Zhou, Bingyuan Xu, and Yu Huang

School of Electronic and Information Engineering,
South China University of Technology, Guangzhou, China
{h.junchu,xu.bingyuan,h.y33}@mail.scut.edu.cn, zhouzh@scut.edu.cn

Abstract. Stochastic gradient descent (SGD) is popular for large scale optimization but has slow convergence. To remedy this problem, stochastic variance reduced gradient (SVRG) is proposed, which adopts average gradient to reduce the effect of variance. Since its expensive computational cost, average gradient is maintained between m iterations, where m is set to the same order of data size. For large scale problems, the efficiency will be decreased due to the prediction on average gradient maybe not accurate enough. We propose a method of using a mini-batch of samples to estimate average gradient, called stochastic mini-batch variance reduced gradient (SMVRG). SMVRG greatly reduces the computational cost of prediction on average gradient, therefore it is possible to estimate average gradient frequently thus more accurate. Numerical experiments show the effectiveness of our method in terms of convergence rate and computation cost.

Keywords: Optimization algorithms · Stochastic gradient descent · Machine learning

1 Introduction

In machine learning, the following empirical risk minimization problem is often encountered. Let $f_i(w)$ be a loss function defined on instance i, where $i = 1, 2, \cdots, n$ and w is the parameter to learn. To improve generalization ability, $f_i(w)$ is often with a regularization term. In this paper, it is assumed that each $f_i(w)$ is derivable. Our objective is to in search of an approximate solution of such optimization problem

$$\min P(w), \ P(w) \triangleq \frac{1}{n} \sum_{i=1}^{n} f_i(w) \tag{1}$$

For example, given a set of labeled instances $(x_1, y_1), \cdots, (x_n, y_n)$, if $f_i(w)$ is defined as $\log(1 + \exp(-y_i x_i^T w)) + \frac{\lambda}{2}\|w\|^2$, problem (1) will be known as the

© Springer International Publishing AG 2017
F. Cong et al. (Eds.): ISNN 2017, Part I, LNCS 10261, pp. 346–353, 2017.
DOI: 10.1007/978-3-319-59072-1_41

logistic regression. To solve multi-classification problem, $f_i(w)$ can be defined as (2), which is known as softmax regression with a L2-norm regularization term.

$$f_i(w) = \sum_{j=1}^{l} -1\{y_j = j\}log\frac{exp(w_j^T x_i)}{\sum_{i=1}^{l} exp(w_j^T x_i)} \tag{2}$$

where $1\{true\} = 1$ and $1\{false\} = 0$. Here l means the number of classes, for example in cifar10 l equals 10. Gradient descent [1] is a basic method to solve this problem. At each iteration t $(t = 1, 2, \cdots)$ w^{t-1} is updated by

$$w^t = w^{t-1} - \eta_t \nabla P(w^{t-1}) = w^{t-1} - \eta_t \sum_{i=1}^{n} \frac{1}{n} \nabla f_i(w^{t-1}) \tag{3}$$

However at each step, it requires to compute derivatives of all instances, which is costly especially when n is extremely large. Stochastic gradient descent (SGD) [1,2] is a popular method to remedy it, where at each step i_t is randomly drawn from $\{1, 2, \cdots, n\}$ and

$$w^t = w^{t-1} - \eta_t \nabla f_{i_t}(w^{t-1}) \tag{4}$$

It is usually assumed that $\nabla f_{i_t}(w^{t-1})$ is an unbiased estimation to $\nabla P(w^{t-1})$ because

$$\mathbb{E}\left[\nabla f_{i_t}(w^{t-1})|w^{t-1}\right] = \nabla P(w^{t-1}) \tag{5}$$

On the one hand, the randomness can greatly reduce the computational cost since it only requires to calculate a single instance. On the other hand it also introduces variance, which slows down convergence rate. That is because $\nabla f_{i_t}(w^{t-1})$ equals $\nabla P(w^{t-1})$ in expectation but each $f_{i_t}(w^{t-1})$ may be different.

In order to improve SGD, several researchers have proposed SGD variants, such as SAG [3–5], SAGA [6], SVRG [7–9], SDCA [10,11] and so on. Moreover, unlike SDCA or SAG, SVRG does not require the storage of gradients, hence will be more suitable in complex problems. SVRG adopts average gradient to reduce the effect of variance: where at each step $t = 1, 2, \cdots$ it can be described as

$$w^t = w^{t-1} - \eta_t(\nabla f_{i_t}(w^{t-1}) - \nabla f_{i_t}(w_s) + \tilde{u}) \tag{6}$$

where $\tilde{u} = \frac{1}{n} \sum_{i=1}^{n} \nabla f_i(w_s)$, which means the average gradient, and w_s is the weight obtained at the end of m iterations (this is option I, option II see SVRG). Since its expensive computational cost, average gradient is maintained between m iterations, where m is set to the same order of n. Hence for large scale problems, the prediction on average gradient will be not accurate enough since m is too large. Due to it, the efficiency of SVRG will be decreased. But it is also not wise to chose a smaller m because it will increase the overhead of gradient average computation.

2　Using Mini-Batch Samples on Estimation of Average Gradient

As discussed above, for large scale problems, the value of m is so large that the estimation of the average gradient maybe not accurate enough, which decreases the efficiency of algorithm. However simply choose a smaller m means more frequent updates of the gradient average, which increases of the overhead of computation. We propose a method of using a mini-batch of samples to estimate average gradient, called stochastic mini-batch variance reduced gradient (SMVRG). Differently, we uniformity selected $\{i_1, i_2, \cdots, i_k\}$ from $\{1, 2, \cdots, n\}$ to estimate average gradient ν between m steps, so

$$\nu = \frac{1}{k} \sum_{j=1}^{k} \nabla f_{i_j}(\tilde{w}_s) \tag{7}$$

where k is the size of mini-batch, and k is far less than n. Note that our method greatly reduces the computational cost of prediction on average gradient, hence it is possible to estimate average gradient frequently thus more accurate. And it is natural to choose m to be the same order of k (for example in our experiment m is set to be k). Therefore w^{t-1} is updated by

$$w^t = w^{t-1} - \eta_t(\nabla f_j(w^{t-1}) - \nabla f_j(\tilde{w}_s) + \nu) \tag{8}$$

where w_s is the weight obtained at the end of m iterations.

Algorithm 1. SMVRG

Parameters: update frequency m, learning rate η and mini-batch size k
Initialize: $\tilde{w}_0$
Iterate: for $s = 1, 2, \cdots$;
　　　　Uniformity select $\{i_1, i_2, \cdots, i_k\}$ from $\{1, 2, \cdots, n\}$;
　　　　$\tilde{w} = \tilde{w}_{s-1}$
　　　　$\nu = \frac{1}{k} \sum_{j=1}^{k} \nabla f_{i_j}(\tilde{w}_s)$
　　　　$\tilde{w}_0 = \tilde{w}$
　　　　Iterate: for $t = 1, 2, \cdots, m$
　　　　Randomly pick $i_t \in \{1, 2, \cdots, n\}$ and update weight
　　　　$\tilde{v} = \nabla f_{i_t}(w^{t-1}) - \nabla f_{i_t}(\tilde{w}_{s-1}) + \nu$
　　　　$w^t = w^{t-1} - \eta_t \tilde{v}$
　　　　end
　　　　option I: set $\tilde{w}_{s-1} = w^m$
　　　　option II: set $\tilde{w}_{s-1} = w^t$ for randomly chosen $t \in \{0, 1, \cdots, m\}$
end

Review the ideas embodied in SVRG, it takes average gradient to reduce the effect of variance hence it outperform the standard SGD. Inspired by it, we adopt the average gradient counted by a mini-batch instances to reduce the variance from a single sample. If we always set $k = n$ in SMVRG, then it can be considered to the original SVRG. And the details of the algorithm are shown in Algorithm 1. Compared with SVRG, SMVRG requires less computation, which from $O(2m + n)$ to $(2m + k)$ between m steps.

Our contributions in this paper are in several folds.

(1) It greatly reduces the computational cost of prediction on average gradient, which from $O(n)$ to $O(k)$.
(2) It is possible to estimate average gradient frequently thus more accurate because we can chose a smaller m.
(3) Improve the efficiency of the algorithm especially in large scale optimization. More valuable is that our method requires less computation, which is no relationship with the size of the data. Hence will be more suitable for large scale problem.

3   Experiment

Two datasets are used for evaluation. They are cifar10 and cifar100, which can be downloaded from the CIFAR-10 and CIFAR-100 datasets website (http://www.cs.toronto.edu/~kriz/cifar.html). Detailed information is shown in Table 1. For cifar100 each image comes with a fine label (means the class to which belongs), and a coarse label (means the superclass to which it belongs).

Table 1. Dataset

Dataset	Training instances	Testing instances	Features	Memory	Class
Cifar10	50000	10000	3072	163 MB	10
Cifar100 (coarse-label)	50000	10000	3072	161 MB	20
Cifar100 (fine-label)	50000	10000	3072	161 MB	100

We chose (2) to evaluate our SMVRG. Because full gradient may be the computationally most intensive operation, for fair comparison, we compare SVRG to SMVRG based on the number of gradient computations. The results are shown in Figs. 1, 2, and 3 where loss is an indicator of the error. That means the lower the loss, the better the performance. All of the experiments in this paper, learning rate η is set to the same value. The difference between Figs. 1, 2, and 3 is the size of the mini-batch, which is set to 400, 600 and 800. According to Figs. 1, 2, and 3, it is clear that SMVRG has a lower computational cost, therefore means our method can be more effectively. Comparing Figs. 1, 2, and 3, the algorithm is not sensitive to the size of the mini-batch, although a larger mini-batch may have a better result.

In order to show the superiority of our algorithm, we also compare SVRG to SMVRG based on the computation time. Again, the difference in Figs. 4, 5, and 6 is the size of the mini-batch. Our method requires less computation, which is no relationship with the size of the data, therefore it is much faster than SVRG just as the results show.

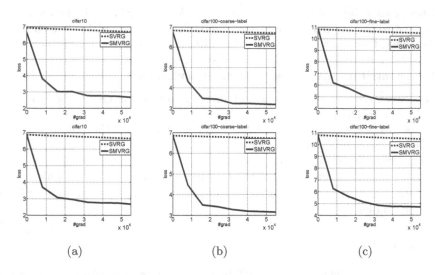

(a)   (b)   (c)

Fig. 1. Contrast between the SVRG and SMVRG based on the number of gradient computations, where k equals 400. (a) Experiments with cifar10: training loss (top), testing loss (down). (b) Experiments with cifar100-coarse-label: training loss (top), testing loss (down). (c) Experiments with cifar100-fine-label: training loss (top), testing loss (down)

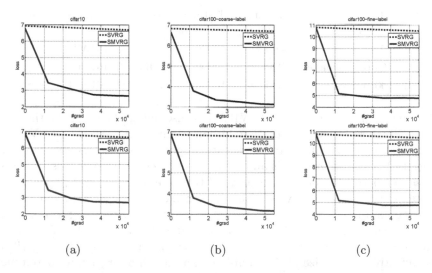

(a)   (b)   (c)

Fig. 2. Contrast between the SVRG and SMVRG based on the number of gradient computations, where k equals 600. (a) Experiments with cifar10: training loss (top), testing loss (down). (b) Experiments with cifar100-coarse-label: training loss (top), testing loss (down). (c) Experiments with cifar100-fine-label: training loss (top), testing loss (down)

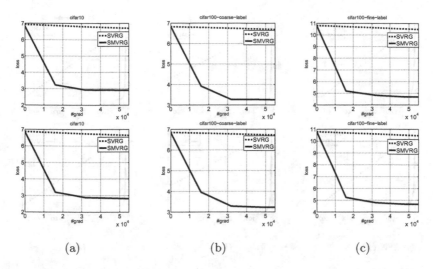

Fig. 3. Contrast between the SVRG and SMVRG based on the number of gradient computations, where k equals 800. (a) Experiments with cifar10: training loss (top), testing loss (down). (b) Experiments with cifar100-coarse-label: training loss (top), testing loss (down). (c) Experiments with cifar100-fine-label: training loss (top), testing loss (down)

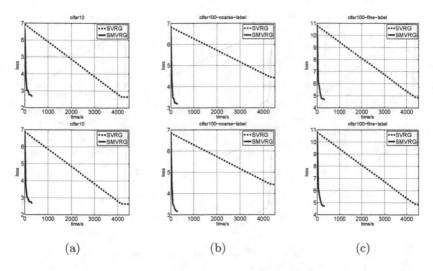

Fig. 4. Contrast between the SVRG and SMVRG based on the computation time, where k equals 400. (a) Experiments with cifar10: training loss (top), testing loss (down). (b) Experiments with cifar100-coarse-label: training loss (top), testing loss (down). (c) Experiments with cifar100-fine-label: training loss (top), testing loss (down)

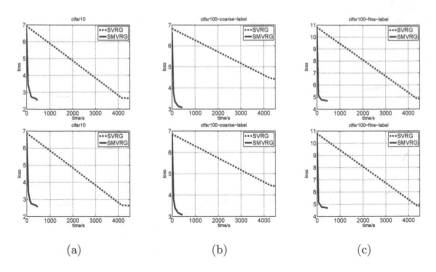

Fig. 5. Contrast between the SVRG and SMVRG based on the computation time, where k equals 600. (a) Experiments with cifar10: training loss (top), testing loss (down). (b) Experiments with cifar100-coarse-label: training loss (top), testing loss (down). (c) Experiments with cifar100-fine-label: training loss (top), testing loss (down)

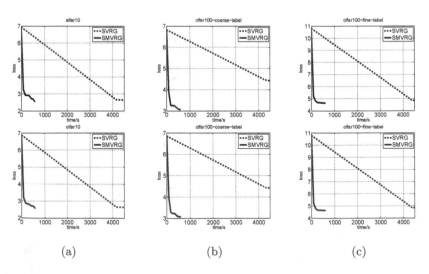

Fig. 6. Contrast between the SVRG and SMVRG based on the computation time, where k equals 800. (a) Experiments with cifar10: training loss (top), testing loss (down). (b) Experiments with cifar100-coarse-label: training loss (top), testing loss (down). (c) Experiments with cifar100-fine-label: training loss (top), testing loss (down)

4   Conclusion

Review the ideas embodied in SVRG, it takes average gradient to reduce the effect of variance hence it outperform the standard SGD. Inspired by it, we adopt the average gradient counted by a mini-batch instances to reduce the variance from a single sample, which greatly reduces the computational cost of average gradient. Hence it is possible to estimate the average gradient more frequently. Numerical experiments show that our method requires less calculation and thus faster than SVRG. More valuable is that SMVRG is no relationship with the size of the data. Hence will be more suitable for solving large scale problem.

Acknowledgements. We thanks Rie Johnson for his advice. And the work is supported by National Natural Science Foundation of China (61372142, U1401252, U1404603), Guangdong Province Science and technology plan (2013B010102004, 2013A011403003), Guangzhou city science and technology research projects (201508010023).

References

1. Zhang, T.: Solving large scale linear prediction problems using stochastic gradient descent algorithms. In: International Conference on Machine Learning, p. 116. Omnipress (2004)
2. Bottou, L.: Large-scale machine learning with stochastic gradient descent. In: Lechevallier, Y., Saporta, G. (eds.) Proceedings of COMPSTAT 2010, pp. 177–186. Physica-Verlag HD, Heidelberg (2010)
3. Roux, N.L., Schmidt, M., Bach, F.: A stochastic gradient method with an exponential convergence rate for finite training sets. Adv. Neural Inf. Process. Syst. **4**, 2663–2671 (2012)
4. Schmidt, M., Roux, N.L., Bach, F.: Minimizing finite sums with the stochastic average gradient. Math. Program. **26**(5), 405–411 (2013)
5. Schmidt, M., Roux, N.L., Bach, F.: Erratum to: minimizing finite sums with the stochastic average gradient. Math. Program. **26**(5), 1 (2016)
6. Defazio, A., Bach, F., Lacostejulien, S.: SAGA: a fast incremental gradient method with support for non-strongly convex composite objectives. Adv. Neural Inf. Process. Syst. **2**, 1646–1654 (2014)
7. Rie, J., Tong, Z.: Accelerating stochastic gradient descent using predictive variance reduction. Adv. Neural Inf. Process. Syst. 315–323 (2013)
8. Wang, C., Chen, X., Smola, A., et al.: Variance Reduction for Stochastic Gradient Optimization. University of Illinois Press, Champaign (2013). pp. 181–189
9. Gresti, P.: Linear convergence of variance-reduced projected stochastic gradient without strong convexity. Comput. Sci. **2014**(2), 648–650 (2014)
10. Shalev-Shwartz, S., Zhang, T.: Stochastic dual coordinate ascent methods for regularized loss minimization. J. Mach. Learn. Res. **14**(1), 2013 (2012)
11. Tseng, P., Yun, S.: A coordinate gradient descent method for nonsmooth separable minimization. Math. Program. **117**(1), 387–423 (2009)

Coexistence and Local Exponential Stability of Multiple Equilibria in Memristive Neural Networks with a Class of General Nonmonotonic Activation Functions

Yujiao Huang[✉], Shijun Chen, Jie Xiao, and Pengyi Hao

College of Computer Science and Technology,
Zhejiang University of Technology, Hangzhou 310023, China
hyj0507@zjut.edu.cn

Abstract. This paper addresses the multistability problem of n-dimensional memristive neural networks with a class of general nonmonotonic activation functions. Sufficient conditions are proposed for checking the existence of $(2l+3)^n$ equilibria, of which $(l+2)^n$ equilibria are locally exponentially stable. The obtained stability results improve and extend the existing ones. One numerical example is given to illustrate the effectiveness of the obtained results.

Keywords: Memristive neural network · Coexistence · Local exponential stability · General nonmonotonic activation function

1 Introduction

Memristor was originally postulated by Chua in 1971 [1], and was realized by scientists at Hewlett-Packard Laboratories in 2008 [2]. The memristor is a contraction of memory and resistor due to its function. It can memorize the past quantity of electric charge by supplying a voltage or current. As predicted in [3], in the short term, memristors are most likely to be used in storage devices, but eventually may be used in neural networks, in applications such as pattern recognition or associative memory. Recently, dynamical behaviors of memristive neural networks have been reported, see [4–8] and references therein.

Existence of many equilibria is a necessary feature in the applications of neural networks to associative memory storage. In the recent years, considerable efforts have been devoted to studying the multistability for recurrent neural networks [9–11,13]. Meanwhile, some efforts have been devoted to investigating the multstbility for memristive recurrent neural networks [14].

It is well known that the activation functions play an important role in the dynamical analysis of recurrent neural networks [12,15–19]. Number of equilibria of neural networks depends on the characteristic of activation functions. In the existing references, a lot of works were based on the assumption that

© Springer International Publishing AG 2017
F. Cong et al. (Eds.): ISNN 2017, Part I, LNCS 10261, pp. 354–362, 2017.
DOI: 10.1007/978-3-319-59072-1_42

the activation functions were nondecreasing [9,11]. In [10], the authors considered Mexican-hat-type activation functions. The activation functions were non-monotonic. [14] considered nonmonotonic piecewise linear activation functions. However, the activation functions in these references have only several corner points. This class of activation functions is special. It is necessary to propose a class of general nonmonotonic activation functions, and to discuss multistability of memristive neural networks with general nonmonotonic activation functions. This class of activation functions has the property of localization in both time and frequency. To the best of the authors' knowledge, there are no result about the multistability of memristive neural networks with general nonmonotonic activation functions.

Motivated by the above discussions, we will investigate multistability of memristive neural networks with general nonmonotonic activation functions. The main contributions of the paper can be summarized as follows. 1. A class of general nonmonotonic activation functions is proposed. 2. By Brouwer's fixed point theorem, sufficient criteria are established to guarantee the existence of multiple equilibria of memristive neural networks. 3. Multistability of memristive neural networks with general nonmonotonic activation functions is analyzed.

2   Paper Preparation

In this paper, we consider the following n-neuron memristive neural networks

$$\dot{x}_i(t) = -d_i x_i(t) + \sum_{j=1}^{n} a_{ij}(x_j(t)) f_j(x_j(t)) + I_i. \quad i = 1, 2 \cdots n, \quad (1)$$

where $x(t) = (x_1(t), x_2(t), \cdots, x_n(t))^T \in \mathscr{R}^n$ is the state vector; $f_j(\cdot)$ denotes the neuron activation function; $I = (I_1, I_2, \cdots, I_n)$ is the control input vector; $d_i > 0$, and $a_{ij}(\cdot)$ represent the memristive synaptic weights. According to the definition of memristive neural networks, suppose

$$a_{ij}(x_j(t)) = \begin{cases} a_{ij}^1, & |x_j(t)| \leq T_j, \\ a_{ij}^2, & |x_j(t)| > T_j. \end{cases}$$

Consider a class of general nonmonotonic continuous activation function as follows:

$$f_i(x) = \begin{cases} u_i, & x \in (-\infty, p_i^0), \\ k_i^m x + b_i^m, & x \in [p_i^{m-1}, p_i^m), \\ v_i, & x \in [p_i^{2l+1}, +\infty), \end{cases} \quad (2)$$

where $u_i < v_i$, $k_i^m > 0$ for $m = 1, 3, 5 \cdots 2l + 1$, $k_i^m < 0$ for $m = 2, 4, 6 \cdots 2l$, $l = 1, 2, 3, \cdots$. Denote $\bar{a}_{ij} = \max\{a_{ij}^1, a_{ij}^2\}$, $\underline{a}_{ij} = \min\{a_{ij}^1, a_{ij}^2\}$, Given a set $\varXi \in \mathscr{R}$, $co[\varXi]$ represents the closure of the convex hull of $\varXi$. Hence,

$$co[a_{ij}(x_j(t))] = \begin{cases} a_{ij}^1, & |x_j(t)| < T_j, \\ [\underline{a}_{ij}, \bar{a}_{ij}], & |x_j(t)| = T_j, \\ a_{ij}^2, & |x_j(t)| > T_j. \end{cases}$$

There exists $\hat{a}_{ij} \in co[a_{ij}(x_j(t))]$ such that

$$\dot{x}_i(t) = -d_i x_i(t) + \sum_{j=1}^{n} \hat{a}_{ij} f_j(x_j(t)) + I_i. \quad \text{for a.e. } t \geq 0, \ i = 1, 2 \cdots n.$$

Denote

$$(-\infty, p_i^0] = (-\infty, p_i^0]^1 \times (p_i^0, p_i^1)^0 \times \cdots \times [p_i^{2l+1}, +\infty)^0,$$
$$(p_i^0, p_i^1) = (-\infty, p_i^0]^0 \times (p_i^0, p_i^1)^1 \times \cdots \times [p_i^{2l+1}, +\infty)^0,$$
$$\cdots$$
$$[p_i^{2l+1}, +\infty) = (-\infty, p_i^0]^0 \times (p_i^0, p_i^1)^0 \times \cdots \times [p_i^{2l+1}, +\infty)^1,$$

then $\mathscr{R}^n$ can be divided into $(2l+3)^n$ parts:

$$\Phi = \{\prod_{i=1}^{n}(-\infty, p_i^0]^{\delta_i^{(0)}} \times (p_i^0, p_i^1)^{\delta_i^{(1)}} \times \cdots \times [p_i^{2l+1}, +\infty)^{\delta_i^{(2l+2)}},$$

$$(\delta_i^{(0)}, \delta_i^{(1)}, \cdots \delta_i^{(2l+2)}) = (1, 0, 0, \cdots, 0) \text{ or } (0, 1, 0, \cdots, 0) \text{ or } \cdots \text{ or } (0, 0, 0, \cdots, 1)\}.$$

Lemma 1 [20]. Assume that activation functions f_i are Lipschitz continuous on $\mathscr{R}$ with Lipschitz constants $\rho_j > 0$. If $f_j(\pm T_j) = 0$ $(j = 1, 2, \cdots, n)$, then $|co[a_{ij}(x_j(t))]f_j(x_j(t)) - co[a_{ij}(y_j(t))]f_j(y_j(t))| \leq A_{ij}\rho_j|x_j(t) - y_j(t)|$ hold for $i, j = 1, 2, \cdots, n$, where $A_{ij} = \max\{|a_{ij}^1|, |a_{ij}^2|\}$.

Lemma 2 (Brouwer's fixed point theorem). Any continuous function from D^n to D^n $(n \geq 2)$, $G : D^n \to D^n$, has a fixed point.

3   Main Result

In this section, we will discuss the existence and stability for memristive neural networks (1) with activation function (2).

Theorem 1. There exist $(2l+3)^n$ equilibria in $\mathscr{R}^n$ for system (1) with activation function (2), if the following conditions hold for $i = 1, 2, \cdots, n$:

$$- d_i p_i^m + \max\{\underline{a}_{ii} f_i(p_i^m), \bar{a}_{ii} f_i(p_i^m)\} + \sum_{j=1, j \neq i}^{n} \max\{\underline{a}_{ij} u_j,$$

$$\underline{a}_{ij} v_j, \bar{a}_{ij} u_j, \bar{a}_{ij} v_j\} + I_i < 0, \ m = 0, 2, 4, \cdots, 2l, \quad (3)$$

$$- d_i p_i^m + \min\{\underline{a}_{ii} f_i(p_i^m), \bar{a}_{ii} f_i(p_i^m)\} + \sum_{j=1, j \neq i}^{n} \min\{\underline{a}_{ij} u_j,$$

$$\underline{a}_{ij} v_j, \bar{a}_{ij} u_j, \bar{a}_{ij} v_j\} + I_i > 0, \ m = 1, 3, 5, \cdots, 2l + 1, \quad (4)$$

Proof. Pick a region arbitrarily marked as $\overline{\Phi}$ from the set Φ, for any point $(x_1 x_2 \cdots x_n)^T \in \overline{\Phi}$, fix $x_1 \cdots x_{i-1}, x_{i+1} \cdots x_n$ except x_i and define

$$F_i(x) = -d_i x + \hat{a}_{ii} f_i(x) + \sum_{j=1, j\neq i}^{n} \hat{a}_{ij} f_j(x_j) + I_i. \tag{5}$$

There exist three possible cases for us to discuss.

Case 1. When $x_i \in (-\infty, p_i^0]$, we can get

$$\begin{aligned}
F_i(p_i^0) = & -d_i p_i^0 + \hat{a}_{ii} f_i(p_i^0) + \sum_{j=1, j\neq i}^{n} \hat{a}_{ij} f_j(x_j) + I_i \\
\leq & -d_i p_i^0 + \max\{\underline{a}_{ii} f_i(p_i^0), \bar{a}_{ii} f_i(p_i^0)\} \\
& + \sum_{j=1, j\neq i}^{n} \max\{\underline{a}_{ij} u_j, \underline{a}_{ij} v_j, \bar{a}_{ij} u_j, \bar{a}_{ij} v_j\} + I_i < 0,
\end{aligned}$$

and $F_i(-\infty) > 0$. Therefore, there exists a point $\bar{x}_i \in (-\infty, p_i^0)$ such that $F_i(\bar{x}_i) = 0$.

Case 2. When $x_i \in [p_i^{m-1}, p_i^m]$, $m = 2, 4, 6 \cdots 2l$, we can get $F_i(p_i^m) < 0$, $F_i(p_i^{m-1}) > 0$ by using similar proof with Case 1. Therefore, there exist points $\bar{x}_i \in (p_i^{m-1}, p_i^m)$, $m = 1, 2 \cdots 2l$ such that $F_i(\bar{x}_i) = 0$.

Case 3. When $x_i \in [p_i^{2l+1}, +\infty)$, we can get $F_i(p_i^{2l+1}) > 0$ by using similar proof with Case 1, and $F_i(+\infty) < 0$. Therefore, there exist one point $\bar{x}_i \in (p_i^{2l}, p_i^{2l+1})$ such that $F_i(\bar{x}_i) = 0$, and one point $\bar{x}_i \in [p_i^{2l+1}, +\infty)$ such that $F_i(\bar{x}_i) = 0$.

Define a map $H : \overline{\Phi} \to \overline{\Phi}$ by $H(x_i, x_2 \cdots x_n) = (\bar{x}_1, \bar{x}_2 \cdots \bar{x}_n)$. It is clear that the map is continuous. From Brouwer's fixed point theorem, there exists one fixed point $\bar{x} = (\bar{x}_1, \bar{x}_2 \cdots \bar{x}_n)$ of H in $\overline{\Phi}$, which is also the equilibrium point of system (1) with activation function (2) in $\overline{\Phi}$. It is noted that the number of such part $\overline{\Phi}$ is $(2l + 3)^n$ in $\overline{\Phi}$. Hence, there exists $(2l + 3)^n$ equilibria for system (1) with activation function (2) in $\mathscr{R}^n$. $\square$

Denote

$$\begin{aligned}
\Omega = \{ & \prod_{i=1}^{n} (-\infty, p_i^0]^{\delta_i^{(0)}} \times (p_i^0, p_i^1)^0 \times [p_i^1, p_i^2]^{\delta_i^{(1)}} \times \cdots \times (p_i^{2l}, p_i^{2l+1})^0 \\
& \times [p_i^{2l+1}, +\infty)^{\delta_i^{(l+1)}}, (\delta_i^{(0)}, \delta_i^{(1)}, \cdots \delta_i^{(l+1)}) = (1, 0, \cdots, 0) \\
& \text{or } (0, 1 \cdots, 0) \text{ or } \cdots \text{ or } (0, 0, \cdots, 1) \}.
\end{aligned}$$

It is obvious that Ω is made up of $(l + 2)^n$ parts and Ω is bounded in Φ.

Now we first establish some positively invariant sets for system (1) and investigate stability of the equilibrium point in each invariant set.

Lemma 3. Assume that conditions (3) and (4) hold, then each Ω is positively invariant under the solution flow generated by system (1) with activation function (2).

Proof. Pick a region arbitrarily $\overline{\Omega} \subset \Omega$, Consider any initial condition $\phi = (\phi_1, \phi_2, \cdots \phi_n) \in \overline{\Omega}$, we claim that $x(t)$ would stay in $\overline{\Omega}$ for all $t \geq 0$.

Case 1. $\phi_i \in (-\infty, p_i^0]$, $i = 1, 2 \cdots n$. If there exists some $t^* \geq 0$ such that $x_i(t^*) = p_i^0$, then

$$
\begin{aligned}
\dot{x}_i(t^*) = & -d_i x_i(t^*) + \hat{a}_{ii} f_i(x_i(t^*)) + \sum_{j=1, j \neq i}^{n} \hat{a}_{ij} f_j(x_i(t^*)) + I_i \\
\leq & -d_i p_i^0 + \max\{\underline{a}_{ii} f_i(p_i^0), \overline{a}_{ii} f_i(p_i^0)\} \\
& + \sum_{j=1, j \neq i}^{n} \max\{\underline{a}_{ij} u_j, \underline{a}_{ij} v_j, \overline{a}_{ij} u_j, \overline{a}_{ij} v_j\} + I_i < 0.
\end{aligned}
$$

Case 2. $\phi_i \in [p_i^{m-1}, p_i^m]$, $i = 1, 2 \cdots n$, $m = 2, 4, 6 \cdots 2l$. If there exists some $t^* \geq 0$ such that $x_i(t^*) = p_i^m$, then $\dot{x}_i(t^*) < 0$. If there exists some $t^* \geq 0$ such that $x_i(t^*) = p_i^{m-1}$, then $\dot{x}_i(t^*) > 0$.

Case 3. $\phi_i \in [p_i^{2l+1}, +\infty)$, $i = 1, 2 \cdots n$. If there exists some $t^* \geq 0$ such that $x_i(t^*) = p_i^{2l+1}$, then $\dot{x}_i(t^*) > 0$. Therefore, the solution $x(t)$ will never escape from $\overline{\Omega}$ for all $t \geq 0$. That is, each Ω is positive invariant under the solution flow generated by system (1) with activation function (2). □

Theorem 2. Suppose that $f_i(\pm T_i) = 0$, and conditions (3) and (4) hold for $i = 1, 2 \cdots n$. Meanwhile,

$$
d_i > \sum_{j=1}^{n} A_{ij} k_j^*, \quad i = 1, 2 \cdots n, \tag{6}
$$

where $A_{ij} = \max\{|a_{ij}^1|, |a_{ij}^2|\}, k_j^* = \max\{|k_j^m|, m = 2, 4 \cdots 2l\}$, $i, j = 1, 2 \cdots n$. Then system (1) with activation function (2) has $(l+2)^n$ locally exponentially stable equilibria in $\mathscr{R}^n$.

Proof. For each $i = 1, 2 \cdots n$, we consider the single-variable function $G_i(\xi) = d_i - \xi - \sum_{j=1}^{n} A_{ij} k_j^*$. Condition (6) implies $G_i(0) > 0$, and there exists a constant $\lambda > 0$ such that $G_i(\lambda) > 0$ for all $i = 1, 2 \cdots n$. Pick a region arbitrarily $\overline{\Omega} \subset \Omega$. According to Theorem 1, we can get that there exists one equilibrium point x^* in $\overline{\Omega}$. We will prove that the equilibrium point x^* is locally exponentially stable. Denote $y_i(t) = x_i(t) - x_i^*$, $i = 1, 2 \cdots n$. Then

$$
\dot{y}_i(t) \in -d_i y_i(t) + \sum_{j=1}^{n} \{co[a_{ij}(x_j(t))] f_j(x_j(t)) - co[a_{ij}(x_j^*)] f_j(x_j^*)\}. \tag{7}
$$

It follows from Lemma 1 that

$$
|co[a_{ij}(x_j(t))] f_j(x_j(t)) - co[a_{ij}(x_j^*)] f_j(x_j^*)| \leq A_{ij} k_j^* |y_j(t)|. \tag{8}
$$

Therefore,

$$\frac{d}{dt}|y_i(t)| \leq -d_i|y_i(t)| + \sum_{j=1}^{n} A_{ij}k_j^*|y_j(t)|.$$

Now, consider the function $z_i(\cdot)$ defined by $z_i(t) = e^{\lambda t}|y_i(t)|$, $i = 1, 2 \cdots n$, Let $\delta > 1$, $K = \max\limits_{1 \leq i \leq n}\{|x_i(0) - x_i^*|\} > 0$. It follows that $z_i(0) < K\delta$, $i = 1, 2 \cdots n$. We shall justify that

$$z_i(t) < K\delta, \text{ for all } t > 0, \ i = 1, 2 \cdots n. \tag{9}$$

Suppose (9) does not hold, then there is a $k \in \{1, 2 \cdots n\}$ and a $t_1 > 0$ for the first time such that $z_i(t) \leq K\delta, t \in [0, t_1], i = 1, 2 \cdots n, i \neq k$; $z_k(t) \leq K\delta, t \in [0, t_1)$; $z_k(t_1) = K\delta, \dot{z}_k(t_1) \geq 0$. In fact

$$\dot{z}_k(t_1) = \lambda e^{\lambda t_1}|y_k(t_1)| + e^{\lambda t_1}\frac{d}{dt}|y_k(t_1)| \leq \lambda z_k(t_1) - d_k z_k(t_1) + \sum_{j=1}^{n} A_{kj}k_j^* z_j(t_1)$$

$$\leq -\{d_k - \lambda - \sum_{j=1}^{n} A_{kj}k_j^*\}K\delta < 0.$$

Therefore, there exits one contradiction. So assertion (9) holds and $z_i(t) \leq K$ for all $t > 0$, $i = 1, 2 \cdots n$, by taking $\delta \to 1^+$. We can obtain that $|x_i(t) - x_i^*| \leq e^{-\lambda t} \max\limits_{1 \leq i \leq n}\{|x_i(0) - x_i^*|\}, t > 0, i = 1, 2 \cdots n$. Therefore, $x(t)$ converges to x^* exponentially. That is, system (1) with activation function (2) has one exponentially stable equilibrium point in $\overline{\Omega}$. Because of the arbitrary of $\overline{\Omega} \subset \Omega$, So system (1) with activation function (2) has $(l + 2)^n$ locally exponentially stable equilibria in $\mathscr{R}^n$. $\square$

Remark 1. In [14], the authors investigated multistability of memristive neural networks with nonmonotonic piecewise linear activation functions. However, the activation functions had only four corner points, which is special. In this paper, the activation functions are general, which have lots of corner points.

4   Illustrative Example

Consider the following two-dimensional memristor-based neural networks

$$\dot{x}_i(t) = -d_i x_i(t) + \sum_{j=1}^{n} a_{ij}(x_j(t))f_j(x_j(t)) + I_i. \quad i = 1, 2, \tag{10}$$

where $d_1 = d_2 = 1.5$, $I_1 = I_2 = -1$,

$$a_{11}(x_1(t)) = \begin{cases} 1.85, & |x_1(t)| \leq 1, \\ 2.2, & |x_1(t)| > 1, \end{cases} \quad a_{12}(x_2(t)) = \begin{cases} -0.02, & |x_2(t)| \leq 1, \\ 0.2, & |x_2(t)| > 1, \end{cases}$$

$$a_{21}(x_1(t)) = \begin{cases} 0.05, & |x_1(t)| \leq 1, \\ 0.01, & |x_1(t)| > 1, \end{cases} \quad a_{22}(x_2(t)) = \begin{cases} 2.4, & |x_2(t)| \leq 1, \\ 1.9, & |x_2(t)| > 1. \end{cases}$$

and the activation function $f_i(x)$ ($i = 1, 2$) are defined as follows:

$$f_i(x) = \begin{cases} -1, & -\infty < x < -\frac{4}{3}, \\ 3(x+1), & -\frac{4}{3} \le x \le -\frac{2}{3}, \\ \frac{3}{5}(1-x), & -\frac{2}{3} < x < \frac{8}{3}, \\ 15x - 41, & \frac{8}{3} \le x \le 3, \\ 4, & 3 < x < +\infty. \end{cases} \tag{11}$$

System (10) satisfies the conditions in Theorem 2. Hence, system (10) with activation function (11) has 9 locally exponentially stable equilibria. The dynamics of system (10) are illustrated in Fig. 1.

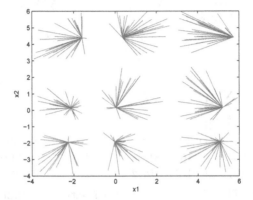

Fig. 1. Transient states of system (10).

5   Conclusion

In this paper, we have investigated multistability of n-dimensional memristive neural networks with a class of general nonmonotonic continuous activation functions. Sufficient criteria have been established to ensure the local existence and local exponential stability of multiple equilibria. Compared with the existing literature about the multistability of memristive neural networks, the activation functions have multiple corner points in this paper. This class of activation functions is general. The effectiveness of the obtained results has been verified by one numerical example. Several extensions would be welcome as future work: such as, studying the basin of attraction of some equilibrium points, considering multistability of time-delay system.

Acknowledgments. This work was supported by the National Natural Science Foundation of China (Grant Nos. 61503338 and 61502422) and the Natural Science Foundation of Zhejiang Province, China (Grant Nos. LQ15F030005, LQ15F020006 and LQ15F020008).

References

1. Chua, L.: Memristor-the missing circuit element. IEEE Trans. Circ. Theory **18**, 507–519 (1971)
2. Strukov, D., Snider, G., Stewart, G., Williams, R.: The missing memristor found. Nature **453**, 80–83 (2008)
3. Anthes, G.: Memristor: pass or fail. Commun. ACM **54**(3), 22–24 (2010)
4. Wu, A., Zeng, Z.: Exponential stabilization of memristive neural networks with time delays. IEEE Trans. Neural Netw. Learn. Syst. **23**, 1919–1929 (2012)
5. Zhang, G., Shen, Y.: Exponential synchronization of delayed memristor based chaotic neural networks via periodically intermittent control. Neural Netw. **55**, 1–10 (2014)
6. Chandrasekar, A., Rakkiyappan, R., Cao, J., Lakshmanan, S.: Synchronization of memristor-based recurrent neural networks with two delay components based on second-order reciprocally convex approach. Neural Netw. **57**, 79–93 (2014)
7. Yang, X., Cao, J., Yu, W.: Exponential synchronization of memristive Cohen-Grossberg neural networks with mixed delays. Cogn. Neurodyn. **8**(3), 239–249 (2014)
8. Wang, Z., Ding, S., Huang, Z., Zhang, H.: Exponential stability and stabilization of delayed memristive neural networks based on quadratic convex combination method. IEEE Trans. Neural Netw. Learn. Syst. **27**, 2337–2350 (2016)
9. Nie, X., Cao, J.: Multistability of second-order competitive neural networks with nondecreasing saturated activation functions. IEEE Trans. Neural Netw. **22**, 1694–1708 (2011)
10. Wang, L., Chen, T.: Multistability of neural networks with Mexican-hat-type activation functions. IEEE Trans. Neural Netw. Learn. Syst. **23**, 1816–1826 (2012)
11. Zeng, Z., Zheng, W.: Multistability of two kinds of recurrent neural networks with activation functions symmetrical about the origin on the phase plane. IEEE Trans. Neural Netw. Learn. Syst. **24**, 1749–1762 (2013)
12. Huang, Y., Wang, X., Long, H., Yang, X.: Synthesization of high-capacity auto-associative memories using complex-valued neural networks. Chin. Phys. B **12**, 120701-1–120701-8 (2016)
13. Liu, P., Zeng, Z.G., Wang, J.: Complete stability of delayed recurrent neural networks with Gaussian activation functions. Neural Netw. **85**, 21–32 (2017)
14. Nie, X., Zheng, W., Cao, J.: Coexistence and local μ-stability of multiple equilibria for memristive neural networks with nonmonotonic piecewise linear activation functions and unbounded time-varying delays. Neural Netw. **84**, 172–180 (2016)
15. Zhang, H., Wang, Y.: Stability analysis of Markovian jumping stochastic Cohen-Grossberg neural networks with mixed time delays. IEEE Trans. Neural Netw. **19**(2), 366–370 (2008)
16. Zhang, H., Wang, Z., Liu, D.: Global asymptotic stability of recurrent neural networks with multiple time-varying delays. IEEE Trans. Neural Netw. **19**(5), 855–873 (2008)
17. Zhang, H., Liu, Z., Huang, G., Wang, Z.: Novel weighting-delay-based stability criteria for recurrent neural networks with time-varying delay. IEEE Trans. Neural Netw. **21**(1), 91–106 (2010)
18. Zheng, Y., Ling, H., Chen, S., Xue, J.: A hybrid neuro-fuzzy network based on differential biogeography-based optimization for online population classification in earthquakes. IEEE Trans. Fuzzy Syst. **23**(4), 1070–1083 (2015)

19. Zheng, Y., Sheng, W., Sun, X., Chen, S.: Airline passenger profiling based on fuzzy deep machine learning. IEEE Trans. Neural Netw. Learn. Syst. (2016). doi:10.1109/TNNLS.2016.2609437
20. Chen, J., Zeng, Z., Jiang, P.: Global Mittag-Leffler stability and synchronization of memristor-based fractional-order neural networks. Neural Netw. **51**, 1–8 (2014)

A Reinforcement Learning Method with Implicit Critics from a Bystander

Kao-Shing Hwang[1], Chi-Wei Hsieh[1], Wei-Cheng Jiang[1(✉)],
and Jin-Ling Lin[2]

[1] Electrical Engineering, National Sun Yat-sen University, Kaohsiung, Taiwan
hwang@ccu.edu.tw, darkdesert18@gmail.com,
enjoysea0605@gmail.com
[2] Information Management, Shih Hsin University, Taipei, Taiwan
jllin@mail.shu.edu.tw

Abstract. In Reinforcement Learning, we train agent many times, so agents can get experience from learning, and then, agent can complete every behavior of different missions. In this paper, we propose architecture to allow agent get experience from environment. We use Adaptive Heuristic Critic (AHC) as a learning architecture and combine an action bias with AHC to solve the problem of continuous action system. On account of the problems of recognition error and state delay, we use Reinforcement Learning which learns from cumulative reward to update the experience of agents.

Keywords: Reinforcement learning · Adaptive Heuristic Critic · Stochastic Real Value · Reward function

1 Introduction

Generally, there are two methods in machine learning, supervised learning and Reinforcement Learning [1, 2]. Agent learns policy to do behaviors by environmental reward in different missions, called reinforcement learning. Supervised learning which is human commanding or teaching them. In real life, an agent interacts with not only environment, but also human. Agents can interactive with human and get some important experience from facial expression, besides getting reward from environment. In fact, human are not the expert, they don't know how to describe with words, so they express their emotions to agent is much easier. In [3–5], Agent has its own emotion, and it can be changed by what it encounters. In [6–9], they propose a framework that let human can train agent, and they combine environment reward with human reward to be a new reward function. When agent gets feedback from human, it may occur delay reward [10]. Therefore, we use cumulative reward as update value instead of immediate reward. We use Adaptive Heuristic Critic (AHC) [11, 12] which is one of the Actor-Critic method [13] to accomplish Reinforcement Learning. Because the agent might not have few actions in real life, in [14, 15], they proposed Stochastic Real Value (SRV) to solve continuous problem. And we use the same concept to combine Action Bias with AHC to become a continuous value to solve this problem.

© Springer International Publishing AG 2017
F. Cong et al. (Eds.): ISNN 2017, Part I, LNCS 10261, pp. 363–370, 2017.
DOI: 10.1007/978-3-319-59072-1_43

The Back-Propagation neural network (BP) [16] is a feed forward network, and uses a supervised learning scheme with a different threshold function and learning rules. Deep Learning uses the concept of BP to train its neural network, and then it learns to extract the feature. There are three main algorithms for deep learning: Convolution Neural Network (CNN) [17], Deep Belief Network (DBN) [18] and Sparse Autoencoder [19, 20].

2 Reinforcement Learning

Reinforcement learning is one of machine learning, and it is a Markov Decision Process (MDP). In the process of training, agent has its target in the environment, and agent interactive with environment for getting experience to update its policy. For training many episodes, agent learns to find the optimal policy. As shown in Fig. 1, Actor-Critic is one of method of Reinforcement Learning, and we use the concept of Actor-Critic in this thesis. Actor Critic methods are TD methods that have a separate memory structure to explicitly represent the policy independent of the value function. The policy structure is known as the *actor*, because it is used to selected actions, and the estimated value function is known as the *critic*, because it criticizes the actions made by the actor.

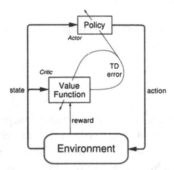

Fig. 1. The actor-critic architecture

3 Actor Critic-Q with Continuous Actions

In this section, the proposed methods include two parts, Actor Critic-Q and action bias. The Actor Critic-Q is for policy learning and action bias is used for output continuous action. The following sections will explain the content in detailed.

3.1 Actor Critic-Q

In classic AHC [11], it can predict the state value by getting external reinforcement signal r from environment in ACE element. Then, ACE element updates its weights by internal reinforcement $\hat{r}$ called TD error. At last, ACE element emits internal reinforcement signal $\hat{r}$ to ASE to update its weights and generate the real value for doing an action. But, there is a weak point that AHC just can solve two direction problems like cart pole system or mountain car. For example, maze system may be multiple direction problems, so classic AHC is failed to solve. In order to solve this problem, we make ASE element be multiple neurons, each of them represents the action value. If agent has four actions, up, down, right, left, in maze system, the ASE element will have four neurons to represent them. The Actor-Critic-Q architecture is shown in Fig. 2.

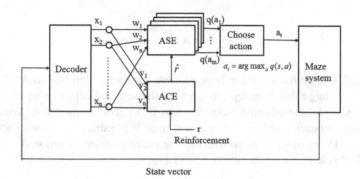

Fig. 2. The Actor-Critic-Q architecture in maze system

The neurons in ASE element are according to the number of action, and we use $\varepsilon - greddy$ policy to choose action. Simply, we choose neuron which has the optimal action value as action, but we have a probability to choose randomly for exploration. In the part of eligibility in ASE, on account of the increasing of neuron, the eligibility can be written as:

$$e_{ij} \leftarrow \delta e_{ij} + (1 - \delta)x_i \tag{1}$$

We update all the weights of ASE each time. But we find that there is not any physical meaning of Q-value in ASE element in this method. Therefore, we combine the output value with eligibility as: For example in service robot, when it service host, there are many different obstacle at home. Therefore, we want to use continuous action instead of discrete action.

$$e_{ij} \leftarrow \delta e_{ij} + (1 - \delta)q(a_t)x_i \tag{2}$$

3.2 Actor-Critic-Q with Continuous Actions

In Actor-Critic-Q, I divide ASE element into four neuron representing action value. For making Actor-Critic-Q have real-valued action function, I combine Actor-Critic-Q with action bias. Action bias $B \sim N(Bm, Bv)$, is a stochastic value generating from normal distribution by using mean *Bm* and standard deviation *Bv*. We evaluate and update the value of mean and standard deviation. An illustration of action bias is shown in Fig. 3.

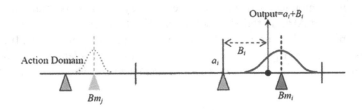

Fig. 3. An illustration of action bias

Each ASE neuron has its action bias, so there are four action bias in ASE element. If the bias is so large that overstep the range of the selected action, it try another bias again. For example, the wheeled robot can turn 0° to 360° as its action. We divide it into four part, and each part has its own action bias. We restrict the range of action bias from −45° to 45° in order to avoid overstep the range of selected action. We show the diagram of wheeled robot with action bias in Fig. 4.

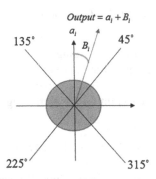

Fig. 4. The diagram of wheeled robot with action bias

- Action Bias Mean

  In SRV, the updating rule of mean is estimate the immediate reward and the difference between the real action output and the mean of normal distribution. In this thesis, we use the same concept to update mean, and we use the TD error with eligibility instead of immediately reward to update. We will show the updating rule in Eqs. 3 and 4.

$$\Delta q = \hat{r}e_{ij} \tag{3}$$

$$m_i \leftarrow m_i + \beta_m(\Delta q)(B(a_i) - Bm(a_i)) \tag{4}$$

Where $\hat{r}$ is TD error, and e_{ij} is an eligibility that remember the state how long it stayed and what action it chose. If $\Delta q > 0$, it says that the learning is right, and we adjust the mean toward the real action output of distribution. If $\Delta q < 0$, we should adjust our mean in opposite direction to the real action output. β_m is a learning rate, and we let:

$$\beta_m = \begin{cases} \beta_p & \Delta q > 0 \\ \beta_n & \Delta q < 0 \end{cases}, where \ \beta_p > \beta_n > 0 \tag{5}$$

- Action Bias Standard deviation

The standard deviation is like an exploration. When learning is finish, the action value is converge, and the $\Delta q \approx 0$. We let standard deviation become zero so that the real action output is according to the mean. The updating rule is shown in Eq. 6.

$$s_i \leftarrow s_i + \beta_v(|\Delta q| - Bv(a_i)) \tag{6}$$

where β_v is a learning rate. When q is too large, it says that learning is not already, so it should let standard deviation be larger to explore. When q approach zero, it says that it learning already, so it should not explore. We show the Actor-Critic-Q with action bias architecture in Fig. 5.

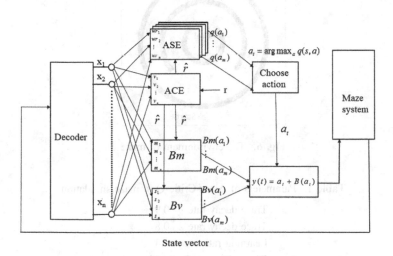

State vector

Fig. 5. The Actor-Critic-Q with continuous actions architecture

4 Simulation

In this section, two methods are compared Actor-Critic-Q with discrete action and continuous actions.

4.1   Compare Discrete Actions with Continuous Actions

The simulation is to command a mobile robot to approach a goal in the maze as shown in Fig. 6. The mobile robot starts from the starting point in the size of 300×300 maze. The goal, which the robot is attempting to reach, is in the center of the maze. The task of the robot is to explore the goal and find a trace to approach the goal, simultaneously. The agent has four options; taking an action toward the up, down, left, or right direction one step at a time. Each direction has $90°$ range; for example, the range of the up direction is from $45°$ to $135°$. The three gray broken rings of the maze are walls. If the agent hits the wall after taking an action, the agent stays in the same position and receives a reward -1. When the agent arrives at the goal, the agent receives a reward 5 and the episode is terminated. Otherwise, the agent moves to the next position and receives a reward -0.01. Since the shapes of the walls are curves, the optimal path of robot is moving along the curves. The proposed continuous actions would control the robot moving along the walls. The parameters of the approaching goal are set as shown in Table 1.

Fig. 6. The environment of maze

Table 1. Parameters of Actor-Critic-Q in maze simulation

Trace decay rate δ	0.9
Trace decay rate λ	0.8
Learning rate α	0.4
Discount rate γ	0.99
Step size	5 cm

The simulation results are shown in Fig. 7. We compared the Actor-Critic-Q with discrete action and with continuous action. The simulations of Actor-Critic-Q and Actor-Critic-Q with action bias in maze are shown in Fig. 7(a) and (b).

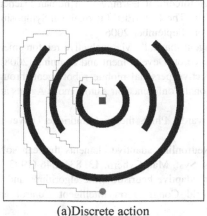

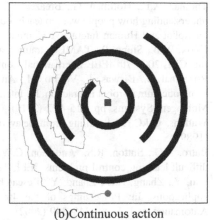

(a)Discrete action   (b)Continuous action

Fig. 7. The learned path of discrete action and continuous actions

5   Conclusion

As science and technology are improving, the reinforcement learning has been applied to more complex environment in the real. In this paper, we propose an Actor-Critic-Q with continuous action to control the robots. According to the simulation results, the proposed method can control to achieve goal by a smooth path. In future, we hope our method will be used in another environment or implemented in real life.

References

1. Sutton, R.S., Barto, A.G.: Reinforcement Learning: An Introduction. MIT Press, Cambridge (1998)
2. Kaebling, L.P., Littman, M.L., Moore, A.W.: Reinforcement learning: a survey. J. Artif. Intell. Res. **4**, 237–285 (1996)
3. Ayesh, A.: Emotionally motivated reinforcement learning based controller. IEEE Int. Conf. Syst. Man Cybernet. **1**, 874–878 (2004)
4. Broekens, J.: Emotion and reinforcement: affective facial expressions facilitate robot learning. In: Huang, T.S., Nijholt, A., Pantic, M., Pentland, A. (eds.) Artifical Intelligence for Human Computing. LNCS, vol. 4451, pp. 113–132. Springer, Heidelberg (2007). doi:10. 1007/978-3-540-72348-6_6
5. Obayashi, M., Takuno, T., Kuremoto, T., Kobayashi, K.: An emotional model embedded reinforcement learning system. In: 2012 IEEE International Conference on Systems, Man, and Cybernetics (2012)
6. Sridharan, M.: Augmented Reinforcement learning for interaction with non-expert humans in agent domains. In: 2011 10th International Conference on Machine Learning and Applications and Workshops (ICMLA), vol. 1 (2011)

7. Thomaz, A.L., Hoffman, G., Breazeal, C.: Reinforcement learning with human teachers: understanding how people want to teach robots. In: The 15th IEEE International Symposium on Robot and Human Interactive Communication, September 2006

8. Knox, W.B., Stone, P.: TAMER: training an agent manually via evaluative reinforcement. In: ICDL 2008 7th IEEE International Conference on Development and Learning (2008)

9. Rosenthal, S., Biswas, J., Veloso, M.: An effective personal mobile robot agent through symbiotic human-robot interaction. In: International Conference on Autonomous Agents and Multiagent Systems, pp. 915–922 (2010)

10. Watkins, C.J.C.H.: Learning from delayed rewards. Ph.D. thesis, Cambridge University (1989)

11. Batro, A.G., Sutton, R.S., Anderson, C.W.: Neuronlike adaptive elements that can solve difficult learning control problems. IEEE Trans. Syst. Man Cybern. **13**, 834–846 (1993)

12. Sun, Y., Zhang, R.B., Zhang, Y.: Research on adaptive heuristic critic algorithms and its applications. In: Proceedings of the 4th World Congress on Intelligent Control and Automation, vol. 1, pp. 345–349 (2002)

13. Konda, V., Tsitsiklis, J.: Actor-critic algorithms. In: Advances in Neural Information Processing Systems (2000)

14. Gullapalli, V.: A stochastic reinforcement learning algorithm for learning real valued functions. Neural Netw. **3**, 671–692 (1990)

15. Gullapalli, V.: Associative reinforcement learning of real valued functions. In: Proceedings of IEEE, System, Man, Cybernetics, Charlottesville, VA, October 1991

16. Widrow, B., Lehr, M.A.: 30 years of adaptive neural networks: perceptron, madaline, and backpropagation. Proc. IEEE **78**, 1415–1442 (1990)

17. Krizhevsky, A., Sutskever, I., Hinton, G.E.: Imagenet classification with deep convolutional neural networks. In: NIPS (2012)

18. Hinton, G., Osindero, S., The, Y.: A fast learning algorithm for deep belief nets. Neural Comput. (2006)

19. Vincent, P., Larochelle, H., Lajoie, I.: Stacked denoising autoencoders: learning useful representations in a deep network with a local denoising criterion. J. Mach. Learn. Res. Arch. **11**, 3371–3408 (2010)

20. Baldi, P.: Autoencoders, unsupervised learning, and deep architectures. JMLR Workshop Conf. Proc. **27**, 37–50 (2012)

The Mixed States of Associative Memories Realize Unimodal Distribution of Dominance Durations in Multistable Perception

Takashi Kanamaru[✉]

Department of Mechanical Science and Engineering, School of Advanced Engineering,
Kogakuin University, 2665-1 Nakano, Hachioji-city, Tokyo 192-0015, Japan
kanamaru@cc.kogakuin.ac.jp

Abstract. We propose a pulse neural network that exhibits chaotic pattern alternations among stored patterns as a model of multistable perception, which is reflected in phenomena such as binocular rivalry and perceptual ambiguity. When we regard the mixed state of patterns as a part of each pattern, the durations of the retrieved pattern obey unimodal distributions. The mixed states of the patterns are essential to obtain the results that are consistent with psychological studies. Based on these results, it is proposed that many pre-existing attractors in the brain might relate to the general category of multistable phenomena, such as binocular rivalry and perceptual ambiguity.

Keywords: Pulse neural network · Chaotic pattern alternations · Multistable perception · Binocular rivalry · Perceptual ambiguity · Dominance duration

1 Introduction

In the perception of visual information, it is well known that multiple stable states compete for perceptual dominance. For example, when two different stimuli are presented to the eyes, the dominant stimulus perceived fluctuates over time, a phenomenon known as binocular rivalry [1,2]. Similarly, when an ambiguous figure such as a Necker cube is presented, the dominant interpretation also fluctuates over time [3]. Research has also indicated that the duration of the dominant state (dominance duration) may be characterized by a unimodal distribution, such as the gamma distribution [2,3] or the log-normal distribution [1].

One possible mechanism for such fluctuations in multistable perception is associated with noise in the visual system, which is generated by small eye movements and microsaccades. On the other hand, the deterministic chaos generated by nonlinear dynamics in the brain may also be responsible for such fluctuations. Several dynamical models in which the state of the network changes chaotically among several patterns have been proposed [4–6]. However, the duration of a pattern in the chaotic networks does not obey a unimodal distribution, but it typically obeys a monotonically decreasing distribution [6].

© Springer International Publishing AG 2017
F. Cong et al. (Eds.): ISNN 2017, Part I, LNCS 10261, pp. 371–378, 2017.
DOI: 10.1007/978-3-319-59072-1_44

In the present study, we report that the pattern alternations caused by chaotic dynamics of a pulse neural network can reproduce the properties of multistable perception. This network is composed of neuronal models which emit pulses when a sufficiently strong input is injected [7–9], while the previous models were composed of conventional neuronal models based on firing rates. By storing several patterns based on the mechanism of associative memory, this network shows chaotic pattern alternations [10,11]. It is observed that the durations of the retrieved pattern obey unimodal distributions when we regard the mixed state of patterns as a part of each pattern.

Based on these results, it is proposed that many pre-existing attractors in the brain might relate to the general category of multistable phenomena, such as binocular rivalry and perceptual ambiguity. In the previous work, we called such a set of pre-existing attractors as "attractor landscape" [11].

This paper is organized as follows. In Sect. 2, we define a pulse neural network composed of excitatory neurons and inhibitory neurons exhibiting synchronized, chaotic firing. This network is referred to as the one-module system. In Sect. 3, we connect eight modules of networks in which three patterns are stored according to the mechanism of associative memory. We show that chaotic dynamics are responsible for alterations in the retrieved patterns over time. It is observed that the durations of the retrieved pattern are shown to obey unimodal distributions. The final section provides conclusions.

2   One-Module System

In Sects. 2 and 3, we introduce a neural network of theta neurons with phases as their internal states [7–9]. When a sufficiently strong input is provided, each neuron yields a pulse by increasing its phase around a circle and returning to its original phase. The network is composed of N_E excitatory neurons and N_I inhibitory neurons governed by the following equations:

$$\dot{\theta}_E^{(i)} = (1 - \cos\theta_E^{(i)}) + (1 + \cos\theta_E^{(i)})(r + \xi_E^{(i)}(t) + g_{int}I_E(t) - g_{ext}I_I(t)), \quad (1)$$

$$\dot{\theta}_I^{(i)} = (1 - \cos\theta_I^{(i)}) + (1 + \cos\theta_I^{(i)})(r + \xi_I^{(i)}(t) + g_{ext}I_E(t) - g_{int}I_I(t)), \quad (2)$$

$$I_X(t) = \frac{1}{2N_X} \sum_{j=1}^{N_X} \sum_k \frac{1}{\kappa_X} \exp\left(-\frac{t - t_k^{(j)}}{\kappa_X}\right), \quad (3)$$

$$\langle \xi_X^{(i)}(t)\xi_Y^{(j)}(t')\rangle = D\delta_{XY}\delta_{ij}\delta(t - t'), \quad (4)$$

where $\theta_E^{(i)}$ and $\theta_I^{(i)}$ are the phases of the ith excitatory neuron and the ith inhibitory neuron, respectively. r is a parameter of the neurons that determines whether the equilibrium of each neuron is stable or not. We used $r = -0.025$ to ensure that each neuron had a stable equilibrium. $X = E$ or I denote the excitatory or inhibitory ensemble, respectively, while $t_k^{(j)}$ is the kth firing time of the jth neuron in the ensemble X, and the firing time is defined as the time at which $\theta_X^{(j)}$ exceeds π in the positive direction. The neurons communicate with each other using the post-synaptic potentials whose waveforms are the exponential functions as shown in Eq. (3). $\xi_X^{(i)}(t)$ represents Gaussian white noise added to the ith neuron in the ensemble X.

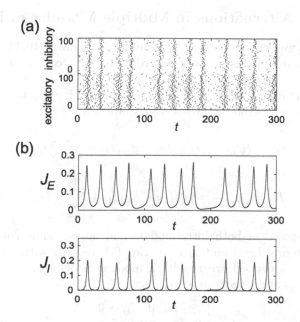

Fig. 1. (a) Chaotic synchronization observed in a module with $D = 0.0032, r = -0.025$, $g_{int} = 4$, and $g_{ext} = 2.5$. Raster plot of spikes of 200 randomly chosen excitatory neurons and inhibitory neurons in a module with $N_E = N_I = 2000$ is shown. (b) Chaotic synchronization in a module with an infinite number of neurons obtained by analysis with Fokker-Planck equations. The values of parameters are the same as those used in (a). Temporal changes in the instantaneous firing rates J_E and J_I are shown.

In the following, this network is referred to as a one-module system, which exhibits various patterns of synchronized firing [9]. We utilized the chaotic synchronization shown in Fig. 1. In Fig. 1(a), a raster plot of spikes of 200 randomly chosen excitatory neurons and inhibitory neurons in a module with $N_E = N_I = 2000$ is shown. This plot allows one to observe the synchronized firing of neurons, and that the intervals of synchronized firing do not remain constant. To analyze this variability, we took the limit of $N_E, N_I \to \infty$ in order to obtain the Fokker-Planck equation, which governs the dynamics of the probability densities $n_E(\theta_E)$ and $n_I(\theta_I)$ of $\theta_E^{(i)}$ and $\theta_I^{(i)}$ as shown in Ref. [10]. The instantaneous firing rates J_E and J_I of the excitatory and inhibitory ensembles obtained from the analysis of the Fokker-Planck equation are shown in Fig. 1(b). The largest Lyapunov exponent of the data in Fig. 1(b) is positive [9], indicating that the dynamics of J_E and J_I are chaotic.

In the following, only the one-module systems with infinite neurons treated in Fig. 1(b) are considered, as the Fokker-Planck equation does not contain noise, allowing for the reproduction of analyses.

3   Pattern Alternations in Multiple Modules of Network

In this section, we define a network with multiple modules [10,11]. Several patterns can be stored in this network according to the mechanism of associative memory.

The synaptic inputs T_{Ei} and T_{Ii} injected to the ith excitatory ensemble Ei and the inhibitory ensemble Ii, respectively, are defined as

$$T_{Ei} = (g_{int} - \gamma\epsilon_{EE})I_{Ei} - g_{ext}I_{Ii} + \sum_{j=1}^{M} \epsilon_{ij}^{E} I_{Ej}, \tag{5}$$

$$T_{Ii} = (g_{ext} - \gamma\epsilon_{IE})I_{Ei} - g_{int}I_{Ii} + \sum_{j=1}^{M} \epsilon_{ij}^{I} I_{Ej}, \tag{6}$$

which are composed of both intra-module and inter-module connections. By replacing the terms $I_E(t)$ and $I_I(t)$ in Eqs. (1) and (2) with T_{Ei} and T_{Ii} in Eqs. (5) and (6), a network with multiple modules is defined.

The strengths of connections are defined as

$$\epsilon_{ij}^{E} = \begin{cases} \epsilon_{EE} K_{ij} & \text{if } K_{ij} > 0 \\ 0 & \text{otherwise} \end{cases}, \tag{7}$$

$$\epsilon_{ij}^{I} = \epsilon_{IE}|K_{ij}|, \tag{8}$$

$$K_{ij} = \frac{1}{Ma(1-a)} \sum_{\mu=1}^{p} \eta_i^{\mu}(\eta_j^{\mu} - a), \tag{9}$$

where $\eta_i^{\mu} \in \{0,1\}$ is the stored value in the ith module for the μth pattern, M is the number of modules, p is the number of patterns, and a is the rate of modules that store the value "1". Note that ϵ_{EE} and ϵ_{IE} scale the strengths of the inter-module connections to the excitatory and inhibitory ensembles, respectively. In the following, we set $M = 8$, $p = 3$, and $a = 0.5$.

Three patterns stored in the network of eight modules are defined as

$$\eta_i^1 = \begin{cases} 1 & \text{if } i \leq M/2 \\ 0 & \text{otherwise} \end{cases}, \tag{10}$$

$$\eta_i^2 = \begin{cases} 1 & \text{if } M/4 < i \leq 3M/4 \\ 0 & \text{otherwise} \end{cases}, \tag{11}$$

$$\eta_i^3 = \begin{cases} 1 & \text{if } i \bmod 2 = 1 \\ 0 & \text{otherwise} \end{cases}. \tag{12}$$

In the following, the dynamics of the network are examined by regulating the inter-module connections ϵ_{IE}, for the fixed values of parameters $\gamma = 0.6$ and $\epsilon_{EE} = 1.25$.

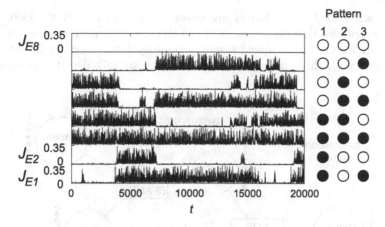

Fig. 2. Chaotic pattern alternations observed for $\epsilon_{IE} = 1.68$.

In Fig. 2, the dynamics of eight modules for $\epsilon_{IE} = 1.68$ are shown. The changes in the instantaneous firing rates J_{Ei} of the excitatory ensemble in the ith module are aligned vertically. It is observed that the retrieved pattern alters over time. The analysis of the network is performed with the Fokker-Planck equation, which does not contain noise because the limit $N_E, N_I \to \infty$ is taken. Therefore, the dynamics shown in Fig. 2 are not caused by noise but by chaos that is inherent in the network. This fact can be confirmed via analysis using Lyapunov spectra [10].

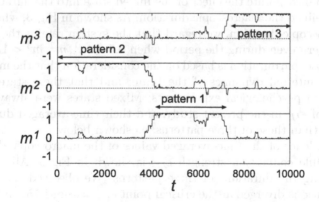

Fig. 3. Changes in three overlaps over time.

In order to investigate the retrieved pattern in the network, it is useful to define the overlap of the network with each pattern, which is similar to the inner product (detailed in Ref. [10]). The overlaps calculated from the dynamics during $0 \le t \le 10000$ shown in Fig. 2 are shown in Fig. 3. Note that m^μ takes values close to 1 when the μth pattern is retrieved.

In Figs. 2 and 3, short bursts are observed around $t \simeq 1000, 6000, 8500$, where the modules that do not store "1" in the retrieved pattern oscillate. Such patterns are referred to as mixed states in the associative memory literature. In our network with three patterns, we can observe six mixed states as shown in Fig. 4.

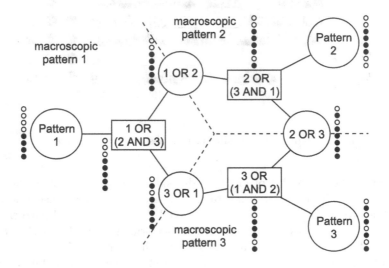

Fig. 4. The relationships among three patterns and their mixed states.

In order to incorporate the effect of the mixed state into the duration of each pattern, we defined the macroscopic duration. As shown in Fig. 3, when examining the macroscopic duration, we regard that the system retains the previously retrieved pattern even during the period when $0.5 \leq m^1, m^2, m^3 < 1$.

The macroscopic duration is based on the consideration that the mixed states represent the internal dynamics of the brain, and that these states are thus unobservable in psychological experiments. Mixed states were always unstable in the range of ϵ_{IE} in the present study, and their time-averaged duration was much shorter than those of three patterns, as shown below.

The dependence of the time-averaged values of the macroscopic duration on the inter-module connection strength ϵ_{IE} is shown in Fig. 5. All values were calculated using the durations of three patterns. We observed that the time-averaged durations diverged at the critical point $\epsilon_{IE} = \epsilon_0 \simeq 1.75$, and monotonically decreased with decreases in ϵ_{IE}. The time-averaged durations of the mixed states were always below 200 and much shorter than those of three patterns (data not shown).

Next we examine the distribution of the duration of each pattern when chaotic pattern alternations occur. The distributions of the macroscopic durations are shown in Fig. 6, in which the solid lines show the fit with the log-normal distribution. In Ref. [1], the dominant durations of binocular rivalry follow a log-normal distribution. Similarly, the distribution of the macroscopic durations in

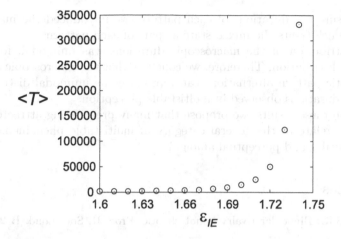

Fig. 5. The dependence of time-averaged macroscopic duration on the inter-module connection strength ϵ_{IE}.

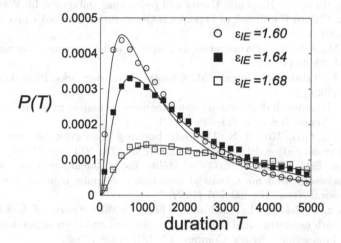

Fig. 6. The distributions of the macroscopic durations. The solid lines indicate the fit with the log-normal distribution.

our system also follows a log-normal distribution, as shown in Fig. 6. Therefore, we conclude that these macroscopic durations are appropriate as models of the dominance durations of binocular rivalry and perceptual ambiguity.

4   Conclusions

We proposed a pulse neural network that exhibits chaotic pattern alternations between three stored patterns as a model of multistable perception, which is reflected in such phenomena as binocular rivalry and perceptual ambiguity.

To measure the durations of each pattern, we introduced the macroscopic duration, which treats the mixed state as part of each pattern.

The distribution of the macroscopic durations was unimodal, following a log-normal distribution. Therefore, we conclude that the macroscopic durations of the chaotic pattern alternations can reproduce the unimodal distribution of dominance durations observed in multistable perception.

Based on these results, we propose that many pre-existing attractors in the brain might relate to the general category of multistable phenomena, such as binocular rivalry and perceptual ambiguity.

References

1. Lehky, S.R.: Binocular rivalry is not chaotic. Proc. R. Soc. Lond. B **259**, 71–76 (1995)
2. Blake, R.: A primer on binocular rivalry, including current controversies. Brain Mind **2**, 5–38 (2001)
3. Alais, D., Blake, R.: Binocular rivalry and perceptual ambiguity. In: Wagemans, J. (ed.) The Oxford Handbook of Perceptual Organization. Oxford University Press, Oxford (2015)
4. Adachi, M., Aihara, K.: Associative dynamics in a chaotic neural network. Neural Netw. **10**, 83–98 (1997)
5. Aihara, K., Takabe, T., Toyoda, M.: Chaotic neural networks. Phys. Lett. A **144**, 333–340 (1990)
6. Tsuda, I.: Dynamic link of memory - chaotic memory map in nonequilibrium neural networks. Neural Netw. **5**, 313–326 (1992)
7. Ermentrout, G.B., Kopell, N.: Parabolic bursting in an excitable system coupled with a slow oscillation. SIAM J. Appl. Math. **46**, 233–253 (1986)
8. Izhikevich, E.M.: Class 1 neural excitability, conventional synapses, weakly connected networks, and mathematical foundations of pulse-coupled models. IEEE Trans. Neural Netw. **10**, 499–507 (1999)
9. Kanamaru, T., Sekine, M.: Synchronized firings in the networks of class 1 excitable neurons with excitatory and inhibitory connections and their dependences on the forms of interactions. Neural Comput. **17**, 1315–1338 (2005)
10. Kanamaru, T.: Chaotic pattern transitions in pulse neural networks. Neural Netw. **20**, 781–790 (2007)
11. Kanamaru, T., Fujii, H., Aihara, K.: Deformation of attractor landscape via cholinergic presynaptic modulations: a computational study using a phase neuron model. PLoS ONE **8**, e53854 (2013)

Possibilities of Neural Networks for Personalization Approaches for Prevention of Complications After Endovascular Interventions

Tatiana V. Lazovskaya[1]([✉]), Dmitriy A. Tarkhov[2], Gelena A. Berezovskaya[3,4], Nikolay N. Petrischev[3,4], and Ildar U. Zulkarnay[5]

[1] CC FEB RAS, 65, Kim Yu Chen Street, 680000 Khabarovsk, Russia
tatianala@list.ru

[2] Peter the Great St. Petersburg Polytechnical University,
29 Politechnicheskaya str., 195251 Saint Petersburg, Russia
dtarkhov@gmail.com

[3] Federal Almazov Medical Research Centre,
2 Akkuratova str., 197341 Saint Petersburg, Russia
berezovgel@mail.ru

[4] Pavlov First Saint Petersburg State Medical University,
6-8 L'va Tolstogo str., 197022 Saint Petersburg, Russia

[5] Bashkir State University, 32 Zaki Validi str., 450076 Ufa, Russia
zulkar@inbox.ru

Abstract. It is known that most of the diseases of the cardiovascular system are accompanied by disorders in the hemostatic system. The hemostatic system is one of the most complex systems. It has a hierarchical structure with a plurality of components. We analyze the results of thrombin generation test (TGT) which allows of estimating the actions of all components of the hemostatic system. The problem is complicated by the presence of too many various clinical cases. The simple statistical methods do not provide global assessments. We suggest the universal neural network approach for building hemostatic system models based on the factors which don't have a statistically significant difference for various types of clinical post surgery cases. The neural network instruments allow of taking into account the nonlinear hierarchical nature of considered system and building individual models for each clinical cases. The aim of our study is to develop the neural network hemostatic system model for forecasting of disease progression and complications after endovascular interventions.

Keywords: Artificial neural networks · Modeling · Forecasting · Hierarchical models · Thrombin generation test · Hemostatic system · Endovascular intervention

1 Introduction

Nowadays artificial neural networks are widely used for processing medical data and medical diagnosis [1–5]. Large volumes of data make possible to study the

© Springer International Publishing AG 2017
F. Cong et al. (Eds.): ISNN 2017, Part I, LNCS 10261, pp. 379–385, 2017.
DOI: 10.1007/978-3-319-59072-1_45

advantages and disadvantages of various learning algorithms and types of the neural networks architecture [2, 3, 5]. At the same time, the medical data obtained from actual histories are often the small size samples. This may be a rare metrics, results of expensive medical tests. In such case, we can speak only about studying the neural network potential to use in medical diagnostics [4].

Requirements to medical studies become more strict every year, but findings and conclusions often depend on design of study and statistical processing methods, chosen for received data analysis. Meanwhile all pathological conditions in the human body represent complicated processes, multiplaned assessment of which needs integrative (total) research methods and mathematical analysis; and the main purpose of their investigation is a possibility of individual complications prediction [13, 14].

Medical diagnostics tasks are the type of problems when all the real conditions cannot be taken into account. The researcher can only distinguish some presumed set of the most important indicators. Built on such data, the algorithm will be imprecise and approximate. Moreover, the rules of construction and finding the answer cannot be clearly defined. In such problems, the use of neural network techniques is justified [14–18].

The present study is aimed at searching for possibilities of individual complications prediction after percutaneous coronary intervention (PCI), which is the most popular treatment method of coronary artery disease (CAD). There are following difficulties in the investigation of complications, developing after PCI are that this process involves alterations in vessels wall and haemostasis system. Haemostasis system indicators can be evaluated by thrombin generation test (TGT), chosen in our study for scrutinizing.

In paper [5], authors explored the applications of artificial neural networks in the prognostic evaluation of post-surgery complications too. But the model suggested there based on the data set with a statistically significant difference and the data volume was large enough. In our work we investigate the artificial neural network possibility to identify patterns and nonlinear connections in the case of a small data sample and the uniform data set in the terms of classical methods of statistics.

2   Materials and Methods

Material in the study is venosus blood data samples, which were drawn before PCI from 66 patients with coronary heart disease at the age of 53 to 77 years. We assessed haemostasis system state by using TGT in two parallel assays, accomplished according to the method, offered by Hemker et al. [6]. Clinical outcomes were considered version of resumption coronary artery disease symptoms, no events (51 observations); disease progression (8 observations); acute coronary syndrome (ACS) or myocardial infraction (MI) (7 observations).

We considered quantitative characteristics of TGT for which classical statistical method applying (correlation, regression analyses and others) had no significant effect. Our medical researchers selected more important characteristics. So, the next measures were investigated as factors.

The index of endogenous thrombin potential (ETP) was suggested by Hemker and coauthors for quantitative expression of thrombin generation [7].

- ETP is the area under the curve of thrombin generation
- Quantitative parameters of thrombogram also include the peak thrombin $PEAK$, which reflects the maximum amount of formed thrombin
- VI_{TM} is the thrombin generation velocity changing
- ΔLT is a percentage of thrombin formation initiation period decreasing.

We used the neural network technique for mathematical modeling of the classification of complications after surgery. The model based on TTG data obtained just before the surgical intervention. During numerical experiments, we considered neural networks with various architecture, numbers of the parameters and neurons (nodes). In general, the result of modeling is a one hidden layer neural network output [8] in the form

$$U(\mathbf{a}, \mathbf{c}; \mathbf{t}) = c_0 + \sum_{i=1}^{n} c_i v(\mathbf{a}_i, \mathbf{t}), \tag{1}$$

where n is the number of neurons (nodes), scalars $\mathbf{c} = (c_0, \ldots, c_n)$ and vectors $\mathbf{a} = (\mathbf{a}_0, \ldots, \mathbf{a}_n)$ are input network weights (parameters); v is selected neural network basis element [8], vector $\mathbf{t}$ is data used. In particular, we denote data of j-th observation as $\mathbf{t}_j$.

After trying various types of basis function we stopped on the sigmoid basis functions (Perceptron) [8,15] with hyperbolic tangent as basis elements in the form

$$v(\mathbf{a}, \mathbf{t}) = \text{th}(a_0 + \sum_{k-1}^{K} a_k t_k), \tag{2}$$

wher K is a number of data for each observation. In our case, four factors are investigated in modeling

$$\mathbf{t} = (ETP, PEAK, \Delta LT, VI_{TM}).$$

So, we can say that the hyperbolic tangent function is best suited for classification tasks.

To avoid an overcomplicating, we constructed the neural network model in several steps. At each step we added and made learned only one neuron. It meant that the learning process went on until getting optimum results.

At first step, the values G_j from the error functional (3) were coded as -3 for the occurrence of the complications after surgery and 1 for the absence. The construction of binary classification was premised on the size of the data explored, in particular that the number of observations with complications is small.

At next steps, G_j-s take the values of the errors of previous stage approximation for every observation. These errors were approximated by new neural network output with one neuron. At the end, we summarize all outputs of neural networks and the number of neurons in the model increased by one. The common number of model parameters is equal to $1 + 6n$, where n means a number of neurons.

At all steps of network learning, the neural network weights are determined by the minimization of the so-called error functional in the discrete form

$$I(a,c) = \sum_{j=1}^{M} \delta_j \big(U(\mathbf{a}, \mathbf{c}; \mathbf{t}_j) - G_j\big)^2, \tag{3}$$

where M is the number of observations, G_j are the values of the approximated function at points $\mathbf{t}_j$, δ_j are positive penalty coefficients.

To solve the minimization problem at every step of the genetic scheme above, we use the optimization algorithm combining RProp and the Particle Swarm methods [8,9]. The Particle Swarm optimization is often used to solve the problems of medical diseases diagnosis and classification [2].

The initial neural networks weights were located by the aim of balancing the scores of each factor.

After solving the global nonlinear optimization problems of all steps, we obtain the output of neural network $U(\mathbf{a}, \mathbf{c}; \mathbf{t}) = U_{a,c}(\mathbf{t})$ with fixed parameters (weights). This function takes real values. We could obtain the classification function by cut-off value selection based on the ROC analysis [10].

The necessary number of neurons is the subject of study. The balance has to be struck between the minimum of the model parameter number on the one hand and the quality of classification model on the other hand.

3 Results of Modeling and Classification

The quality of neural network classification model was estimated by the receiver operating characteristics (ROC) analysis [5,10]. Let us compare the results for model with the increasing number of neurons at each step of modeling. The Table 1 is presented the characteristics of the area under ROC curve (AUC). AUC does not depend on the relations of different type errors. The ideal classification corresponds to AUC equaled 1. This measure is often used to compare different models of classification.

Table 1. The characteristics of the area under ROC curve (AUC), models in the form (1)

Number of neurons	AUC	Std. error	Asimpt. sign	Asimpt. 95-conf. interval
1	0.695	0.074	0.022	0.551–0.840
2	0.778	0.060	0.001	0.661–0.895
3	0.824	0.064	0.000	0.697–0.950
4	0.886	0.051	0.000	0.785–0.987

The standard errors of AUC were estimated based on nonparametric method. Admittedly, the quality of neural network classification models is good even in the case of three neurons. If the number of neurons is equal to four the classification quality is very good [10].

Table 2 illustrates the quality of classifications constructed after choosing the cut-off value. The cut-off values for all our models were equal zero. Here are presented such quality characteristics as: Sensitivity is the percentage of people having the complications after surgery intervention who are correctly identified; Specificity is the percentage of people not having the complications who are correctly identified (the true negative rate). Often, for classification quality describing used are Efficiency (the average of Specificity and Sensitivity), True Positive Rate (TPR) is the percentage of people identified as having the complications who are correctly identified, and True Negative (TNR) Rates.

Table 2. ROC analysis classification quality, models in the form (1)

Number of neurons	1	2	3	4
Sensitivity, perc	60,0	80,0	80,0	86,7
Specificity, perc	72,5	62,7	74,5	80,4
Efficiency, perc	66,3	71,4	77,3	83,5
TPR, perc	39,1	38,7	48,0	56,5
TNR, perc	86,0	91,4	92,67	95,3

The model with one neuron is not enough sensitive. The second neuron adding improves the recognition of the observation with the complications but the specificity falls. The three and four-neuron models have enough level of the sensitivity and specificity.

4   Discussion of Results

Results of the analysis have confirmed the assumption that quantitative TGT indicators (ETP, $PEAK$), reflecting intensity of its formation affect on resumption CAD symptoms independently of relapse version. Received data coincide with conclusions of other investigators researches, who used standard mathematical methods to find relationships between TGT indicators and the death from cardiovascular causes after urgent PCI [11].

Earlier expressed hypothesis about protein C role in complications development after PCI and possibility of using indicators describing activity level of this system by changing thrombin generation velocity (VI_{TM}) and percentage of thrombin formation initiation period decreasing (ΔLT) after thrombomodulin adding in reaction mixture also was confirmed. The role of protein C system in stent thrombosis after PCI was confirmed too in another research [12], where activity of this system was also estimated with TGT.

5   Conclusion

Thus, indicators of TGT accomplished before PCI can be used for predicting of the resumption of CAD symptoms after PCI and development of the individual plan of patient management. It's expected that it would improve the effectiveness of this type of CAD treatment and reduce the risk of complications after PCI. The estimation of such risks is particularly true in the conditions of the Arctic zone.

Our study illustrates the ability of neural network modeling to identify hidden dependencies and complex functional nature of such processes as the human pathological conditions. We used the genetic algorithms of neural network model construction and learning. The dependencies above could reveal even in the case of the uniform data set in the terms of classical methods of statistics. The upgrowth in the number of neurons allows us to talk about increasing of the neural network classification quality. However, it is obvious, when constructing the real good model we must use the big data source. A special feature of neural network models is the ability to finish learning [14–18] based on new information, which could include the new observations and additional factors.

Acknowledgements. The work was supported by the Russian Foundation for Basic Research, project number 14-38-00009.

References

1. Moein, S.: Medical Diagnosis Using Artificial Neural Networks. IGI Global, Hershey (2014)
2. Beheshti, Z., Beheshti, E.: Enhancement of artificial neural network learning using centripetal accelerated particle swarm optimization for medical diseases diagnosis. Soft. Comput. Methodol. Appl. **18**(11), 2253–2270 (2013)
3. Ince, T., Kiranyaz, S., Pulkkinen, J., Gabbouj, M.: Evaluation of global and local training techniques over feed-forward neural network architecture spaces for computer-aided medical diagnosis. Expert Syst. Appl. **37**, 8450–8461 (2010)
4. Gorbachenko, V.I., Kuznetsova, O., Silnov, D.S.: Investigation of neural and fuzzy neural networks for diagnosis of endogenous intoxication syndrome in patients with chronic renal failure. Int. J. Appl. Eng. Res. **11**(7), 5156–5162 (2016)
5. Souza, C., Pizzolato, E., Mendes, R., Borghi-Silva, A.: Artificial neural networks prognostic evaluation of post-surgery complications in patients underwent to coronary artery bypass graft surgery. In: International Conference on Machine Learning and Applications (2009)
6. Hemker, H.C., Giesen, P., Al Dieri, R., Regnault, V., de Smedt, E., Wagenvoord, R., Lecompte, T., Beguin, S.: Calibrated automated thrombin generation measurement in clotting plasma. Pathophysiol. Haemost. Thromb. **33**, 4–15 (2003)
7. Hemker, H.C., Wielders, S., Kessels, H., Beguin, S.: Continuous registration of thrombin 10 generation in plasma, its use for the determination of the thrombin potential. Thromb. Haemost. **70**, 617–624 (1993)
8. Tarkhov, D.A., Vasilyev, A.N.: Neural Network Modeling. Principles. Algorithms. Applications. SPbSPU Publishing House, Saint-Petersburg (2009). (in Russian)

9. Riedmiller, M., Braun, H.: A direct adaptive method for faster backpropagation learning: the RPROP algorithm. In: Proceedings of the IEEE International (1993)
10. Fawcett, T.: An introduction to ROC analysis. Pattern Recogn. Lett. **27**(8), 861–874 (2006)
11. Attanasio, M., Marcucci, R., Gori, A.M., Paniccia, R., Valente, S., Balzi, D., Barchielli, A., Carrabba, N., Valenti, R., Antoniucci, D., Abbate, R., Gensini, G.F.: Residual thrombin potential predicts cardiovascular death in acute coronary syndrome patients undergoing percutaneous coronary intervention. Thromb. Res. **147**, 52–57 (2016)
12. Loeffen, R., Godschalk, T.C., van Oerle, R., Spronk, H.M., Hackeng, C.M., ten Berg, J.M., ten Cate, H.: The hypercoagulable profile of patients with stent thrombosis. Heart **101**(14), 1126–1132 (2015)
13. Bolgov, I., Kaverzneva, T., Kolesova, S., Lazovskaya, T., Stolyarov, O., Tarkhov, D.: Neural network model of rupture conditions for elastic material sample based on measurements at static loading under different strain rates. J. Phys: Conf. Ser. **772**, 012032 (2016). doi:10.1088/1742-6596/772/1/012032
14. Filkin, V., Kaverzneva, T., Lazovskaya, T., Lukinskiy, E., Petrov, A., Stolyarov, O., Tarkhov, D.: Neural network modeling of conditions of destruction of wood plank based on measurements. J. Phys: Conf. Ser. **772**, 012041 (2016). doi:10.1088/1742-6596/772/1/012041
15. Kaverzneva, T., Lazovskaya, T., Tarkhov, D., Vasilyev, A.: Neural network modeling of air pollution in tunnels according to indirect measurements. J. Phys: Conf. Ser. **772**, 012035 (2016). doi:10.1088/1742-6596/772/1/012035
16. Gorbachenko, V.I., Lazovskaya, T.V., Tarkhov, D.A., Vasilyev, A.N., Zhukov, M.V.: Neural network technique in some inverse problems of mathematical physics. In: Cheng, L., Liu, Q., Ronzhin, A. (eds.) ISNN 2016. LNCS, vol. 9719, pp. 310–316. Springer, Cham (2016). doi:10.1007/978-3-319-40663-3_36
17. Tarasenko, F.D., Tarkhov, D.A.: Basis functions comparative analysis in consecutive data smoothing algorithms. In: Cheng, L., Liu, Q., Ronzhin, A. (eds.) ISNN 2016. LNCS, vol. 9719, pp. 482–489. Springer, Cham (2016). doi:10.1007/978-3-319-40663-3_55
18. Blagoveshchenskaya, E.A., Dashkina, A.I., Lazovskaya, T.V., Ryabukhina, V.V., Tarkhov, D.A.: Neural network methods for construction of sociodynamic models hierarchy. In: Cheng, L., Liu, Q., Ronzhin, A. (eds.) ISNN 2016. LNCS, vol. 9719, pp. 513–520. Springer, Cham (2016). doi:10.1007/978-3-319-40663-3_59

Relief R-CNN: Utilizing Convolutional Features for Fast Object Detection

Guiying Li[1], Junlong Liu[1], Chunhui Jiang[1], Liangpeng Zhang[1], Minlong Lin[2], and Ke Tang[1(✉)]

[1] School of Computer Science and Technology,
University of Science and Technoloy of China,
Hefei 230027, Anhui, People's Republic of China
{lgy147,junlong,beethove,udars}@mail.ustc.edu.cn, ketang@ustc.edu.cn
[2] Tencent Company, Shenzhen 518057, People's Republic of China
minlonglin@tencent.com

Abstract. R-CNN style methods are sorts of the state-of-the-art object detection methods, which consist of region proposal generation and deep CNN classification. However, the proposal generation phase in this paradigm is usually time consuming, which would slow down the whole detection time in testing. This paper suggests that the value discrepancies among features in deep convolutional feature maps contain plenty of useful spatial information, and proposes a simple approach to extract the information for fast region proposal generation in testing. The proposed method, namely Relief R-CNN (R^2-CNN), adopts a novel region proposal generator in a trained R-CNN style model. The new generator directly generates proposals from convolutional features by some simple rules, thus resulting in a much faster proposal generation speed and a lower demand of computation resources. Empirical studies show that R^2-CNN could achieve the fastest detection speed with comparable accuracy among all the compared algorithms in testing.

Keywords: Object detection · R-CNN · CNN · Convolutional features · Deep learning · Deep neural networks

1 Introduction

One type of the state-of-the-art deep learning methods for object detection is R-CNN [8] and its derivative models [7,18]. R-CNN consists of two main stages: the category-independent region proposals generation and the proposal classification. The region proposals generation produces the rectangular Regions of Interest (RoIs) [7,18] that may contain object candidates. In the proposal classification stage, the generated RoIs are fed into a deep CNN [15], which will classify these RoIs as different categories or the background.

However, R-CNN is time inefficient in testing, especially when running on hardwares with limited computing power like mobile phones. The time cost of R-CNN comes from three parts: (1) the iterative RoIs generation process [12];

© Springer International Publishing AG 2017
F. Cong et al. (Eds.): ISNN 2017, Part I, LNCS 10261, pp. 386–394, 2017.
DOI: 10.1007/978-3-319-59072-1_46

(2) the deep CNN with a huge computation requirement [10,15,22]; and (3) the naive combination of RoIs and the deep CNN [8]. Many attempts on these three parts have been made to speed up R-CNN in testing. For RoI generation, Faster R-CNN [18] trains a Region Proposal Network (RPN) to predict RoIs in images instead of traditional data-independent methods that iteratively generate RoIs from images like Objectness [1], Selective Search [21], EdgeBox [3] and Bing [2]. For the time consuming deep CNN, some practical approaches [9,14] have been proposed to simplify the CNN structure. For the combination of RoIs and the deep CNN, SPPnet [11] and Fast R-CNN [7], which are the most popular approaches, reconstruct the combination of RoIs and CNN by directly mapping the RoIs to a specific pooling layer inside the deep CNN model. However, all these methods still cannot be efficiently deployed on low-end hardwares, since they still require considerable computing.

In this paper, we propose Relief R-CNN (R^2-CNN), which aims to speed up the deployment of RoI generation for a trained R-CNN without any extra training. For a trained R-CNN style model in deployment phase, R^2-CNN abandons the original RoIs generation process used in training, and directly extracts RoIs from the trained CNN. R^2-CNN is inspired by the analogy between relief sculptures in real life and feature maps in CNN. Visualization of convolutional layers [16,20] has shown that convolutional features with high values in a trained CNN directly map to the recognizable objects on input images. Therefore, R^2-CNN utilizes these convolutional features for region proposal generation. That is done by directly extracting the local region wrapping features with high values as RoIs. This approach is faster than many other methods, since a considerably large part of its computations are comparison operations instead of time consuming multiplication operations. Furthermore, R^2-CNN uses the convolutional features produced by CNN for RoI generation, while most of the methods need additional feature extraction from raw images for RoIs. In short, R^2-CNN could reduce much more computations in RoI generation phase compared with other methods discussed above.

The rest of the paper is organized as follows: Sect. 2 describes the details of Relief R-CNN. Section 3 presents the experimental results about R^2-CNN and relevant methods. Section 4 concludes the paper.

2   Relief R-CNN

In this section we present the details of R^2-CNN. Figure 1 shows the brief structure of R^2-CNN.

General Idea. The value discrepancies among features in a feature map of CNN are sorts of edge details. These details are similar to the textures on sculpture reliefs, which describe the vision by highlighting the height discrepancies of objects. Intuitively speaking, two nearby features that have significant value discrepancy may indicate they are on the boundary of objects, which is a type of edge details. There comes the basic assumption of R^2-CNN: region proposals can be generated from the object boundaries, which consist of enough edge

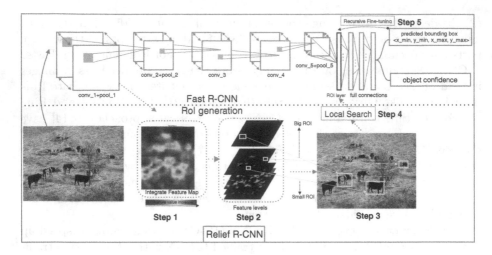

Fig. 1. Overview of Relief R-CNN. (Step 1) First is generating an **Integrate Feature Map** $f_{integrate}$ based on feature maps in pool1 layer of Alexnet [15], (Step 2) followed by separating features of $f_{integrate}$ into different **Feature Levels**. (Step 3) Then extracting **Big RoIs** and **Small RoIs** and using (Step 4, 5) additional proposal refinement techniques for better performance. The process conducted by solid lines is the procedure of Fast R-CNN, while the process along with dotted lines is the special work flow of R^2-CNN

details described by significant value discrepancies in CNN feature maps, with some simple rules based on the characteristics of convolutional feature maps.

The idea above comes from the observations on convolutional feature maps [16,20], and the similarity between the feature maps and sculpture relief, so that the proposed method is called Relief R-CNN. In testing phase, by searching the regions have significant more salient features than nearby context features in convolutional feature maps of a trained CNN, R^2-CNN can locate the objects in the source image by utilizing these region. R^2-CNN can be summarized into 5 steps as follows, in which steps 1–4 replace the RoI generator in the original trained models and step 5 boosts the performance of the fast generated RoIs in classification phase.

Step 1: Integrate Feature Map Generation. A synthetic feature map called **Integrate Feature Map**, denoted as $f_{integrate}$, is generated by adding all feature maps up to one map. $f_{integrate}$ brings two advantages, the first is dramatically reducing the number of feature maps, the second is eliminating noisy maps. The generation of $f_{integrate}$ consists of two steps:

1 Each feature map is normalized by dividing by its maximal feature value.
2 A $f_{integrate}$ is generated by adding all the normalized feature maps together in element-wise.

Step 2: Separating Feature Levels by Feature Interrelationship. Once the $f_{integrate}$ is ready, feature levels in $f_{integrate}$ should be formulated. As wrote in

General Idea, R^2-CNN tries to locate objects by a special sort of edge details, which is depicted by feature value discrepancies. However, it is hard to define how large the discrepancy between two features indicates a part of a boundary. To overcome this obstacle, we propose to separate features into different feature levels, and features in different feature levels are considered to be discriminative. Therefore, the contours formed by nearby features in a feature level directly represent the boundaries.

In this paper, feature levels in a $f_{\text{integrate}}$ are generated by dividing the value range of all the features into several subranges. Each subrange is a specific level which covers a part of features in the $f_{\text{integrate}}$. The number of subranges is a hyper-parameter, denoted as l. R^2-CNN uniformly divides the $f_{\text{integrate}}$ into l feature levels, see Algorithm 1. The step 2 in Fig. 1 shows some samples of feature levels generated from the first pooling layer of CaffeNet model (CaffeNet is a caffe implementation of AlexNet [15]).

Algorithm 1. Feature Level Separation

Input: $(f_{\text{integrate}}, l)$ ▷Integrate Feature Map and Feature Level Number
1: Finding the maximal value $value_{\max}$ and minimal value $value_{\min}$ in $f_{\text{integrate}}$
2: ▷uniformly dividing the value range into l subranges
3: $stride = (value_{\max} - value_{\min})/l$
4: ▷$feature_{\text{level\_i}}$ is the feature level i for $f_{\text{integrate}}$
5: **for** $i = 1 \rightarrow l$ **do**
6:     Finding features bigger than $value_{\min} + (i - 1) * stride$ and smaller than $value_{\min} + i * stride$ in $f_{\text{integrate}}$ as $feautre_{\text{level\_i}}$
7: **end for**
8: **return** $< feature_{\text{level\_1}}, ..., feature_{\text{level\_l}} >$

Step 3: RoIs Generation. The approach R^2-CNN adopted for RoIs generation is, as be mentioned in step 2, finding the contours formated by nearby features in a feature level, which needs the help of some deep network structure related observations. As the step 3 shown in Fig. 1, the neighboring features, which are surely belong to the same object, can form a small RoI. Furthermore, a larger RoI can be assembled from several small RoIs, in case of some large objects be consisted of small ones. Here's the summarized operations:

- Small RoIs: Firstly, it searches for the feature clusters (namely the neighboring features) in the given $feature_{\text{level\_i}}$, and then mapping the feature clusters to the input image as **Small RoIs**.
- Big RoI: For the purpose of simplicity (avoiding the combinatorial explosion), only one **Big RoI** is generated in a feature level by assembling all the small RoIs.

Step 4: Local Search. Convolutional features from source image are not produced by seamless sampling. As a result, RoIs extracted in convolutional feature maps might be quite coarse. Local Search in width and height is applied to tackle this problem. For each RoI, which its width and height are denoted as (w, h), local search algorithm needs two scale ratios α and β to generate 4 more RoIs: $(\beta * w, \beta * h), (\beta * w, \alpha * h), (\alpha * w, \alpha * h), (\alpha * w, \beta * h)$. In experiments, α was fixed to 0.8 and β was fixed to 1.5. The Local Search can give about 1.8 mAP improvement in detection performance.

Step 5: Recursive Fine-Tuning. Previous steps provide a fast RoI generation for testing. However, the accuracy of testing is restricted because of the different proposals distribution between training and testing. Owing to this fact, we propose the method called recursive fine-tuning to boost the detection performance during the classification phase of RoIs.

The recursive fine-tuning is a very simple step. It does not need any changes to existing R-CNN style models, but just a recursive link from the output of a trained box regressor back to its input. Briefly speaking, it is a trained box regressor wrapped up into a closed-loop system from a R-CNN style model. This step aims at making full use of the box regressor, by recursively refining the RoIs until their performance have been converged.

It should be noticed that there exists a similar method called Iterative Localization [5]. It needs a bounding box regressor be trained in another settings and starts the refinement from the proposals generated by Selective Search, while the recursive fine-tuning bases on the regressor in a unified trained R-CNN and starts refinement from the RoIs generated by above steps (namely Step 1–4). Furthermore, recursive fine-tuning does not reject any proposals but only improve them if possible, while iterative localization drops the proposals below a threshold at the beginning.

3 Experiments

3.1 Setup

In this section, we compared our R^2-CNN with some state-of-the-art methods for accelerating trained R-CNN style models. The proposals of Bing, Objectness, EdgeBoxes and Selective Search were the pre-generated proposals published by [12], since the the algorithm settings were the same. The evaluation code used for generating Fig. 2 was also published by [12].

The baseline of R-CNN style model is Fast R-CNN with CaffeNet. The Fast R-CNN model was trained with Selective Search just the same as in [7]. The Faster R-CNN [18] used in experiments was based on project py-faster-rcnn [6]. Despite the difficulty of Faster R-CNN for low power devices, RPN of Faster R-CNN is still one of the state-of-the-art proposal methods. Therefore, RPN was still adopted in experiments using the same Fast R-CNN model consistent with other methods for detection. The RPN in experiments was trained on the first stage of Faster R-CNN training phases. This paradigm is the unshared Faster R-CNN model mentioned in [18]. For the R^2-CNN model, the number of recursive loops was set as 3, and the number of feature levels was 10.

All experiments were tested on PASCAL VOC 2007 [4]. Deep CNNs in this section got support from Caffe [13], a famous open source deep learning framework. All the proposal generation methods were running on CPU (inc. R^2-CNN and RPN) while the deep neural networks of classification were running on GPU. All the deep neural networks had run on one NVIDIA GTX Titan X, and the CPU used in the experiments was Intel E5-2650V2 with 8 cores, 2.6 GHZ.

3.2   Speed and Detection Performance

Table 1 contains the results of comparison about time in testing. The testing time is separated into proposal time and classification time. The proposal time is the time cost for proposal generation, and the classification time is the time cost for verifying all the proposals.

Table 1. Testing time & performance comparison. The object detection model used here is Fast R-CNN. The R^2-CNN needs recursive fine-tuning which makes classification be time-consuming. "Total Time" is the sum of values in "Proposal Time" and "Classification Time". "*" indicates the runtime reported in [12]. "RPN" is the proposal generation model used in Faster R-CNN. **Bold** items are the results of R^2-CNN. R^2-CNN presents the fastest speed and comparable detection performance.

Methods	Proposal time (sec.)	Proposals	Classification time (sec.)	Total time (sec.)	mAP	Mean precision (%)
R^2-CNN	**0.00048**	**760.19**	**0.146**	**0.14648**	**53.8**	9.2
Bing	0.2*	2000	0.115	0.315	41.2	2
EdgeBoxes	0.3*	2000	0.115	0.415	55.5	4.2
RPN	1.616	2000	0.115	1.731	55.2	3.5
Objectness	3*	2000	0.115	3.115	44.4	1.7
Selective search	10*	2000	0.115	10.115	57.0	5.9

Table 1 has also shown the detection performances of R^2-CNN and other comparison methods. Precision [17] is a well known metric to evaluate the precision of predictions, mAP (abbreviation of mean Average Precision) is a highly accepted evaluation in the object detection task [19].

The empirical results in Table 1 reveal that R^2-CNN could achieve a very competitive detection performance compared with state-of-the-art Selective Search, EdgeBoxes and Faster R-CNN with a much more fast CPU speed, which means it's a more suitable RoI method for deploying trained R-CNN style models on low-end hardwares.

3.3   Proposal Quality

To evaluate the quality of proposals, the evaluation metric [12] Recall-to-IoU curve was adopted, see Fig. 2. The metric *IoU* (abbreviation of *intersection*

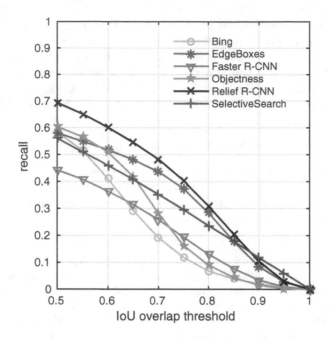

Fig. 2. Recall to *IoU* threshold with 200 proposals in count. R^2-CNN had nearly dominated other methods.

over union) [19], is an evaluation criterion to measure how similar two regions are. A larger *IoU* indicates more similar regions.

In Fig. 2, it could be found that R^2-CNN had nearly dominated other methods in *IoU* threshold between 0.5–0.9, and became the secondary best in *IoU* threshold 0.9–1.0.

It should be noticed that R^2-CNN could not control the number of proposals, but it got the best results with hundreds of proposals while others need thousands. The experiments in this section have shown that R^2-CNN could get a very good performance in the situation of limit proposals with a high speed, which is also a good character for platforms with limited computation resources.

4   Conclusion

This paper presents a unified object detection model called Relief R-CNN (R^2-CNN). By directly extracting region proposals from convolutional feature discrepancies, namely the location information of salient features in local regions, R^2-CNN reduces the RoI generation time required for a trained R-CNN style model in testing phase. Hence, R^2-CNN is more suitable to be deployed on low-end hardwares than existing R-CNN variants. Moreover, R^2-CNN introduces no additional training budget. Empirical studies demonstrated that R^2-CNN was faster than previous works with competitive detection performance.

Acknowledgments. This work was supported in part by the National Natural Science Foundation of China under Grant 61329302 and Grant 61672478, and in part by the Royal Society Newton Advanced Fellowship under Grant NA150123.

References

1. Alexe, B., Deselaers, T., Ferrari, V.: What is an object?. In: 2010 IEEE Computer Society Conference on Computer Vision and Pattern Recognition, pp. 73–80 (2010)
2. Cheng, M.M., Zhang, Z., Lin, W.Y., Torr, P.: BING: binarized normed gradients for objectness estimation at 300fps. In: 2014 IEEE Conference on Computer Vision and Pattern Recognition, pp. 3286–3293 (2014)
3. Dollár, P., Zitnick, C.L.: Fast edge detection using structured forests. IEEE Trans. Pattern Anal. Mach. Intell. **37**(8), 1558–1570 (2015)
4. Everingham, M., Eslami, S.M.A., Van Gool, L., Williams, C.K.I., Winn, J., Zisserman, A.: The pascal visual object classes challenge: a retrospective. Int. J. Comput. Vis. **111**(1), 98–136 (2015)
5. Gidaris, S., Komodakis, N.: Object detection via a multi-region and semantic segmentation-aware CNN model. In: The IEEE International Conference on Computer Vision (ICCV), pp. 1134–1142 (2015)
6. Girshick, R.: Project of Faster R-CNN (python implementation). https://github.com/rbgirshick/py-faster-rcnn
7. Girshick, R.: Fast R-CNN. In: 2015 IEEE International Conference on Computer Vision (ICCV), pp. 1440–1448 (2015)
8. Girshick, R., Donahue, J., Darrell, T., Malik, J.: Rich feature hierarchies for accurate object detection and semantic segmentation. In: 2014 IEEE Conference on Computer Vision and Pattern Recognition, pp. 580–587 (2014)
9. Han, S., Mao, H., Dally, W.J.: Deep compression: compressing deep neural networks with pruning, trained quantization and huffman coding. In: International Conference on Learning Representations (ICLR) (2016)
10. He, K., Zhang, X., Ren, S., Sun, J.: Deep residual learning for image recognition. In: 2016 IEEE Conference on Computer Vision and Pattern Recognition (CVPR), pp. 770–778 (2016)
11. He, K., Zhang, X., Ren, S., Sun, J.: Spatial pyramid pooling in deep convolutional networks for visual recognition. In: Fleet, D., Pajdla, T., Schiele, B., Tuytelaars, T. (eds.) ECCV 2014. LNCS, vol. 8691, pp. 346–361. Springer, Cham (2014). doi:10.1007/978-3-319-10578-9_23
12. Hosang, J., Benenson, R., Dollár, P., Schiele, B.: What makes for effective detection proposals? IEEE Trans. Pattern Anal. Mach. Intell. **38**(4), 814–830 (2016)
13. Jia, Y., Shelhamer, E., Donahue, J., Karayev, S., Long, J., Girshick, R., Guadarrama, S., Darrell, T.: Caffe: convolutional architecture for fast feature embedding. In: ACM Multimedia, pp. 675–678. ACM (2014)
14. Kim, Y.D., Park, E., Yoo, S., Choi, T., Yang, L., Shin, D.: Compression of deep convolutional neural networks for fast and low power mobile applications. In: International Conference on Learning Representations (ICLR) (2016)
15. Krizhevsky, A., Sutskever, I., Hinton, G.E.: ImageNet classification with deep convolutional neural networks. In: Pereira, F., Burges, C.J.C., Bottou, L., Weinberger, K.Q. (eds.) Advances in Neural Information Processing Systems, vol. 25, pp. 1097–1105. Curran Associates, Inc. (2012)

16. Mahendran, A., Vedaldi, A.: Understanding deep image representations by inverting them. In: 2015 IEEE Conference on Computer Vision and Pattern Recognition (CVPR), pp. 5188–5196 (2015)
17. Özdemir, B., Aksoy, S., Eckert, S., Pesaresi, M., Ehrlich, D.: Performance measures for object detection evaluation. Pattern Recogn. Lett. **31**(10), 1128–1137 (2010)
18. Ren, S., He, K., Girshick, R., Sun, J.: Faster R-CNN: towards real-time object detection with region proposal networks. In: Cortes, C., Lawrence, N.D., Lee, D.D., Sugiyama, M., Garnett, R. (eds.) Advances in Neural Information Processing Systems, vol. 28, pp. 91–99. Curran Associates, Inc. (2015)
19. Russakovsky, O., Deng, J., Su, H., et al.: ImageNet large scale visual recognition challenge. Int. J. Comput. Vis. **115**(3), 211–252 (2015)
20. Zeiler, M.D., Fergus, R.: Visualizing and understanding convolutional networks. In: Fleet, D., Pajdla, T., Schiele, B., Tuytelaars, T. (eds.) ECCV 2014. LNCS, vol. 8689, pp. 818–833. Springer, Cham (2014). doi:10.1007/978-3-319-10590-1_53
21. Uijlings, J.R.R., van de Sande, K.E.A., Gevers, T., Smeulders, A.W.M.: Selective search for object recognition. Int. J. Comput. Vis. **104**(2), 154–171 (2013)
22. Simonyan, K., Zisserman, A.: Very deep convolutional networks for large-scale image recognition. In: ICLR (2015)

The Critical Dynamics in Neural Network Improve the Computational Capability of Liquid State Machines

Xiumin Li[1,2], Qing Chen[1,2], Fangzheng Xue[1,2], and Hongjun Zhou[3]([✉])

[1] Key Laboratory of Dependable Service Computing in Cyber Physical Society of Ministry of Education, Chongqing University, Chongqing 400044, China
{xmli,chenqing,xuefangzheng}@cqu.edu.cn
[2] College of Automation, Chongqing University, Chongqing 400044, China
[3] School of Economics and Business Administration, Chongqing University, Chongqing 400044, China
hjzhou@cqu.edu.cn

Abstract. In recent years, increasing studies have shown that the networks in the brain can reach a critical state where dynamics exhibit a mixture of synchronous and asynchronous firing activity. It has been hypothesized that the homeostatic level balanced between stability and plasticity of this critical state may be the optimal state for performing diverse neural computational tasks. Motivated by this, the role of critical state in neural computation based on liquid state machines (LSM), which is one of the neural network application model of liquid computing, has been investigated in this note. Different from a randomly connect structure in liquid component of LSM in most studies, the synaptic weights among neurons in proposed liquid are refined by spike-timing-dependent plasticity (STDP); meanwhile, the degrees of neurons excitability are regulated to maintain a low average activity level by Intrinsic Plasticity (IP). The results have shown that the network yield maximal computational performance when subjected to critical dynamical states.

Keywords: Computation capability · LSM · Critical dynamic · STDP · IP

1 Introduction

Recently, many studies have been advanced to study the critical state of the network in the brain [1–3]. A remarkable phenomena that critical state exhibits is power law distributions of the spontaneous neuronal avalanches sizes approximately with a slope of -1.5 [4]. The functional rule of this dynamical criticality can bring about optimal transmission [1], storage of information [5] and sensitivity to external stimuli [6]. The influences of network structures on critical state have been widely researched considering from the perspective of complex

© Springer International Publishing AG 2017
F. Cong et al. (Eds.): ISNN 2017, Part I, LNCS 10261, pp. 395–403, 2017.
DOI: 10.1007/978-3-319-59072-1_47

network, such as scale-free network [7,8], small-world network [9,10] and hierarchical modular network [11,12]. However, critical dynamics are rarely used in computational neuroscience.

In this paper, the influences of critical state on the computational performance of LSM for real-time computing have been studied. As shown in Fig. 1(a), LSMs include three components: input component, liquid component, readout component [13]. Synaptic inputs, which are integrated from the input parts, are send to the neurons in liquid component and then it can be described in a higher dimensional state known as liquid state. As to specific assignments, the outputs of liquid component are projecting to the readout component, which plays a role as a memory-less function. In the process of computations, all the connections in the liquid component will always keep unchanged once the structure is setted up, except that readouts are trained through linear regression algorithm. As a result, many researchers are concentrating on studying the network dynamics under a predefined topological network. Considering the flexibility of neuronal connectivity in the brain, it is more reasonable to consider self-organizing neural networks based on neural plasticity.

One of the widely known forms of synaptic plasticity is the spike-timing-dependent plasticity [14]. In our previous work [15], we have given a novel liquid component of LSM refined by STDP. Compared with the LSM with tradition random liquid, LSM with new liquid has better computational performance on complex input streams. Besides, recent experimental results show that the intrinsic excitability of individual biological neurons can be adjusted to match the synaptic input by the activity of their voltage gated channels [17]. This adaption of neuronal intrinsic excitability called intrinsic plasticity (IP) has been observed in cortical areas and plays an important role on cortical functions of neural circuits [18]. It is hypothesized that IP can keep the mean firing activity of neuronal population in a homeostatic level [19], which is essential for avoiding highly intensive and synchronous firing caused by the STDP learning. Therefore, it is necessary to investigate it in combination with existing network learning algorithms to maximize the information capacity.

In this paper, we have refined the liquid component of LSM though STDP and IP learning. Therein, the synaptic weights among neurons in liquid are updated by STDP; while IP learning regulates the degrees of neurons excitability. The influence of critical dynamics on the computational performance of proposed LSM has been investigated. Results demonstrate that the network yield maximal computational performance when subjected to critical dynamical states. These results may be very significant in finding out the relationship between network learning and efficiency of information processing.

2   Network Description

2.1   Network Architecture

In this paper, as described in Fig. 1(a), we have added four different inputs to four equivalently divided groups in the liquid component. Each input is made

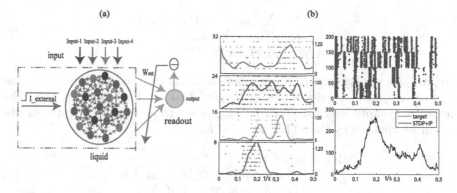

Fig. 1. (a) Network structure. Neurons marked with different colors are subjected to different inputs. (b) Left: Four independent input. Right: The response of neurons in STDP+IP network(up) and the corresponding output of the readouts (bottom) according to the target signal $r_1 + r_3$. (Color figure online)

of eight independent signal streams and generated by the Poisson process with randomly varying rates as $r_i(t)$, i $= 1, ...4$ (Fig. 1(b)-left), which are chosen as follows [20]. The baseline firing rates for input 1 and 2 are chosen to be 5 Hz, with randomly distributed bursts of 120 Hz for 50 ms. The rates for input 3 and 4 are periodically updated, by randomly drawn from the two values of 30 Hz and 90 Hz. The curves in Fig. 1(b)-left represents the final firing rates. Figure 1(b)-right show the responses of the neurons in liquid networks and the corresponding outputs of LSM compared with the target teaching signal $r_1 + r_3$. The results show that input signal can be expressed well as the high-dimensional liquid state, where information can be encoded into the intrinsic dynamical of neuronal population, thus the high precision of computational capability can be realized.

2.2 Neuron Model

The network used in this article is composed of 200 Izhikevich neuron [21] described by

$$\dot{v}_i = 0.04v_i^2 + 5v_i + 140 - u_i + I + I_i^{syn}$$
$$\dot{u}_i = a(bv_i - u_i) + D\xi_i \tag{1}$$

$$\text{if } v_i > 30 \text{ mV, then } \begin{cases} v_i \leftarrow c \\ u_i \leftarrow u_i + d \end{cases} \tag{2}$$

where $i = 1, 2, ..., 200$. v_i and u_i is the membrane potential and membrane recovery variable of the neurons, respectively. The parameters a, b, c, d are constant. Choosing different values of these parameters can obtain various firing dynamic [21]. The parameter ξ_i stand for the independent Gaussian noise with zero mean and intensity D is the noisy background. I is the external current. I_i^{syn} is the total synaptic current through neuron i and is governed by the dynamics of the synaptic variable s_j:

$$I_i^{syn} = -\sum_{1(j\neq i)}^{N} g_{ji} s_j (v_i - v_{syn})$$
$$\dot{s}_j = \alpha(v_j)(1 - s_j) - s_j/\tau \qquad (3)$$
$$\alpha(v_j) = \alpha_0/(1 + e^{-v_j/v_{shp}})$$

here, $\alpha(v_j)$ is the synaptic recovery function. If the presynaptic neuron is in the silent state $v_j < 0$, s_j reduces to $\dot{s}_j = -s_j/\tau$; if not, s_j jumps quickly to 1. The excitatory synaptic reversal potential v_{syn} is set to be 0. The synaptic weight g_{ij} will be updated by the STDP function F:

$$\Delta g_{ij} = g_{ij} F(\Delta t)$$
$$F(\Delta t) = \begin{cases} A_+ \exp(-\Delta t/\tau_+) & \text{if } \Delta t > 0 \\ -A_- \exp(\Delta t/\tau_-) & \text{if } \Delta t < 0 \end{cases} \qquad (4)$$

where $\Delta t = t_j - t_i$, t_i and t_j is the spike time of the presynaptic and postsynaptic neuron, respectively. τ_+ and τ_- determine the temporal window for synaptic modification. $F(\Delta t) = 0$ when $\Delta t = 0$. A_+ and A_- determine the maximum amount of synaptic modification. Here, $\tau_- = \tau_+ = 20$, $A_+ = 0.05$ and $A_-/A_+ = 1.05$. The synaptic weights are distribute in $[0, g_{max}]$, where $g_{max} = 0.015$ is the maximum value.

Particularly, parameter b has a significant influence on the neurons excitability. To get a heterogeneous network, the initial values of b are randomly distributed in $[0.12, 0.2]$. The neurons with larger value b can exhibit stronger excitability, thus fire with a higher frequency. As a result we consider plastic modifications of b as a representative scheme describing IP mechanisms. The model we proposed is based on neurons' inter-spike interval (ISI), in which a function ϕ_i is used to determine the amount of excitability modification:

$$\Delta b_i = b_i \phi_i$$
$$\phi_i = \begin{cases} -\eta_{IP} \cdot exp(\frac{T_{min}-ISI_i}{T_{min}}) & \text{if } ISI_i < T_{min} \\ \eta_{IP} \cdot exp(\frac{ISI_i-T_{max}}{T_{max}}) & \text{if } ISI_i > T_{max} \\ 0, others \end{cases} \qquad (5)$$

where η_{IP} is learning rate. The neuronal inter-spike interval (ISI) is $ISI_i^k = t_i^{k+1} - t_i^k$, where t_i^k is the kth firing time of neuron i; T_{min} and T_{max} are thresholds, they determine the expected ranges of ISI. During the learning process, the most recent ISI is examined every t_{ck} time and used to adjust the neuronal excitability: If ISI_i is larger than the threshold T_{max}, the neuronal excitability is strengthened to make the neuron more sensitive to input stimuli; if ISI_i is less than the threshold T_{min}, the neuronal excitability is weakened to make the neuron less sensitive to input stimuli. The histogram of firing rate response during IP learning for a randomly driven network is shown in Fig. 2, from which an normal distribution of firing rate is observed, and this result is consistent with the theory that the maximum-entropy distribution is Gaussian if the desired $(p(x) = \frac{exp[-(x-\mu)^2/2\sigma^2]}{\sigma\sqrt{2\pi}})$ variance is fixed. It indicates that our IP model is reasonable. Additionally, the values of other parameters are $\alpha_0 = 3$, $\tau = 2$, $V_{shp} = 5$, $a = 0.02$, $c = -65$, $d = 8$, $D = 0.1$, $T_{max} = 110$, $T_{min} = 90$.

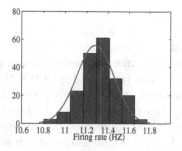

Fig. 2. Histogram of firing rate response by IP learning and its Gassian fit: $\mu = 11.2795$, $\sigma = 0.1736$.

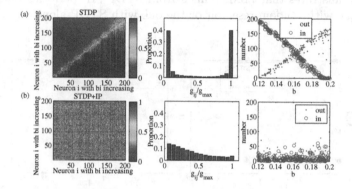

Fig. 3. Network structures obtained from learning rules. Left: STDP alone; Right: STDP plus IP. (a) Schematic diagram of the normalized synaptic matrix. (b) Histogram of the normalized synaptic weights. (c) Scatter plots of neurons strong synaptic weights for in-degrees and out-degrees.

At the beginning of the learning, each neuron in liquid network is bidirectionally connected to each other with the same synaptic weight of $g_{max}/2$ and the same external current of 6. After sufficient time the updated network structure by STDP alone or STDP+IP is shown in Fig. 3. Figure 3(a) indicates the active-neuron dominant structure obtained by STDP learning, where the strong connections are mainly distributed to the synapses from neurons with large values of b to inactive ones with small values of b, and most of the synapses are rewired to be either 0 or 1; while IP strengthens the competition among different neurons and makes the connectivity structure more complex and the distribution is not bimodal, but rather is skewed toward smaller values. The degree distribution for different networks are also examined, the out-degree(out) and in-degree(in) are defined as in [16]. It is demonstrated that neurons with larger values of b have larger out-degrees and smaller in-degrees in STDP condition, while only neurons with intermediate sensitivity keep this principle when IP is switched on.

3   Results

In this section, lists of real-time computational tasks were conducted to investigate the influence of critical dynamic on the computational perfromance of LSM updated by STDP+IP. To characterize and quantify the computational performance of networks systematically, we purposely tested the sensitivity of different types of LSM by varying the external current I. The results of average MSEs shown in Fig. 4(a) were obtained from 20 times independent simulations. The results of the three network are non-monotonic, which reaches the minimal value when the external stimulus current is about 5. The computational performance becomes worse when the external stimulus current I is too strong or too weak. Besides, it illustrates that LSMs refined from STDP+IP performs much better than the one with random reservoir or the one with STDP alone.

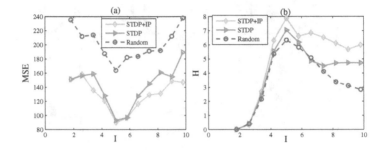

Fig. 4. (a) Computational capability of networks with different topologies. (b) Entropy of network activity for networks with different topologies. i.e. STDP network, random network and STDP+IP network.

In order to get an insight into the potential advantages of the turning point, we have specially investigated the influence of stimulus external current on network activity. Figure 5 has shown the network activities of different network with different stimulus. It can be seen that the synchronization degree of network activity has been increased with the increase of external stimulus. Particularly, the activity exhibit a mixture of synchronous and asynchronous firing activity when the stimulus current is about 5, indicating the highly complexity of network activity. To further quantify the complexity of network activity, we have computed the information entropy of network activity, which measures the complexity of activity patterns in a neural network and defined as

$$H = -\sum_{i=1}^{n} p_i \log_2 p_i \tag{6}$$

where, n is the number of unique binary patterns. p_i is the probability that pattern i occurs [22]. For calculation convenience, neuronal activities are measured in pattern units consisting of a certain number of neurons. In each time bin, if

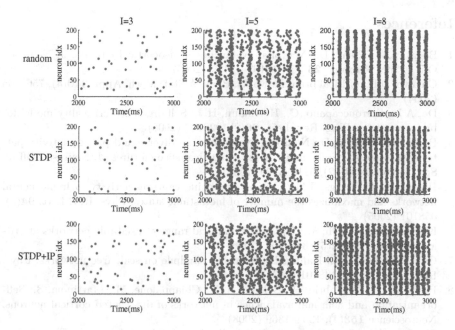

Fig. 5. Firing activity of the three network with different stimulates (I = 3, 5, 8).

any neuron of the unit is firing then the event of this unit is active; otherwise it is inactive. Surprisingly, the results have shown that the maximal entropy has been reached when the current is about 5 (see Fig. 4(b)) where networks have the optimal computational performance. Therefore, these results demonstrate that the critical state with dynamics between synchronized firings and unsynchronized firings makes the system have maximal dynamical complexity and thus achieve optimal computational performance.

4   Conclusion

In this paper, the effect of critical dynamics on computational capability of liquid state machine updated by STDP+IP has been investigated. Our results have shown that the critical dynamic can remarkable improve the computation performance of liquid state machine. At the critical state, the information entropy of network activity is maximized indicating the complexity of activity patterns are maximized, which can encode the rich dynamics of different neurons. These results may be very significant in finding out the relationship between network learning and efficiency of information processing.

Acknowledgments. This work was supported by the National Natural Science Foundation of China (Nos. 61473051 and 61304165).

References

1. Beggs, J.M., Plenz, D.: Neuronal avalanches in neocortical circuits. J. Neurosci. Off. J. Soc. Neurosci. **23**(35), 11167–11177 (2003)
2. Chialvo, D.R.: Critical brain networks. Phys. A Stat. Mech. Appl. **340**(4), 756–765 (2004)
3. De, A.L., Perronecapano, C., Herrmann, H.J.: Self-organized criticality model for brain plasticity. Phys. Rev. Lett. **96**(2), 028107 (2006)
4. Beggs, J.M., Plenz, D.: Neuronal avalanches are diverse and precise activity patterns that are stable for many hours in cortical slice cultures. J. Neurosci. Off. J. Soc. Neurosci. **24**(22), 5216–5229 (2004)
5. Haldeman, C., Beggs, J.M.: Critical branching captures activity in living neural networks and maximizes the number of metastable states. Phys. Rev. Lett. **94**(5), 058101 (2005)
6. Kinouchi, O., Copelli, M.: Optimal dynamical range of excitable networks at criticality. Nat. Phys. **2**(5), 348–351 (2006)
7. Goh, K.I., Lee, D.S., Kahng, B., Kim, D.: Sandpile on scale-free networks. Phys. Rev. Lett. **91**(14), 148701 (2003)
8. Pasquale, V., Massobrio, P., Bologna, L.L., Chiappalone, M., Martinoia, S.: Self-organization and neuronal avalanches in networks of dissociated cortical neurons. Neuroscience **153**(4), 1354–1369 (2008)
9. Lin, M., Chen, T.: Self-organized criticality in a simple model of neurons based on small-world networks. Phys. Rev. E **71**(1), 016133 (2005)
10. Pajevic, S., Plenz, D.: Efficient network reconstruction from dynamical cascades identifies small-world topology of neuronal avalanches. PLoS Comput. Biol. **5**(1), e1000271 (2009)
11. Wang, S.J., Zhou, C.: Hierarchical modular structure enhances the robustness of self-organized criticality in neural networks. New J. Phys. **14**(2), 023005 (2012)
12. Wang, S.J., Hilgetag, C., Zhou, C.: Sustained activity in hierarchical modular neural networks: self-organized criticality and oscillations. Front. Comput. Neurosci. **5**, 30 (2011)
13. Natschläger, T., Maass, W., Markram, H.: The "liquid computer": a novel strategy for real-time computing on time series. In: Special issue on Foundations of Information Processing of TELEMATIK, vol. 8 (LNMC-ARTICLE-2002-005), pp. 39–43 (2002)
14. Markram, H., Lübke, J., Frotscher, M., Sakmann, B.: Regulation of synaptic efficacy by coincidence of postsynaptic APs and EPSPs. Science **275**(5297), 213–215 (1997)
15. Xue, F., Hou, Z., Li, X.: Computational capability of liquid state machines with spike-timing-dependent plasticity. Neurocomputing **122**, 324–329 (2013)
16. Li, X., Small, M.: Enhancement of signal sensitivity in a heterogeneous neural network refined from synaptic plasticity. New J. Phys. **12**(8), 083045 (2010)
17. Daoudal, G., Debanne, D.: Long-term plasticity of intrinsic excitability: learning rules and mechanisms. Learn. Mem. **10**(6), 456–465 (2003)
18. Marder, E., Abbott, L.F., Turrigiano, G.G., Liu, Z., Golowasch, J.: Memory from the dynamics of intrinsic membrane currents. Proc. Nat. Acad. Sci. **93**(24), 13481–13486 (1996)
19. Triesch, J.: Synergies between intrinsic and synaptic plasticity in individual model neurons. In: NIPS, pp. 1417–1424 (2004)

20. Maass, W., Joshi, P., Sontag, E.D.: Computational aspects of feedback in neural circuits. PLoS Comput. Biol. **3**(1), e165 (2007)
21. Izhikevich, E.M.: Simple model of spiking neurons. IEEE Trans. Neural Netw. **14**(6), 1569–1572 (2003)
22. Shew, W.L., Yang, H., Yu, S., Roy, R., Plenz, D.: Information capacity and transmission are maximized in balanced cortical networks with neuronal avalanches. J. Neurosci. **31**(1), 55–63 (2011)

Exponential Stability of the Coupled Neural Networks with Different State Dimensions

Jieyin Mai, Manchun Tan$^{(\boxtimes)}$, Yunfeng Liu, and Desheng Xu

College of Information Science and Technology,
Jinan University, Guangzhou 510632, China
tanmc@jnu.edu.cn

Abstract. In this paper, the exponential stability is studied for a class of coupled neural networks, in which the model has nodes of different dimensions, and has different internal time-delays and coupling delays. Based on Lyapunov stability theory and linear matrix inequality technique, some sufficient conditions are derived for ensuring the exponential stability of the equilibrium of system. Finally, a numerical example is given to show the effectiveness of our results.

Keywords: Coupled neural networks · Exponential stability · Different state dimensions · Different coupling time delays

1 Introduction

During the past decade, coupled neural networks (CNNs) have received increasing attention of researchers, due to the fact that coupled networks may exhibit more complicated and unpredictable behaviors than a single neural network (see [1–4] and references therein). Since the properties of dynamic behaviors are important in design and application of neural networks, stability is one of the hot topics [5,6]. Various different coupled neural networks can be found in the literature, e.g. coupled term with or without time delays, or hybrid both (see [7–14]). Some global stability criteria for arrays of linearly coupled delayed neural networks with nonsymmetric coupling are established on the basis of linear matrix inequality (LMI) method in [7]. By applying the theory of Kronecker product of matrices, the Wirtinger-based inequality, and the method of reciprocally convex combination, the authors derived some delay-dependent synchronization conditions for CNNs in terms of linear matrix inequalities in [1].

Although there are lots of researches for the exponential stability or stabilization problems of neural networks with time delay, most of them concern with the same dimension of the states. If a network is constructed by nodes with different state dimension, the network will exhibit different dynamical behaviours (see [15–17] and references therein). Dimensions of nodes are actually different in many practical situations, so such coupled complex networks need more in-depth study. The global stability of a class of CNNs with nodes of different dimensions is studied in [2], as an extension of this work, we discuss the exponential stability of CNNs.

© Springer International Publishing AG 2017
F. Cong et al. (Eds.): ISNN 2017, Part I, LNCS 10261, pp. 404–412, 2017.
DOI: 10.1007/978-3-319-59072-1_48

2   System Description and Preliminaries

Consider the coupled neural networks (CNNs) with nodes of different dimensions:

$$\frac{dx_i(t)}{dt} = -D_i x_i(t) + A_i f_i(x_i(t)) + B_i f_i(x_i(t - \tau_{i1})) + \alpha_i \sum_{j=1}^{N} g_{ij} C_{ij} x_j(t)$$

$$+\beta_i \sum_{j=1}^{N} g_{ij} \Gamma_{ij} x_j(t - \tau_{i2}), \tag{1}$$

where $i = 1, 2, \cdots, N$; $x_i(t) = (x_{i1}(t), x_{i2}(t), \cdots, x_{in_i}(t))^T \in \mathbb{R}^{n_i}$ is the state vector of the ith node; $A_i, B_i \in \mathbb{R}^{n_i \times n_i}$ are constant matrices representing the feedback matrix without and with time delays respectively; $D_i > 0$ is a constant and diagonal matrix; $f_i(*)$ is the activation function; $G = (g_{ij})_{N \times N}$ is an outer coupling matrix representing the coupling strength and the topological structure of the neural networks; $C_{ij}, \Gamma_{ij} \in \mathbb{R}^{n_i \times n_j}$ are inner coupling matrices representing the inner-linking strengths between the cells without and with time delays respectively; α_i, β_i are the strengths of the constant coupling and delayed coupling, respectively; τ_{i1}, τ_{i2} are the constant internal delays and coupling delays, respectively; and $0 \leq \tau_{i1} < \tau$, $0 \leq \tau_{i2} < \tau$.

Remark 1. The dimension of isolated node network is different from each other in this paper. Most results in the literature (e.g., [1,9,10,14,18]) concern with the CNNs with nodes of the same dimension.

Remark 2. The zero-row-sum condition $g_{ii} = - \sum\limits_{j=1, j \neq i}^{N} g_{ij}$ for G is required in some references (see [7,17]). However, the condition is removed in this paper.

The initial condition associated with (1) is given as follows:

$$x_{ij}(s) = \varphi_{ij}(s) \in \mathbb{C}([-\tau, 0], \mathbb{R}), \tag{2}$$

where $\tau_i = \max\{\tau_{i1}, \tau_{i2}\}$, $\tau = \max\{\tau_1, \tau_2, \cdots, \tau_N\}$, $i = 1, 2, \cdots, N$, $j = 1, 2, \cdots, n_i$.

Assumption 1. The activation function $f_i(x_i(t)) = (f_{i1}(x_{i1}(t)), f_{i2}(x_{i2}(t)), \cdots, f_{in_i}(x_{in_i}(t)))^T$ is Lipschitz continuous, i.e., there exist constants $w_{il} > 0$, such that $|f_{il}(\xi_1) - f_{il}(\xi_2)| \leq w_{il}|\xi_1 - \xi_2|$ holds for any $\xi_1, \xi_2 \in \mathbb{R}$, and $\xi_1 \neq \xi_2$, where $i = 1, 2, \cdots, N$; $l = 1, 2, \cdots, n_i$.

For convenience, the notations are givens as follows:
$M = n_1 + n_2 + \cdots + n_N$; I_{n_i} denotes $n_i \times n_i$ identity matrix; I denotes $M \times M$ identity matrix; $diag(\cdots)$ denotes a block-diagonal matrix; $\|y\| = \sqrt{y^T y}$ denotes the norm of y; $X(t) = (x_1^T(t), x_2^T(t), \cdots, x_N^T(t))^T$; $D = diag(D_1, D_2, \cdots, D_N)$; $A = diag(A_1, A_2, \cdots, A_N)$; $B = diag(B_1, B_2, \cdots, B_N)$;

$W_i = diag(w_{i1}, w_{i2}, \cdots, w_{in_i}); \quad W = diag(W_1, W_2, \cdots, W_N);$
$F(X(t - \tau_1)) = (f_1^T(x_1(t - \tau_{11})), f_2^T(x_2(t - \tau_{21})), \cdots, f_N^T(x_N(t - \tau_{N1})))^T;$
$X(t - \tau_2) = (x_1^T(t - \tau_{12}), x_2^T(t - \tau_{22}), \cdots, x_N^T(t - \tau_{N2}))^T;$
$\lambda_{\max}(*)$ and $\lambda_{\min}(*)$ respectively denote the maximum and minimum eigen-
value of $*$; $\begin{bmatrix} X & Y \\ * & Z \end{bmatrix}$ is defined as a matrix in form of $\begin{bmatrix} X & Y \\ Y^T & Z \end{bmatrix}$.

Hence, Eq. (1) can be rewritten as

$$\frac{dX(t)}{dt} = -DX(t) + AF(X(t)) + BF(X(t - \tau_1)) + HX(t) + KX(t - \tau_2), (3)$$

where

$$H = \begin{bmatrix} \alpha_1 g_{11} C_{11} & \alpha_1 g_{12} C_{12} & \cdots & \alpha_1 g_{1N} C_{1N} \\ \vdots & \vdots & \ddots & \vdots \\ \alpha_N g_{N1} C_{N1} & \alpha_N g_{N2} C_{N2} & \cdots & \alpha_N g_{NN} C_{NN} \end{bmatrix},$$

$$K = \begin{bmatrix} \beta_1 g_{11} \Gamma_{11} & \beta_1 g_{12} \Gamma_{12} & \cdots & \beta_1 g_{1N} \Gamma_{1N} \\ \vdots & \vdots & \ddots & \vdots \\ \beta_N g_{N1} \Gamma_{N1} & \beta_N g_{N2} \Gamma_{N2} & \cdots & \beta_N g_{NN} \Gamma_{NN} \end{bmatrix}.$$

Assume that X^* is the equilibrium point of the CNNs (1), then it satisfies

$$-DX^* + Af(X^*) + Bf(X^*) + HX^* + KX^* = 0, \tag{4}$$

where $X^* = (x_1^{*T}, x_2^{*T}, \cdots, x_N^{*T})^T$ and $x_i^* = (x_{i1}^*, x_{i2}^*, \cdots, x_{in_i}^*)^T$.
Define the linear coordinate transformation $E(t) = X(t) - X^*$, then the new
dynamical systems can be described as follows:

$$\frac{de_i(t)}{dt} = -D_i e_i(t) + A_i \phi_i(e_i(t)) + B_i \phi_i(e_i(t - \tau_{i1}))$$

$$+ \alpha_i \sum_{j=1}^{N} g_{ij} C_{ij} e_j(t) + \beta_i \sum_{j=1}^{N} g_{ij} \Gamma_{ij} e_j(t - \tau_{i2}). \tag{5}$$

That is

$$\frac{dE(t)}{dt} = -DE(t) + A\Phi(E(t)) + B\Phi(E(t - \tau_1)) + HE(t) + KE(t - \tau_2), \tag{6}$$

where $e_i(t) = (e_{i1}(t), e_{i2}(t), \cdots, e_{in_i}(t))^T = (x_{i1}(t) - x_{i1}^*, \cdots, x_{in_i}(t) - x_{in_i}^*)^T$,
$\phi_i(e_i(t)) = f_i(x_i(t)) - f_i(x_i^*), \quad \phi_i(e_i(t - \tau_{i1})) = f_i(x_i(t - \tau_{i1})) - f_i(x_i^*),$
$E(t) = (e_1^T(t), e_2^T(t), \cdots, e_N^T(t))^T, \quad E(t) = X(t) - X^*,$
$\Phi(E(t)) = (\phi_1^T(e_1(t)), \phi_2^T(e_2(t)), \cdots, \phi_N^T(e_N(t)))^T = F(X(t)) - F(X^*),$
$\Phi(E(t - \tau_1)) = F(X(t - \tau_1)) - F(X^*),$
$E(t - \tau_2) = (e_1^T(t - \tau_{12}), e_2^T(t - \tau_{22}), \cdots, e_N^T(t - \tau_{N2}))^T.$

Definition 1 [13]. For given $k > 0$, the dynamical system (1) is said to be exponentially stable, if there exist constant $Z > 0$ such that the following inequality

holds: $\|E(t)\|^2 \le Z \|\varphi\|^2 e^{-kt}$, for all initial conditions $e_{ij}(s)(i = 1, \cdots, N; j = 1, \cdots, n_i)$ of system (5) and any $t \ge T_0$ (sufficiently large $T_0 > 0$), where
$$\|\varphi\| = \sup_{-\tau \le s \le 0} \sqrt{\sum_{i=1}^{N} \sum_{j=1}^{n_i} |e_{ij}(s)|^2}.$$

Lemma 1 [6]. For any $x, y \in \mathbb{R}^n$ and positive definite matrix $Q \in \mathbb{R}^{n \times n}$, the following matrix inequality holds: $2x^T y \le x^T Q x + y^T Q^{-1} y$.

Lemma 2 (Schur Complement) [20]. The linear matrix inequality (LMI) $\begin{bmatrix} Q(x) & S(x) \\ S^T(x) & R(x) \end{bmatrix} > 0$, where $Q^T(x) = Q(x)$, $R^T(x) = R(x)$, is equivalent to $R(x) > 0$, and $Q(x) - S(x)R^{-1}(x)S^T(x) > 0$.

Lemma 3 [19]. If Q, R are real symmetric matrices, and $Q > 0$, $R \ge 0$, there exists a positive constant σ, such that the inequality holds: $-Q + \sigma R < 0$.

3 Exponential Stability Analysis

Theorem 1. Under the Assumption 1, the CNNs (1) is exponentially stable, if there exist $M \times M$ positive definite diagonal matrices P, Q, R, such that the following linear matrix inequality holds:

$$\Xi = \begin{bmatrix} \psi & PA & PB & PH & PK \\ * & Q & 0 & 0 & 0 \\ * & * & R & 0 & 0 \\ * & * & * & Q & 0 \\ * & * & * & * & Q \end{bmatrix} > 0, \tag{7}$$

where $\psi = PD + D^T P - WQW - WRW - 2Q$,
$P_i = diag(p_{i1}, p_{i2}, \cdots, p_{in_i})$, $P = diag(P_1, P_2, \cdots, P_N)$,
$Q_i = diag(q_{i1}, q_{i2}, \cdots, q_{in_i})$, $Q = diag(Q_1, Q_2, \cdots, Q_N)$,
$R_i = diag(r_{i1}, r_{i2}, \cdots, r_{in_i})$, $R = diag(R_1, R_2, \cdots, R_N)$.

Proof. From Lemmas 2 and 3 and inequality (7), we know that there exists a positive constant λ, such that

$$\Xi_1 = \begin{bmatrix} \psi^* & PA & PB & PH & PK \\ * & -Q & 0 & 0 & 0 \\ * & * & -R & 0 & 0 \\ * & * & * & -Q & 0 \\ * & * & * & * & -Q \end{bmatrix} < 0,$$

where $\psi^* = -\psi + \lambda P + (e^{\lambda \tau} - 1)(Q + WRW)$.

Consider the Lyapunov functional as

$$V(t) = V_1(t) + V_2(t) + V_3(t), \tag{8}$$

where $V_1(t) = e^{\lambda t} \sum_{i=1}^{N} e_i^T(t) P_i e_i(t)$, $V_2(t) = \sum_{i=1}^{N} \int_{t-\tau_{i2}}^{t} e^{\lambda(s+\tau)} e_i^T(s) Q_i e_i(s) ds$,

$V_3(t) = \sum_{i=1}^{N} \int_{t-\tau_{i1}}^{t} e^{\lambda(s+\tau)} \phi_i^T(e_i(s)) R_i \phi_i(e_i(s)) ds$.

Calculating the time derivatives of $V_1(t)$, $V_2(t)$ and $V_3(t)$, we have

$$\dot{V}_1(t) = e^{\lambda t}[\lambda E^T(t) P E(t) - E^T(t)(PD + D^T P)E(t)] + 2e^{\lambda t} E^T(t) P[A\Phi(E(t))$$
$$+ B\Phi(E(t-\tau_1)) + HE(t) + KE(t-\tau_2)], \tag{9}$$
$$\dot{V}_2(t) \leq e^{\lambda(t+\tau)} E^T(t) Q E(t) - e^{\lambda t} E^T(t-\tau_2) Q E(t-\tau_2), \tag{10}$$
$$\dot{V}_3(t) \leq e^{\lambda(t+\tau)} \Phi^T(E(t)) R\Phi(E(t)) - e^{\lambda t} \Phi^T(E(t-\tau_1)) R\Phi(E(t-\tau_1)). \tag{11}$$

From Lemma 1 and Assumption 1, we obtain

$$2E^T(t) PA\Phi(E(t)) \leq E^T(t) PAQ^{-1} A^T P E(t) + \Phi^T(E(t)) Q\Phi(E(t))$$
$$\leq E^T(t) PAQ^{-1} A^T P E(t) + E^T(t) WQW E(t),$$
$$2E^T(t) PB\Phi(E(t-\tau_1)) \leq E^T(t) PBR^{-1} B^T P E(t)$$
$$+ \Phi^T(E(t-\tau_1)) R\Phi(E(t-\tau_1)),$$
$$2E^T(t) PKE(t-\tau_2) \leq E^T(t) PKQ^{-1} K^T P E(t) + E^T(t-\tau_2) Q E(t-\tau_2),$$
$$2E^T(t) PHE(t) \leq E^T(t) PHQ^{-1} H^T P E(t) + E^T(t) Q E(t). \tag{12}$$

Substituting (9)–(12) into $\dot{V}(t)$, we have

$$\dot{V}(t) \leq e^{\lambda t}\{[\lambda E^T(t) P E(t) - E^T(t)(PD + D^T P)E(t)] + E^T(t) PAQ^{-1} A^T P E(t)$$
$$+ E^T(t) WQW E(t) + E^T(t) PBR^{-1} B^T P E(t)$$
$$+ \Phi^T(E(t-\tau_1)) R\Phi(E(t-\tau_1)) + E^T(t) PHQ^{-1} H^T P E(t) + E^T(t) Q E(t)$$
$$+ E^T(t) PKQ^{-1} K^T P E(t) + E^T(t-\tau_2) Q E(t-\tau_2)$$
$$+ e^{\lambda\tau} E^T(t) Q E(t) - E^T(t-\tau_2) Q E(t-\tau_2)$$
$$+ e^{\lambda\tau} E^T(t) WRW E(t) - \Phi^T(E(t-\tau_1)) R\Phi(E(t-\tau_1))\}$$
$$= e^{\lambda t}\{E^T(t) \Delta E(t)\}, \tag{13}$$

where $\Delta = \lambda P - PD - D^T P + WQW + e^{\lambda\tau} Q + e^{\lambda\tau} WRW + Q + PAQ^{-1} A^T P + PBR^{-1} B^T P + PHQ^{-1} H^T P + PKQ^{-1} K^T P$.

From Lemma 2, we know that $\Xi_1 < 0$ is equivalent to $\Delta < 0$. Therefore, $\dot{V}(t) \leq 0$ is negative-definite, and $V(t)$ is decreasing from $t=0$, and $V(t) \leq V(0)$.

From the initial conditions (2) of the coupled dynamical system (1), we obtain the initial conditions of the new dynamical system (5) are $e_{ij}(s) = \varphi_{ij}(s) - x^*_{ij}(s) \in \mathbb{C}\,([-\tau, 0], \mathbb{R})$. It follows from (8) that

$$V(0) = E^T(0) P E(0) + \sum_{i=1}^{N} \int_{-\tau_{i2}}^{0} e^{\lambda(s+\tau)} e_i^T(s) Q_i e_i(s) ds$$

$$+ \sum_{i=1}^{N} \int_{-\tau_{i1}}^{0} e^{\lambda(s+\tau)} \phi_i^T(e_i(s)) R_i \phi_i(e_i(s)) ds, \tag{14}$$

and

$$\sum_{i=1}^{N} \int_{-\tau_{i2}}^{0} e^{\lambda(s+\tau)} e_i^T(s) Q_i e_i(s) ds \le e^{\lambda\tau} \int_{-\tau}^{0} e^{\lambda s} E^T(s) Q E(s) ds$$

$$\le e^{\lambda\tau} \lambda_{\max}(Q) \int_{-\tau}^{0} e^{\lambda s} ds \cdot \|\varphi\|^2 = \frac{e^{\lambda\tau}-1}{\lambda} \cdot \lambda_{\max}(Q)\|\varphi\|^2, \tag{15}$$

$$\sum_{i=1}^{N} \int_{-\tau_{i2}}^{0} e^{\lambda(s+\tau)} \phi_i^T(e_i(s)) R_i \phi_i(e_i(s)) ds \le e^{\lambda\tau} \int_{-\tau}^{0} e^{\lambda s} \Phi^T(E(s)) R \Phi(E(s)) ds$$

$$\le e^{\lambda\tau} \int_{-\tau}^{0} e^{\lambda s} E^T(s) W R W E(s) ds \le \frac{e^{\lambda\tau}-1}{\lambda} \cdot \lambda_{\max}(WRW)\|\varphi\|^2. \tag{16}$$

Substituting (15)–(16) into (14), we get

$$V(0) \le \left\{ \lambda_{\max}(P) + \frac{e^{\lambda\tau}-1}{\lambda} \cdot [\lambda_{\max}(Q) + \lambda_{\max}(WRW)] \right\} \|\varphi\|^2. \tag{17}$$

From $e^{\lambda t} \cdot \lambda_{\min}(P) \|E(t)\|^2 \le V_1(t) \le V(t)$ and $V(t) \le V(0)$, we have

$$e^{\lambda t} \cdot \lambda_{\min}(P) \|E(t)\|^2 \le V(0). \tag{18}$$

Combining (17) with (18), we obtain $\|E(t)\|^2 \le Z \|\varphi\|^2 e^{-\lambda t}$, where $Z = \frac{1}{\lambda_{\min}(P)} \cdot \{\lambda_{\max}(P) + \frac{e^{\lambda\tau}-1}{\lambda} \cdot [\lambda_{\max}(Q) + \lambda_{\max}(WRW)]\}$. That is, the CNNs (1) is exponentially stable. This completes the proof.

4   Numerical Examples

Example 1. Consider the CNNs (1) that are composed of two isolated node networks, in which the parameters are given as follows:

$$N = 2, \quad n_1 = 2, \quad n_2 = 3, \quad f(x_{il}) = 2\cos(x_{il}) + 1, \quad i = 1, 2, \quad l = 1, ..., n_i,$$

$$G = \begin{bmatrix} 0.8 & -0.7 \\ -0.1 & 0.4 \end{bmatrix}, \quad D_1 = \begin{bmatrix} 35 & 0 \\ 0 & 36 \end{bmatrix}, \quad A_1 = \begin{bmatrix} 1 & -1 \\ 5 & -4 \end{bmatrix}, \quad B_1 = \begin{bmatrix} -3 & 1 \\ 4 & 5 \end{bmatrix},$$

$$D_2 = \begin{bmatrix} 33 & 0 & 0 \\ 0 & 21 & 0 \\ 0 & 0 & 34 \end{bmatrix}, \quad A_2 = \begin{bmatrix} -3 & 1 & 4 \\ 1 & 1 & -1 \\ -2 & 1 & 1 \end{bmatrix}, \quad B_2 = \begin{bmatrix} -5 & -5 & 4 \\ 1 & 2 & 4 \\ 2 & -4 & 3 \end{bmatrix},$$

$$C_{11} = \begin{bmatrix} -5 & -1 \\ 1 & -2 \end{bmatrix}, \quad C_{12} = \begin{bmatrix} 4 & -2 & -4 \\ -2 & -4 & 2 \end{bmatrix}, \quad \Gamma_{11} = \begin{bmatrix} -1 & -2 \\ -1 & 2 \end{bmatrix},$$

$$\Gamma_{12} = \begin{bmatrix} 3 & -5 & -5 \\ -1 & -3 & 1 \end{bmatrix}, \quad C_{21} = \begin{bmatrix} 1 & 4 \\ -5 & -2 \\ 2 & 4 \end{bmatrix}, \quad C_{22} = \begin{bmatrix} -4 & 2 & 1 \\ 3 & -5 & 4 \\ -4 & 4 & 4 \end{bmatrix},$$

$$\Gamma_{21} = \begin{bmatrix} 3 & -3 \\ 2 & 1 \\ -3 & 5 \end{bmatrix}, \quad \Gamma_{22} = \begin{bmatrix} -4 & 3 & -3 \\ -5 & 2 & 3 \\ -2 & 4 & 3 \end{bmatrix}, \quad \alpha_1 = 0.1, \quad \alpha_2 = 0.4,$$

and $\beta_1 = 0.9$, $\beta_2 = 0.6$, $\tau_{11} = 0.5$, $\tau_{12} = 1$, $\tau_{21} = 0.8$, $\tau_{22} = 0.4$.

It is easy to derive that $w_{il} = 2$, and $W = 2 * I_5$, for $i = 1, 2; l = 1, 2, \cdots, n_i$. Solving the linear matrix inequalities $\Xi > 0$ in Theorem 1, we obtain the following feasible solutions:

$P = diag\,(2.5743, 2.6871, 3.0825, 3.6538, 5.3390),$
$Q = diag\,(12.7886, 13.4662, 15.2123, 17.0441, 17.4001),$
$R = diag\,(11.6732, 13.6302, 17.5218, 19.9204, 24.7094).$

According to Theorem 1, the CNNs (1) can achieve the exponential stability. Given the random initial condition, the simulation results are plotted in Fig. 1, in which all state trajectories converge to the equilibrium point.

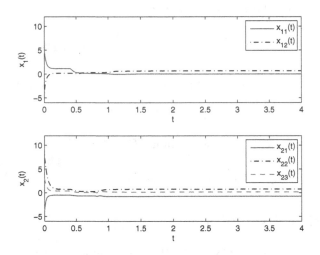

Fig. 1. The state trajectories of coupled neural networks

5   Conclusions

In this paper, we study the exponential stability of the coupled neural networks. The state dimensions in each isolated network can be different, so the derived results will have wider applicability. The criteria of exponential stability are derived on basis of the linear matrix inequality (LMI) and Lyapunov functional. The effectiveness of the proposed theoretical results has been demonstrated by a numerical simulation example.

Acknowledgments. This work was partly supported by grants from the National Natural Science Foundation of China (No.61572233, No.11471083), and the Fundamental Research Funds for the Central Universities (No. 21612443).

References

1. Zhang, J., Gao, Y.B.: Synchronization of coupled neural networks with time-varying delay. Neurocomputing **219**, 154–162 (2017)
2. Tan, M.C.: Stabilization of coupled time-delay neural networks with nodes of different dimensions. Neural Process. Lett. **43**(1), 255–268 (2016)
3. Liu, X.Y., Cao, J.D., Yu, W.W.: Nonsmooth finite-time synchronization of switched coupled neural networks. IEEE Trans. Cybern. **46**(10), 2360–2371 (2016)
4. Wang, J.L., Wu, H.N., Huang, T.W.: Pinning control for synchronization of coupled reaction-diffusion neural networks with directed topologies. IEEE Trans. Syst. Man. Cybern. Syst. **46**(8), 1109–1120 (2016)
5. Mahmoud, M.S., Selim, S.Z., Shi, P.: Global exponential stability criteria for neural networks with probabilistic delays. IET Control Theory Appl. **4**(11), 2405–2415 (2010)
6. Tan, M.C., Zhang, Y.N.: New sufficient conditions for global asymptotic stability of Cohen-Grossberg neural networks with time-varying delays. Nonlinear Anal. RWA **10**, 2139–2145 (2009)
7. Wang, Z.S., Zhang, H.G.: Synchronization stability in complex interconnected neural networks with nonsymmetric coupling. Neurocomputing **108**, 84–92 (2013)
8. Song, Q.K.: Synchronization analysis in an array of asymmetric neural networks with time-varying delays and nonlinear coupling. Appl. Math. Comput. **216**(5), 1605–1613 (2010)
9. Wang, G., Yin, Q., Shen, Y.: Exponential synchronization of coupled fuzzy neural networks with disturbances and mixed time-delays. Neurocomputing **106**, 77–85 (2013)
10. Lu, J.Q., Ho, D.W.C., Cao, J.D., Kurths, J.: Exponential synchronization of linearly coupled neural networks with impulsive disturbances. IEEE Trans. Neural Netw. **22**(2), 329–335 (2011)
11. Zhang, H.G., Gong, D.W., Wang, Z.S., Ma, D.Z.: Synchronization criteria for an array of neutral-type neural networks with hybrid coupling: a novel analysis approach. Neural Process. Lett. **35**(1), 29–45 (2012)
12. Cao, J.D., Chen, G.R., Li, P.: Global synchronization in an array of delayed neural networks with hybrid coupling. IEEE Trans. Syst. Man. Cybern. B **38**(2), 488–498 (2008)
13. Zhang, G.B., Wang, T., Li, T., Fei, S.M.: Exponential synchronization for delayed chaotic neural networks with nonlinear hybrid coupling. Neurocomputing **85**, 53–61 (2012)
14. Chen, G.R., Zhou, J., Liu, Z.R.: Global synchronization of coupled delayed neural networks and applications to chaotic CNN models. Int. J. Bifurcat. Chaos **14**(7), 2229–2240 (2004)
15. Wang, Y.H., Fan, Y.Q., Wang, Q.Y., Zhang, Y.: Stabilization and synchronization of complex dynamical networks with different dynamics of nodes via decentralized controllers. IEEE Trans. Circ. Syst. I **59**(8), 1786–1795 (2012)
16. Fan, Y.Q., Wang, Y.H., Zhang, Y., Wang, Q.R.: The synchronization of complex dynamical networks with similar nodes and coupling time-delay. Appl. Math. Comput. **219**(12), 6719–6728 (2013)
17. Tan, M.C., Tian, W.X.: Finite-time stabilization and synchronization of complex dynamical networks with nonidentical nodes of different dimensions. Nonlinear Dyn. **79**(1), 731–741 (2015)

18. Hua, C.C., Wang, Q.G., Guan, X.P.: Exponential stabilization controller design for interconnected time delay systems. Automatica **44**(10), 2600–2606 (2008)
19. Tan, M.C., Zhang, Y., Su, W.L., Zhang, Y.N.: Exponential stability of neural networks with variable delays. Int. J. Bifurcat. Chaos **20**(5), 1541–1549 (2010)
20. Thuan, M.V., Hien, L.V., Phat, V.N.: Exponential stabilization of non-autonomous delayed neural networks via Riccati equations. Appl. Math. Comput. **246**, 533–545 (2014)

Critical Echo State Networks that Anticipate Input Using Morphable Transfer Functions

Norbert Michael Mayer[✉]

Department of Electrical Engineering and AIM-HI,
National Chung Cheng University, Min-Hsiung, Chia-Yi, Taiwan
mikemayer@ccu.edu.tw

Abstract. The paper investigates a new type of truly critical echo state networks where individual transfer functions for every neuron can be modified to anticipate the expected next input. Deviations from expected input are only forgotten slowly in power law fashion. The paper outlines the theory, numerically analyzes a one neuron model network and finally discusses technical and also biological implications of this type of approach.

1 Introduction

Recurrent neural networks (RNNs) with input are examples for non-autonomous dynamical systems. One fundamental property is their dependence on their initial states (i.e. the initial settings of the recurrent layer neurons) with regard to one given input sequence. On one hand and for obvious reasons, networks that sensitively and for all future states depend on the setting of the initial state, will not work very well. On the other hand, if the network forgets too fast information about the past, it essentially works in the same way as a feed-forward network, and if that is good enough for the given task it is much easier to replace the recurrent network with the feed-forward solution. In the field of reservoir computing [1,2] and particularly in the case of echo state networks (ESNs) [3–5] much efforts have been undertaken in order to quantify to which extent an RNN is sensitive to the initial state. As a result several methods exist to detect the fine line between network parameters that – in combination with a given input sequence – finally result in a forgetting of the past within the network versus such parameter values for which essentially differences in the initial settings prevail in all the future. More interestingly, heuristics show that parameter settings that are near the border line, however, on the side of the forgetting type of networks, show the best performance for certain relevant tasks [6–8]. These networks are called near critical networks. An important notice from experimental biology is that also the statistics of dynamics of neurons in brain slices hint towards a near critical or even critical tuning of biological neurons in the brain [9]. Practical state of the art near critical networks usually require a certain margin towards the critical state because by design unexpected input deviations may push the state of the network over the critical point, in which case the performance deteriorates. In contrast to near critical networks, a relatively new study [10,11]

© Springer International Publishing AG 2017
F. Cong et al. (Eds.): ISNN 2017, Part I, LNCS 10261, pp. 413–420, 2017.
DOI: 10.1007/978-3-319-59072-1_49

brought up the idea to train the synaptic weights of the recurrent layer in the way that certain points (so-called epi-critical points, ECPs) within the transfer function are hit. If the network receives unexpected input, these special points are missed and result in an under-critical behavior. Given an expected input, the resulting network is tuned exactly to the critical point; other network features are power law forgetting of an unexpected input if it is succeeded by a sequence of expected input. Although that approach lines up a complete and new concept of designing critical ESNs, for practical purposes, there are still some problems. Most important, the proposed learning algorithm does not guarantee for a good performance of the network for many tasks. Different from that approach, the present work does not apply learning to the input weights and the recurrent weights. Rather, it proposes adaptive transfer functions for each neuron where the ECPs are always shifted towards the next expected transition point.

2   Echo State Networks and Criticality

The system is intended to resemble the dynamics of a biological recurrent neural network. We follow here the notation of Jäger:

$$\mathbf{x}_{lin,t} = \mathbf{W}\mathbf{x}_{t-1} + \mathbf{w}^{in}\mathbf{u}_t \tag{1}$$

$$\mathbf{x}_t = \theta\left(\mathbf{x}_{lin,t}\right) = f(\mathbf{x}_{t-1}, \mathbf{u}_t), \tag{2}$$

where $\mathbf{u}_t \in \mathbb{R}^n$ are items that form a left infinite time series that in total are called $\bar{\mathbf{u}}^\infty$ Supervised learning is done by linear regression using $\mathbf{x}_t$ as input [3], $\mathbf{x}_t \in \mathbb{R}^k$ represents activity in the hidden layer. $\mathbf{W}$ and $\mathbf{w}^{in}$ are matrices that represent (constant) synaptic weights. In principal, the complete time series $\mathbf{x}_t$ is determined by the tuple of the initial state $\mathbf{x}_0$ and a time series $\bar{\mathbf{u}}^\infty$. Comparing any two time series $\mathbf{y}_t = f(\mathbf{y}_{t-1}, \mathbf{u}_t)$ and $\mathbf{x}_t = f(\mathbf{x}_{t-1}, \mathbf{u}_t)$, such that $\mathbf{y}_0 \neq \mathbf{x}_0$, one can quantify how the difference develops over time. Important for the definition of echo state networks is the concept of state contraction that is if

$$\lim_{t \to \infty} ||\mathbf{y}_t - \mathbf{x}_t||_2 = 0, \tag{3}$$

i.e. the Euclidean distance converges to zero. In combination with the assumption that the processing of the neural network is acting in a time invariant manner, the concept has been named uniformly state contracting system [3]. Uniformly state contracting networks are echo state networks and thus capable to learn by linear regression. There has been some confusion about the definition of uniformly state contracting networks. Some researches define it to describe the dynamics of a network with regard to a specific input sequence (cf. [12]). Within this paper networks are called uniformly state contracting only if for a given network the relation of Eq. 3 holds for **any** input. Some calculus shows that a network with the dynamics of Eq. 2 is always an ESN if the recurrent connectivity matrix is orthogonal ($\mathbf{W} \in O(k)$) and the derivative of the transfer function is in the range $0 \leq \dot{\theta}(.) < 1$. Single inflection points where $\dot{\theta}(.) = 1$ may be permissible [3,11]. These points are important in the following considerations, and are called epi-critical points (ECPs). In analogy to calculating the Lyapunov exponent in autonomous dynamic systems, one can define also a Lyapunov exponent for ESNs with regard to a certain input sequence $\bar{\mathbf{u}}^\infty$ [11,13]

$$\Lambda_{\bar{\mathbf{u}}^\infty} = \lim_{|\mathbf{y}_0 - \mathbf{x}_0| \to 0} \lim_{t \to \infty} \frac{1}{t} \log \frac{\|\mathbf{y}_t - \mathbf{x}_t\|_2}{\|\mathbf{y}_0 - \mathbf{x}_0\|_2}. \tag{4}$$

If the Lyapunov exponent for all $\bar{\mathbf{u}}^\infty$ is negative, the network is shown to be uniformly state contracting and thus is an ESN. If the Lyapunov exponent for any input time series is positive the network is not an ESN. In addition, it is worthwhile to introduce the following definition: A network that for some input sequences has a Lyapunov exponent of zero but for which Eq. 3 still holds for any input sequence shall be called a **critical** ESN. According to this definition, an ESN is critical with respect to a particular input sequence. Technically, this can be achieved by training the network towards a setting where for some input sequences $\dot{\theta}(\mathbf{x}_{lin,t}) = 1$. A deeper insight into the theory behind that formula can be found in [10,11]. That is the aim of the training is to direct the input to those single inflection points of the transfer function.

3   Adaptive Transfer Functions

Instead of implementing plasticity on $\mathbf{W}$ and $\mathbf{w}^{in}$, the proposal for the current work is to implement the plasticity on the shape of the transfer function θ. Assuming that each neuron has an intrinsic mechanism to predict several possible values of $x_{lin,t}$ (that is one item in the vector $\mathbf{x}_{lin,t}$) the prediction happens before the input $\mathbf{u}_t$ is perceptible to the neuron (compare [14,15] for another scenario for self-prediction in ESNs). The neuron does not need to restrict itself to one prediction, instead a list of those values are possible. So, the ECPs Π_i should be shifted towards the predicted values of the linear response. The transfer function can thus be defined as

$$\theta(x_{lin}) = \tanh(x_{lin} - \Pi_i) + \tanh(\Pi_i), \tag{5}$$

for all values x_{lin} near any of the Π_i. Else, the transfer function is

$$\theta(x_{lin}) = \tanh(x_{lin}). \tag{6}$$

Figure 1 depicts two possible resulting transfer functions. The transfer function is designed in the way that $\dot{\theta}(\Pi_i) = 1$.

4   Synthetic One Neuron Reservoirs

In order to illustrate the proposed methodology, we designed a reservoir with one neuron that expects an alternating input of $u_t = 1$s and -1s. The following update equation can then be used

$$x_t = \theta\left(-\alpha x_{t-1} + (1 - \alpha \tanh(1))u_t\right), \tag{7}$$

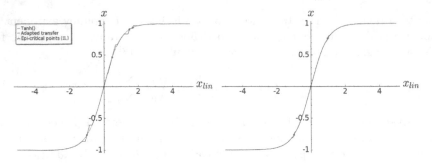

Fig. 1. The plots show examples of two versions of a transfer function $x = \theta(x_{lin})$ according to Eqs. 5–6 as blue curve where the ECPs can be organized in an adaptive way. The green curve depicts the underlying tanh function on which all ECPs are located. Green dots indicate ECPs for this particular example. The version at the right side is the version that is used in Sect. 3. Here areas of the transfer function where $\dot{\theta} = 0$ have been avoided, which leads to better results in the following graphs. Since the derivative of tanh(0) is 1, the point $(0,0)$ is marked as an additional ECP in both plots. (Color figure online)

where the factor α takes the role of a $\mathbf{W} \in R^{1 \times 1}$ matrix and the factor $(1 - \alpha \tanh(1))$ takes the role of $\mathbf{w}^{in}$. Note that if $\alpha = 1$ one may call $\mathbf{W} = O(1)$ a one dimensional orthogonal matrix. For any α and in case the network receives the expected alternating input the linear response converges to $x_{lin} = \pm 1$. So, two ECPs may be used $\Pi_0 = -1$ and $\Pi_1 = 1$. With regard to the resulting transfer function (that includes the ECPs) one can now measure the Lyapunov exponent according to Eq. 4 and for differing values of α. Figure 2 (left) depicts the results. One can see that although independent from α for the predicted input, the dynamics are always alternating 1s and -1s; this dynamic is only stable for the range of α between 0 and 1. In this range the network is an echo state network that is under-critical if $\alpha < 1$. Further numerical tests and also theoretical considerations show that at the point $\alpha = 1$ the network is still an ESN; however, it is critical. For values of $\alpha > 1$, the network is not an ESN anymore. The purpose of this work is to propose ESNs of $\alpha = 1$ as optimal critical ESNs. For comparison one may consider a one neuron version of a traditional near edge of chaos approach which basically relates to the common experience that the given theoretical limits for the ESNs can be significantly overtuned for many practical time series. Those overtuned ESNs in many cases show a much better performance than those that actually obey Jäger initial limit. So recently researchers came up with theoretical insights with regard to ESNs that are subject to a network and a particular input statistic [12] which fundamentally relate a network **and** an input statistic to a limit. One might assume that those approaches show similar properties as the one that has been presented above. However, for a good reason those approaches all are coined as 'near edge of chaos' approaches. In order to illustrate the problems that arise from those approaches one may consider what happens if those overtuned ESNs

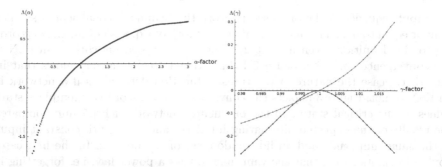

Fig. 2. Left: Depicted is the Lyapunov exponent for the example system of Eq. 7 for different values of α. At $\alpha = 1$ the Lyapunov exponent crosses zero if the input sequence is the expected alternating sequence of 1s and -1s. Right: Lyapunov exponents for two one neuron networks: In both cases the amplitude of the alternating input u_t is varied in a series of measurements of the Lyapunov exponent. In the case of the network according to Eq. 7 (blue) one can see that the Lyapunov exponent never becomes larger than 0. In the case of Eq. 8 (green) positive Lyapunov exponents occur if the amplitude of the input is larger than the critical value. (Color figure online)

are set exactly to the critical point. Here, just for the general understanding one may consider again a one neuron network and a tanh as a transfer function, so

$$x_t = \tanh(-bx_{t-1} + u_t). \tag{8}$$

Note that the ESC limit as outlined above requires that the recurrent connectivity should be $b < 1$. From the previous section one can now take the input time series $u_t = (-1)^t \pi/4$. Slightly tedious but basically simple calculus results in a critical value of $b \approx 2.344$ for the input time series, where $x_t \approx (-1)^t \times 0.757$. For the following results, the value of b is always set to the critical value. In this situation one can test for convergence of two slightly different initial conditions and one can get a power law decay of the difference. However, setting up the amplitude of the input just a tiny bit higher is going to result in two diverging time series x_t and y_t. In other words, if the conditions of the ESN are chosen to be exactly at the critical point it is possible that a not trained input sequence very near to the trained input sequence can turn the ESN into a state where Jäger's echo state condition is not fulfilled anymore, i.e. the Lyapunov exponent is positive for the given network in combination with these input sequences. In order to illustrate this difference numerical experiments (cf. Fig. 2 right side) have been done where both the networks according to Eq. 7 and Eq. 8 receive input with a slightly higher or lower input amplitude, i.e. an input sequence $\tilde{u}_t = \gamma \cdot u_t$ is perceived, where γ is a constant factor and u_t in both cases is the expected input that produces the critical behavior. Here, the amplitude γ is used as an example as an arbitrary continuous parameter that defines properties of the input sequences. If γ is equal to one, the resulting input sequences for both the examples of Eqs. 7 and 8 result in a critical dynamics with a Lyapunov exponent of 1. The difference between the two networks is that in the case of Eq. 8 positive Lyapunov are possible for γ-factors larger than one whereas for any input sequence of the proposed network the Lyapunov coefficient is smaller than or equal to one. This means that the network of Eq. 8 is not an ESN according to the definition given in Sect. 2, while the proposed network is an ESN if the convergence condition of Eq. 3 holds. Analytic calculus [11] shows that in the critical case the nature of the transfer function determines if a network is an ESN or not. Figure 3 depicts the convergence process of two exemplary start values at the critical state and the one neuron network of Eq. 7 and compares the results of the expected alternating input (1s and −1s) with constant input of the same amplitude and an iid. random set of 1s and −1s. In the first case, double log plots reveal that the vanishing follows a power law, i.e. forgetting is a slow process. Thus,

$$d(x_t, y_t) \propto t^{-c_a}, \tag{9}$$

with a constant c_a. The other type of input statistics result in faster forgetting. Here, every input value may be seen as an event that demands memory capacity. The result is effectively an exponential forgetting, i.e.

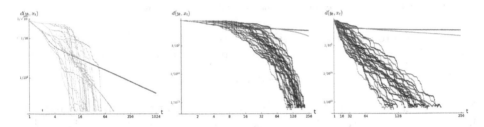

Fig. 3. The graphs depict different versions of the same data. Each red curve is the forgetting curve of the initial difference between 2 networks if the input sequence is alternating between 1s and −1s. The orange curves depict the forgetting curve for constant input with amplitude 1. The other curves show several iid. random sequences of 1s and −1s with equal probability. The left plot is a log-log plot with a focus on the alternating and constant inputs. The red curve converges towards a straight line, which indicates an underlying power law of this data. The curve resulting from constant input shows large values even at later time steps. However, the convergence appears to be faster than a power law, hence the curve bends toward the bottom. Finally random input shows the fastest convergence. The middle and right side graphs clearly indicate that all except for the alternating input show (roughly) an exponential decay. Both the middle and the right side plot show a scale down to 10^{-15}, which is about the limit of precision of double precision floating point numbers. Once a difference between two initial states reaches zero, it is beyond the logarithmic scale and not plotted anymore. So, the curve ends at that iteration. (Color figure online)

$$d(x_t, y_t) \asymp c_b{}^t, \tag{10}$$

with a constant c_b. This is the same result that one would also expect for all memory decay in under-critical networks. Exponential decay appears as a straight line in **semi-logarithmic** plots. The single neuron network simulations have been done by using double precision floating point variables, i.e. in 64 bits. Since the experimental setting in Fig. 3 organizes the initial difference between the 2 networks in an identical way as the randomness of the following inputs (that are identical to both networks) one would expect that the differences vanish over the time of 64 iterations. So one expects that the difference between the 2 network vanishes roughly in about 64 iterations. Considering the result of Fig. 3, one can see that indeed the difference vanishes in about 64 to 200 iterations. The fact that the forgetting process is slower than 64 iterations may indicate that several variant input histories can result in the same identical reservoir state.

5 Discussion

ESNs can be tuned to the critical value on the spot. At the same time it can be guaranteed that no input can push the network over the permissible limit. The setting of the ECPs leads to new insights into the network dynamics and relate those to information theory. If the next input is predictable, the next state of the network is going to hit one ECP exactly. One may interpret the resulting network in the way that predictable input is always directed to the ECPs and in this way prevented from consuming too much space (i.e. entropy) in the reservoir. Instead, **deviations** from predicted input materialize in the reservoir as distances to the ECPs. These deviations prevail than in a power-law fashion. This is true for both the present approach and the approach proposed in [10]. Different from [10], the approach here focuses on an adaptive transfer function. Overall there is very limited literature about adaptive transfer functions in neural networks (e.g. [16]). With regard to reservoir computing investigations into adaptive transfer functions may be promising. In the present approach, one target of the investigation was to find a way of training where the position of the transition point $\theta(\mathbf{x}_{lin,t})$ was unchanged, only the environment around it was transformed in the way that $\dot{\theta}(\mathbf{x}_{lin,t}) = 1$. This method in some sense changes the topology of the reservoir: By design, in every transition reservoirs lose information about previous inputs, however this information loss is not homogeneous and independent from the input time series. Rather it varies depending on features of the network, on the current input value and other parameters. Using the method of ECPs, the reservoir transforms then into a magnifying glass around those predicted states, which allows the network to look deep into the past if the incidence of aberrations from the predicted values are rare. So, aberrations from the predicted states can leave traces in the reservoir for very long times – if they are rare. In this sense the input-driven network turns into an **event-driven** network, i.e. a system that reacts strongly to an unpredicted event in contrast to the everyday and usual input. This can thought of as a lossy memory compression of a sliding window with an infinite but more and more lossy reproducibility of the far past.

Acknowledgements. This manuscript has been posted at arxiv.org. The authors thanks MOST of Taiwan for financial support and O. Obst for all his help.

References

1. Lukoševičius, M., Jäger, H.: Reservoir computing approaches to recurrent neural network training. Comput. Sci. Rev. **3**(3), 127–149 (2009)
2. Schrauwen, B., Verstraeten, D., Van Campenhout, J.: An overview of reservoir computing: theory, applications and implementations. In: Proceedings of the 15th European Symposium on Artificial Neural Networks. Citeseer (2007)
3. Jäger, H.: The "echo state" approach to analysing and training recurrent neural networks - with an erratum note. In: GMD Report 148, GMD German National Research Insitute for Computer Science (2010). http://www.gmd.de/People/Herbert.Jaeger/Publications.html
4. Jäger, H.: Adaptive nonlinear system identification with echo state networks. In: Proceedings of NIPS 2002, AA14 (2003)
5. Jäger, H., Maass, W., Principe, J.: Special issue on echo state networks and liquid state machines. Neural Netw. **20**(3), 287–289 (2007)
6. Natschläger, T., Bertschinger, N., Legenstein, R.: At the edge of chaos: real-time computations and self-organized criticality in recurrent neural networks. In: Advances in Neural Information Processing Systems, vol. 17 (2005)
7. Hajnal, M.A., Lörincz, A.: Critical echo state networks. In: Kollias, S.D., Stafylopatis, A., Duch, W., Oja, E. (eds.) ICANN 2006. LNCS, vol. 4131, pp. 658–667. Springer, Heidelberg (2006). doi:10.1007/11840817_69
8. Boedecker, J., Obst, O., Lizier, J., Mayer, N., Asada, M.: Information processing in echo state networks at the edge of chaos. Theory Biosci. **131**, 205–213 (2012)
9. Beggs, J., Plenz, D.: Neuronal avalanches in neocortical curcuits. J. Neurosci. **24**(22), 5216–5229 (2004)
10. Mayer, N.M.: Adaptive critical reservoirs with power law forgetting of unexpected input events. Neural Comput. **27**, 1102–1119 (2015)
11. Mayer, N.M.: Critical echo state networks that anticipate input using adaptive transfer functions (2016). http://arxiv.org/abs/1606.03674
12. Manjunath, G., Jaeger, H.: Echo state property linked to an input: exploring a fundamental characteristic of recurrent neural networks. Neural Comput. **25**(3), 671–696 (2013)
13. Wainrib, G., Galtier, M.N.: A local echo state property through the largest lyapunov exponent. Neural Netw. **76**, 39–45 (2016)
14. Mayer, N.M., Browne, M.: Self-prediction in echo state networks. In: Proceedings of the First International Workshop on Biological Inspired Approaches to Advanced Information Technology (BioAdIt2004), Lausanne (2004)
15. Mayer, N.M., Asada, M.: Is self-prediction a useful paradigm for echo state networks that are driven by robotic sensory input? In: 20th Neural Information Processing Systems Conference (NIPS 2006): Workshop on Echo State Networks and Liquid State Machines, H. Jaeger, W. Maass, Jose C. Principe (Organisers), December 2006
16. Wang, L., Chen, X., Li, S., Cai, X.: General adaptive transfer functions design for volume rendering by using neural networks. In: King, I., Wang, J., Chan, L.-W., Wang, D.L. (eds.) ICONIP 2006. LNCS, vol. 4233, pp. 661–670. Springer, Heidelberg (2006). doi:10.1007/11893257_74

INFERNO: A Novel Architecture for Generating Long Neuronal Sequences with Spikes

Alex Pitti[✉], Philippe Gaussier, and Mathias Quoy

Laboratoire ETIS, CNRS UMR 8051, University of Cergy-Pontoise,
ENSEA, Cergy, France
alexandre.pitti@u-cergy.fr
http://www.etis.ensea.fr

Abstract. Human working memory is capable to generate dynamically robust and flexible neuronal sequences for action planning, problem solving and decision making. However, current neurocomputational models of working memory find hard to achieve these capabilities since intrinsic noise is difficult to stabilize over time and destroys global synchrony. As part of the principle of free-energy minimization proposed by Karl Friston, we propose a novel neural architecture to optimize the free-energy inherent to spiking recurrent neural networks to regulate their activity. We show for the first time that it is possible to stabilize iteratively the long-range control of a recurrent spiking neurons network over long sequences. We identify our architecture as the working memory composed by the Basal Ganglia and the Intra-Parietal Lobe for action selection and we make some comparisons with other networks such as deep neural networks and neural Turing machines. We name our architecture INFERNO for *Iterative Free-Energy Optimization for Recurrent Neural Network. abstract* environment.

Keywords: Free-energy · Predictive coding · Working memory · Neuronal sequences · Spiking neurons · STDP · Basal ganglia · Cortico-basal loops · Habit learning

1 Introduction

Hierarchical plans and tree structures are a hallmark for human language and cognition [2]. But how the brain does to construct and retrieve them dynamically? For instance, the spontaneous activity within the network rapidly perturbs the neural dynamics and it is rather difficult then to maintain any stability for controlling long-range synchrony.

Making an analogy with the butterfly effect in chaos theory, small perturbations can destroy the dynamics even after few iterations. At reverse, exploiting this intrinsic noise can serve to make to converge neural dynamics to attractors, as a chaotic itinerancy [12]. In spiking neural networks, we propose that the tiny control of the neurons' sub-threshold activity (small events) can drive at another order of magnitude the generation of spikes (big events).

© Springer International Publishing AG 2017
F. Cong et al. (Eds.): ISNN 2017, Part I, LNCS 10261, pp. 421–428, 2017.
DOI: 10.1007/978-3-319-59072-1_50

As a novel mechanism, we propose to exploit this intrinsic noise to regulate the neural activity and the neurons' firing; an idea in line with the free-energy minimization principle proposed by Friston [3]. The minimization of the free energy means to predict for one particular policy its expected state and to optimize it over time in order to minimize future errors [4]. Our neural model is based on this principle of *Iterative Free-Energy Optimization for Recurrent Neural Networks*, and we named it INFERNO [9], see Fig. 1.

Moreover, this architecture is supported by several proposals and observations that consider the functional organization between the cortex with the subcortical regions (the basal ganglia); c.f. [1,6–8,10,11].

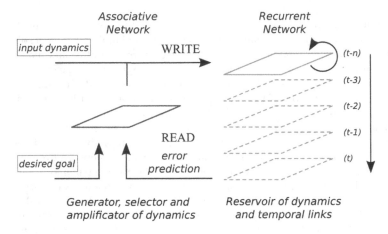

Fig. 1. Neural architecture INFERNO for iterative free-energy optimization of recurrent neural networks. This neural architecture is based on the coupled system formed by an associative memory (AM) and a recurrent neural network (RNN). INFERNO generates, selects and stores a set of rules in AM to assemble dynamically a neuronal sequence from a reservoir of dynamics in RNN toward a desired goal and based on free-energy minimization. It has some similarities with a turing machine that a table of instructions, write and read heads to generate a code from an infinite tape. (Color figure online)

2   Methods

We use the Rank-Order Coding (ROC) algorithm to model spiking neurons and the learning mechanism of Spike Timing-Dependent Plasticity [13]. The neurons' output y is computed by multiplying the rank order of the sensory signal vector x, $f(x) = \frac{1}{rank(x)}$, by the synaptic weights w so that $y = \sum wf(x)$. The function $rank(x)$ corresponds to the *argsort* function in Matlab and the synaptic weights of the neurons Δw are updated at each iteration.

The first neural architecture consists of one recurrent neural network of spiking neurons RNN arranged as in Fig. 1(a) in red. The second neural network consists on one associate neural network ANN as in Fig. 1(a) in blue. The output vector y of RNN is compared to one desired goal activity y^* to compute the error prediction $e = y^* - y$. Based on the variational error Δe, a stochastic descent gradient is used to generate the input vector x that will minimize error e on the long-term. The ANN learns to reconstruct the RNN input vector x based on the error prediction e on the output vector y so that $x = \sum vf(y)$. The coupled system composed of ANN and RNN attempts to minimize error dynamically. The former learns to control the latter system to generate a temporal sequence directly based on the feeded back activity. We describe below the stochastic search algorithm (Table 1).

Table 1. Free-energy optimization based on stochastic gradient descent to minimize prediction error.

Stochastic optimization as
Accumulation Evidences Process

#01 At time $t = 0$, initialize V, V^*, I
#02 choose randomly I_{search}
#03 compute $V_{search}(t)$ from $V(t)$, $I + I_{search}$
#04 **While** $t \leq horizon\_time$, repeat:
#05   compute $V_{search}(t + 1)$ from $V_{search}(t)$
#06   **If** $V^* - V \geq V^* - V_{search}(t + 1)$:
#07 $I = I + I_{search}$
#08 $V = V_{search}$
#09   **break**
#10 $t = t + 1$
#11 **Goto** *#02*

Prediction error E on the output vector V is used as a reinforcement signal to control the level of noise I_{search} to inject in the input dynamics I in order to explore local or global minima toward V^*.

3   Results

We propose at first to explain how ANN controls the neural dynamics of RNN with respect to one goal vector and error prediction relative to it. We plot in Fig. 2 the dynamics of RNN after several iteration of error descent gradient and explorative search till discovery of the solution (a). After few iterations, ANN finds the input dynamics that makes to converge RNN (b). That is, the coupled system self-organizes itself to minimize error toward a goal dynamically (c–d).

We propose to use the ANN-RNN architecture for controlling one 3 DOF arm motion toward goals that we give on the fly, see Fig. 3. Only three RNN neurons

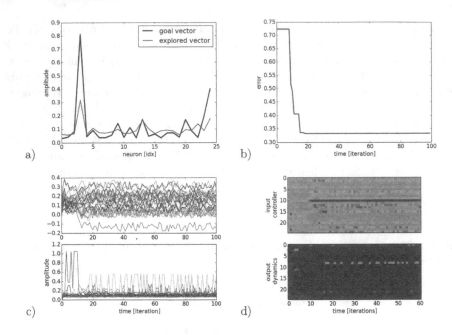

Fig. 2. Explored dynamics toward a goal. (a) ANN makes to converge the RNN neurons (in blue) to some desired dynamics (in black) thanks to prediction error. (b) the ANN prediction error diminishes to reach a local minimal value during the first 20 iterations. (c–d) dynamics of the ANN to control the dynamics of the RNN. (Color figure online)

control the normalized angles of the robot. We emphasize also hat ANN has no information about it, just about distance error between the location of the end effector and the location of the target. We change dynamically the target place and the arm is searching for a new configuration that minimizes error. ANN dynamics are changing everytime the target is placed at a new location (middle, upper chart), as do the three neurons of RNN, which converge to angles that reach the goal (middle, lower chart). Over time, each ANN neuron learns the dynamics that control the RNN, see Fig. 3. This neural architecture is capable to generate neuronal sequences based on habit learning once ANN has learned to control RNN (exploitation), before this happens, ANN searches to minimize error in a supervised manner toward a desired goal (exploration).

We can let the two coupled networks to self-organize their dynamics so that ANN triggers one specific neural trajectory in RNN, which triggers back one ANN neuron, the most problable one from the generated neural trajectory, see Fig. 4. For each neuron in ANN, ANN triggers one specific rule to direct RNN dynamics. The most probable ANN rule is then selected depending on the y vector found in RNN. In this way, the two systems control themselves to generate serial neuronal sequence several for hundreds of iterations without error. We emphasize that these dynamics are not completely learned but generated off-the-shelf.

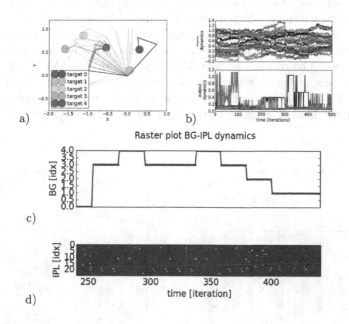

Fig. 3. RNN arm control based on predictive coding. On the left, RNN controls the three joint angles of one planar robot. ANN controls RNN based on error prediction toward targets. In the middle, ANN and RNN dynamics when switching dynamically to the euclidean distance to the goal location furnishes a reward to the motor neurons. On the right, ANN controls the neural activity of RNN over time.

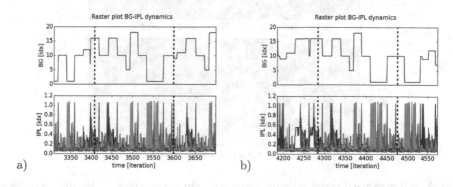

Fig. 4. ANN-RNN long-range neuronal sequences. Each neuron of ANN can control the neural dynamics of RNN for a relative long period (20 iterations in our case), which in return selects the correct ANN neuron to pursue the sequence. By doing so, the coupled system can produce the serial ordering of chunks toward multistep computation. The same sequence is generated within the dashed lines.

In an experience of serial ordering computation, ANN learns simple connecting rules to trigger the RNN dynamics. During this task, eventhough error and variability occur they are minimized dynamically to retrieve the goal vector. The three trajectories of spiking neurons present similar dynamics, mixing variability and robustness. The coupled system ANN-RNN formed by INFERNO presents some of the properties of a working memory to be robust and flexible at the same time. To some point, INFERNO appears to overcome the exploration and exploitation dilemna of machine learning algorithms thanks to predictive coding (Fig. 5).

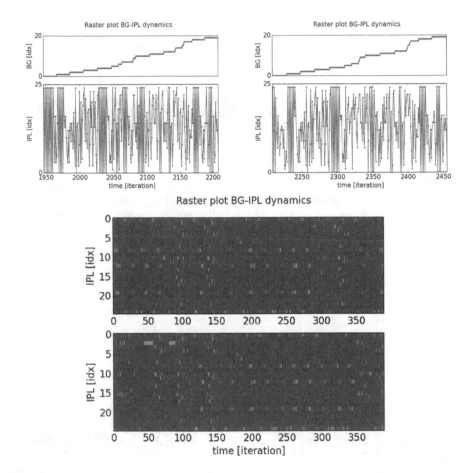

Fig. 5. Self-organized serial ordering sequence. Three examples of dynamic sequence ordering show that self-organization is not rigid and that variability occurs during time. The error minimization serves to rebind the two systems from each other.

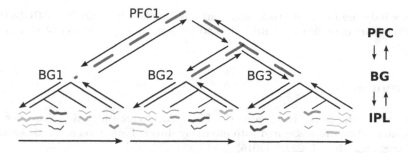

Fig. 6. Working memory for tree-like sequences. Unwrapped in time, INFERNO generates tree-like trajectories as a A* algorithm and as a virtually deep feed-forward neural network. RNN has neuronal primitives that ANN can selects, amplifies. This is similar to cortico-basal loops, having the basal ganglia to control the dynamics of the cortical maps and learns context-dependent rules depending on prefrontal cortex goal state.

4 Discussion

We propose a framework based on a coupled recurrent spiking neuronal system that achieves to perform long sequential planning by controlling the amplitude level of the spiking neurons through reinforcement signals. The control done is weak so that the propagated reinforced signals let the working memory plastic enough to converge to the desired internal states from various trajectories. Used in a robotic simulation, the neural dynamics can drive a three d.o.f. arm to reach online different locations.

The neural control is done by controlling tiny variations injected into the recurrent network that can iteratively change its dynamics to make it to converge to attractors. To this respect, our framework embodies some aspects of the free-energy optimization principle proposed by [4].

INFERNO generates, selects and stores a set of rules to assemble dynamically a neuronal sequence from a reservoir of dynamics toward a desired goal and based on free-energy minimization.

While the RNN working memory provides, stores, and manipulates representations; the ANN maps current states to courses of action. ANN can serve for selection of complex, sequenced actions at RNN. Thus, it can be interpreted as a repertoire of if-then rules or a set of stimulus-response associations to select appropriate cortical chains. To some points, we think it has some similarities with a Turing machine with a table of instructions, Write and Read heads to generate a code from an infinite tape [5,14]. Unwrapped in time, INFERNO generates tree-like trajectories as a A* algorithm and as a virtually deep feed-forward neural network, see Fig. 6. With INFERNO, we make a parallel with the cortico-basal system to construct a working memory. Iteratively, the basal ganglia forms 'habits' or rules that select cortical primitives in order to generate neuronal sequences based on a desired goal provided by the prefrontal cortex. The reinforcement signals given by the dopaminergic neuromodulator is similar to error prediction optimization or to free-energy minimization.

Acknowledgments. This work was partially funded by EQUIPEX-ROBOTEX (CNRS), chaire dexcellence CNRS-UCP and project Labex MME-DII (ANR11-LBX-0023-01).

References

1. Benedek, M., Jauk, E., Beaty, R., Fink, A., Koschutnig, K., Neubauer, A.: Brain mechanisms associated with internally directed attention and self-generated though. Sci. Rep. **6**, 22959 (2016)
2. Dehaene, S., Meyniel, F., Wacongne, C., Wang, L., Pallier, C.: The neural representation of sequences: from transition probabilities to algebraic patterns and linguistic trees. Neuron **88**, 2–19 (2015)
3. Friston, K.: A theory of cortical responses. Philos. Trans. R. Soc. Lond. Ser. B Biol. Sci. **360**(1456), 815–836 (2005)
4. Friston, K., Kilner, J.: A free energy principle for the brain. J. Physiol. Paris **100**, 70–87 (2006)
5. Graves, A.: Hybrid computing using a neural network with dynamic external memory. Nature **538**, 471–476 (2016)
6. Guthrie, M., Leblois, A., Garenne, A., Boraud, T.: Interaction between cognitive and motor cortico-basal ganglia loops during decision making: a computational study. J. Neurophysiol. **109**, 3025–3040 (2013)
7. Koechlin, E.: Prefrontal executive function and adaptive behavior in complex environments. Curr. Opin. Neurobiol. **37**, 1–6 (2016)
8. Miller, E.: The "working" of working memory. Dialog. Clin. Neurosci. **15**(4), 411–418 (2015)
9. Pitti, A., Gaussier, P., Quoy, M.: INFERNO: iterative free-energy optimization for recurrent neural networks. PLoS ONE **12**(3), e0173684 (2017)
10. Seger, C., Miller, E.: Category learning in the brain. Annu. Rev. Neurosci. **33**, 203–219 (2010)
11. Topalidou, M., Rougier, N.: [Re] interaction between cognitive and motor cortico-basal ganglia loops during decision making: a computational study. ReScience **1**(1), 1–6 (2015)
12. Tsuda, I., Fujii, H., Tadokoro, S., Yasuoka, T., Yamaguti, Y.: Chaotic itinerancy as a mechanism of irregular changes between synchronization and desynchronization in a neural network. J. Integr. Neurosci. **3**, 159–182 (2004)
13. Van Rullen, R., Gautrais, J., Delorme, A., Thorpe, S.: Face processing using one spike per neurone. BioSystems **48**, 229–239 (1998)
14. Zylberberg, A., Dehaene, S., Roelfsema, P., Sigman, M.: The human turing machine: a neural framework for mental programs. Trends Cogn. Sci. **7**, 293–300 (2011)

Global Exponential Stability for Matrix-Valued Neural Networks with Time Delay

Călin-Adrian Popa[✉]

Department of Computer and Software Engineering,
Polytechnic University Timişoara,
Blvd. V. Pârvan, No. 2, 300223 Timişoara, Romania
calin.popa@cs.upt.ro

Abstract. Complex-, quaternion-, and Clifford-valued neural networks can all be generalized to matrix-valued neural networks, which have matrix states. This paper derives a sufficient criterion given in the form of linear matrix inequalities that guarantees the global exponential stability of the equilibrium point for matrix-valued Hopfield neural networks with time delay. A simulation example demonstrates the effectiveness of the theoretical results.

Keywords: Matrix-valued neural networks · Global stability · Linear matrix inequality · Time delay

1 Introduction

Multidimensional neural networks have received an increasing interest during recent years. The simplest type of multidimensional neural networks are the complex-valued ones, which were proposed in the 1970's (see, for example, [22]). Due to their wide range of applications, including the ones in telecommunications, complex-valued signal processing, and image processing (see, for example, [5]), they have gained more interest in the past decade. Hyperbolic numbers, which also form a 2-dimensional algebra, represent the basis for hyperbolic-valued neural networks, which are another type of multidimensional networks, see [10,14]. The 4-dimensional algebra of quaternion numbers represents the domain of quaternion-valued neural networks. First developed in the 1990's, see [1], they have been applied since then, for example, to the 4-bit parity problem, chaotic time series prediction, and quaternion-valued signal processing.

First defined in [15,16], and later discussed, for example, in [2,10], were the Clifford-valued neural networks. Clifford algebras have dimension 2^n, where $n \geq 1$, and represent a generalization of the complex, hyperbolic, and quaternion algebras. Their applications in engineering and physics make them appealing as novel types of data representation also for the neural network domain.

The complex, hyperbolic, quaternion, and Clifford numbers all have a matrix representation. For instance, the complex number $a + ib$, $i = \sqrt{-1}$, can be

© Springer International Publishing AG 2017
F. Cong et al. (Eds.): ISNN 2017, Part I, LNCS 10261, pp. 429–438, 2017.
DOI: 10.1007/978-3-319-59072-1_51

represented as $\begin{pmatrix} a & b \\ -b & a \end{pmatrix}$, the hyperbolic number $a + ub$, $u^2 = 1$, $u \neq \pm 1$, as $\begin{pmatrix} a & b \\ b & a \end{pmatrix}$, and the quaternion $a + ib + jc + kd$, $i^2 = j^2 = k^2 = ijk = -1$, as $\begin{pmatrix} a & b & c & d \\ -b & a & -d & c \\ -c & d & a & -b \\ -d & -c & b & a \end{pmatrix}$.

This means that each of these algebras can be represented as a subalgebra of the square matrix algebra. This observation gave rise to the idea of generalizing the above-discussed neural networks to matrix-valued neural networks, first in their feedforward variant, see [17], and then in the recurrent Hopfield variant, see [18]. Due to the degree of generality given by their definition, these neural networks have the potential to be applied in the future to problems at which the traditional neural networks have performed poorly or failed.

On the other hand, at the beginning of the 1980's, Hopfield introduced an energy function with the purpose of studying the dynamical behavior of recurrent neural networks, see [6,7]. Starting then, recurrent Hopfield neural networks had numerous applications to image processing, speech processing, the synthesis of associative memories, control systems, pattern matching, signal processing, etc.

Multidimensional Hopfield neural networks were proposed in recent years. The complex-valued Hopfield neural networks have been discussed in [3,11,19, 20], the hyperbolic-valued ones in [8,10], the quaternion-valued ones in [13,21], and the Clifford-valued ones in [9,12,23].

All these facts led to the idea of introducing matrix-valued recurrent neural networks, which, as stated earlier, generalize all the before-mentioned models. Their potential applications include pattern matching and image processing, where data can be represented as matrices, and the synthesis of matrix-valued associative memories. Time delays unavoidably appear in real life implementations of neural networks, and they can lead to unwanted behavior such as oscillations and chaos. Because of this, we consider, in the present paper, matrix-valued Hopfield neural networks with time delay, and study the global exponential stability of their equilibrium point.

Thus, the outline of the rest of the paper is the following. Hopfield neural networks with matrix values are introduced, and one assumption and one useful lemma are presented in Sect. 2. A sufficient condition for the global exponential stability of matrix-valued Hopfield neural networks with time delay is established in Sect. 3. A numerical example that proves the effectiveness of the theoretical results is presented in Sect. 4. The conclusions are given in Sect. 5.

Notations: $\mathbb{R}$ denotes the real number set and $\mathbb{R}^n$ denotes the n dimensional Euclidean space. The algebra of real square matrices of order n is $\mathcal{M}_n$. A^T represents the transpose of matrix A. I_n denotes the identity matrix of order n. $A > 0$ $(A < 0)$ means that matrix A is positive definite (negative definite). $\lambda_{\min}(P)$ is defined as the smallest eigenvalue of positive definite matrix P. $\| \cdot \|$ is the vector Euclidean norm or the matrix Frobenius norm.

2   Preliminaries

A matrix-valued recurrent Hopfield neural network model has all the states, weights, and outputs given in the form of square matrices, i.e., they all belong to $\mathcal{M}_n$. The following system of differential equations describes this type of network:

$$\dot{X}_i(t) = -d_i X_i(t) + \sum_{j=1}^{N} A_{ij} f_j(X_j(t)) + \sum_{j=1}^{N} B_{ij} g_j(X_j(t-\tau)) + U_i, \qquad (1)$$

for $i \in \{1, \ldots, N\}$, where $X_i(t) \in \mathcal{M}_n$ is the state of the ith neuron at time t, $d_i \in \mathbb{R}$, $d_i > 0$, is the self-feedback weight of the ith neuron, $A_{ij} \in \mathcal{M}_n$ is the weight from the jth neuron to the ith neuron without time delay, $B_{ij} \in \mathcal{M}_n$ is the weight from the jth neuron to the ith neuron with time delay, $f_j : \mathcal{M}_n \to \mathcal{M}_n$ constitutes the nonlinear activation function of the jth neuron without time delay, $g_j : \mathcal{M}_n \to \mathcal{M}_n$ constitutes the nonlinear activation function of the jth neuron with time delay, $\tau \in \mathbb{R}$ is the time delay and we assume $\tau > 0$, and $U_i \in \mathcal{M}_n$ is the external input of neuron i, $\forall i, j \in \{1, \ldots, N\}$.

The derivative is simply the matrix whose elements are the derivatives of the entries of the state $X_i(t)$ with respect to t:

$$\dot{X}_i(t) = \frac{dX_i(t)}{dt} := \left(\frac{d([X_i(t)]_{ab})}{dt}\right)_{1 \leq a, b \leq n} = \left([\dot{X}_i(t)]_{ab}\right)_{1 \leq a, b \leq n}.$$

The activation functions f_j, g_j are each formed of n^2 functions f_j^{ab}, g_j^{ab} : $\mathcal{M}_n \to \mathbb{R}$, $1 \leq a, b \leq n$, $\forall j \in \{1, \ldots, N\}$:

$$f_j(X) = \left(f_j^{ab}(X)\right)_{1 \leq a, b \leq n}, \; g_j(X) = \left(g_j^{ab}(X)\right)_{1 \leq a, b \leq n}.$$

In order to study the stability of the above defined network, we need to make an assumption about the activation functions.

Assumption 1. *The following Lipschitz conditions are satisfied by the matrix-valued activation functions f_j, g_j, for any $X, X' \in \mathcal{M}_n$:*

$$\|f_j(X) - f_j(X')\| \leq l_j^f \|X - X'\|,$$

$$\|g_j(X) - g_j(X')\| \leq l_j^g \|X - X'\|,$$

where $l_j^f > 0$, $l_j^g > 0$ are the Lipschitz constants, $\forall j \in \{1, \ldots, N\}$. Furthermore, we denote $\overline{L_f} = diag(l_1^f I_{n^2}, l_2^f I_{n^2}, \ldots, l_N^f I_{n^2})$, $\overline{L_g} = diag(l_1^g I_{n^2}, l_2^g I_{n^2}, \ldots, l_N^g I_{n^2})$.

Firstly, we will transform the matrix-valued set of differential Eq. (1) into a real-valued one. For this, we expand each equation in (1) into n^2 real-valued equations:

$$[\dot{X}_i(t)]_{ab} = -d_i[X_i(t)]_{ab} + \sum_{j=1}^{N} \sum_{c=1}^{n} [A_{ij}]_{ac} f_j^{cb}(X_j(t))$$

$$+ \sum_{j=1}^{N} \sum_{c=1}^{n} [B_{ij}]_{ac} g_j^{cb}(X_j(t-\tau)) + [U_i]_{ab}, \qquad (2)$$

for $1 \leq a, b \leq n$, $i \in \{1, \ldots, N\}$. Now, using the vectorization operation, we can write

$$\text{vec}(\dot{X}_i(t)) = -d_i I_{n^2} \text{vec}(X_i(t)) + \sum_{j=1}^{N}(I_n \otimes A_{ij})\text{vec}(f_j(X_j(t)))$$

$$+ \sum_{j=1}^{N}(I_n \otimes B_{ij})\text{vec}(g_j(X_j(t - \tau))) + \text{vec}(U_i), \qquad (3)$$

for $i \in \{1, \ldots, N\}$. Finally, by denoting

$$Y(t) = (\text{vec}(X_1(t))^T, \text{vec}(X_2(t))^T, \ldots, \text{vec}(X_N(t))^T)^T,$$

$$\overline{D} = \text{diag}(d_1 I_{n^2}, d_2 I_{n^2}, \ldots, d_N I_{n^2}), \quad \overline{A} = \begin{bmatrix} I_n \otimes A_{11} & I_n \otimes A_{12} & \cdots & I_n \otimes A_{1N} \\ I_n \otimes A_{21} & I_n \otimes A_{22} & \cdots & I_n \otimes A_{2N} \\ \vdots & \vdots & \ddots & \vdots \\ I_n \otimes A_{N1} & I_n \otimes A_{N2} & \cdots & I_n \otimes A_{NN} \end{bmatrix},$$

$$\overline{B} = \begin{bmatrix} I_n \otimes B_{11} & I_n \otimes B_{12} & \cdots & I_n \otimes B_{1N} \\ I_n \otimes B_{21} & I_n \otimes B_{22} & \cdots & I_n \otimes B_{2N} \\ \vdots & \vdots & \ddots & \vdots \\ I_n \otimes B_{N1} & I_n \otimes B_{N2} & \cdots & I_n \otimes B_{NN} \end{bmatrix},$$

$\overline{f}(Y(t)) = (\text{vec}(f_1(X_1(t)))^T, \text{vec}(f_2(X_2(t)))^T, \ldots, \text{vec}(f_N(X_N(t)))^T)^T$, $\overline{g}(Y(t - \tau)) = (\text{vec}(g_1(X_1(t - \tau)))^T, \text{vec}(g_2(X_2(t - \tau)))^T, \ldots, \text{vec}(g_N(X_N(t - \tau)))^T)^T$, $\overline{U} = (\text{vec}(U_1)^T, \text{vec}(U_2)^T, \ldots, \text{vec}(U_N)^T)^T$, with the simplifying notations $Y = Y(t)$ and $Y^\tau = Y(t - \tau)$, system (1) becomes

$$\dot{Y} = -\overline{D}Y + \overline{A}\,\overline{f}(Y) + \overline{B}\,\overline{g}(Y^\tau) + \overline{U}. \qquad (4)$$

Now shifting to the origin the equilibrium point of (4), the system (4) becomes

$$\dot{\tilde{Y}} = -\overline{D}\tilde{Y} + \overline{A}\,\tilde{f}(\tilde{Y}) + \overline{B}\,\tilde{g}(\tilde{Y}^\tau), \qquad (5)$$

where $\tilde{f}(\tilde{Y}) = \overline{f}(\tilde{Y} + \hat{Y}) - \overline{f}(\hat{Y})$ and $\tilde{g}(\tilde{Y}^\tau) = \overline{g}(\tilde{Y}^\tau + \hat{Y}) - \overline{g}(\hat{Y})$.

Remark 1. The system (5) is equivalent with the system (1), which means that any property proved about system (5) will also hold for system (1). For this reason, from now on we will only study stability properties for the origin of system (5).

Remark 2. It can be clearly seen from the above derivation that the Hopfield neural network with matrix values, defined in (1), is not equivalent with a general nN-dimensional real-valued Hopfield neural network, because, for such a network, the matrices $\overline{A}$ and $\overline{B}$ would be general unconstrained matrices, and wouldn't have the particular form given above.

We will also need the following lemma:

Lemma 1 [4]. *The following inequality holds for any positive definite matrix* $M \in \mathcal{M}_{n^2 N}$ *and vector function* $Y : [a, b] \to \mathbb{R}^{n^2 N}$:

$$\left(\int_a^b Y(u)du \right)^T M \left(\int_a^b Y(u)du \right) \leq (b-a) \int_a^b Y^T(u)MY(u)du,$$

in which the integrals are well defined.

3   Main Results

We give a sufficient condition that ensures the global exponential stability of the origin of system (5).

Theorem 1. *If Assumption 1 holds, then the origin of system (5) is globally exponentially stable if there are positive definite matrices* P, Q_1, Q_2, Q_3, S_1, S_2, S_3, S_4, *positive block-diagonal matrices* R_1, R_2, R_3, R_4, *all from* $\mathcal{M}_{n^2 N}$, *and* $\varepsilon > 0$, *so that the following linear matrix inequality (LMI) is true*

$$(\Pi)_{9 \times 9} < 0, \tag{6}$$

where $\Pi_{1,1} = 2\varepsilon P - P\overline{D} - \overline{D}P + Q_1 + \tau S_2 - \tau^{-1}e^{-2\varepsilon\tau}S_1 + \tau\overline{D}S_1\overline{D} + \overline{L_f}^T R_1\overline{L_f} + \overline{L_g}^T R_3\overline{L_g}$, $\Pi_{1,2} = \tau^{-1}e^{-2\varepsilon\tau}S_1$, $\Pi_{1,3} = P\overline{A} - \tau\overline{D}S_1\overline{A}$, $\Pi_{1,6} = P\overline{B} - \tau\overline{D}S_1\overline{B}$, $\Pi_{2,2} = -e^{-2\varepsilon\tau}Q_1 - \tau^{-1}e^{-2\varepsilon\tau}S_1 + \overline{L_f}^T R_2\overline{L_f} + \overline{L_g}^T R_4\overline{L_g}$, $\Pi_{3,3} = Q_2 + \tau S_3 - R_1 + \tau\overline{A}^T S_1\overline{A}$, $\Pi_{3,6} = \tau\overline{A}^T S_1\overline{B}$, $\Pi_{4,4} = -e^{-2\varepsilon\tau}Q_2 - R_2$, $\Pi_{5,5} = Q_3 + \tau S_4 - R_3$, $\Pi_{6,6} = -e^{-2\varepsilon\tau}Q_3 - R_4 + \tau\overline{B}^T S_1\overline{B}$, $\Pi_{7,7} = -\tau^{-1}e^{-2\varepsilon\tau}S_2$, $\Pi_{8,8} = -\tau^{-1}e^{-2\varepsilon\tau}S_3$, $\Pi_{9,9} = -\tau^{-1}e^{-2\varepsilon\tau}S_4$.

Proof. Consider the Lyapunov-Krasovskii functional

$$\begin{aligned}
V(\tilde{Y}(t)) = {} & e^{2\varepsilon t}\tilde{Y}^T(t)P\tilde{Y}(t) \\
& + \int_{t-\tau}^t e^{2\varepsilon u}\tilde{Y}^T(u)Q_1\tilde{Y}(u)du \\
& + \int_{t-\tau}^t e^{2\varepsilon u}\tilde{f}^T(\tilde{Y}(u))Q_2\tilde{f}(\tilde{Y}(u))du \\
& + \int_{t-\tau}^t e^{2\varepsilon u}\tilde{g}^T(\tilde{Y}(u))Q_3 g(\tilde{Y}(u))du \\
& + \int_{-\tau}^0 \int_{t+\theta}^t e^{2\varepsilon u}\dot{\tilde{Y}}^T(u)S_1\dot{\tilde{Y}}(u)du d\theta \\
& + \int_{-\tau}^0 \int_{t+\theta}^t e^{2\varepsilon u}\tilde{Y}^T(u)S_2\tilde{Y}(u)du d\theta \\
& + \int_{-\tau}^0 \int_{t+\theta}^t e^{2\varepsilon u}\tilde{f}^T(\tilde{Y}(u))S_3\tilde{f}(\tilde{Y}(u))du d\theta \\
& + \int_{-\tau}^0 \int_{t+\theta}^t e^{2\varepsilon u}\tilde{g}^T(\tilde{Y}(u))S_4 g(\tilde{Y}(u))du d\theta.
\end{aligned}$$

Its derivative with respect to t for system (5) is

$$\dot{V}(\tilde{Y}) = e^{2\varepsilon t} \Big[2\varepsilon \tilde{Y}^T P \tilde{Y} + \dot{\tilde{Y}}^T P \tilde{Y} + \tilde{Y}^T P \dot{\tilde{Y}} + \tilde{Y}^T Q_1 \tilde{Y} - e^{-2\varepsilon\tau} \tilde{Y}^{\tau T} Q_1 \tilde{Y}^\tau$$

$$+ \tilde{f}^T(\tilde{Y}) Q_2 \tilde{f}(\tilde{Y}) - e^{-2\varepsilon\tau} \tilde{f}^T(\tilde{Y}^\tau) Q_2 \tilde{f}(\tilde{Y}^\tau) + \tilde{g}^T(\tilde{Y}) Q_3 \tilde{g}(\tilde{Y})$$

$$- e^{-2\varepsilon\tau} \tilde{g}^T(\tilde{Y}^\tau) Q_3 \tilde{g}(\tilde{Y}^\tau) + \tau \dot{\tilde{Y}}^T S_1 \dot{\tilde{Y}} - \int_{t-\tau}^t e^{2\varepsilon(u-t)} \dot{\tilde{Y}}^T(u) S_1 \dot{\tilde{Y}}(u) du$$

$$+ \tau \tilde{Y}^T S_2 \tilde{Y} - \int_{t-\tau}^t e^{2\varepsilon(u-t)} \tilde{Y}^T(u) S_2 \tilde{Y}(u) du + \tau \tilde{f}^T(\tilde{Y}) S_3 \tilde{f}(\tilde{Y})$$

$$- \int_{t-\tau}^t e^{2\varepsilon(u-t)} \tilde{f}^T(\tilde{Y}(u)) S_3 \tilde{f}(\tilde{Y}(u)) du + \tau \tilde{g}^T(\tilde{Y}) S_4 \tilde{g}(\tilde{Y})$$

$$- \int_{t-\tau}^t e^{2\varepsilon(u-t)} \tilde{g}^T(\tilde{Y}(u)) S_4 g(\tilde{Y}(u)) du \Big]$$

$$\leq e^{2\varepsilon t} \Big[2\varepsilon \tilde{Y}^T P \tilde{Y} + (-\overline{D}\tilde{Y} + \overline{A}\,\tilde{f}(\tilde{Y}) + \overline{B}\,\tilde{g}(\tilde{Y}^\tau))^T P \tilde{Y} + \tilde{Y}^T P(-\overline{D}\tilde{Y}$$

$$+ \overline{A}\,\tilde{f}(\tilde{Y}) + \overline{B}\,\tilde{g}(\tilde{Y}^\tau)) + \tilde{Y}^T Q_1 \tilde{Y} - e^{-2\varepsilon\tau} \tilde{Y}^{\tau T} Q_1 \tilde{Y}^\tau + \tilde{f}^T(\tilde{Y}) Q_2 \tilde{f}(\tilde{Y})$$

$$- e^{-2\varepsilon\tau} \tilde{f}^T(\tilde{Y}^\tau) Q_2 \tilde{f}(\tilde{Y}^\tau) + \tilde{g}^T(\tilde{Y}) Q_3 \tilde{g}(\tilde{Y}) - e^{-2\varepsilon\tau} \tilde{g}^T(\tilde{Y}^\tau) Q_3 \tilde{g}(\tilde{Y}^\tau)$$

$$+ \tau \dot{\tilde{Y}}^T S_1 \dot{\tilde{Y}} - \tau^{-1} e^{-2\varepsilon\tau} \left(\int_{t-\tau}^t \dot{\tilde{Y}}(u) du \right)^T S_1 \left(\int_{t-\tau}^t \dot{\tilde{Y}}(u) du \right)$$

$$+ \tau \tilde{Y}^T S_2 \tilde{Y} - \tau^{-1} e^{-2\varepsilon\tau} \left(\int_{t-\tau}^t \tilde{Y}(u) du \right)^T S_2 \left(\int_{t-\tau}^t \tilde{Y}(u) du \right)$$

$$+ \tau \tilde{f}^T(\tilde{Y}) S_3 \tilde{f}(\tilde{Y}) - \tau^{-1} e^{-2\varepsilon\tau} \left(\int_{t-\tau}^t \tilde{f}(\tilde{Y}(u)) du \right)^T S_3 \left(\int_{t-\tau}^t \tilde{f}(\tilde{Y}(u)) du \right)$$

$$+ \tau \tilde{g}^T(\tilde{Y}) S_4 \tilde{g}(\tilde{Y}) - \tau^{-1} e^{-2\varepsilon\tau} \left(\int_{t-\tau}^t \tilde{g}(\tilde{Y}(u)) du \right)^T S_4 \left(\int_{t-\tau}^t \tilde{g}(\tilde{Y}(u)) du \right) \Big], \quad (7)$$

where we used Lemma 1 for the inequality.

Assumption 1 can be written as

$$\| f_j(X) - f_j(X') \| \leq l_j^f \| X - X' \|$$

$$\Leftrightarrow \| \text{vec}(f_j(X)) - \text{vec}(f_j(X')) \| \leq l_j^f \| \text{vec}(X) - \text{vec}(X') \|,$$

for $j \in \{1, \ldots N\}$. Taking into account this inequality (and the analogous one for the functions g_j), and the above notations, there exist positive block-diagonal matrices $R_1 = \text{diag}(r_1^1 I_{n^2}, r_2^1 I_{n^2}, \ldots, r_N^1 I_{n^2})$, $R_2 = \text{diag}(r_1^2 I_{n^2}, r_2^2 I_{n^2}, \ldots, r_N^2 I_{n^2})$, $R_3 = \text{diag}(r_1^3 I_{n^2}, r_2^3 I_{n^2}, \ldots, r_N^3 I_{n^2})$, $R_4 = \text{diag}(r_1^4 I_{n^2}, r_2^4 I_{n^2}, \ldots, r_N^4 I_{n^2})$, such that

$$0 \leq \tilde{Y}^T \overline{L_f}^T R_1 \overline{L_f} \tilde{Y} - \tilde{f}^T(\tilde{Y}) R_1 \tilde{f}(\tilde{Y}), \quad 0 \leq \tilde{Y}^{\tau T} \overline{L_f}^T R_2 \overline{L_f} \tilde{Y}^\tau - \tilde{f}^T(\tilde{Y}^\tau) R_2 \tilde{f}(\tilde{Y}^\tau),$$
$$(8)$$

$$0 \leq \tilde{Y}^T \overline{L_g}^T R_3 \overline{L_g} \tilde{Y} - \tilde{g}^T(\tilde{Y}) R_3 \tilde{g}(\tilde{Y}), \quad 0 \leq \tilde{Y}^{\tau T} \overline{L_g}^T R_4 \overline{L_g} \tilde{Y}^\tau - \tilde{g}^T(\tilde{Y}^\tau) R_4 \tilde{g}(\tilde{Y}^\tau).$$
$$(9)$$

Combining (8) and (9) with (7), yields

$$\dot{V}(\tilde{Y}) \le e^{2\varepsilon t} \zeta^T \Pi \zeta, \tag{10}$$

where

$$\zeta = \left[\tilde{Y}^T\ \tilde{Y}^{\tau T}\ \tilde{f}^T(\tilde{Y})\ \tilde{f}^T(\tilde{Y}^\tau)\ \tilde{g}^T(\tilde{Y})\tilde{g}^T(\tilde{Y}^\tau) \right.$$
$$\left. \left(\int_{t-\tau}^t \tilde{Y}(u)du \right)^T \left(\int_{t-\tau}^t \tilde{f}(\tilde{Y}(u))du \right)^T \left(\int_{t-\tau}^t \tilde{g}(\tilde{Y}(u))du \right)^T \right]^T,$$

and Π is given by (6). Also from condition (6) we have that $\Pi < 0$, so (10) becomes $\dot{V}(\tilde{Y}) < 0$, from which we infer that $V(\tilde{Y}(t))$ is strictly decreasing for $t \ge 0$. This fact, together with the definition of $V(\tilde{Y}(t))$, imply that

$$e^{2\varepsilon t}\lambda_{\min}(P)\|\tilde{Y}(t)\|^2 \le e^{2\varepsilon t}\tilde{Y}^T(t)P\tilde{Y}(t) \le V(t) \le V_0, \ \forall t \ge T,\ T \ge 0,$$

where $V_0 = \max\limits_{0 \le t \le T} V(t)$. Consequently, we have that

$$\|\tilde{Y}(t)\|^2 \le \frac{V_0}{e^{2\varepsilon t}\lambda_{\min}(P)} \Leftrightarrow \|\tilde{Y}(t)\| \le Me^{-\varepsilon t},\ \forall t \ge 0,$$

for $M = \sqrt{\frac{V_0}{\lambda_{\min}(P)}}$. Thus, we obtained the global exponential stability for the origin of system (5).

4   Numerical Example

Next, we give a numerical example to prove the correctness of the above-derived criterion.

Example 1. Let us consider the following delayed matrix-valued Hopfield neural network with two neurons:

$$\begin{cases} \dot{X}_1(t) = -d_1 X_1(t) + \sum_{j=1}^2 A_{1j} f_j(X_j(t)) + \sum_{j=1}^2 B_{1j} g_j(X_j(t-\tau)) + U_1, \\ \dot{X}_2(t) = -d_2 X_2(t) + \sum_{j=1}^2 A_{2j} f_j(X_j(t)) + \sum_{j=1}^2 B_{2j} g_j(X_j(t-\tau)) + U_2, \end{cases} \tag{11}$$

where $d_1 = d_2 = 11$,

$$A_{11} = \begin{bmatrix} 1 & 1 \\ 2 & 2 \end{bmatrix},\ A_{12} = \begin{bmatrix} 1 & -1 \\ -1 & 1 \end{bmatrix},\ A_{21} = \begin{bmatrix} -1 & 1 \\ 2 & -2 \end{bmatrix},\ A_{22} = \begin{bmatrix} 1 & 2 \\ 2 & 1 \end{bmatrix},$$

$$B_{11} = \begin{bmatrix} -1 & 1 \\ 2 & -2 \end{bmatrix},\ B_{12} = \begin{bmatrix} 1 & -1 \\ -1 & 1 \end{bmatrix},\ B_{21} = \begin{bmatrix} 1 & 1 \\ 2 & 2 \end{bmatrix},\ B_{22} = \begin{bmatrix} 1 & 2 \\ 2 & 1 \end{bmatrix},$$

$$U_1 = \begin{bmatrix} 5 & -10 \\ 15 & 10 \end{bmatrix},\ U_2 = \begin{bmatrix} 5 & 15 \\ -10 & -15 \end{bmatrix},$$

$$f\left(([X]_{ab})_{1\leq a,b\leq n}\right) = \left(\frac{1}{1+e^{-[X]_{ab}}}\right)_{1\leq a,b\leq n},$$

$$g\left(([X]_{ab})_{1\leq a,b\leq n}\right) = \left(\frac{1-e^{-[X]_{ab}}}{1+e^{-[X]_{ab}}}\right)_{1\leq a,b\leq n},$$

from which we get that $l_1^f = l_2^f = \frac{1}{2}$ and $l_1^g = l_2^g = 2$.

The state trajectories of the elements of matrices X_1 and X_2 are given in Fig. 1, for four initial values. The time delay was taken to be $\tau = 0.5$.

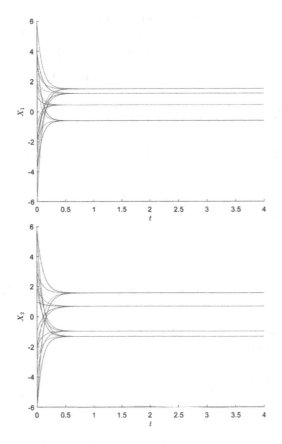

Fig. 1. State trajectories of elements of X_1 and X_2 in Example 1

Solving the LMI condition (6) in Theorem 1, we obtain that the equilibrium point of system (11) is globally exponentially stable for $\varepsilon = 0.2$, $R_1 = \text{diag}(3.8444I_4, 4.3276I_4)$, $R_2 = \text{diag}(0.4434I_4, 0.3695I_4)$, $R_3 = \text{diag}(1.8953I_4, 1.1550I_4)$, $R_4 = \text{diag}(0.3570I_4, 0.2839I_4)$,

$$P = \begin{bmatrix} 1.2130 & 0.0891 & 0 & 0 & 0.0534 & -0.0414 & 0 & 0 \\ 0.0891 & 1.0538 & 0 & 0 & -0.0239 & 0.0199 & 0 & 0 \\ 0 & 0 & 1.2130 & 0.0891 & 0 & 0 & 0.0534 & -0.0415 \\ 0 & 0 & 0.0891 & 1.0538 & 0 & 0 & -0.0239 & 0.0199 \\ 0.0534 & -0.0239 & 0 & 0 & 0.8634 & -0.1098 & 0 & 0 \\ -0.0414 & 0.0199 & 0 & 0 & -0.1098 & 0.7639 & 0 & 0 \\ 0 & 0 & 0.0534 & -0.0239 & 0 & 0 & 0.8634 & -0.1098 \\ 0 & 0 & -0.0415 & 0.0199 & 0 & 0 & -0.1098 & 0.7639 \end{bmatrix}.$$

(The values of the other matrices are not given due to space limitations.)

5   Conclusions

A sufficient criterion expressed as a linear matrix inequality was given, which assures that the equilibrium point of delayed Hopfield neural networks with matrix values is globally exponentially stable, by making the assumption that the Lipschitz conditions are satisfied by the activation functions. The correctness of the proposed criterion was showed by providing a numerical example.

Because of their degree of generality, encompassing the complex-, hyperbolic-, quaternion-, and Clifford-valued particular cases, matrix-valued recurrent neural networks provide the potential for further study, both in terms of stability properties as well as in applications.

References

1. Arena, P., Fortuna, L., Muscato, G., Xibilia, M.: Multilayer perceptrons to approximate quaternion valued functions. Neural Netw. **10**(2), 335–342 (1997)
2. Buchholz, S., Sommer, G.: On Clifford neurons and Clifford multi-layer perceptrons. Neural Netw. **21**(7), 925–935 (2008)
3. Chen, X., Zhao, Z., Song, Q., Hu, J.: Multistability of complex-valued neural networks with time-varying delays. Appl. Math. Comput. **294**, 18–35 (2017)
4. Gu, K.: An integral inequality in the stability problem of time-delay systems. In: Proceedings of the 39th IEEE Conference on Decision and Control, pp. 2805–2810 (2000)
5. Hirose, A.: Complex-Valued Neural Networks, Studies in Computational Intelligence, vol. 400. Springer, Heidelberg (2012)
6. Hopfield, J.J.: Neural networks and physical systems with emergent collective computational abilities. Proc. Nat. Acad. Sci. U.S.A. **79**(8), 2554–2558 (1982)
7. Hopfield, J.J.: Neurons with graded response have collective computational properties like those of two-state neurons. Proc. Nat. Acad. Sci. U.S.A. **81**(10), 3088–3092 (1984)
8. Kobayashi, M.: Hyperbolic Hopfield neural networks. IEEE Trans. Neural Netw. Learn. Syst. **24**(2), 335–341 (2013)
9. Kuroe, Y.: Models of Clifford recurrent neural networks and their dynamics. In: International Joint Conference on Neural Networks (IJCNN), pp. 1035–1041. IEEE (2011)

10. Kuroe, Y., Tanigawa, S., Iima, H.: Models of Hopfield-type clifford neural networks and their energy functions - hyperbolic and dual valued networks -. In: Lu, B.-L., Zhang, L., Kwok, J. (eds.) ICONIP 2011. LNCS, vol. 7062, pp. 560–569. Springer, Heidelberg (2011). doi:10.1007/978-3-642-24955-6_67

11. Liu, X., Chen, T.: Global exponential stability for complex-valued recurrent neural networks with asynchronous time delays. IEEE Trans. Neural Netw. Learn. Syst. **27**(3), 593–606 (2016)

12. Liu, Y., Xu, P., Lu, J., Liang, J.: Global stability of Clifford-valued recurrent neural networks with time delays. Nonlinear Dyn. **84**(2), 767–777 (2016)

13. Liu, Y., Zhang, D., Lu, J., Cao, J.: Global μ-stability criteria for quaternion-valued neural networks with unbounded time-varying delays. Inf. Sci. **360**, 273–288 (2016)

14. Nitta, T., Buchholz, S.: On the decision boundaries of hyperbolic neurons. In: International Joint Conference on Neural Networks (IJCNN), pp. 2974–2980. IEEE (2008)

15. Pearson, J., Bisset, D.: Back propagation in a Clifford algebra. Int. Conf. Artif. Neural Netw. **2**, 413–416 (1992)

16. Pearson, J., Bisset, D.: Neural networks in the Clifford domain. In: International Conference on Neural Networks, vol. 3, pp. 1465–1469. IEEE (1994)

17. Popa, C.-A.: Matrix-valued neural networks. In: Matoušek, R. (ed.) Mendel 2015. AISC, vol. 378, pp. 245–255. Springer, Cham (2015). doi:10.1007/978-3-319-19824-8_20

18. Popa, C.-A.: Matrix-valued hopfield neural networks. In: Cheng, L., Liu, Q., Ronzhin, A. (eds.) ISNN 2016. LNCS, vol. 9719, pp. 127–134. Springer, Cham (2016). doi:10.1007/978-3-319-40663-3_15

19. Song, Q., Yan, H., Zhao, Z., Liu, Y.: Global exponential stability of complex-valued neural networks with both time-varying delays and impulsive effects. Neural Netw. **79**, 108–116 (2016)

20. Song, Q., Zhao, Z.: Stability criterion of complex-valued neural networks with both leakage delay and time-varying delays on time scales. Neurocomputing **171**, 179–184 (2016)

21. Valle, M.: A novel continuous-valued quaternionic Hopfield neural network. In: Brazilian Conference on Intelligent Systems (BRACIS), pp. 97–102. IEEE, October 2014

22. Widrow, B., McCool, J., Ball, M.: The complex LMS algorithm. Proc. IEEE **63**(4), 719–720 (1975)

23. Zhu, J., Sun, J.: Global exponential stability of Clifford-valued recurrent neural networks. Neurocomputing **173**, Part 3, 685–689 (2016)

Global Asymptotic Stability
for Octonion-Valued Neural Networks
with Delay

Călin-Adrian Popa[✉]

Department of Computer and Software Engineering,
Polytechnic University Timişoara,
Blvd. V. Pârvan, No. 2, 300223 Timişoara, Romania
calin.popa@cs.upt.ro

Abstract. Over the last few years, neural networks with values in multi-dimensional domains have been intensely studied. This paper introduces octonion-valued neural networks with delay, for which the states and weights are octonions. The octonion algebra represents a non-associative normed division algebra which generalizes the complex and quaternion algebras and doesn't fall into the category of Clifford algebras, which are associative. A sufficient criterion is derived in terms of linear matrix inequalities that ensures the existence, uniqueness, and global asymptotic stability of the equilibrium point for the proposed networks. Finally, a simulation example illustrates the effectiveness of the theoretical results.

Keywords: Octonion-valued neural networks · Global stability · Linear matrix inequality · Time delay

1 Introduction

In recent years, neural networks with values in multidimensional domains have been studied with increasing interest. Complex-valued neural networks are the most popular form of multidimensional neural networks. Although they were first introduced in the 1970's (see, for example, [22]), they have received more attention in the 1990's and in the past decade, especially due to their numerous applications, ranging from those in complex-valued signal processing to those in telecommunications and image processing (see, for example, [6,13]).

Next, came the neural networks defined on the 4-dimensional quaternion algebra, which started gaining interest in the recent years. Quaternion-valued neural networks were first introduced in the 1990's, first as a generalization of the complex-valued neural networks, see [1,2]. But later, an increasing number of applications of quaternion-valued neural networks appeared in chaotic time series prediction, the 4-bit parity problem, and, very recently, in quaternion-valued signal processing.

One of the most general types of multidimensional neural networks are the ones defined on Clifford algebras, which have dimension 2^n, $n \geq 1$, and represent a generalization of both the complex and quaternion algebras. The numerous

© Springer International Publishing AG 2017
F. Cong et al. (Eds.): ISNN 2017, Part I, LNCS 10261, pp. 439–448, 2017.
DOI: 10.1007/978-3-319-59072-1_52

applications in physics and engineering of Clifford or geometric algebras, made them appealing also for the field of neural networks. Clifford-valued neural networks were defined in [15,16], and later discussed, for example, in [3]. Possible applications of neural networks with values in Clifford algebras include processing different geometric objects and applying different geometric models to data, because of the Clifford algebras' underlying connection with geometry.

A different generalization of the complex and quaternion numbers, which doesn't fall into the Clifford algebra category, is the 8-dimensional algebra of octonions. The easiest way to see this is by considering the fact that Clifford algebras are associative, whereas the octonion algebra is not. Nonetheless, it has an important property, especially for applications, which Clifford algebras don't have: the octonion algebra is a normed division algebra, which means that a norm and a multiplicative inverse can be defined on it. In fact, complex, quaternion and octonion algebras are the only normed division algebras that can be defined over the field of real numbers.

Octonions have many applications in physics and geometry (see [4,14]), and they have also been successfully applied to signal processing in the very recent years (see [18]). Taking all the above facts into consideration, defining octonion-valued neural networks seemed a promising idea, first in the form of feedforward networks [17]. Octonion-valued neural networks may be applied in signal processing and all other areas related to higher-dimensional object processing.

On the other hand, at the beginning of the 1980's, Hopfield had the idea of introducing an energy function in order to study the dynamics of fully connected recurrent neural networks, see [7,8]. He showed that combinatorial problems can be solved by using this type of networks. Since then, Hopfield neural networks have been applied to the synthesis of associative memories, image processing, speech processing, control systems, signal processing, pattern matching, etc.

Generalizations of the Hopfield neural networks to multidimensional domains appeared over the last few years. Complex-valued Hopfield networks were discussed in [10,19,20], quaternion-valued Hopfield networks in [12,21], and Clifford-valued Hopfield networks in [11,23]. Taking these facts into account, in this paper, we introduce octonion-valued neural networks with delay, which could be applied to solve octonion optimization problems.

The rest of the paper is organized as follows: Sect. 2 gives the definition of octonion-valued neural networks with delay, and an assumption and useful lemmas. A sufficient condition for the existence, uniqueness, and global asymptotic stability of the equilibrium point of these networks is given in Sect. 3. The effectiveness of the theoretical results is illustrated by a numerical example in Sect. 4. Section 5 presents the conclusions of the study.

Notations: $\mathbb{R}$ denotes the set of real numbers, $\mathbb{R}^n$ denotes the n dimensional Euclidean space, and $\mathbb{R}^{n \times n}$ the set of real matrices of dimension $n \times n$. A^T denotes the transpose of matrix A and $*$ denotes the symmetric terms in a matrix. I_n denotes the identity matrix of dimension n. $|| \cdot ||$ stands for the Euclidean vector norm or the induced matrix 2-norm. $A > 0$ $(A < 0)$ means that A is a positive definite (negative definite) matrix.

2 Preliminaries

We start by defining the algebra of octonions and highlighting some of its properties.

The algebra of octonions is defined as

$$\mathbb{O} := \left\{ x = \sum_{p=0}^{7} [x]_p e_p \,\middle|\, [x]_0, [x]_1, \dots, [x]_7 \in \mathbb{R} \right\},$$

where e_p represent the octonion units, $0 \le p \le 7$, and satisfy the following multiplication table

×	e_0	e_1	e_2	e_3	e_4	e_5	e_6	e_7
e_0	e_0	e_1	e_2	e_3	e_4	e_5	e_6	e_7
e_1	e_1	$-e_0$	e_3	$-e_2$	e_5	$-e_4$	$-e_7$	e_6
e_2	e_2	$-e_3$	$-e_0$	e_1	e_6	e_7	$-e_4$	$-e_5$
e_3	e_3	e_2	$-e_1$	$-e_0$	e_7	$-e_6$	e_5	$-e_4$
e_4	e_4	$-e_5$	$-e_6$	$-e_7$	$-e_0$	e_1	e_2	e_3
e_5	e_5	e_4	$-e_7$	e_6	$-e_1$	$-e_0$	$-e_3$	e_2
e_6	e_6	e_7	e_4	$-e_5$	$-e_2$	e_3	$-e_0$	$-e_1$
e_7	e_7	$-e_6$	e_5	e_4	$-e_3$	$-e_2$	e_1	$-e_0$

The addition of octonions is defined by $x + y = \sum_{p=0}^{7}([x]_p + [y]_p)e_p$, and the multiplication is given by the multiplication of the unit octonions shown in the above table. Scalar multiplication is given by $\alpha x = \sum_{p=0}^{7}(\alpha[x]_p)e_p$, and thus $\mathbb{O}$ is a real algebra. It can be verified using the multiplication table that $e_i e_j = -e_j e_i \ne e_j e_i$, $\forall i \ne j$, $0 < i, j \le 7$, which means that $\mathbb{O}$ is not commutative, and that $(e_i e_j)e_k = -e_i(e_j e_k) \ne e_i(e_j e_k)$, for i, j, k distinct, $0 < i, j, k \le 7$, or $e_i e_j \ne \pm e_k$, which shows that $\mathbb{O}$ is also not associative.

The conjugate of an octonion x is defined by $\bar{x} = [x]_0 e_0 - \sum_{p=1}^{7}[x]_p e_p$. Using the conjugate, the norm of an octonion can be defined as $||x|| = \sqrt{x\bar{x}} = \sqrt{\sum_{p=0}^{7}[x]_p^2}$, and the inverse of an octonion as $x^{-1} = \frac{\bar{x}}{||x||^2}$. Thus, $\mathbb{O}$ is a normed non-associative division algebra, unlike the 8-dimensional Clifford algebras, which are associative algebras, but not division algebras. In fact, the only three real division algebras that can be defined are the complex, quaternion, and octonion algebras.

We can now introduce octonion-valued Hopfield neural networks for which the states and weights are from $\mathbb{O}$. The following set of differential equations describes this type of networks:

$$\dot{x}_i(t) = -d_i x_i(t) + \sum_{j=1}^{N} a_{ij} f_j(x_j(t)) + \sum_{j=1}^{N} b_{ij} g_j(x_j(t - \tau)) + u_i, \tag{1}$$

for $i \in \{1, \ldots, N\}$, where $x_i(t) \in \mathbb{O}$ is the state of neuron i at time t, $d_i \in \mathbb{R}$, $d_i > 0$, is the self-feedback connection weight of neuron i, $a_{ij} \in \mathbb{O}$ is the weight connecting neuron j to neuron i without delay, $b_{ij} \in \mathbb{O}$ is the weight connecting neuron j to neuron i with delay, $f_j : \mathbb{O} \to \mathbb{O}$ is the nonlinear octonion-valued activation function of neuron j without delay, $g_j : \mathbb{O} \to \mathbb{O}$ is the nonlinear octonion-valued activation function of neuron j with delay, $\tau \in \mathbb{R}$ is the delay and we assume $\tau > 0$, and $u_i \in \mathbb{O}$ is the external input of neuron i, $\forall i, j \in \{1, \ldots, N\}$.

The derivative $\frac{dx_i(t)}{dt}$ is considered to be the octonion formed by the derivatives of each element $[x_i(t)]_p$ of the octonion $x_i(t)$ with respect to t: $\dot{x}_i(t) = \frac{dx_i(t)}{dt} := \sum_{p=0}^{7} \frac{d([x_i]_p)}{dt} e_p$. Thus, the above set of differential equations has values in $\mathbb{O}$, and the multiplication between the weights and the values of the activation functions is the octonion multiplication.

We need to make an assumption about the activation functions, in order to study the stability of the above defined network.

Assumption 1. *The octonion-valued activation functions f_j and g_j satisfy the following Lipschitz conditions:*

$$\|f_j(x) - f_j(x')\| \leq l_j^f \|x - x'\|, \ \forall x, x' \in \mathbb{O},$$

$$\|g_j(x) - g_j(x')\| \leq l_j^g \|x - x'\|, \ \forall x, x' \in \mathbb{O},$$

where $l_j^f > 0$ and $l_j^g > 0$ are the Lipschitz constants, $\forall j \in \{1, \ldots, N\}$. Moreover, we denote $\overline{L_f} = diag((l_1^f)^2 I_8, (l_2^f)^2 I_8, \ldots, (l_N^f)^2 I_8)$, $\overline{L_g} = diag((l_1^g)^2 I_8, (l_2^g)^2 I_8, \ldots, (l_N^g)^2 I_8)$.

We will first transform the octonion-valued differential Eq. (1) into real-valued ones. To do so, we will detail each equation in (1) into 8 real-valued equations:

$$[\dot{x}_i(t)]_p = -d_i [x_i(t)]_p + \sum_{j=1}^{N} \sum_{q=0}^{7} [a_{ij}]_{pq} [f_j(x_j(t))]_q$$

$$+ \sum_{j=1}^{N} \sum_{q=0}^{7} [b_{ij}]_{pq} [g_j(x_j(t - \tau))]_q + [u_i]_p, \tag{2}$$

for $0 \leq p \leq 7$, $i \in \{1, \ldots, N\}$, where $[x]_{pq}$ is an element of the matrix $mat(x)$, defined by

$$mat(x) := \begin{bmatrix} [x]_0 & -[x]_1 & -[x]_2 & -[x]_3 & -[x]_4 & -[x]_5 & -[x]_6 & -[x]_7 \\ [x]_1 & [x]_0 & -[x]_3 & [x]_2 & -[x]_5 & [x]_4 & [x]_7 & -[x]_6 \\ [x]_2 & [x]_3 & [x]_0 & -[x]_1 & -[x]_6 & -[x]_7 & [x]_4 & [x]_5 \\ [x]_3 & -[x]_2 & [x]_1 & [x]_0 & -[x]_7 & [x]_6 & -[x]_5 & -[x]_4 \\ [x]_4 & [x]_5 & [x]_6 & [x]_7 & [x]_0 & -[x]_1 & -[x]_2 & -[x]_3 \\ [x]_5 & -[x]_4 & [x]_7 & -[x]_6 & [x]_1 & [x]_0 & [x]_3 & -[x]_2 \\ [x]_6 & -[x]_7 & -[x]_4 & [x]_5 & [x]_2 & -[x]_3 & [x]_0 & [x]_1 \\ [x]_7 & [x]_6 & -[x]_5 & -[x]_4 & [x]_3 & [x]_2 & -[x]_1 & [x]_0 \end{bmatrix}.$$

Now, if we denote $\text{vec}(x) = ([x]_0, [x]_1, \ldots, [x]_7)^T$, the Eq. (2) can be written as

$$\text{vec}(\dot{x}_i(t)) = -d_i I_8 \text{vec}(x_i(t)) + \sum_{j=1}^{N} \text{mat}(a_{ij}) \text{vec}(f_j(x_j(t)))$$

$$+ \sum_{j=1}^{N} \text{mat}(b_{ij}) \text{vec}(g_j(x_j(t - \tau))) + \text{vec}(u_i), \tag{3}$$

for $i \in \{1, \ldots, N\}$. Denoting $w(t) = (\text{vec}(x_1(t))^T, \text{vec}(x_2(t))^T, \ldots, \text{vec}(x_N(t))$ $^T)^T$, $\overline{D} = \text{diag}(d_1 I_8, d_2 I_8, \ldots, d_N I_8)$, $\overline{A} = (\text{mat}(a_{ij}))_{1 \le i,j \le N}$, $\overline{B} = (\text{mat}(b_{ij}))_{1 \le i,j \le N}$, $\overline{f}(w(t)) = (\text{vec}(f_1(x_1(t)))^T, \text{vec}(f_2(x_2(t)))^T, \ldots, \text{vec}(f_N(x_N(t)))^T)^T$, $\overline{g}(w(t - \tau)) = (\text{vec}(g_1(x_1(t - \tau)))^T, \text{vec}(g_2(x_2(t - \tau)))^T, \ldots, \text{vec}(g_N(x_N(t - \tau)))^T)^T$, $\overline{u} = (\text{vec}(u_1)^T, \text{vec}(u_2)^T, \ldots, \text{vec}(u_N)^T)^T$, with the simplifying notations $w = w(t)$ and $w^\tau = w(t - \tau)$, system (1) becomes

$$\dot{w} = -\overline{D}w + \overline{A}\,\overline{f}(w) + \overline{B}\,\overline{g}(w^\tau) + \overline{u}. \tag{4}$$

Remark 1. The system (4) is equivalent with the system (1), which means that any property proven about system (4) will also hold for system (1). Because of this, from now on we will only study the existence, uniqueness, and global asymptotic stability of the equilibrium point of system (4).

We will also need the following lemmas:

Lemma 1 [5]. *If $H(w) : \mathbb{R}^{8N} \to \mathbb{R}^{8N}$ is a continuous map that satisfies the following conditions:*

(i) $H(w)$ is injective on $\mathbb{R}^{8N}$,
(ii) $\|H(w)\| \to \infty$ as $\|w\| \to \infty$,

then $H(w)$ is a homeomorphism of $\mathbb{R}^{8N}$ onto itself.

Lemma 2 [9]. *For any vectors x, $y \in \mathbb{R}^{8N}$, positive definite matrix $P \in \mathbb{R}^{8N \times 8N}$, and real constant $\varepsilon > 0$, the following linear matrix inequality (LMI) holds:*

$$2x^T y \le \varepsilon x^T P x + \frac{1}{\varepsilon} y^T P^{-1} y.$$

3 Main Results

In this section, we give an LMI-based sufficient condition for the existence, uniqueness, and global asymptotic stability of the equilibrium point for (4).

Theorem 1. *If Assumption 1 holds, then system (4) has a unique equilibrium point which is globally asymptotically stable if there exist real numbers $\varepsilon_1 > 0$ and $\varepsilon_2 > 0$, and positive definite matrix $P \in \mathbb{R}^{8N \times 8N}$ such that the following LMI holds*

$$\begin{bmatrix} P\overline{D} + \overline{D}P - \varepsilon_1\overline{L_f} - \varepsilon_2\overline{L_g} & P\overline{A} & P\overline{B} \\ * & \varepsilon_1 I_{8N} & 0 \\ * & * & \varepsilon_2 I_{8N} \end{bmatrix} > 0. \tag{5}$$

Proof. Define the function $H : \mathbb{R}^{8N} \to \mathbb{R}^{8N}$,

$$H(w) = -\overline{D}w + \overline{A}\overline{f}(w) + \overline{B}\overline{g}(w) + \overline{u}. \tag{6}$$

We will first prove that H is injective. Assume by contradiction that there exist $w, w' \in \mathbb{R}^{8N}$, $w \neq w'$, such that $H(w) = H(w')$. This equality is equivalent with

$$-\overline{D}(w - w') + \overline{A}(\overline{f}(w) - \overline{f}(w')) + \overline{B}(\overline{g}(w) - \overline{g}(w')) = 0. \tag{7}$$

By left multiplying this relation by $2(w - w')^T P$, we get that

$$2(w - w')^T P(-\overline{D}(w - w') + \overline{A}(\overline{f}(w) - \overline{f}(w')) + \overline{B}(\overline{g}(w) - \overline{g}(w'))) = 0, \tag{8}$$

which can be rewritten as

$$(w - w')^T(-P\overline{D} - \overline{D}P)(w - w') + 2(w - w')^T P\overline{A}(\overline{f}(w) - \overline{f}(w'))$$
$$+2(w - w')^T P\overline{B}(\overline{g}(w) - \overline{g}(w')) = 0, \tag{9}$$

From Assumption 1, we can deduce that

$$(f(w) - f(w'))^T(f(w) - f(w')) \leq (w - w')^T \overline{L_f}(w - w'), \tag{10}$$

$$(g(w) - g(w'))^T(g(w) - g(w')) \leq (w - w')^T \overline{L_g}(w - w'). \tag{11}$$

Now, taking into account Lemma 2 and inequalities (10) and (11), we have from (9) that

$$(w - w')^T(-P\overline{D} - \overline{D}P)(w - w') + 2(w - w')^T P\overline{A}(\overline{f}(w) - \overline{f}(w'))$$
$$+2(w - w')^T P\overline{B}(\overline{g}(w) - \overline{g}(w'))$$
$$\leq (w - w')^T(-P\overline{D} - \overline{D}P)(w - w') + \varepsilon_1(\overline{f}(w) - \overline{f}(w'))^T(\overline{f}(w) - \overline{f}(w'))$$
$$+\varepsilon_1^{-1}(w - w')^T P\overline{A}\,\overline{A}^T P(w - w') + \varepsilon_2(\overline{g}(w) - \overline{g}(w'))^T(\overline{g}(w) - \overline{g}(w'))$$
$$+\varepsilon_2^{-1}(w - w')^T P\overline{B}\,\overline{B}^T P(w - w')$$
$$\leq (w - w')^T(-P\overline{D} - \overline{D}P)(w - w') + \varepsilon_1(w - w')^T \overline{L_f}(w - w')$$
$$+\varepsilon_1^{-1}(w - w')^T P\overline{A}\,\overline{A}^T P(w - w') + \varepsilon_2(w - w')^T \overline{L_g}(w - w')$$
$$+\varepsilon_2^{-1}(w - w')^T P\overline{B}\,\overline{B}^T P(w - w')$$
$$= -(w - w')^T(P\overline{D} + \overline{D}P - \varepsilon_1\overline{L_f} - \varepsilon_2\overline{L_g} - \varepsilon_1^{-1}P\overline{A}\,\overline{A}^T P$$
$$-\varepsilon_2^{-1}P\overline{B}\,\overline{B}^T P)(w - w'). \tag{12}$$

Using Schur's complement, from condition (5), we get that

$$P\overline{D} + \overline{D}P - \varepsilon_1\overline{L_f} - \varepsilon_2\overline{L_g} - \varepsilon_1^{-1}P\overline{A}\,\overline{A}^T P - \varepsilon_2^{-1}P\overline{B}\,\overline{B}^T P > 0, \tag{13}$$

which, plugged back into (12), finally yields $H(w) - H(w') < 0$, which is a contradiction with our initial assumption. We deduce that H is injective.

Next, we prove that $||H(w)|| \to \infty$ as $||w|| \to \infty$. To this end, we deduce from (13) that there exists a sufficiently small $\varepsilon > 0$, such that $-P\overline{D} - \overline{D}P + \varepsilon_1\overline{L_f} + \varepsilon_2\overline{L_g} + \varepsilon_1^{-1}P\overline{A}\,\overline{A}^T P + \varepsilon_2^{-1}P\overline{B}\,\overline{B}^T P < -\varepsilon I_{8N}$. Considering $w' = 0$ in (12), we have

$$2w^T P(H(w) - H(0)) \le w^T(-P\overline{D} - \overline{D}P + \varepsilon_1\overline{L_f} + \varepsilon_2\overline{L_g} + \varepsilon_1^{-1}P\overline{A}\,\overline{A}^T P$$
$$+\varepsilon_2^{-1}P\overline{B}\,\overline{B}^T P)w < -\varepsilon||w||^2. \tag{14}$$

By applying the Cauchy-Schwarz inequality in relation (14), we get that

$$2||w||\,||P||(||H(w)|| + ||H(0)||) > \varepsilon||w||^2,$$

from which we conclude that $||H(w)|| \to \infty$ when $||w|| \to \infty$.

Now we can use Lemma 1 to deduce that H is a homeomorphism of $\mathbb{R}^{8N}$ onto itself. This means that the equation $H(w) = 0$ has a unique solution, and so system (4) also has a unique equilibrium point, which we will denote by $\hat{w}$.

We shift this equilibrium point to the origin, and thus system (4) is equivalent with

$$\dot{\tilde{w}} = -\overline{D}\tilde{w} + \overline{A}\,\tilde{f}(\tilde{w}) + \overline{B}\,\tilde{g}(\tilde{w}^\tau), \tag{15}$$

where $\tilde{w} = w - \hat{w}$, $\tilde{w}^\tau = w^\tau - \hat{w}$, $\tilde{f}(\tilde{w}) = \overline{f}(\tilde{w} + \hat{w}) - \overline{f}(\hat{w})$, and $\tilde{g}(\tilde{w}^\tau) = \overline{g}(\tilde{w}^\tau + \hat{w}) - \overline{g}(\hat{w})$. Construct the following Lyapunov-Krasovskii functional:

$$V(\tilde{w}(t)) = \tilde{w}^T(t)P\tilde{w}(t) + \int_{t-\tau}^t \tilde{w}(s)^T Q\tilde{w}(s)ds,$$

where $Q \in \mathbb{R}^{8N \times 8N}$, $Q > 0$.

The derivative of $V(\tilde{w}(t))$ with respect to t along the trajectories of system (15) is computed as

$$\dot{V}(\tilde{w}) = \dot{\tilde{w}}^T P\tilde{w} + \tilde{w}^T P\dot{\tilde{w}} + \tilde{w}^T Q\tilde{w} - \tilde{w}^{\tau T}Q\tilde{w}^\tau$$
$$= \tilde{w}^T P(-\overline{D}\tilde{w} + \overline{A}\,\tilde{f}(\tilde{w}) + \overline{B}\,\tilde{g}(\tilde{w}^\tau))$$
$$(-\overline{D}\tilde{w} + \overline{A}\,\tilde{f}(\tilde{w}) + \overline{B}\,\tilde{g}(\tilde{w}^\tau))^T P\tilde{w} + \tilde{w}^T Q\tilde{w} - \tilde{w}^{\tau T}Q\tilde{w}^\tau$$
$$= \tilde{w}^T(-P\overline{D} - \overline{D}P)\tilde{w} + \tilde{w}^T P\overline{A}\,\tilde{f}(\tilde{w}) + \tilde{f}^T(\tilde{w})\overline{A}^T P\tilde{w} + \tilde{w}^T P\overline{B}\,\tilde{g}(\tilde{w}^\tau)$$
$$+\tilde{g}^T(\tilde{w}^\tau)\overline{B}^T P\tilde{w} + \tilde{w}^T Q\tilde{w} - \tilde{w}^{\tau T}Q\tilde{w}^\tau. \tag{16}$$

If we multiply relations (10) and (11) by $\varepsilon_1 > 0$ and $\varepsilon_2 > 0$, we obtain

$$0 \le \varepsilon_1(\tilde{w}^T\overline{L_f}\tilde{w} - \tilde{f}^T(\tilde{w})\tilde{f}(\tilde{w})), \tag{17}$$

$$0 \le \varepsilon_2(\tilde{w}^{\tau T}\overline{L_g}\tilde{w}^\tau - \tilde{g}^T(\tilde{w}^\tau)\tilde{g}(\tilde{w}^\tau)). \tag{18}$$

Adding inequalities (17) and (18) to (16), gives

$$\dot{V}(\tilde{w}) \le \xi^T \Omega \xi, \tag{19}$$

where

$$\xi = \left[\tilde{w}^T \; \tilde{w}^{\tau T} \; \tilde{f}^T(\tilde{w}) \; \tilde{g}^T(\tilde{w}^\tau)\right]^T,$$

$$\Omega = \begin{bmatrix} -P\overline{D} - \overline{D}P + Q + \varepsilon_1 \overline{L_f} & 0 & P\overline{A} & P\overline{B} \\ * & -Q + \varepsilon_2 \overline{L_g} & 0 & 0 \\ * & * & -\varepsilon_1 I_{8N} & 0 \\ * & * & * & -\varepsilon_2 I_{8N} \end{bmatrix}.$$

Now, we have $\Omega < 0$ if and only if $Q > \varepsilon_2 \overline{L_g}$ and

$$\begin{bmatrix} -P\overline{D} - \overline{D}P + Q + \varepsilon_1 \overline{L_f} & P\overline{A} & P\overline{B} \\ * & -\varepsilon_1 I_{8N} & 0 \\ * & * & -\varepsilon_2 I_{8N} \end{bmatrix} < 0. \tag{20}$$

Together, the linear matrix inequalities (20) and $Q > \varepsilon_2 \overline{L_g}$ are equivalent with condition (5), which means that (19) becomes $\dot{V}(\tilde{w}) < 0$, from which we can conclude that the equilibrium point of (4) is globally asymptotically stable, thus ending the proof of the theorem.

4   Numerical Example

A numerical example is given to demonstrate the effectiveness of our results.

Example 1. Consider the following two-neuron octonion-valued recurrent neural network with time delay:

$$\begin{cases} \dot{x}_1(t) = -d_1 x_1(t) + \sum_{j=1}^2 a_{1j} f_j(x_j(t)) + \sum_{j=1}^2 b_{1j} g_j(x_j(t-\tau)) + u_1, \\ \dot{x}_2(t) = -d_2 x_2(t) + \sum_{j=1}^2 a_{2j} f_j(x_j(t)) + \sum_{j=1}^2 b_{1j} g_j(x_j(t-\tau)) + u_2, \end{cases} \tag{21}$$

where $d_1 = 50$, $d_2 = 40$,

$$\text{vec}(a_{11}) = (1,1,2,2,1,-1,-1,1)^T, \; \text{vec}(a_{12}) = (2,1,1,-2,2,1,-2,2)^T$$

$$\text{vec}(a_{21}) = (2,-2,2,1,2,-2,1,2)^T, \; \text{vec}(a_{22}) = (1,2,2,-2,1,1,2,-2)^T,$$

$$\text{vec}(b_{11}) = (2,1,2,1,-2,2,-1,2)^T, \; \text{vec}(b_{12}) = (-2,2,-2,2,1,2,-2,2)^T,$$

$$\text{vec}(b_{21}) = (1,-2,2,-2,1,2,2,2)^T, \; \text{vec}(b_{22}) = (1,2,2,1,2,-2,-2,1)^T,$$

$$\text{vec}(u_1) = (10,-20,30,-40,50,-70,80,-90)^T,$$

$$\text{vec}(u_2) = (90,-40,10,-60,30,-80,50,-20)^T,$$

$$f_j([x]_p) = \frac{1}{1+e^{-[x]_p}}, \; g_j([x]_p) = \frac{1-e^{-[x]_p}}{1+e^{-[x]_p}}, \; p \in \{0,1,\ldots,7\}, \; j \in \{1,2\},$$

from which we deduce that $l_1^f = l_2^f = \frac{1}{\sqrt{2}}$ and $l_1^g = l_2^g = 2\sqrt{2}$. The time delay is taken to be $\tau = 0.5$.

By solving the LMI condition (5) in Theorem 1, we get that system (21) has a unique equilibrium point which is globally asymptotically stable for $\varepsilon_1 = 1.3270$, $\varepsilon_2 = 0.7199$. (The value of P is not given due to space limitations.) The state trajectories of the elements of octonions x_1 and x_2 are given in Fig. 1, for four initial values.

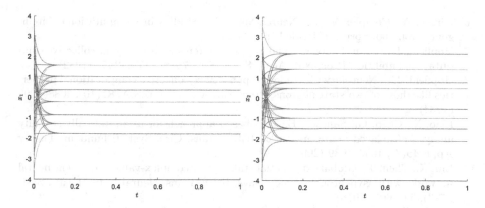

Fig. 1. State trajectories of elements of x_1 and x_2 in Example 1

5 Conclusions

The existence and uniqueness of the global equilibrium point for octonion-valued recurrent neural networks with time delay was proved using the homeomorphism theory. A sufficient criterion was derived in terms of linear matrix inequalities, which assures that the equilibrium point is also globally asymptotically stable, by making the assumption that the activation functions satisfy the Lipschitz condition. The effectiveness of the proposed criterion was illustrated by giving a numerical example.

It is very likely that the future will bring even more applications for the complex- and quaternion-valued neural networks, and also for their generalization, namely the Clifford-valued neural networks. Especially due to the property of being a normed division algebra that the octonion algebra has, octonion-valued neural networks can represent an alternative to networks defined on Clifford algebras of dimension 8.

References

1. Arena, P., Fortuna, L., Muscato, G., Xibilia, M.: Multilayer perceptrons to approximate quaternion valued functions. Neural Netw. **10**(2), 335–342 (1997)
2. Arena, P., Fortuna, L., Occhipinti, L., Xibilia, M.: Neural networks for quaternion-valued function approximation. In: International Symposium on Circuits and Systems (ISCAS), vol. 6, pp. 307–310. IEEE (1994)
3. Buchholz, S., Sommer, G.: On Clifford neurons and Clifford multi-layer perceptrons. Neural Netw. **21**(7), 925–935 (2008)
4. Dray, T., Manogue, C.: The Geometry of the Octonions. World Scientific, Singapore (2015)
5. Forti, M., Tesi, A.: New conditions for global stability of neural networks with application to linear and quadratic programming problems. IEEE Trans. Circ. Syst. I: Fundam. Theory Appl. **42**(7), 354–366 (1995)

6. Hirose, A.: Complex-Valued Neural Networks, Studies in Computational Intelligence, vol. 400. Springer, Heidelberg (2012)
7. Hopfield, J.J.: Neural networks and physical systems with emergent collective computational abilities. Proc. Nat. Acad. Sci. U.S.A. **79**(8), 2554–2558 (1982)
8. Hopfield, J.J.: Neurons with graded response have collective computational properties like those of two-state neurons. Proc. Nat. Acad. Sci. U.S.A. **81**(10), 3088–3092 (1984)
9. Liao, X., Chen, G., Sanchez, E.: LMI-based approach for asymptotically stability analysis of delayed neural networks. IEEE Trans. Circ. Syst. I: Fundam. Theory Appl. **49**(7), 1033–1039 (2002)
10. Liu, X., Chen, T.: Global exponential stability for complex-valued recurrent neural networks with asynchronous time delays. IEEE Trans. Neural Netw. Learn. Syst. **27**(3), 593–606 (2016)
11. Liu, Y., Xu, P., Lu, J., Liang, J.: Global stability of Clifford-valued recurrent neural networks with time delays. Nonlinear Dyn. **84**(2), 767–777 (2016)
12. Liu, Y., Zhang, D., Lu, J., Cao, J.: Global μ-stability criteria for quaternion-valued neural networks with unbounded time-varying delays. Inf. Sci. **360**, 273–288 (2016)
13. Mandic, D.P., Goh, V.S.L.: Complex Valued Nonlinear Adaptive Filters: Noncircularity, Widely Linear and Neural Models. Wiley-Blackwell, Hoboken (2009)
14. Okubo, S.: Introduction to Octonion and Other Non-Associative Algebras in Physics. Cambridge University Press, Cambridge (1995)
15. Pearson, J., Bisset, D.: Back propagation in a Clifford algebra. In: International Conference on Artificial Neural Networks, vol. 2, pp. 413–416 (1992)
16. Pearson, J., Bisset, D.: Neural networks in the Clifford domain. In: International Conference on Neural Networks, vol. 3, pp. 1465–1469. IEEE (1994)
17. Popa, C.-A.: Octonion-valued neural networks. In: Villa, A., Masulli, P., Pons Rivero, A. (eds.) ICANN 2016. LNCS, vol. 9886, pp. 435–443. Springer, Cham (2016). doi:10.1007/978-3-319-44778-0_51
18. Snopek, K.M.: Quaternions and octonions in signal processing - fundamentals and some new results. Przeglad Telekomunikacyjny + Wiadomosci Telekomunikacyjne **6**, 618–622 (2015)
19. Song, Q., Yan, H., Zhao, Z., Liu, Y.: Global exponential stability of complex-valued neural networks with both time-varying delays and impulsive effects. Neural Netw. **79**, 108–116 (2016)
20. Song, Q., Zhao, Z.: Stability criterion of complex-valued neural networks with both leakage delay and time-varying delays on time scales. Neurocomputing **171**, 179–184 (2016)
21. Valle, M.: A novel continuous-valued quaternionic Hopfield neural network. In: Brazilian Conference on Intelligent Systems (BRACIS), pp. 97–102. IEEE, October 2014
22. Widrow, B., McCool, J., Ball, M.: The complex LMS algorithm. Proc. IEEE **63**(4), 719–720 (1975)
23. Zhu, J., Sun, J.: Global exponential stability of Clifford-valued recurrent neural networks. Neurocomputing **173**, Part 3, 685–689 (2016)

Convolutional Neural Networks for Thai Poem Classification

Nuttachot Promrit[✉] and Sajjaporn Waijanya[✉]

Department of Computing, Faculty of Science, Silpakorn University,
Sanam Chandra Palace Campus, Muang District, Nakhon Pathom, Thailand
{promrit_n,waijanya_s}@silpakorn.edu

Abstract. In this work, we propose a Convolutional Neural Networks (CNNs) that able to be unsupervised feature learning to classify Thai poem (Klon-8) categories and Thai poem sentiment analysis. Thai poem has prosody, syllable rhyme and rhythm, there are challenges and different from prose text classification. The input of model representation by the vector (word2vec) generated from Thai-Text corpus 5.9 Million words. We perform the experiments by comparing with Support Vector Machine (SVM) and Naïve Bayes. CNNs showed the performance of poem categories 83% and performance of sentiment analysis 61%. CNNs showed a good performance, although unused knowledge about the composition of the poem for feature extraction.

Keywords: Poem classification · Poem sentiment analysis · Convolutional neural networks · Thai poem · Klon-8 · Word2Vec

1 Introduction

Klon-8 is the poem that has been most popular in Thailand since 200 years ago [1]. Nevertheless, writing klon-8 by corrects prosody and beautiful melodious must practice very hard. Since the poem is the language art that has exquisiteness and beautiful language. To continue the cultural heritage as Thai Poem Klon-8 by computer and information technology. Then we started the project for developing Artificial intelligence to compose Klon-8. At the beginning, machine is able to understand klon-8 by poem category classification and poem sentiment analysis.

The poem classification have to use feature extraction same with other text classification. However, the stringent prosody of the poem such as a number of syllables, rhyme position and words rhymes to be the cause of incomplete sentence. Incomplete sentence in klon-8 has no subject and object in sometimes. The example of klon-8 1 unit (1 baat) in Thai is "ดินน้ำไฟฝนฟ้าลมอากาศ – พืชแร่ธาตุข้าวปลาธัญญาหาร" the phonetic alphabet is "din^1 nam^4 <u>fai^1 fon^5 fa;^4</u> lom^1 <u>?a;^1ka;d^2</u> - phv;d^3 rx;^3**tha;d^3** kha;w^3 <u>pla;^1</u> than^1ja;^1ha;n^5". Then translate to English is "earth, water, fire, rain, sky, wind, air. - plants, minerals, fish, rice, cereals". From the example the underline word is internal rhyme and bold is external rhyme. All words from the example are only "noun". Feature extractions such as syntactic feature and entity feature need the quite complete sentence. By these reason we will not use the

F. Cong et al. (Eds.): ISNN 2017, Part I, LNCS 10261, pp. 449–456, 2017.
DOI: 10.1007/978-3-319-59072-1_53

syntactic feature and entity feature for poem classification and sentiment analysis. Then we use Word2Vec that is embedding feature extraction for Thai poem klon-8 instead.

The appearance of above feature extraction, the embedding feature is effortless method and the knowledge of Thai-poem klon-8 is not required. This research selected the embedding feature to extract the feature of klon-8. The result of feature extraction to be the input of convolutional neural networks (CNNs) for classifications. We use Word2Vec that is embedding feature extraction to classify poem into 7 categories include "Royalty", "Festival", "Parent+Teacher", "Advise", "Happy in Love", "Broken" and "Depress". And sentiment analysis 3 emotional include "good", "normal" and "bad". We estimated and expected the result by CNNs will be better than Support Vector Machine and Naïve Bayes.

2   Related Works

Text Classification is one important task in Natural Language Processing (NLP). Key task of most researches text classification is feature extraction. Text engineering for feature extraction including syntactical parsing, entity extraction, statistical features and word embedding. The popular methodologies of prose and poem text category classification are Support Vector Machine (SVM), Latent Dirichlet Allocation (LDA) and Term Frequency–Inverse Document Frequency (TF-IDF) [2–4]. Besides category classification, the sentiment analysis is another important task. Traditional NLP research often use Naive Bayes and SVM to analyze user's text such as reviews and social post [5–7]. For poem sentiment analysis we found only 1 work, they use weighted personalized page rank and lexical network to analyze sentiment of classical Chinese poem [8].

Text feature extraction technique such as TF-IDF and LDA must have huge training dataset to build bag of word. On the other hand, Word2Vec [9] has use bag of word from text corpus instead of training data. Normally, collecting Thai text corpus is easier than collecting Thai Poem training set. For other text feature extractions such as syntactical parsing or named entity recognition, they are necessary for prose text and completed sentence but the characteristic of Thai poem is not similar. Thai poem has quite complexity prosody and most of them have the incomplete sentence.

This research used Word2Vec instead of TF-IDF to embedding words in the poem. Word2Vec is used for learning vector representations of words. We applied the continuous skip-gram model in Word2Vec that used the current word to predict the surrounding window. Then the position of words in vector space can represent the semantic of words.

Since, Thai Poem text feature extraction used Word2Vec represent word not whole sentence. The effortless way to extract feature be applied by Convolutional Neural Networks (CNNs) [10, 11]. CNNs is popular methodology in many domain [12–14]. However we never found researcher uses CNNs for Poem Category Classification and Poem Sentiment Analysis before.

3 Model and Methodology

3.1 Process Overview

This research use Thai language to be input of process. We separated processes into 2 main process groups include (1) the preparation process and (2) the word embedding and classification process has shown in Fig. 1. The preparation process including with Thai word segmentation and Thai poem transformation. Because of the sentence in Thai language has no spaces, that mean it's no stop word. Thus, we must have word segmentation process. Thai word segmentation will separate words from 2 types of content Thai-poem and Thai-prose (news, encyclopedia, books, etc.). The output of word segmentation will be Thai Text corpus and only output of word segmentation from Thai-poem will send to Thai poem transformation process and creating to Thai poem training data and test data.

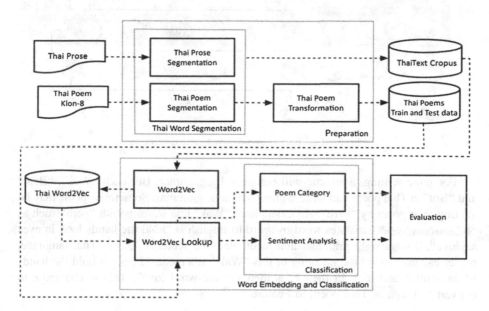

Fig. 1. Process overview of convolutional neural networks for Thai poem classification model

In the word embedding and classification process, Word2Vec process use data from ThaiText corpus and the result of this process is Thai word embedding (Thai-Word2Vec), it is input for Word2Vec lookup process. The result of Word2Vec lookup is input to classification process both poem category classification and poem sentiment analysis.

The evaluation process is last step to evaluate the performance of our model versus SVM and Naïve Bayes.

3.2   Thai Poem (Input)

To understand what is klon-8 poem, the rhyme and prosody term of klon-8 has shown in Fig. 2, it has prosody for number of syllable in line. In each line are 7 to 9 syllables allowed. If one line is having more than 9 or less than 7 syllables, an error is implicated in the length of the line. Moreover, Thai poem klon-8 has rhyme. The rhyme of klon-8 means syllables in "rhyme positions" must have same vowel sound and same spelling-sound such as "rak^4" (รัก: love) and "nak^2" (หนัก: hard) but its phoneme must not duplicated.

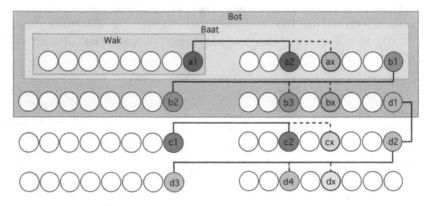

Fig. 2. Thai poem Klon-8 structure and prosody

For prose writing, Syntactic will complete by grammar. But each "Wak", "Baat" and "Bot" in Thai poem can write without syntactic grammar. Sometimes poets may be starting their poem by "verb". An example of "Wak" that starting with "verb" such as "จับมือมองจ้องตาเอ่ยว่ารัก" translates word by word to English is "hold the hands look in eyes saylove". It's not right grammar and incomplete the sentence, but in Thai language, reader can understand the meaning of this "Wak". It's mean "2 people hold the hands of each other and look in their eyes then say the word love". This is challenge to convert text data as Thai poem to Feature.

3.3   Thai Word Segmentation

Refer to Thai poem Klon-8 structure in Fig. 2, Syllables are very important in prosody of Thai Poetry. Each "Wak" has a rule for number of syllables. The relation between "Wak" and "Bot" has to check the sound of the syllable. But the syllable has no meaning by itself because it is sound. The unit with meaning is word and word consists of one syllable or many syllables.

In this work used Thai word segmentation API [15] to cut words from Thai Poem both training set and test set by using longest matching with dictionary base technique. We also cut word from Thai prose content in preparation process to be ThaiText corpus.

3.4   Word2Vec

To preparing Word Vectors to be input data for Convolutional Neural Network model, we selected Word2Vec by define size of word vectors 200 dimension and train Word2Vec by skip-gram model using ThaiText corpus 5.9 Million words from 5 online resources including with (1) "BEST I Corpus" by NECTEC (2) Contemporary Poets Society-www.kawethai.com (3) www.wannakadee.com (4) www.thaipoem.com and (5) www.aromklon.com

The result of Word2Vec from Thai Text corpus in Fig. 3 shows a vector of continuous value that represents semantic attributes of the words with t-SNE [16]. The example of words with similar meaning in Thai are "คุณ" = "you" and "ท่าน" = "you". The example of words with semantic relation in Thai are "พูด" = "speak", "คิด" = "think", "ถาม" = "ask", " ร้อง"="call".

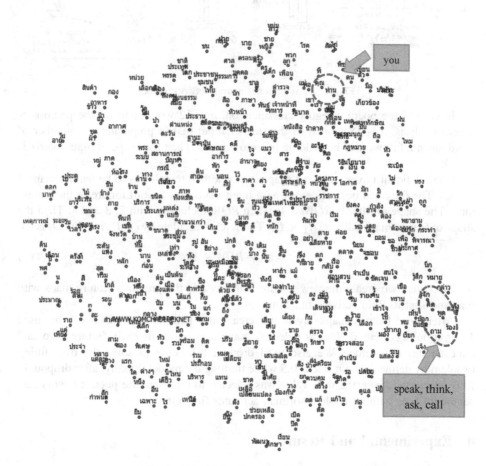

Fig. 3. Word2Vec from ThaiText corpus

3.5 Convolutional Neural Network

Our model in Fig. 4 shows the process of Convolutional Neural Network model. The input of this model is Thai Poem Word Embedding. Each word in the poem is represented by each vector. Size of vector is k-dimension and the number of words is n.

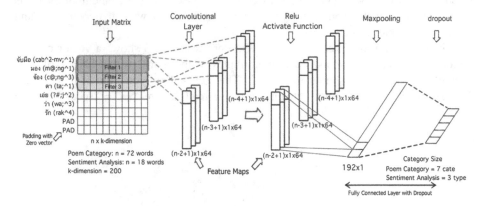

Fig. 4. Convolutional neural network model

In case of the poem category classification, n is 72 words and for the poem sentiment analysis, n is 18 words. In the step of input matrix preparation, if number of word has not full we will be padding by zero vectors. Then the shape of input matrix is $n \times k$.

Convolutional Layer has input shape is $n \times k$ then transforms to be feature maps 3 shapes. The filter had been slide down on input matrix by 1 word for create the feature map. The shape of filter is $ROW \times k$, when ROW including with 2, 3 and 4. Then the shape of feature maps is $W_1 \times H_1 \times D_1$. The W_1 can calculate by (1) H_1 is 1 and D_1 is 64.

$$W_1 = (n - ROW + 1) \tag{1}$$

Next step, we create the new feature maps by add the previous feature maps with bias and sent to Relu activate function.

After new feature maps layer, we create the matrix shape **192 × 1** by used **1_maxpooling** method. It was selected maximum value in each feature map and concatenates each other. We also used dropout technique to solve the over fitting problem by define dropout rate is 0.5 while training the model. Finally, after dropout in fully connected layer, the max value was selected to represent the poem category and sentiment type without applied softmax activates function.

4 Experimental and Result

The experiments in this work had focusing on 2 main tasks poem category classification and sentiment analysis. Training data for the poem category classification is 500 poems and testing data is 55 poems. Each poem has its length 2 "bot" (see in Fig. 2). Training data for the poem sentiment analysis is 2000 "baat" and testing data is 220 "baat".

We selected the Word2Vec parameter (k-dimension) and CNNs parameter (number of feature map) by adjusting in an experimental. The results of poem category classification and poem sentiment analysis by CNNs as shown in Table 1. Then we compare with SVM and Naïve Bayes. The results of poem category classification and poem sentiment analysis by SVM and Naïve Bayes as shown in Table 2.

Table 1. The results of poem category classification by CNNs

k-dimension	Number of feature-maps	Accuracy of poem category classification	Accuracy of poem sentiment analysis
128	32	68.00%	54.00%
128	64	70.00%	52.00%
200	32	80.00%	57.20%
200	64	83.00%	61.00%

Table 2. The results of poem sentiment analysis by SVM and Naive Bayes

Models	Accuracy of poem category classification	Accuracy of poem sentiment analysis
CNNs (k = 200, feature maps = 64)	83.00%	61.00%
SVM	16.36%	24.54%
Naïve Bayes	14.54%	41.36%

The results from the experiment, k-dimension = 200 and number of feature = 64 perform the best prediction result is 83% in poem category classification and 61% in poem sentiment analysis. Although the poem sentiment analysis has 3 classes which smaller than number of classes of poem categories. But the accuracy of sentiment analysis is not better than poem category classification. It may be because of Thai poems often using the metaphors that compose with multi emotional in a sentence (baat). The example is "กระแสลมแปรปรวนชวนตะลึง มันน่าทึ่งที่ยังรู้ทิศทาง" translated to English is "*The turbulence winds make people feel stunning, But It's amazing to know the wind direction*".

Comparing with SVM and Naïve Bayes, CNNs had shown the accuracy better than both models because only Word2Vec feature cannot present the concept of sentence such as name entity and path of speech.

5   Conclusion and Future Work

In this research, we used CNNs model and adjusted parameters for Thai poem category classification and sentiment analysis. This is first research which classified the poem by CNNs. We used the Word2Vec embedding to be input for CNNs. The Word2Vec had built amount of Thai Text corpus 5.9 million words. The large size of the corpus can build words volume in the bag of word to be enough for the word embedding process. For training the model, even though we use small training dataset but the propose model can performed the best over SVM and Naïve Bayes. The results have shown that

CNNs method can develop the machine's ability to classify the categories and analyses the sentiments the Klon-8.

To continue the developing artificial intelligence to compose Klon-8, in the future we will compare text feature extraction such as syntactic feature, entity feature, statistical feature and word embedding for Thai Poem domain. Moreover we will apply them to the next task of Thai poem.

References

1. Thailand's Shakespeare? Sunthorn Phu | ThingsAsian. http://thingsasian.com/story/thailands-shakespeare-sunthorn-phu
2. Kumar, V., Minz, S.: Poem classification using machine learning approach. In: Babu, B.V., Nagar, A., Deep, K., Pant, M., Bansal, J.C., Ray, K., Gupta, U. (eds.) Proceedings of the Second International Conference on Soft Computing for Problem Solving (SocProS 2012), December 28-30, 2012. AISC, vol. 236, pp. 675–682. Springer, New Delhi (2014). doi:10.1007/978-81-322-1602-5_72
3. Jamal, N., Mohd, M., Noah, S.A.: Poetry classification using support vector machines. J. Comput. Sci. **8**, 1441–1446 (2012)
4. Multilabel Subject-Based Classification of Poetry - Research Publication, http://researchr.org/publication/LouIT15
5. Vanzo, A., Croce, D., Basili, R.: A context-based model for Sentiment Analysis in Twitter. In: COLING (2014)
6. Yessenov, K., Misailovic, S.: Sentiment analysis of movie review comments. Methodology, 1–17 (2009)
7. Liparas, D., HaCohen-Kerner, Y., Moumtzidou, A., Vrochidis, S., Kompatsiaris, I.: News articles classification using random forests and weighted multimodal features. In: Lamas, D., Buitelaar, P. (eds.) IRFC 2014. LNCS, vol. 8849, pp. 63–75. Springer, Cham (2014). doi:10.1007/978-3-319-12979-2_6
8. Hou, Y., Frank, A.: Analyzing sentiment in classical Chinese poetry. In: LaTeCH 2015, p. 15 (2015)
9. Mikolov, T., Chen, K., Corrado, G., Dean, J.: Efficient Estimation of Word Representations in Vector Space. arXiv:13013781 Cs (2013)
10. Le, Q.V., Brain, G., Inc, G.: A Tutorial on Deep Learning Part 1: Nonlinear Classifiers and the Backpropagation Algorithm (2015)
11. Le, Q.V., Brain, G., Inc, G.: A Tutorial on Deep Learning Part 2: Autoencoders, Convolutional Neural Networks and Recurrent Neural Networks (2015)
12. Rios, A., Kavuluru, R.: Convolutional neural networks for biomedical text classification: application in indexing biomedical articles. In: Proceedings of the 6th ACM Conference on Bioinformatics, Computational Biology and Health Informatics, pp. 258–267. ACM, New York (2015)
13. Zhang, Y., Wallace, B.: A Sensitivity Analysis of (and Practitioners' Guide to) Convolutional Neural Networks for Sentence Classification. arXiv:151003820 Cs (2015)
14. Weston, J., Chopra, S., Adams, K.: #TAGSPACE: semantic embeddings from hashtags. Presented at the 2014 Conference on Empirical Methods in Natural Language Processing (EMNLP), Doha, Qatar (2014)
15. Veer Sattayamas: GitHub - veer66/PhlongTaIam: PHP Thai word breaker (2014)
16. van der Maaten, L., Hinton, G.: Visualizing data using t-SNE. J. Mach. Learn. Res. **9**, 2579–2605 (2008)

A Quaternionic Rate-Based Synaptic Learning Rule Derived from Spike-Timing Dependent Plasticity

Guang Qiao[1,2], Hongyue Du[1], and Yi Zeng[2,3,4(✉)]

[1] Harbin University of Science and Technology, Harbin, China
[2] Institute of Automation, Chinese Academy of Sciences, Beijing, China
[3] Center for Excellence in Brain Science and Intelligence Technology,
Chinese Academy of Sciences, Shanghai, China
[4] University of Chinese Academy of Sciences, Beijing, China
{qiaoguang2014,yi.zeng}@ia.ac.cn

Abstract. Most of the differential Hebbian rules derived from Spike-Timing Dependent Plasticity (STDP) focus on the rates of change of post-synaptic activity that carries the information about the future and enables the neural network to predict. And the current model mainly consider three factors for the adjustment of synaptic weight, namely, the rate of pre- and post-synaptic activity and the rate of change of post-synaptic activity. We argue that the rate of change of pre-synaptic activity also plays an important role on the adjustment of synaptic weight. Hence, this paper proposes a quaternionic rate-based synaptic learning rule that depends on four elements, namely, the instantaneous firing rates of both pre- and post-synaptic neurons and their time derivatives.

Keywords: Spike-Timing Dependent Plasticity · Quaternionic rate-based synaptic learning rule · Instantaneous firing rate

1 Introduction

Synaptic connectivity and its dynamics play a key role in the brain and artificial neural networks that implement various functions. Spike-Timing Dependent Plasticity (STDP) depends the exact timing difference of pre- and post-synaptic neuron's spikes and is believed to be the major form of synaptic changes in neurons [7,15,20]. Many investigations have been done to explore its biological mechanisms [5,21] and to infer its biological functions [9,16]. Inspired by the biological explorations, synaptic dynamics and their interpretations have been investigated as part of a learning procedure in artificial neural networks [1,11].

Because the spikes occur discretely and couple with time tightly, the most direct research method based on the original discrete forms of STDP would be computer simulation which is time consuming and noise sensitive. Rate-based rules, beyond overcoming those shortcomings, have more favourable mathematical analysis properties to facilitate the analysis of the dynamics of networks

© Springer International Publishing AG 2017
F. Cong et al. (Eds.): ISNN 2017, Part I, LNCS 10261, pp. 457–465, 2017.
DOI: 10.1007/978-3-319-59072-1_54

and to reveal more substantive characteristics, which make it significant to convert the spike-based STDP to a rated-based rule. There have many work which aim at connecting the STDP rule and rate-based rules. Both the BCM learning rule and the standard rate-based Hebbian learning rule have been considered to be identical or equivalent to the STDP learning rule under certain conditions [12,13]. The differential Hebbian rules which take the changes of the firing rate into account can catch more temporal information and hence are shown to be the better approximations to the STDP rule [2,17,18,24]. Problematically, for most of differential Hebbian rules which were derived from the STDP rule, they emphasized the effect of the time derivative of the post-synaptic firing rate, and neglected the importance of the time derivative of the pre-synaptic firing rate, which takes the imbalance into the final formulas. The original STDP rule seems to be symmetric for the importance of pre- and post-synaptic spikes and a shift of spike time of post-synaptic neuron can be compensated by the shift of pre-synaptic spike, which implies maybe the corresponding rate-base rule is also symmetric and the time derivative of the post-synaptic firing rate and the time derivative of the pre-synaptic firing rate should share the same importance. The lack of rate of change of pre-synaptic activity has been noticed and a symmetric learning rule has been proposed [19]. This paper proves the STDP rule can be converted to a rate-based rule that depends on the instantaneous firing rate of both pre- and post-synaptic neurons and their time derivatives, reveals the important functions of the time derivative of the pre-synaptic firing rate in synaptic weights updating, and hence provides an evidence that the symmetric learning rule is reasonable.

2   Spike-Timing Dependent Plasticity

Spike-Timing Dependent Plasticity (STDP) refers to the phenomenons observed in biological experiments that the changes in the synaptic strength depends on the spike time difference $\Delta t = t_{post} - t_{pre}$ between the post-synaptic spike and the pre-synaptic spike. According to [3], "repetitive post-synaptic spiking within a time window of 20 msec after pre-synaptic activation resulted in Long-Term Potentiation (LTP), whereas post-synaptic spiking within a window of 20 msec before the repetitive pre-synaptic activation led to long-term depression (LTD)" [3]. This is referred as biphasic rule. The triphasic STDP function are also reported [4,23]. Equations 1 and 2 formulate the biphasic and the triphasic STDP respectively [4].

$$\Delta W = \begin{cases} A^+ exp(-\Delta t/\tau^+) & \Delta t > 0 \\ -A^- exp(\Delta t/\tau^-) & \Delta t < 0 \end{cases} \tag{1}$$

$$\Delta W = A^+ exp\left(-\frac{(\Delta t - 15)^2}{\tau^+}\right) - A^- exp\left(-\frac{(\Delta t - 20)^2}{\tau^-}\right) \tag{2}$$

It is a common requirement to determine the weight changes in the time window (t_1, t_2) through a given spike train of pre- and post-synaptic neurons. Let

$W_{ij}(t)$ denote the synaptic weight from pre-synaptic neuron j to post-synaptic neuron i at time t. $x_i(t) = \sum_n \delta(t - T_i^n)$, where δ is the Dirac delta function and T_i^n is the nth spike time of the ith neuron. If all pre- and post-synaptic spike pairs make contributions to the changes of synaptic strength and all delays are ignored so that the change happens immediately when a spike occurs, then

$$W_{ij}(t_2) - W_{ij}(t_1) = \int_{t_1}^{t_2} A_{ij}(t) + B_{ij}(t)dt, \tag{3}$$

where

$$A_{ij}(t) = \int_{-\infty}^{0} f(u)x_j(t)x_i(t+u)du, \quad B_{ij}(t) = \int_{0}^{+\infty} f(u)x_j(t-u)x_i(t)du.$$

The first term $A_{ij}(t)$ counts the effect of all pairs consisting of the pre-synaptic spike which occurred in time t and the post-synaptic spike which occurred before the pre-synaptic spike and the spike time difference is negative. The term $B_{ij}(t)$ counts the effect of all pairs consisting of the post-synaptic spike which occurred in time t and the pre-synaptic spike which occurred before the post-synaptic spike. Hence any spike pair which would change the synaptic weight in the time window (t_1, t_2) has been counted in the two terms.

3 Relating Spike-Based Learning Rules to Rate-Based Learning Rules

Since the spikes occurred discretely, it is difficult to analyze the properties of the spike-based result (described as Eq. 3). It is more favourable to convert it to a continuous rate-based rule for further analysis. A rate-based updating rule also has its biological significance more than computational convenience. The brains are full of noise and uncertainty, including "a few microseconds of jitter over length of about 10 cm to each spike", "unreliable release of vesicles in to the synaptic cleft", "variant amount of neurotransmitter in each vesicle" [10,14,22]. The brains also have spontaneous spikes and uncertain synaptic delays and need to face various circumstances, which make themselves eager for a robust rule. The STDP which depends the exact timing difference of pre- and post-synaptic neurons' spikes at the micro level should work through a rate-based way in a bigger scope so that it could meet the requirement. An intuitive explanation for how the spike-based STDP can influence the synaptic efficiency through a rate-based way can be found in [2].

3.1 The Definition of Instantaneous Firing Rate

The average firing rate of neuron is well-defined, while the instantaneous rate is not. If the pre-synaptic firing rates are known variables, the post-synaptic firing rate could be defined by relating the firing rate v to the membrane potential u by a nonlinear monotonically increasing function $v = g(u)$, and the membrane potential can be calculated from the pre-synaptic firing rate and the synaptic weights [8]. That definition works well in artificial neural networks for the pre-synaptic firing rates which represent the features of samples are normally defined manually. If for the aim to directly set the instantaneous firing rate of certain neuron based on its spikes train, a possible way is to convolve the spiking train with a smoothing filter such as a Gaussian profile [6],

$$\phi(t) = \frac{1}{\sqrt{2\pi\tau}} exp\left(-\frac{t^2}{2\tau^2}\right).$$ (4)

Let $v(t)$ represent the instantaneous firing rate at time t, then we can define an estimate of the instantaneous firing rate via the convolution [6]

$$v(t) = \int_{-\infty}^{+\infty} x(s)\phi(t-s)ds.$$ (5)

The parameter τ in Eq. 4 controls the size of the time window over which spike times are averaged, and

$$\lim_{\tau \to 0} v(t) = x(t).$$

The firing rate of neuron $v(t)$ can also be considered as some probability measures of neuron to spike. A strong motivation to convert spike-based rule to rate-based rule is for convenience of further computations and analysis about the neural network dynamics. In comparison with utilizing Eq. 5 to compute the instantaneous firing rate of a given spikes train, we prefer to take $v(t)$ as a primitive quantity in theoretical analysis. An ideal instantaneous firing rate $v(t)$ should be continuous and smooth enough to apply some mathematical tools like derivation. The integral of $v(t)$ over any effective interval should be equal to the amount of spikes occurred in this interval approximately. For the rest of this paper, we assume $v(t)$ is provided with such properties.

3.2   Rate-Based Aspect of STDP

Spikes that are generated by neurons are with certain degree of randomness, here we aim to avoid the randomness to some extent. The correlation between the pre-synaptic neuron's spikes and the post-synaptic neuron's spikes is ignored. It makes sense since there are many pre-synaptic neurons to connect to the post-synaptic neuron. A spike from one of the pre-synaptic neurons has little impact on post-synaptic neuron, and for the existence of the spike jitters and spontaneous spikes, there cannot be a consistent time difference between the spikes which come from any specified one of the pre-synaptic neurons and the post-synaptic neuron. If the correlation is ignored, the spike-based learning rule (as shown in Eq. 3) can be rewritten to the rate-based learning rule through replacing $x(t)$ by $v(t)$,

$$W_{ij}(t_2) - W_{ij}(t_1) = \int_{t_1}^{t_2} A_{ij}(t) + B_{ij}(t)dt,$$ (6)

where

$$A_{ij} = \int_{-\infty}^{0} f(u)v_j(t)v_i(t+u)du, \quad B_{ij} = \int_{0}^{+\infty} f(u)v_i(t)v_j(t-u)du.$$

Performing Taylor expansion, $v(t+u) = v(t) + u\dot{v}(t) + o(u)$ is obtained. Since $f(u)$ has significant value only in a small time window around 0, the value of $\int f(u) \cdot o(u)du$ can be ignored,

$$\Delta W_{ij} = \alpha \int_{t_1}^{t_2} v_i(t)v_j(t)dt + \beta \int_{t_1}^{t_2} v_j(t)\dot{v}_i(t)dt + \gamma \int_{t_1}^{t_2} v_i(t)\dot{v}_j(t)dt \quad (7)$$

where

$$\alpha = \int_{-\infty}^{+\infty} f(u)du, \quad \beta = \int_{-\infty}^{0} uf(u)du, \quad \gamma = -\int_{0}^{+\infty} uf(u)du.$$

Equation 7 clearly shows there are mainly three items which take effects on the changes of the synaptic weight, including the product of pre- and post-synaptic neurons' firing rate, the product of pre-synaptic neuron's firing rate and the time derivative of post-synaptic neuron's firing rate, and the product of post-synaptic neuron's firing rate and the time derivative of pre-synaptic neuron's firing rate with their coefficients. The three items take effects together and can be thought to share the same importance if ignoring their coefficients which depend on the STDP function they choose. If the STDP function is perfectly antisymmetric or has the same size between the area which is below x-axis and the area which is above x-axis, the coefficient α would be zero and the first term could be omitted. If $\alpha \neq 0$, it may be difficult for the synaptic weight to reach its stationary point. Since the firing rate is always non-negative and the first item will always have impact on the synaptic weight to decay or strengthen, depending on whether the value of α is negative or positive. This indicates perhaps it needs some other mechanisms to compensate (e.g. to avoid continuous potentiation for very long time). Inspired by the methods in [24], our model extends the original conclusion mainly by taking the influence of the rate of change of pre-synaptic activity into account. Hence it makes a more comprehensive consideration about the elements which influences the synaptic weight.

3.3   Experimental Validation

Different STDP functions and firing rate functions have been tested to verify the reliability of the rate-based rule (as shown in Eq. 7). Both biphasic rule (as shown in Eq. 1) and triphasic rule (as shown in Eq. 2) are tested. In the setting of biphasic rule, two groups of parameters are used, including $A^+ = 0.06$, $A^- = 0.06$, $\tau^+ = 20$, $\tau^- = 20$ and $A^+ = 0.06$, $A^- = 0.03$, $\tau^+ = 20$, $\tau^- = 20$, referred as balanced biphasic rule and imbalanced biphasic rule respectively [4]. The parameters of triphasic rule chosen here are the same as in [4], $A^+ = 0.23$, $A^- = 0.15$, $\tau^+ = 200$, $\tau^- = 2000$. The spike train and the corresponding firing rate are generated by the following procedures: Set the initial firing rate function or invoke the generating function $p(t)$ at first, then generate the spiking train accordingly. The spikes are generated at each discrete time t with probability proportional to $p(t)$. The firing rate $v(t)$ of the generated spike train is computed with Eq. 5. Although the spike train was generated according to the initial firing rate function $p(t)$, the early tests showed it worked well when $p(t)$ was large if let $p(t)$ be $v(t)$ directly, but failed when $p(t)$ was small. We infer that if $p(t)$ was small, the spike train would very sparse along the time axis, which makes the firing rate no longer be $p(t)$, but larger when there is a spike and smaller when there is not. Smaller $p(t)$ means smaller size of spike samples, which introduces the contingency and reduces the irrelevance of the pre- and post-synaptic spikes. Synaptic efficiency is calculated by the spike-based rule and the rate-based rule respectively. In spike-based rule, all pairs of pre-synaptic and post-synaptic spikes are taken into account, and numerically only the pairs which the time difference is less than tens of milliseconds have a significant influence on the synaptic efficiency.

The experimental results show that the weight changes calculated by the rate-based rule can match the spike-based rule well, as shown in Fig. 1. The unbounded amplitude of weight is to show the trends of weight change. All the four subfigures showed in Fig. 1 are with the balanced biphasic rule. Figure 1(a) depicts how the weight changes with both spike-based and rate-based rule in the setting of both pre-synaptic and post-synaptic neuron having constant initial firing rate. The constant firing rate would not drive the weight away from the initial point too much with the balance rule. The pre-synaptic neuron's firing rate in (b) is sinusoidal function, and hence the line in (b) shows some rhythmical fluctuations, which clearly emphasizes the importance of the change of the pre-synaptic neuron's firing rate in synaptic weight changes. If the changes of firing rate are synchronized in pre- and post-synaptic neurons, the synaptic weight would not show significant changes. In (c), both pre- and post-synaptic neurons have the firing rate of sinusoidal functions with same phases, but the synaptic weights don't have significant changes though the firing rates have significant fluctuations. The unexpected potentiation in the last segment of (c) possibly is because of noises. Note that if the changes of firing rate have different phase in pre- and post-synaptic neurons just like (d) in which the pre-synaptic firing rate is sinusoidal function and the post-synaptic firing rate is cosinoidal function, the synaptic weight would have significant potentiation or depression. This possibly indicates if two stimuli happened at the same time, the synaptic weight would not have significant change. If they happened with fixed delay, the synaptic weight would be potentiated or depressed. This is a rate-based STDP rule working in large time scale, which is triggered by stimuli, not the spikes. The imbalanced biphasic and triphasic rule also show the accordance between the rate-based rule and the spike-based rule, which is not shown in the figure.

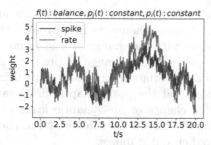

(a) Balanced biphasic STDP with constant pre- and post-synaptic firing rates

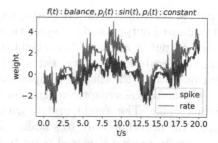

(b) Balanced biphasic STDP with sinusoidal pre-synaptic firing rate and constant post-synaptic firing rate

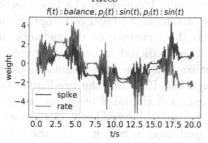

(c) Balanced biphasic STDP with sinusoidal pre- and post-synaptic firing rates

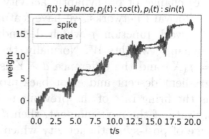

(d) Balanced biphasic STDP with cosinoidal pre-synaptic firing rate and sinusoidal post-synaptic firing rate

Fig. 1. Change of synaptic efficiency with various settings. The blue line represents the synaptic efficiency calculated by spike-based rule, and red line is calculated by the rate-based rule. (Color figure online)

4   Discussion

In this paper, we provide a theoretic framework to convert the spike-based synaptic efficiency updating rule to the corresponding rate-based rule. Experimental validations show that the pre- and post-synaptic neurons' firing rates and their time derivatives all make contributions to the changes of synaptic efficiency. This result is obtained under the assumption that the pre-synaptic neurons' spike train has no relation or weak relation to the post-synaptic neurons' train. The experiments testify the spike-based and the rate-based rules are basically consistent with each other. The rate-based rules have more favourable mathematical analysis properties to facilitate the analysis of the dynamics of networks, and could be a favourable substitution for spike-based rules.

In comparison to the rate of change of post-synaptic firing rate which has been shown a great deal of importance to synaptic weight updates, the rate of change of pre-synaptic firing rate do not possess enough attention in most of existing rate-based rules derived from STDP. Undoubtedly, the rate of change of pre-synaptic firing rate is closely related to synaptic weight updating in the brain, its effect are different from the one of the rate of change of post-synaptic firing rate. The change of post-synaptic firing rate indicates the change of desired output. A potentiation on the post-synaptic firing rate indicates a potentiation on the desired output, and the synaptic weight can enhance itself to produce the desired output in advance. The weight update with the change of post-synaptic firing rate enables the neural network to predict. The change of pre-synaptic firing rate means the change of input, and a potentiation on pre-synaptic firing rate will make the synaptic weight decay so that the post-synaptic firing rate remains stable. The weight update with the change of pre-synaptic firing rate enables the neural network to stabilize. The two factors coordinate and restrain each other to reach a balanced point to predict and stabilize.

It is very reasonable to believe the rate of change of pre-synaptic firing rate should also play a big part in the field of artificial neural networks where the rate of change of pre-synaptic activity is usually ignored. In the basic setting of feedforward networks, we have features of sample X, label of sample d, and the network function f which depends on the structure of the network and the synaptic weights W. Normally, there is a predicted value Y computed by $Y = f(X)$ and the difference $d - Y$ is utilized to tune the weights by means of gradient descent and error back-propagation. If taking the predicted value Y as the firing rate of the previous moment, and the label of sample d as the firing rate of the next moment, then $d - Y$ could be considered as the rate of change of post-synaptic activity which is well used in training. For the setting of some networks, the features X remain unchanged, which makes the rate of change of pre-synaptic activity be zero. In some other networks, X are changing while training, but the changes are still dependent on the rate of change of post-synaptic activity, and the rate of change of pre-synaptic activity has no direct influence on synaptic weights. If the rate of change of pre-synaptic activity can be constructed and utilized, it could possibly accelerate the training process or enhance the capacity of networks. The traditional network structures and training procedures maybe restrain the utilization of the rate of change of pre-synaptic activity. How to extend the existing network structures based on the idea proposed in this paper needs more further explorations.

Acknowledgement. This study was funded by the Strategic Priority Research Program of the Chinese Academy of Sciences (XDB02060007), and Beijing Municipal Commission of Science and Technology (Z151100000915070, Z161100000216124). This research is conducted at Institute of Automation, Chinese Academy of Science.

References

1. Bengio, Y., Lee, D.H., Bornschein, J., Lin, Z.: Towards biologically plausible deep learning. arXiv preprint arXiv:1502.04156 (2015)
2. Bengio, Y., Mesnard, T., Fischer, A., Zhang, S., Wu, Y.: STDP as presynaptic activity times rate of change of postsynaptic activity. arXiv preprint arXiv:1509.05936 (2015)
3. Bi, G.Q., Poo, M.M.: Synaptic modifications in cultured hippocampal neurons: dependence on spike timing, synaptic strength, and postsynaptic cell type. J. Neurosci. **18**(24), 10464–10472 (1998)
4. Chrol-Cannon, J., Grüning, A., Jin, Y.: The emergence of polychronous groups under varying input patterns, plasticity rules and network connectivities. In: The 2012 International Joint Conference on Neural Networks (IJCNN), pp. 1–6 (2012)

5. Froemke, R.C., Letzkus, J.J., Kampa, B.M., Hang, G.B., Stuart, G.J.: Dendritic synapse location and neocortical spike-timing-dependent plasticity. Front. Synaptic Neurosci. **2**, 29 (2010)
6. Gabbiani, F., Cox, S.J.: Mathematics for neuroscientists. Academic Press, Cambridge (2010)
7. Gerstner, W., Kempter, R., van Hemmen, J.L., Wagner, H.: A neuronal learning rule for sub-millisecond temporal coding. Nature **383**(6595), 76–78 (1996)
8. Gerstner, W., Kistler, W.M.: Mathematical formulations of hebbian learning. Biol. Cybern. **87**(5–6), 404–415 (2002)
9. Guyonneau, R., VanRullen, R., Thorpe, S.J.: Neurons tune to the earliest spikes through STDP. Neural Comput. **17**(4), 859–879 (2005)
10. Henneman, E., Mendell, L.M.: Functional organization of motoneuron pool and its inputs. In: Comprehensive Physiology, pp. 423–507 (2011). http://onlinelibrary. wiley.com/doi/10.1002/cphy.cp010211/abstract;jsessionid=DC8038A8B928C8490 64AFF986290D55A.f04t01
11. Iakymchuk, T., Rosado-Muñoz, A., Guerrero-Martínez, J.F., Bataller-Mompeán, M., Francés-Víllora, J.V.: Simplified spiking neural network architecture and STDP learning algorithm applied to image classification. EURASIP J. Image Video Process. **2015**(1), 1–11 (2015)
12. Izhikevich, E.M., Desai, N.S.: Relating STDP to BCM. Neural Comput. **15**(7), 1511–1523 (2003)
13. Kempter, R., Gerstner, W., Van Hemmen, J.L.: Spike-based compared to rate-based hebbian learning. Adv. Neural Inf. Process. Syst. **11**, 125–131 (1999)
14. Lass, Y., Abeles, M.: Transmission of information by the axon: I. Noise and memory in the myelinated nerve fiber of the frog. Biol. Cybern. **19**(2), 61–67 (1975)
15. Markram, H., Sakmann, B.: Action potentials propagating back into dendrites trigger changes in efficacy of single-axon synapses between layer V pyramidal neurons. Soc. Neurosci. Abstr. **21**, 2007 (1995)
16. Nessler, B., Pfeiffer, M., Maass, W.: STDP enables spiking neurons to detect hidden causes of their inputs. In: Advances in Neural Information Processing Systems, pp. 1357–1365 (2009)
17. Rao, R.P., Sejnowski, T.J.: Spike-timing-dependent hebbian plasticity as temporal difference learning. Neural Comput. **13**(10), 2221–2237 (2001)
18. Roberts, P.D.: Computational consequences of temporally asymmetric learning rules: I. Differential hebbian learning. J. Comput. Neurosci. **7**(3), 235–246 (1999)
19. Scellier, B., Bengio, Y.: Equilibrium propagation: bridging the gap between energy-based models and backpropagation. arXiv preprint arXiv:1602.05179 (2016)
20. Senn, W., Pfister, J.P.: Spike-timing-dependent plasticity, learning rules. In: Jaeger, D., Jung, R. (eds.) Encyclopedia of Computational Neuroscience, pp. 1–10. Springer, Heidelberg (2014)
21. Sjöström, P.J., Rancz, E.A., Roth, A., Häusser, M.: Dendritic excitability and synaptic plasticity. Physiol. Rev. **88**(2), 769–840 (2008)
22. Stevens, C.F., Wang, Y., et al.: Changes in reliability of synaptic function as a mechanism for plasticity. Nature **371**(6499), 704–707 (1994)
23. Waddington, A., Appleby, P.A., De Kamps, M., Cohen, N.: Triphasic spike-timing-dependent plasticity organizes networks to produce robust sequences of neural activity. Front. Comput. Neurosci. **6**, 88 (2012)
24. Xie, X., Seung, H.S.: Spike-based learning rules and stabilization of persistent neural activity. Adv. Neural Inf. Process. Syst. **12**, 199–208 (2000)

Cognitive Load Recognition Using Multi-channel Complex Network Method

Jian Shang[1], Wei Zhang[2], Jiang Xiong[2], and Qingshan Liu[1(✉)]

[1] School of Automation, Huazhong University of Science and Technology,
Wuhan 430074, China
{shangjian,qsliu}@hust.edu.cn
[2] College of Computer Science and Engineering,
Chongqing Three Gorges University, Chongqing 404000, China
xjcq123@sohu.com, cqec126@126.com

Abstract. Modeling the cognitive events of human beings is an interesting task, but finding effective representations from electroencephalogram (EEG) data is one of the challenges. Recently, complex network analysis has gained considerable attention in the time series analysis, but most of the analysis is devoted to investigating single time series or just the time domain statistical features. Herein, we propose a novel approach using the frequency domain features to construct connections between different EEG channels to generate a multi-channel network. First, we transform the EEG time series to a frequency domain feature using the spectrogram of three frequency bands. Next, we generate a multi-channel network using the space distance and the classification is based on the network structural features. The results indicate that the proposed method gets good performance and is more efficient than the deep learning method to some degrees.

Keywords: Cognitive events · Electroencephalogram (EEG) · Frequency domain features · Multi-channel complex network

1 Introduction

Recognizing individual's cognitive load is important in Brain-Computer Interfaces (BCI) and daily life. Cognitive load beyond individual's capacity could lead to overload state that may put too much pressure on the brain causing confusion and lower the learning ability [1]. EEG is a widely used signal which meatures changes of electrical voltage on the scalp directly for its high temporal resolution, non-invasion and relatively low cost. Here, we explore the capabilities of EEG for reflecting the cognitive activities. In fact, there's numerous research using continuous EEG time series and applying supervised learning algorithms such as support vector machines (SVMs) [2–4]. Deep neural networks

---

This work was supported in part by the National Natural Science Foundation of China under Grant 61473333.

F. Cong et al. (Eds.): ISNN 2017, Part I, LNCS 10261, pp. 466–474, 2017.
DOI: 10.1007/978-3-319-59072-1_55

have achieved great success in many recognition tasks such as computer vision, speech and text recognition [5–8]. Convolutional and recurrent neural networks have shown their potential in extracting representations from EEG signals [9–11]. Spampinato et al. [12] propose to read the mind and transfer human visual capabilities to computer vision methods using Recurrent Neural Network (RNN) and Convolutional Neural Network (CNN). In [4] Bashivan et al. propose a novel approach to learn robust representations from EEG data by transforming the time series into $2 - D$ images based on deep learning. They adopt CNN and Long Short-Term Memory (LSTM) network to deal with the single-frame image approach and the multi-frame approach and reach the state-of-art classification error on the cognitive load classification task dataset. But it costs a lot making images from EEG time series and sometimes lack of the necessary spatial coordinates of eletrodes.

There has been a growing interest in using networks to analyse time series over the last decade and some impressive results in recognition tasks have been achieved using EEG data, such as sleep stage classification, motor imagery and Alzheimer's disease recognition [13–15]. Several approaches have been proposed to transform time series into networks. He et al. [15] propose to analysis motor imagery EEG signals based on probabilistic graphical models [16], using bayesian network with gaussian distribution. Zhang and Small [17] have constructed complex networks using a single node to represent each cycle from pseudo-periodic time series. Wang et al. [14] use limited penetrable visibility graph (LPVG) and phase space method to map single EEG series into networks and study the difference between different regions. Diykh and Li [13] divide each EEG segment into several sub-segments and extract different statistical features to construct networks. The topology of networks can help better understand the relationships between time series. However, these studies are devoted to investigating single time series or just using the time domain features rather than the frequency domain. And the threshold to determine the connection of networks has to be set manually according to the classification performance, which is not automatic or smart.

In this paper, we present a novel automatic cognitive load classification method, which uses the frequency domain features to construct connections between different EEG channels to generate a multi-channel network. Our approach is different from the previous which attempts to use time domain features to represent the single channel EEG data. Instead, the frequency domain features such as power spectrum of each channel will be extracted as representations of all the 64 channels' EEG data. Here, the memory operations related three frequency bands, theta (4–7 Hz), alpha (8–12 Hz) and beta (13–30 Hz), will be used as separate measurement of each electrode. Then we adopt an adaptive threshold which is equal to the mean space distance to determine the connections of every two channels, and construct the multi-channel networks automatically. The network structural features such as degree distribution, clustering coefficient of each complex network will be extracted and fed to a classifier.

2 Main Method

This paper proposes an efficient automatic method to classify different cognitive load levels. Figure 1 illustrates the whole procedure of the proposed method. To evaluate the effectiveness of the method, the extracted features of multi-channel network and the original three bands' frequency features arc forwarded to the SVM classifiers. We also compare the classification error with the deep learning method in [4]. The details are explained in the following sections.

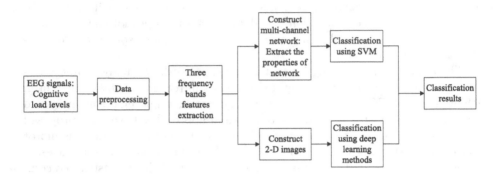

Fig. 1. Block diagram of the automatic cognitive load levels classification.

2.1 EEG Dataset

The dataset [4] used in this paper was collected during a working memory experiment which fifteen participants performed. The dataset was recorded from 64 electrodes which were placed over the scalp at the distances of 10% (the standard 10–10 location). We chose 13 sujects' datasets which were not damaged by the excessive noise and artifacts in the recording process. And the sampling frequency was 500 Hz. There were 2670 samples from all the 13 subjects, and every sample corresponded to a trial with the length of 3.5 s. Four classes 1, 2, 3, 4 define the different cognitive load levels, and each of the trial conditions contains 2, 4, 6, 8 characters corresponding to the four load levels respectively. The participants needed to indicated whether the test character was among the 2, 4, 6 or 8 characters to mark the load levels.

2.2 Signal Preprocessing

Our purposed method is to map the EEG time series to a network. But in the previous works, the time domain features were usually extracted to generate the network. Here, we propose a novel approach to extract the frequency domain of every channel to generate a multi-channel network. First, fast Fourier transform is performed on every channel of the EEG time series to estimate the power spectrum of the data. Then, we extract the three frequency bands, theta (4–7 Hz), alpha (8–12 Hz) and beta (13–30 Hz), which are related to the memory operations to represent every channel, and we use the sum of squared absolute values within each band as the features of every electrode. So every channel has a representation of three frequency band features.

2.3 Generate a Multi-channel Complex Network from the Frequency Domain Representations

A network can describe the relationship between the channels, and the nodes correspond to each channel, while the connection between each pair of nodes refers to the relationship. In our work, we employ the frequency domain representations to generate a network to classify the four different load levels. Each channel consisting three frequency bands' features can be considered as a data point or a basic node in the network. Then we have to determine the connections between every two nodes. The structural properties of each network can be calculated and forwarded to a classifier. As for the connection between two nodes, according to [17], we employ the space distance. Let $d(x_i, x_j)$ represents the space distance between two nodes x_i, x_j. If the distance is less or equals to a predetermined threshold, there is a connection between the two nodes; i.e.,

$$\text{if} \quad d(x_i, x_j) \leq D \tag{1}$$

where D is the predetermined threshold. Here, we choose an adaptive value of D as

$$D = \frac{\sum d(x_i, x_j)}{n * (n - 1)} \tag{2}$$

where n represents the number of nodes in the network.

The threshold is equal to the mean space distance, so we can have almost half of the connections existing, and the threshold is adaptive to every trial of the datasets. We don't need to choose the threshold manually, and the performance is good enough in the classification results. After the network is constructed, we get the adjacent matrix A from the connections of the network using

$$A(x_i, x_j) = \begin{cases} 1, & \text{if} \quad d(x_i, x_j) \leq D \\ 0, & \text{otherwise} \end{cases} \tag{3}$$

Figure 2 shows an example of constructed complex networks consisting of 64 nodes. The lines between every two nodes represent the connections of the network.

2.4 Network Structural Features

In order to analyze the structural features of networks, we calculate the degree distribution, clustering coefficient and average clustering coefficient as three different topological characteristics.

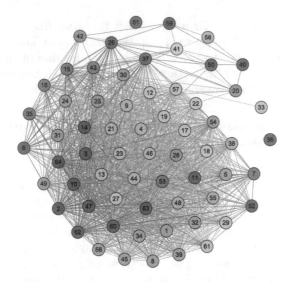

Fig. 2. The constructed complex network with 64 nodes. The number of each node refer to the corresponding electrode which implies the spatial position. Different color means different degree, and the nodes in the same color have the same degree. (Color figure online)

Degree Distribution. The degree refers to the number of connections of a node linking other nodes in the network. Here we investigate each node's degree and use the degree distribution as a feature for classification. As for the average degree (AD) which is popularly used in the previous work, because of the adaptive threshold we choose, the networks we get from different EEG time series will almost have the same size and average degrees, so average degree is not the most significant feature in our work. The average degree is evaluated in the experiment section by

$$\bar{\mathbf{D}} = \frac{1}{n}\sum D_i \tag{4}$$

where D_i is the degree of the ith node, and n is the number of nodes in the network.

Clustering Coefficient. The clustering coefficient of a node in a network quantifies how close its neighbours are. By analyzing the properties of its neighbor nodes, clustering coefficient measures the degree of local interconnections of a node, and it is a significant local feature for the network. In an undirected graph, if node i has k_i neighbours, there's $k_i * (k_i - 1)/2$ connections among the nodes within the neighbourhood. Thus, the clustering coefficient can be defined as

$$\mathbf{C}_i = \frac{2 * e_i}{k_i * (k_i - 1)} \tag{5}$$

where e_i is the number of links in the neighborhood of the node i. And the average clustering coefficient is the clustering coefficient over all the nodes

$$\bar{C} = \frac{1}{n} \sum C_i \tag{6}$$

2.5 Classification Methods

In this work, in order to demonstrate the effectiveness of our approach, we employ the standard Support Vector Machines (SVM) to classify the structural features of the network. And we compare our approach against the state-of-the-art deep learning method in experiments. The k-fold cross validation and the test error are used to evaluate the performance. Let $accuracy_k$ represents the k-th fold classification accuracy, then the performance of cross validation method can be evaluated using

$$\textbf{performance} = \frac{1}{k} \sum accuracy_k \tag{7}$$

where k is the number of folds which the dataset are divided to, and performance is the average accuracy of k folds.

3   Experimental Results

In this section, we evaluate the efficiency of our method using 10-fold cross validation and test error. We divide the dataset into 10 folds of equal size and every time choose one fold to test the performance of the classifier trained by all the other subsets, repeating 10 times. For comparisons, the original frequency features and the structural features of networks are forwarded to a standard SVM. We also compare the test error of the proposed approach with the popular deep learning method.

3.1   Multi-channel Network of Different Cognitive Load Levels

The method is based on the concept that the structure of different cognitive load level's network topology is different. It has been manifested that different brain activities or some mental illness may lead to the difference of the constructed complex network, but recognizing individual's different cognitive load levels is to identify heathy individual's same activity. So we need to figure out whether the network method is suitable for this task. Figure 3 illustrates the group differences of adjacency matrixes of typical networks of different cognitive loads. The adjacency matrixes have been transformed into grayscale image for visualization. The horizontal axis and vertical axis show the 64 nodes. Every white point means the connection between the corresponding nodes. On the contrary, the black means no connection. As you can see, although the number of connections may seem to be almost the same, the connection distribution shows strong difference apparently. In other words, the degree distribution could be the significant feature to identify the different cognitive load levels.

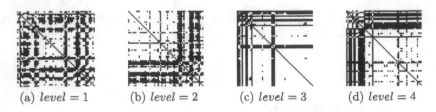

(a) *level* = 1   (b) *level* = 2   (c) *level* = 3   (d) *level* = 4

Fig. 3. Differences of adjacency matrixes of typical networks with different cognitive load levels.

3.2   Performance Comparison Among Different Features

In the experiment, we first compare the performance of different network structural features, the degree distribution and the clustering coefficient. Four factors need to be evaluated, the degree distribution, average degree, clustering coefficient and average clustering coefficient. The average accuracy of 10-fold cross validation is 33.86%, 32.28%, 85.66%, 78.65% and 86.33% respectively, as is shown in Fig. 4.

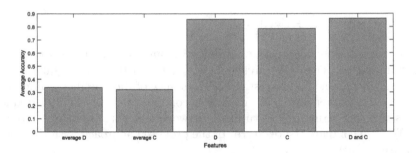

Fig. 4. The average accuracy among different features.

We can find that the average degree and average clustering coefficient could not recognize the right load levels. The reason is that we choose the adaptive threshold as the mean space distance, so almost half of the connections retained lead to the same average degree for different cognitive load levels. The two most important features, the degree distribution and the clustering coefficient, are extracted for the final classification.

3.3   Performance Comparison Among Different Models

As mentioned above, for the classification experiment, we extract the degree distribution and the clustering coefficient as features of the multi-channel networks using SVM classifier. We compare our results with the popular deep learning method in [4] using two measures (i.e., test error and number of parameters).

Table 1. Comparisons of test error (%) among different models

Models	Test error	Number of parameters
Multi-channel networks + SVM	10.36	–
ConvNet + 1D−Conv [4]	11.32	441 k
ConvNet + LSTM [4]	10.54	1.34 mil
ConvNet + LSTM/1D−Conv [4]	8.89	1.62 mil

The results are shown in Table 1, we can find that the proposed method is almost getting the same performance as the deep learning method. The performance of multi-channel network is equal or even better than two of the methods in the literature, but much lower than the CNN + LSTM with 1D−Conv method which uses multi-frame continuous images containing more information in the time domain. The number of parameters is tremendous and needs a long time to train the deep learning model and to generate the images from EEG time series, but the proposed multi-channel network method without any parameter is more efficient to some degrees.

4 Conclusions

This paper proposes a novel approach using the frequency domain features to construct connections between different EEG channels to construct a multi-channel network. First, we transform the EEG time series to a frequency domain feature using the spectrogram of three frequency bands. Next, we generate a multi-channel network using the space distance and the classification is based on the network structural features. The results indicate that the proposed method is more efficient than the deep learning method to some degrees. The proposed approach can be used to identify individual's cognitive load or even the mental states. However, in this paper the multi-channel network method only uses the frequency domain features to construct networks, but the time domain features are not considered. In the future work, we will explore the probability of using different parts of the time series to construct several networks and connect them together to learn more robust representations and make full use of the time domain information.

References

1. Sweller, J., Van Merrienboer, J.J., Paas, F.G.: Cognitive architecture and instructional design. Educ. Psychol. Rev. **10**, 251–296 (1998)
2. Subasi, A., Gursoy, M.I.: EEG signal classification using PCA, ICA, LDA and support vector machines. Expert Syst. Appl. **37**, 8659–8666 (2010)
3. Lotte, F., Congedo, M., Lécuyer, A., Lamarche, F., Arnaldi, B.: A review of classification algorithms for EEG-based brain-computer interfaces. J. Neural Eng. **4**, R1–R13 (2007)

4. Bashivan, P., Rish, I., Yeasin, M., Codella, N.: Learning representations from EEG with deep recurrent-convolutional neural networks. arXiv preprint arXiv:1511. 06448 (2015)
5. Graves, A., Liwicki, M., Bunke, H., Schmidhuber, J., Fernández, S.: Unconstrained on-line handwriting recognition with recurrent neural networks. In: Advances in Neural Information Processing Systems, pp. 577–584 (2008)
6. Karpathy, A., Toderici, G., Shetty, S., Leung, T., Sukthankar, R., Fei-Fei, L.: Large-scale video classification with convolutional neural networks. In: Advances in Neural Information Processing Systems, pp. 1725–1732 (2014)
7. Krizhevsky, A., Sutskever, I., Hinton, G.E.: Imagenet classification with deep convolutional neural networks. In: Proceedings of the Neural Information Processing Systems Conference and Workshop, pp. 1097–1105 (2012)
8. Zhang, X., LeCun, Y.: Text understanding from scratch. arXiv preprint arXiv:1502. 01710 (2015)
9. Mirowski, P., Madhavan, D., LeCun, Y., Kuzniecky, R.: Classification of patterns of EEG synchronization for seizure prediction. Clin. Neurophysiol. **120**, 1927–1940 (2009)
10. Cecotti, H., Graser, A.: Convolutional neural networks for P300 detection with application to brain-computer interfaces. IEEE Trans. Pattern Anal. Mach. Intell. **33**, 433–445 (2011)
11. Güler, N.F., Übeyli, E.D., Güler, I.: Recurrent neural networks employing Lyapunov exponents for EEG signals classification. Expert Syst. Appl. **29**, 506–514 (2005)
12. Spampinato, C., Palazzo, S., Kavasidis, I., Giordano, D., Shah, M., Souly, N.: Deep learning human mind for automated visual classification. arXiv preprint arXiv:1609.00344 (2016)
13. Diykh, M., Li, Y.: Complex networks approach for EEG signal sleep stages classification. Expert Syst. Appl. **63**, 241–248 (2016)
14. Wang, J., Yang, C., Wang, R., Yu, H., Cao, Y., Liu, J.: Functional brain networks in Alzheimers disease: EEG analysis based on limited penetrable visibility graph and phase space method. Physica A: Stat. Mech. Appl. **460**, 174–187 (2016)
15. He, L., Liu, B., Hu, D., Wen, Y., Wan, M., Long, J.: Motor imagery EEG signals analysis based on Bayesian network with Gaussian distribution. Neurocomputing **188**, 217–224 (2016)
16. Koller, D., Friedman, N.: Probabilistic Graphical Models: Principles and Techniques. MIT Press, Cambridge (2009)
17. Zhang, J., Small, M.: Complex network from pseudoperiodic time series: topology versus dynamics. Phys. Rev. Lett. **96**, 238701 (2006)

Event-Triggering Sampling Based Synchronization of Delayed Complex Dynamical Networks: An M-matrix Approach

Yang Tang[✉]

The Key Laboratory of Advanced Control and Optimization
for Chemical Processes, Ministry of Education,
East China University of Science and Technology, Shanghai 200237, China
tangtany@gmail.com

Abstract. In this technical note, the synchronization problem is investigated for delayed complex dynamical networks. A novel distributed event-triggered sampling rule is proposed, i.e., one node can decide its own event time via its own state value and the state values of its neighbor agents as long as the locally-computed error exceeds the given state-dependent threshold. The aim here is to design controllers and some required events such that the considered complex dynamical networks can achieve synchronization. Then the M-matrix method is applied to derive some criteria in the form of eigenvalue-based inequality for achieving the synchronization, and the Zeno behavior can be avoided as well. Finally, a numerical example is presented for demonstrating the availability and effectiveness of the main results.

Keywords: Event-triggered control mechanism · Synchronization · Delayed complex dynamical network · M-matrix · Directed spanning tree · Eigenvalue-based inequality

1 Introduction

In recent decades, much more researches are paying close attention to the research progress of complex dynamic networks, and much more excellent work has been done to meet the demand and realize the desired goal in real-world applications. It has been recognized that, complex dynamical networks where each node can be regarded as a nonlinear dynamical system, can be used to model many real-world systems, such as self-organizing biological swarms, metabolic and gene networks [7,16]. Given this, the synchronization problem of complex dynamical networks has been a research hotspot during the past decades, which can be typically applied to a variety of realms such as secure communication, formation control of mobile robots and information processing [14,15]. To analyze the synchronization problem of the complex dynamical networks, the foremost is designing the appropriate control protocol with different methods, including the sliding-mode control method, the pinning control method, the continuous-time

© Springer International Publishing AG 2017
F. Cong et al. (Eds.): ISNN 2017, Part I, LNCS 10261, pp. 475–482, 2017.
DOI: 10.1007/978-3-319-59072-1_56

feedback control techniques and the mixed optimization approaches [14,17,18]. And an increasing number of excellent investigations of the synchronization and their wide applications in various fields have been presented [11].

Considering the actual benefits for reducing the unnecessary resource consumption, designing the distributed control algorithm depending on the state of each node is more in line with the requirement, then the event-triggered sampling control mechanism can be used [1,4,9], where the number of control updates can be reduced can then the resource can be saved to a certain degree. That is, with appropriately designed triggering events, it is divinable that not only the laxation of bandwidth occupation but also some desired properties including stability and convergence of the close-loop system can be all guaranteed. More specifically, by designing the state-based trigger conditions, the error can grow without destroying the stability of the error system, and the latest states will be transmitted to the controller at the transmission instant decided by the event-trigger condition is violated, which can surely reduces the transmission communication and lengthy computations. In recent years, various event-triggered control schemes have been widely investigated [2,3,12,13], just to name a few. [12] considers the sampled-data synchronization control problem of dynamical networks, where the sampling period is time-varying switching between two different values, which is not flexible enough compared with the event-triggering sampling where the sampling instants are depending on the states of nodes. Moreover, the synchronization conditions of [12] derived by Gronwall's inequality and Jenson inequality are complicated to analyze. In addition, [2,3,10,13] respectively investigate the synchronization of multi-agent systems by means of the different event-triggering sampling mechanisms. However, all these investigations above mentioned are not considering time delays. Therefore, to release the conservativeness of the strong connectivity for the network, and to improve the form of the synchronization condition and to consider time delays as well, this paper investigates the synchronization of delayed complex dynamic networks without confining the strong connectivity of the network but with the directed spanning tree based on the designed event-triggering sampling control mechanism.

2   Problem Statement

The algebraic graph theory can be checked in [3,10,11]. In this paper, the general complex networks with time delay is considered, which can be described as follows:

$$\dot{x}_i(t) = f(t, x_i(t), x_i(t - \tau)) + c \sum_{j=1}^{N} a_{ij}(x_j(t) - x_i(t)), \quad i = 1, 2, \cdots, N, \quad (1)$$

where $\tau > 0$ denotes the time delay of node i. $x_i(t) = (x_{i1}(t), x_{i2}(t), \cdots, x_{in}(t))^T \in \mathbb{R}^n$. $f(t, x_i, x_i(t-\tau)) = \left(f_1(t, x_i, x_i(t-\tau)), \cdots, f_n(t, x_i, x_i(t-\tau))\right)^T \in \mathbb{R}^n$ is a continuous vector-valued function with time delay τ. The initial condition of network (1) is given as $x_i(r) = \phi_i(r) \in \mathcal{C}([-\tau, 0], \mathbb{R}^n)$, $i = 1, 2, \cdots, N$, in which $\mathcal{C}([-\tau, 0], \mathbb{R}^n)$ denotes a class of continuous functions mapping $[-\tau, 0]$ into $\mathbb{R}^n$.

The virtual leader for complex network (1) can be described as follows:

$$\dot{s}(t) = f(t, s(t), s(t - \tau)), \tag{2}$$

where $x_0(t) = s(t) = \left(s_1(t), s_2(t), \cdots, s_n(t)\right)^T \in \mathbb{R}^n$.

The control goal is to let all nodes follow the virtual leader asymptotically: $x_i(t) \to s(t)$ as $t \to \infty$. To introduce the decentralized event-triggered control strategies for (1), define the triggering time sequence of node i as $t_0 = t_0^i, t_1^i, \cdots, t_k^i, \cdots$, at which node i measures its own state and obtain the neighbors' states $x_j(t), j \in N_i$ at $t = t_k^i, k \in \mathbb{Z}_0^+$. As the target points, the measurements $x_i(t_k^i)$ and $x_j(t_k^i), j \in \mathcal{N}_i$ remain unchanged until the next triggering time instant t_{k+1}^i comes at which the individual triggering event for i occurs. Thus the distributed event-triggered control strategies for network (1) can be described as follows, for $i = 1, 2, \cdots, N$,

$$\dot{x}_i(t) = f(t, x_i(t), x_i(t - \tau)) + c \sum_{j=1}^{N} a_{ij}\left(x_j(t_k^i) - x_i(t_k^i)\right) + cb_i\left(s(t_k^i) - x_i(t_k^i)\right), \tag{3}$$

Before the synchronization is achieved, for any $t \in [t_k^i, t_{k+1}^i), i = 1, 2, \cdots, N$, some necessary measurement errors are defined as: $e_{xi}(t) = x_i(t_k^i) - x_i(t)$, $e_{xi}(t) \in \mathbb{R}^{Nn}$, $e_{xij}(t) = x_j(t_k^i) - x_j(t)$, $e_{xij}(t) \in \mathbb{R}^{Nn}$, $\tilde{e}_{xi}(t) = \left(e_{xi1}^T(t), e_{xi2}^T(t), \cdots, e_{xiN}^T(t)\right)^T \in \mathbb{R}^{N^2n}$, where $e_{xij}(t) \equiv 0$ if $j \notin \mathcal{N}_i$. If $i = 0$ for the leader, $s(t) = x_0(t)$ is defined for the following analysis. Define $\hat{x}_i(t) = x_i(t) - s(t), i = 1, 2, \cdots, N, \hat{\mathcal{A}} = \text{diag}\{\mathcal{A}_1, \cdots, \mathcal{A}_N\} \in \mathbb{R}^{N \times N^2}, D = \text{diag}\{d_1, \cdots, d_N\} \in \mathbb{R}^N$ with $d_i = \sum_{j \in \mathcal{N}_i} a_{ij}$. Let $\hat{x}(t) = [\hat{x}_1^T(t), \cdots, \hat{x}_N^T(t)]^T \in \mathbb{R}^{Nn}$, $\tilde{e}_x(t) = [\tilde{e}_{x1}^T(t), \cdots, \tilde{e}_{xN}^T(t)]^T \in \mathbb{R}^{N^2n}$, $e_x(t) = [e_{x1}^T(t), \cdots, e_{xN}^T(t)]^T \in \mathbb{R}^{Nn}$, $e_{x0}(t) = [e_{x10}^T(t), \cdots, e_{xN0}^T(t)]^T \in \mathbb{R}^{Nn}$ $F(t, x(t)) = \left[f^T(t, x_1(t)), \cdots, f^T(t, x_N(t))\right]^T \in \mathbb{R}^{Nn}$, thereby error dynamical system (2) and (3) can be rewritten as the following compact matrix vector form:

$$\dot{\hat{x}}(t) = F(t, x(t), x(t - \tau)) - I_N \otimes f(t, s(t), s(t - \tau)) - \left[c(L + B) \otimes I_n\right]\hat{x}(t)$$
$$+ c(\hat{\mathcal{A}} \otimes I_n)\tilde{e}_x(t) - \left[c(D + B) \otimes I_n\right]e_x(t) + c(B \otimes I_n)e_{x0}(t). \tag{4}$$

For every node i, design the decentralized sampling event in the following:

$$E_i(t) = c(b_i + d_i)\|e_{xi}(t)\| + c\|\mathcal{A}_i\|\|\tilde{e}_{xi}(t)\| + cb_i\|e_{xi0}(t)\| - \beta_i H_i(t) = 0, \quad \beta_i > 0, \tag{5}$$

where $\mathcal{A}_i$ denotes the ith row of matrix $\mathcal{A}$ and $H_i(t) = \sqrt{H_{i1}(t) + H_{i2}(t)}$, $d_i = \sum_{j \in \mathcal{N}_i} a_{ij}$, $H_{i1}(t) = \sum_{j \in \mathcal{N}_i} \|a_{ij}(x_j(t) - x_i(t))\|^2$, $H_{i2}(t) = b_i\|x_0(t) - x_i(t)\|^2$.

3   Main Results

In the following, some lemmas and an assumption are presented for deriving the main results.

Lemma 1 [6]. *For a nonsingular matrix $A = (a_{ij}) \in \mathbb{R}^{n \times n}$ with $a_{ij} \leq 0$ $(i \neq j)$, the following statements are equivalent: (1) A is an M-matrix; (2) All eigenvalues of A have positive real parts, i.e., $\mathcal{R}(\lambda_i(A)) > 0$ for all $i = 1, 2, \cdots, n$; (3) A^{-1} exists, and $A^{-1} \geq 0$; (4) There exists a positive definite diagonal matrix $\Xi = diag(\xi_1, \xi_2, \cdots, \xi_n)$ such that $\Xi A + A^T \Xi > 0$, which indicates that $(\Xi A)_s > 0$.*

Lemma 2 [11]. *The matrix $L + B$ is an M-matrix if and only if $\tilde{\mathcal{G}}$ has a directed spanning tree.*

Assumption 1. *For the nonlinear vector-valued function $f = (f_1, f_2, \cdots, f_n) \in \mathbb{R}^n$, there exist two constant matrices $W = (w_{ij})_{n \times n}$ and $M = (m_{ij})_{n \times n}$ where w_{ij}, $m_{ij} \geq 0$ such that $|f_i(t, x(t), x(t - \tau)) - f_i(t, y(t), y(t - \tau))| \leq \sum_{j=1}^{n} \left(w_{ij}|x_j(t) - y_j(t)| + m_{ij}|x_j(t - \tau) - y_j(t - \tau)| \right)$, $i = 1, 2, \cdots, n$, for all $x = (x_1, x_2, \cdots, x_n)^T$, $y = (y_1, y_2, \cdots, y_n)^T \in \mathbb{R}^n$.*

Theorem 1. *Consider complex networks (1), (2) and (3) with event-triggered sampling rule (5). Assume that Assumption 1 holds and $\tilde{\mathcal{G}}$ has a directed spanning tree. If for any initial condition $x_i(t_0) \in \mathbb{R}^n$, the following inequality holds: $\min_{1 \leq i \leq N} \mathcal{R}(\lambda_{\min}(L + B)) > \rho$, where $\rho = \max \left\{ \left(\sqrt{2\lambda_{\max}(W^T W)} + \frac{1}{2}\sqrt{2\lambda_{\max}(M^T M)} + k\beta h \right) / (c - 1), \frac{1}{2}\sqrt{2\lambda_{\max}(M^T M)} \right\}$ with $c > 1$, $k = \max \left\{ \frac{\|(B+D) \otimes I_n\|}{(b_i + d_i)_m}, \frac{\|\hat{A} \otimes I_n\|}{\|A_m\|}, \frac{b_M \|I_N \otimes I_n\|}{b_m} \right\}$, $(b_i + d_i)_m = \min_{1 \leq i \leq N} \left\{ b_i + d_i, b_i + d_i > 0 \right\}$, $\beta = \max_{1 \leq i \leq N}\{\beta_i\}$, $h = \sqrt{2(a^*)^2 \cdot (N + \mathcal{N}^*) + b_M}$, $\mathcal{N}^* = \max_{1 \leq i \leq N} \{|\mathcal{N}_i|\}$, $b_M = \max_{1 \leq i \leq N}\{b_i\}$, $a^* = \max_{1 \leq i \leq N}\{a_{iM}\}$, $a_{iM} = \max_{j \in \mathcal{N}_i}\{a_{ij}\}$, $b_m = \min_{1 \leq i \leq N} \left\{ b_i, b_i > 0 \right\}$ and $\|A_m\| = \min_{1 \leq i \leq N} \left\{ \|A_i\|, \|A_i\| > 0 \right\}$, then complex network (1) can be globally asymptotically synchronized to the homogeneous trajectory (2). Furthermore, the Zeno-behavior can be avoided.*

Proof. Define λ_i as eigenvalues of $(L + B)$, $i = 1, 2, \cdots, N$. According to Lemmas 1 and 2, the fact that $(L + B)$ is an M-matrix can be guaranteed by the presented condition that $\tilde{\mathcal{G}}$ has a directed spanning tree, which results in $\mathcal{R}(\lambda_i(L + B)) > 0$. One can easily get that $(c\lambda_i - \alpha)$ is an eigenvalue of $((L + B)) - \rho I_N$ and $\mathcal{R}(\lambda_i - \rho) > 0$, deriving that $\left(c(L + B)) - \alpha I_N \right)$ is an M-matrix, where $\rho = \max \left\{ \left(2\lambda_{\max}(W^T W) + \lambda_{\max}(M^T M) + k\beta h \right) / (c - 1), \lambda_{\max}(M^T M) \right\}$, $i = 1, 2, \cdots, N$. Then, according to Lemma 1, it can be checked that there exists $\Xi = diag(\xi_1, \xi_1, \cdots, \xi_N) > 0$ such that $\left[\Xi \left((L + B) - \rho I_N \right) \right]_s > 0$.

Construct the following Lyapunov-Krasovskii functional $V(t) = \frac{1}{2}\hat{x}^T(t)(\Xi \otimes I_n)\hat{x}(t) + \int_{t-\tau}^{t} \hat{x}^T(\zeta)\Big(\Xi(L+B) \otimes I_n\Big)\hat{x}(\zeta)d\zeta$. For $\hat{x}(t) \neq 0$, $V(t) > 0$ can be easily guaranteed by the fact that $\tilde{\mathcal{G}}$ has a directed spanning tree according to Lemmas 1 and 2.

Calculating the time derivative of $V(t)$ along the trajectory (4), one can get $\dot{V}(t) = \hat{x}^T(t)(\Xi \otimes I_n)\Big(F(t, x(t), x(t-\tau)) - I_N \otimes f(t, s(t), s(t-\tau))\Big) - \hat{x}^T(t-\tau)\Big(\Xi(L+B) \otimes I_n\Big)\hat{x}(t-\tau) + \hat{x}^T(t)\Big(\Xi(L+B) \otimes I_n - c\Xi(L+B) \otimes I_n\Big)\hat{x}(t) + c\hat{x}^T(t)(\Xi \otimes I_n)Q(t)$, where $Q(t) = (\hat{A} \otimes I_n)\tilde{e}_x(t) - \Big[(D+B) \otimes I_n\Big]e_x(t) + (B \otimes I_n)e_{x0}(t)$.

For $Q(t) = (\hat{A} \otimes I_n)\tilde{e}_x(t) - \Big[(D+B) \otimes I_n\Big]e_x(t) + (B \otimes I_n)e_{x0}(t)$, we have the following analysis. Note that decentralized event-triggered rule (5) implies the following three inequalities hold for all $t \geq t_0$, $c(b_i + d_i)\|e_{xi}(t)\| \leq \alpha_{i1}(t)\beta_i H_i(t)$, $c\|\mathcal{A}_i\|\|\tilde{e}_{xi}(t)\| \leq \alpha_{i2}(t)\beta_i H_i(t)$, and $cb_i\|e_{xi0}(t)\| \leq \alpha_{i3}(t)\beta_i H_i(t)$, where $\alpha_{i1}(t) + \alpha_{i2}(t) + \alpha_{i3}(t) = 1$ and $\alpha_{i1}(t), \alpha_{i2}(t), \alpha_{i3}(t) > 0$. It can be easily checked that the event is triggered if and only if the three equalities hold simultaneously, at which the node i samples its own states and the neighbors' states at time instant $t = t_k^i$, $k \in \mathbb{Z}_0^+$. By the fact that $\|x - y\|^2 \leq \|x\|^2 + \|y\|^2 + 2\|x\|\|y\| \leq 2\|x\|^2 + 2\|y\|^2$, $\forall x, y \in \mathbb{R}^n$, it can be obtained that $H_i^2(t) = H_{i1}(t) + H_{i2}(t) \leq 2a_{iM}^2 \sum_{j \in \mathcal{N}_i} \Big(\|\hat{x}_j\|^2 + \|\hat{x}_i(t)\|^2\Big) + b_M\|\hat{x}_i(t)\|^2$, in which $a_{iM} = \max_{j \in \mathcal{N}_i}\{a_{ij}\}$ and $b_M = \max_{1 \leq i \leq N}\{b_i\}$. Then it can be checked that $c^2(b_i + d_i)^2\|e_{xi}(t)\|^2 \leq \alpha_{i1}^2(t)\beta_i^2 H_i^2(t) \leq \alpha_{i1}^2(t)\beta_i^2\Big(2a_{iM}^2 \sum_{j \in \mathcal{N}_i} \big(\|\hat{x}_j\|^2 + \|\hat{x}_i(t)\|^2\big) + b_M\|\hat{x}_i(t)\|^2\Big)$, i.e., $c^2(b_i + d_i)_m^2 \sum_{i=1}^{N}\|e_{xi}(t)\|^2 \leq \beta^2 \cdot \Big(2(a^*)^2 \cdot (N + \mathcal{N}^*) + b_M\Big)\|\hat{x}(t)\|^2$, in which $\beta = \max_{1 \leq i \leq N}\{\beta_i\}$, $a^* = \max_{1 \leq i \leq N}\{a_{iM}\}$, $\mathcal{N}^* = \max_{1 \leq i \leq N}\{|\mathcal{N}_i|\}$ and $(b_i + d_i)_m = \min_{1 \leq i \leq N}\{b_i + d_i, b_i + d_i > 0\}$. Then we have $\sum_{i=1}^{N}\|e_{xi}(t)\|^2 \leq \frac{1}{c^2(b_i+d_i)_m^2}\beta^2 h^2\|\hat{x}(t)\|^2$, where $h = \sqrt{2(a^*)^2 \cdot (N + \mathcal{N}^*) + b_M}$, which implies that $\|(B+D) \otimes I_n\| \cdot \|e_x(t)\| \leq \frac{\|(B+D) \otimes I_n\|}{c(b_i+d_i)_m}\beta h\|\hat{x}(t)\|$. By similar procedures above, one can obtain that $\|\hat{A} \otimes I_n\|\|\tilde{e}_x(t)\| \leq \frac{\|\hat{A} \otimes I_n\|}{c\|\mathcal{A}_m\|}\beta h\|\hat{x}(t)\|$ and $\|B \otimes I_n\|\|e_{x0}(t)\| \leq \frac{b_M\|I_N \otimes I_n\|}{cb_m}\beta h\|\hat{x}(t)\|$. Let $k = \max\left\{\frac{\|(B+D) \otimes I_n\|}{(b_i+d_i)_m}, \frac{\|\hat{A} \otimes I_n\|}{\|\mathcal{A}_m\|}, \frac{b_M\|I_N \otimes I_n\|}{b_m}\right\}$, which implies that $\|(B+D) \otimes I_n\| \cdot \|e_x(t)\| + \|\hat{A} \otimes I_n\|\|\tilde{e}_x(t)\| + \|B \otimes I_n\|\|e_{x0}(t)\| \leq \frac{1}{c}k\beta h\|\hat{x}(t)\|$. Then we can easily get that $c\hat{x}^T(t)Q(t) \leq k\beta h \cdot \hat{x}^T(t)\hat{x}(t)$. By Assumption 1, one gets that $\hat{x}^T(t)(F(t, x(t), x(t-\tau)) - I_N \otimes f(t, s(t), s(t-\tau))) \leq \Big(\sqrt{2\lambda_{\max}(W^T W)} + \frac{1}{2}\sqrt{2\lambda_{\max}(M^T M)}\Big)\hat{x}^T(t)\hat{x}(t) + \frac{1}{2}\sqrt{2\lambda_{\max}(M^T M)}\hat{x}^T(t-\tau)\hat{x}(t-\tau)$. Combined with analysis above, one can easily get $\dot{V}(t) \leq -(c-1)\hat{x}^T(t)\Big(\Big[\Xi\big((L+B) - \rho I_N\big)\Big]_s \otimes I_n\Big)\hat{x}(t) - \hat{x}^T(t-\tau)\Big(\Big[\Xi((L+B) - \rho I_N)\Big]_s \otimes I_n\Big)\hat{x}(t-\tau)$. Then $\dot{V}(t) \leq 0$ can be guaranteed, and $\dot{V}(t) = 0$ if and only if $\hat{x}(t) = 0$. That is, the set $\tilde{S} = \{\hat{x}(t)|\hat{x}(t) = 0\}$ is the largest invariant set of the set $\tilde{D} = \{\hat{x}(t)|\dot{V}(t) = 0\}$ for network (4). By LaSalle's invariance principle

[8], for network (4) with any initial condition, every solution of it approaches $\hat{S}$ as $t \to \infty$, indicating that $\hat{x}_i^T(t) \to 0$. Therefore, the synchronization can be globally achieved under decentralized event-triggered sampling rule (5).

It remains to show the Zeno-behavior can be avoided, i.e., the set $\Delta_{k+1}^i = \{t_{k+1}^i - t_k^i | i = 1, 2, \cdots, N, k \in \mathbb{Z}_0^+\}$ has a lower bound which is strictly positive. Let $\{t_k^i\}_0^\infty$ be a sequence of trigger time of agent i at which one can get $e_{x_i}(t_k^i) = 0$, $e_{x_{ij}}(t_k^i) = 0$. Suppose $\|\dot{x}_i(t)\| \leq \omega_{\dot{x}}$, where $\omega_{\dot{x}} > 0$. For any agent i and $t \in [t_k^i, t_{k+1}^i)$, one can obtain that $\|e_{x_i}(t)\| = \|\int_{t_k^i}^t \dot{e}_{x_i}(s)ds\| \leq \omega_{\dot{x}}(t - t_k^i)$, $\|e_{x_{ij}}(t)\| = \|\int_{t_k^i}^t \dot{e}_{x_{ij}}(s)ds\| \leq \omega_{\dot{x}}(t - t_k^i)$, then, one can derive that $\|e_{xi0}(t)\| \leq \omega_{\dot{x}}(t - t_k^i)$, $\|\tilde{e}_{xi}(t)\| = \sqrt{\sum_{j \in \mathcal{N}_i} \|e_{x_{ij}}(t)\|^2} \leq \sqrt{|\mathcal{N}_i|}\omega_{\dot{x}}(t - t_k^i)$. Then, one can obtain $c(b_i + d_i)\|e_{xi}(t)\| + c\|\mathcal{A}_i\|\|\tilde{e}_{xi}(t)\| + cb_i\|e_{xi0}(t)\| \leq c(2b_i + d_i + \|\mathcal{A}_i\|\sqrt{|\mathcal{N}_i|})\omega_{\dot{x}}(t - t_k^i)$. Event triggered rule (5) shows that the next event will be triggered once the trigger function $E_i(t) = 0$, i.e., agent i is sampled at the time instant t_{k+1}^i when $0 = E_i(t) = c(b_i + d_i)\|e_{xi}(t)\| + c\|\mathcal{A}_i\|\|\tilde{e}_{xi}(t)\| + cb_i\|e_{xi0}(t)\| - \beta_i H_i(t)$. In the case where events need to be necessarily triggered, that is $\beta_i H_i(t) > 0$, there exists a constant $\omega_1 > 0$ such that $\beta_i H_i(t) \geq \omega_1 > 0$ before synchronization is arrived. Then one can get $c(2b_i + d_i + \|\mathcal{A}_i\|\sqrt{|\mathcal{N}_i|})\omega_{\dot{x}}(t - t_k^i) \geq \omega_1 > 0$, that is, $\Delta_{k+1}^i = t_{k+1}^i - t_k^i \geq \frac{\omega_1}{c(2b_i + d_i + \|\mathcal{A}_i\|\sqrt{|\mathcal{N}_i|})\omega_{\dot{x}}} > 0$, which implies that the Zeno-behavior can be avoided before the synchronization is reached. Therefore, the proof of Theorem 1 is completed.   □

4   Example

In this example, based on event-triggering sampling rule (5), consider the synchronization of a linearly coupled delayed complex network consisting a host system as the leader and four subordinate systems, i.e., $\dot{x}_i(t) = f(t, x_i(t), x_i(t - \tau)) + c\sum_{j=1}^N a_{ij}(x_j(t) - x_i(t))$, $i = 1, 2, 3, 4$, $\dot{x}_0(t) = s(t) = f(t, s(t), s(t - \tau))$, where $x_i(t) = (x_{i1}(t), x_{i2}(t))^T$ denotes the state variable of the ith node. Choose $c = 10$ and $\tau = 1.0$, and the nonlinear function f is defined by [5] and [11], i.e., $f\left(t, x_i(t), x_i(t - \tau)\right) = -x_i(t) + Ag(x_i(t)) + Bg(x_i(t - \tau))$ with $g(x_i) = 0.5\left(|x_{i1} + 1| - |x_{i1} - 1|, |x_{i2} + 1| - |x_{i2} - 1|\right)^T$, $A = \begin{pmatrix} 1 + \frac{\pi}{4} & 20 \\ 0.1 & 1 + \frac{\pi}{4} \end{pmatrix}$, $B = \begin{pmatrix} -1.3\sqrt{2}\frac{\pi}{4} & 0.1 \\ 0.1 & -1.3\sqrt{2}\frac{\pi}{4} \end{pmatrix}$. Let the weighted adjacency matrix be $\mathcal{A} = \begin{pmatrix} 0 & 1.35 & 0 & 1.1 \\ 2.2 & 0 & 1 & 2.1 \\ 2.25 & 0 & 0 & 2.25 \\ 1.6 & 1.25 & 0.8 & 0 \end{pmatrix}$ and leader adjacency matrix be $B = $ diag$\{4.5, 4.5, 4.75, 4.75\}$. It can be verified that the eigenvalues of $(L + B)$ are 2.2020, 5.4686, 7.4829, 9.4465, i.e., $\min_{1 \leq i \leq N} \mathcal{R}(\lambda_i(L + B)) = 2.2020$. By $d_i = \sum_{j \in \mathcal{N}_i} a_{ij}$, one can get $d_1 = 2.45$, $d_2 = 5.3$, $d_3 = 4.5$, $d_3 = 3.65$. Similarly, $a_{1M} = 1.35$, $a_{2M} = 2.2$, $a_{3M} = 2.25$, $a_{4M} = 1.6$, $b_m = 4.5$, $b_M = 4.75$,

$\|\mathcal{A}_1\| = 1.7414, \|\mathcal{A}_2\| = 3.2016, \|\mathcal{A}_3\| = 3.1820, \|\mathcal{A}_4\| = 2.1823$. Then, one can get $\mathcal{N}^* = \max_{1 \leq i \leq N}\{|\mathcal{N}_i|\} = 3$, $(b_i + d_i)_m = \min_{1 \leq i \leq N}\{b_i + d_i, b_i + d_i > 0\} = 6.95$, $a^* = \max_{1 \leq i \leq N}\{a_{iM}\} = 2.25$, $\|\mathcal{A}_m\| = \min_{1 \leq i \leq N}\{\|\mathcal{A}_i\|, \mathcal{A}_i > 0\} = 1.7414$, $h = 8.6963$, $k = 1.8961$. Choose $\beta_i = 0.42$, i.e., $\beta = \max_{1 \leq i \leq N}\{\beta_i\} = 0.42$.

In addition, it is easy to check that the function f given above satisfies Assumption 1 with the following two matrices: $W = \begin{pmatrix} 3.5998 & 1.0472 \\ 1.4571 & 2.3099 \end{pmatrix}$, $M = \begin{pmatrix} 1.6998 & 1.0472 \\ 1.4571 & 1.9999 \end{pmatrix}$, where $\lambda_{\max}(W^T W) = 18.9077$ and $\lambda_{\max}(M^T M) = 9.5740$. By simple calculations, one can obtain $\left(\sqrt{2\lambda_{\max}(W^T W)} + \frac{1}{2}\sqrt{2\lambda_{\max}(M^T M)} + k\beta h\right)/(c-1) = 1.6959 < 2.1879 = \frac{1}{2}\sqrt{2\lambda_{\max}(M^T M)}$, which shows that $\rho = 2.1879 < 2.2020 = \min_{1 \leq i \leq N} \mathcal{R}(\lambda_i(L+B))$. Hence, the conditions in Theorem 1 are satisfied as well. Choose the initial states as: $\begin{pmatrix} x_0(0) & x_1(0) & x_2(0) & x_3(0) & x_4(0) \end{pmatrix} = \begin{pmatrix} 1.00 & 1.16 & 1.22 & 1.26 & 1.12 \\ 1.32 & 0.87 & 0.34 & 1.25 & 0.79 \end{pmatrix}$. Therefore, the synchronization of the considered complex network can be reached under event-triggering sampling rule (5).

5   Conclusion

Based on the graph theory and M-matrix approach, this paper investigates the synchronization for delayed complex dynamical networks. The decentralized event-triggering sampling rule is designed where the communication of information among agents is reduced by decreasing the frequency of controller updates. Based on the designed event-triggering sampling rule, each node measures its own state and obtain the neighbors' states once the event-trigger error exceeds the given state-based and error-based threshold. By combining with the Lyapunov function and M-matrix approach, some event-triggering based conditions are established to ensure the synchronization of the delayed complex dynamical networks with the assumption of the directed spanning tree, including the exclusion of the Zeno behavior. In addition, a numerical example is given to verify the feasibility and availability of the main results.

Acknowledgments. This paper was supported by the National Natural Science Foundation of China (Grant No. 61673176).

References

1. Abdelrahim, M., Postoyan, R., Daafouz, J., Nešić, D.: Robust event-triggered output feedback controllers for nonlinear systems. Automatica **75**, 96–108 (2017)
2. Dimarogonas, D.V., Frazzoli, E., Johansson, K.H.: Distributed event-triggered control for multi-agent systems. IEEE Trans. Autom. Control **57**(5), 1291–1297 (2012)
3. Fan, Y., Feng, G., Wang, Y., Song, C.: Distributed event-triggered control of multi-agent systems with combinational measurements. Automatica **49**(2), 671–675 (2013)

4. Forni, F., Galeani, S., Nešić, D., Zaccarian, L.: Event-triggered transmission for linear control over communication channels. Automatica **50**(2), 490–498 (2014)
5. Gilli, M.: Strange attractors in delayed cellular neural networks. IEEE Trans. Circuits Syst. I Fundam. Theory Appl. **40**(11), 849–853 (1993)
6. Hom, R.A., Johnson, C.R.: Topics in Matrix Analysis. Cambridge UP, New York (1991)
7. Kauffman, S.A.: Metabolic stability and epigenesis in randomly constructed genetic nets. J. Theor. Biol. **22**(3), 437–467 (1969)
8. Khalil, H.K., Grizzle, J.: Nonlinear Systems, vol. 3. Prentice Hall, New Jersey (1996)
9. Postoyan, R., Bragagnolo, M.C., Galbrun, E., Daafouz, J., Nešić, D., Castelan, E.B.: Event-triggered tracking control of unicycle mobile robots. Automatica **52**, 302–308 (2015)
10. Qin, J., Yu, C., Gao, H.: Coordination for linear multiagent systems with dynamic interaction topology in the leader-following framework. IEEE Trans. Ind. Electron. **61**(5), 2412–2422 (2014)
11. Song, Q., Liu, F., Cao, J., Yu, W.: Pinning-controllability analysis of complex networks: an M-matrix approach. IEEE Trans. Circuits Syst. I Regul. Pap. **59**(11), 2692–2701 (2012)
12. Shen, B., Wang, Z., Liu, X.: Sampled-data synchronization control of dynamical networks with stochastic sampling. IEEE Trans. Automat. Control **57**(10), 2644–2650 (2012)
13. Tabuada, P.: Event-triggered real-time scheduling of stabilizing control tasks. IEEE Trans. Automat. Control **52**(9), 1680–1685 (2007)
14. Tang, Y., Gao, H., Zhang, W., Kurths, J.: Leader-following consensus of a class of stochastic delayed multi-agent systems with partial mixed impulses. Automatica **53**, 346–354 (2015)
15. Tang, Y., Xing, X., Karimi, H.R., Kocarev, L., Kurths, J.: Tracking control of networked multi-agent systems under new characterizations of impulses and its applications in robotic systems. IEEE Trans. Ind. Electron. **63**(2), 1299–1307 (2016)
16. Topaz, C.M., Bertozzi, A.L.: Swarming patterns in a two-dimensional kinematic model for biological groups. SIAM J. Appl. Math. **65**(1), 152–174 (2004)
17. Wong, W.K., Zhang, W., Tang, Y., Wu, X.: Stochastic synchronization of complex networks with mixed impulses. IEEE Trans. Circuits Syst. I Regul. Pap. **60**(10), 2657–2667 (2013)
18. Zhang, W., Tang, Y., Wu, X., Fang, J.A.: Synchronization of nonlinear dynamical networks with heterogeneous impulses. IEEE Trans. Circuits Syst. I Regul. Pap. **61**(4), 1220–1228 (2014)

Learning Human-Understandable Description of Dynamical Systems from Feed-Forward Neural Networks

Sophie Tourret[1(✉)], Enguerrand Gentet[1,2], and Katsumi Inoue[1,3]

[1] National Institute of Informatics, Tokyo, Japan
{tourret,inoue}@nii.ac.jp
[2] Paris-Sud University, Orsay, France
[3] Tokyo Institute of Technology, Tokyo, Japan

Abstract. Learning the dynamics of systems, the task of interest in this paper, is a problem to which artificial neural networks (NN) are naturally suited. However, for a non-expert, a NN is not a convenient tool. There are two reasons for this. First, the creation of an accurate NN requires fine-tuning its architecture and training parameters. Second, even the most accurate NN prediction gives no insight on the rules governing the system. These two issues are addressed in this paper, that presents a method to automatically fine-tune a NN to accurately predict the evolution of a dynamical system and to extract human-understandable rules from it. Experimental results on Boolean systems are presented. They show the relevance of this approach and open the way to many extensions naturally supported by NNs, such as the handling of noisy data, continuous variables or time delayed systems.

1 Introduction

Artificial neural networks (NNs) have been successfully applied to solve a large variety of predictive learning and function approximation problems [1]. Often, the motivation behind their use is their inherent ability to generalize observations and to handle noisy data [2]. As such, it is no wonder that the NN community has been actively researching means of understanding what happens inside NNs since nearly as long as NNs have existed [3]. To do so, the usual method is to extract a symbolic reasoning system from the NN, which can be made of, e.g., logic rules [4–8] or decision trees [9]. To render this extraction possible a method to build a NN with a specific architecture is usually devised first [4,5,8] but standalone extraction methods from trained NN have also been studied [7,10]. Such techniques are not only profitable to NN researchers seeking to understand what is captured by their NNs, but also for people in the field of Inductive Logic Programming (ILP) [11], aiming at constructing logic programs generalizing the observed behavior of systems given in a background theory.

This paper presents a method named NN-LFIT that uses NNs in an ILP learning context. It differs from the neural-symbolic approaches previously mentioned in that it is applied not to a standard classification problem but to the

© Springer International Publishing AG 2017
F. Cong et al. (Eds.): ISNN 2017, Part I, LNCS 10261, pp. 483–492, 2017.
DOI: 10.1007/978-3-319-59072-1_57

modeling of the relational dynamics of a system, i.e., of logic rules that describe the evolution of the system through time, and in that it builds NNs and rules using only the measures of the system. Examples of application include cellular automata studied by physicians and several AI sub-domains such as planning (e.g., discovering action rules), multi-agent systems (e.g., studying social networks evolutions) and systems biology [12,13] (e.g., understanding gene-protein interactions, a key component in the design of better drugs). This work is of interest to the NN community because on the one hand it enhances the methods of automatic generation and tuning of feed-forward NNs for classification tasks from [6,14] in order to deal with dynamical systems in the case of Boolean inputs and on the other hand it gives an explicit method relying on a state-of-the-art symbolic reasoning tool for the extraction of easy-to-understand rules from NNs. Moreover, the experimental results show the relevance of the neural approach which, thanks to the generalization power of NNs, is more accurate than its purely symbolic counterpart LFIT [15]. This suggests extensions such as the handling of continuous data and delayed effects that are very costly for symbolic systems like LFIT but naturally suited to NNs.

In Sect. 2, we present a formal description of the problem. In Sect. 3, the NN-LFIT algorithm is detailed. Section 4 contains the experimental results and their analysis and Sect. 5 concludes this paper. A short version of this article was presented at *ILP 2016* [16] but not included in the formal proceedings.

2   Problem Description

We adopt the representation of dynamical systems used in [15]. The standard terminology and notations of propositional logic (PL) are used[1], e.g., when referring to literals (variables or negation of variables), terms (conjunctions of literals) and formulæ. We are especially concerned with formulæ in disjunctive normal form (DNF), i.e., disjunctions of terms. In this framework a dynamical system is a finite state vector evolving through time $\mathbf{x}(t) = (x_1(t), x_2(t), ..., x_{n_{var}}(t))$ where each $x_i(t)$ is a Boolean variable. In systems biology these variables can represent, e.g., the presence or absence of some genes or proteins inside a cell. The aim of NN-LFIT is to output a normal logic program P that satisfies the condition $\mathbf{x}(t+1) = T_P(\mathbf{x}(t))$ for any t, where T_P is the immediate consequence operator for P [15]. The rules of P are of the form $\forall t, x_i(t+1) \leftarrow F(\mathbf{x}(t))$ for all i in $\{1..., n_{var}\}$ where F is a Boolean formula in DNF. Note that this formalism allows us to describe only the simplest of dynamical systems, meaning those purely Boolean and without delays i.e. where $\mathbf{x}(t+1)$ depends only of $\mathbf{x}(t)$.

Example 1. Figure 1 is an example of application of NN-LFIT. On the left-hand side is the input problem, made of a set of observed transitions of the system. For example, the transition $(1, 0, 1) \rightarrow (0, 0, 1)$ indicates that if at time t, $p = 1$, $q = 0$ and $r = 1$, then at time $t + 1$, $p = 0$, $q = 0$ and $r = 1$. On the right-hand side is the logic program outputed by NN-LFIT. For instance, the first rule of

---

[1] An introduction to logic is available in, e.g., [17].

Input
transitions:
$(p(t), q(t), r(t)) \rightarrow (p(t+1), q(t+1), r(t+1))$

$(1,1,1) \rightarrow (1,1,1) \mid (1,0,1) \rightarrow (0,0,1)$
$(0,1,1) \rightarrow (1,0,0) \mid (0,0,1) \rightarrow (0,0,0)$
$(1,1,0) \rightarrow (1,1,1) \mid (0,1,0) \rightarrow (1,0,0)$
$(1,0,0) \rightarrow (0,1,1) \mid (0,0,0) \rightarrow (0,0,0)$

$\xrightarrow{\quad NN-LFIT \quad}$

Output
logic program:

$p(t+1) \leftarrow q(t)$
$q(t+1) \leftarrow$
$\quad (p(t) \wedge \neg r(t))$
$\quad \vee (p(t) \wedge q(t))$
$r(t+1) \leftarrow p(t)$

Fig. 1. An application of NN-LFIT

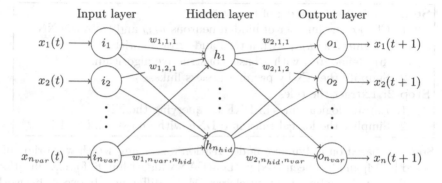

Fig. 2. NN architecture and notations used in NN-LFIT

this program mean that p is true at time $t+1$ iff q is true at time t and the second rule means that q is true at time $t+1$ iff either p is true and r is false at time t or p and q are true at time t.

The type of NN used in NN-LFIT reflects the simplicity of the systems considered. We use feed-forward NNs [2] and we furthermore restrict ourselves to using only one hidden layer, i.e. a total of three layers, because it simplifies a lot the architecture of the NN and its treatment. This does not limit the accuracy of the NN as long as there are enough neurons in the hidden layer [18]. The user is assumed to be familiar with the notion of feed-forward NN, and the notations used in this paper are introduced in Fig. 2. The state vector $\mathbf{x}(t)$ describing the dynamical system is directly fed to the input layer and the output layer predicts the values of the next state $\mathbf{x}(t+1)$. This fixes the number of neurons on the input and output layer to the number of variables in the system. The activation function of the neurons is a sigmoid and the training method used is standard: backward propagation with an adaptive rule on the gradient step and L2 regularization to avoid over-fitting the training data. The errors made by the trained NN on the training, validation[2] and test sets are written respectively by E_{train}, E_{val}, and E_{test} and denote as usual the ratio of incorrect predictions made by each output neuron averaged on all output neurons. The only parameter remaining to choose is the number of neurons on the hidden layer n_{hid},

---

[2] Note that, as is usual, the validation set is made of 20% of the training set.

which is automatically tuned by NN-LFIT to suit each problem as described in the following section.

3   The NN-LFIT Algorithm

This section introduces the details of the NN-LFIT algorithm. This algorithm automatically constructs a model of a system from the observation of its state transitions and generate transition rules which describe the dynamic of the system. The main steps of NN-LFIT are listed bellow:

Step 1: Create the model of the system.
1. Choose the number of hidden neurons n_{hid} and train the NN.
   (a) Initialize n_{hid} with a trial and error algorithm.
   (b) Refine n_{hid} with a basic constructive algorithm.
2. Simplify the NN by pruning useless links.

Step 2: Extract the rules
1. Extract logical rules in DNF by querying the NN.
2. Simplify the logical rules into DNF with an external tool.

Step 1 is based on a dynamic node creation algorithm, which was originally proposed in [14] and has been used in the REANN algorithm [6] for classification tasks with a small number of output classes. Major differences between this work and REANN are explicitly indicated in the following description. Step 2 is an original contribution.

Step 1 - Creation of the Model. The first building step is to generate a fully connected NN with a well fitted architecture to learn the dynamics of the observed system. We first use an initialization algorithm and then we refine the architecture with a constructive algorithm.

Initialization algorithm. The initial number of neurons on the hidden layer n_{hid} is chosen using a simple trial and error algorithm. It consists in training the NN using several architectures with an incremental initial number of hidden neurons starting from one and stopping when E_{val} no longer decreases after a few tries. Every time we try a new architecture, we randomly initialize all the weights. In REANN, this step is skipped. The constructive algorithm is directly used on a randomly initialized NN with only one neuron in the hidden layer. For real problems, one or two hidden neurons are unlikely to be enough. Thus the initialization algorithm speeds up the training process by identifying roughly the number of neurons needed before the constructive algorithm, of which the training converges more slowly, fixes this number.

Constructive algorithm. The architecture is improved by using a basic constructive algorithm. It uses the same principle as the initialization algorithm except that every time a hidden neuron is added, the trained weights attached to the other neurons are left unchanged.

Pruning algorithm. The purpose of this step is to remove useless links. To do so we introduce the notion of link efficiency. To compute the efficiency of a specific link, we multiply its weight by the weights of every other link starting from (or ending to) the same hidden neuron it ends to (or starts from). In other words, the efficiency of a link quantifies the best contribution among all the paths going through this link. It is therefore logical to remove links with low efficiency because they have less effects on the predictions compared to others. We use a simple dichotomous search to remove as many links as possible without increasing E_{train}. After the pruning algorithm has been run, if some hidden neurons have lost all their links to the output layer or all their links from the input layer, they can be removed. Due to the presence of biases in the neurons activation functions, it is not possible to simply delete unreachable hidden neurons, because even without inputs they can still influence the output neurons they are connected to. To remove an unreachable hidden neuron h with a bias b_h, it is thus necessary to update the bias of each of the output neurons under its influence by adding to it the product of its output value (computed from b_h alone) with the weight linking the two neurons before deleting the hidden neuron. On the contrary, hidden neurons with no connection to the output layer can be removed without care since they do not influence the output of the NN.

The REANN algorithm, that handles non-Boolean inputs, includes a discretization step which is unneeded here.

Example 2. Figure 3 shows the NNs obtained after applying each sub-step of Step 1 on the system described in Example 1. The weights are omitted to improve the readability. The error rate on the validation set is given at each step.

Step 2 - Extraction of the Rules. To extract the rules underlying the transition system from the NN, each output neuron o_i is considered independently. First the sub-NN $\mathcal{N}_i$, made of o_i plus all the input and hidden neurons that can reach o_i and their connections to each other, is extracted from the main NN. Then, $\mathcal{N}_i$ is used as a black box to construct the rules. All possible input vectors are fed to $\mathcal{N}_i$ and only those that activate o_i are kept. The union of these vectors is converted into a DNF formula F that is then simplified by computing a prime implicant cover of it using a tool called `primer` [19]. Formally, a prime implicant of F is a term D such that $D \models F$ and for any D' such that $D' \models F$, if $D \models D'$ then $D' \models D$. This means that if a term D'' is such that $D'' \subseteq D$ and $D'' = D$ then $D'' \not\models F$. The notion of a prime implicate is dual to that of a prime implicant. It is a clause C such that $F \models C$ and if there exists another clause C' such that $F \models C'$ and $C \models C'$ then $C' \models C$.

Intuitively, prime implicants and prime implicates can be seen respectively as the most specific conditions and the most general consequences of a formula. When handling a formula in DNF, a formula syntactically simpler than but semantically equivalent is obtained by replacing each term of the formula by a prime implicant that subsumes it. To simplify F, we rely on `primer` to compute a prime implicate cover of the CNF formula $\tilde{F}$ that is called the dual of F. It is obtained by swapping conjunctions and disjunctions in formulæ, hence transforming DNFs in CNFs and vice versa. This is done because primer only accepts

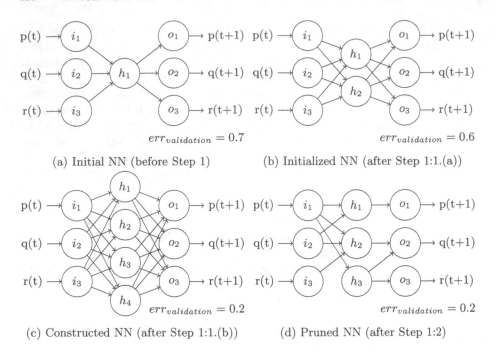

(a) Initial NN (before Step 1)   (b) Initialized NN (after Step 1:1.(a))

(c) Constructed NN (after Step 1:1.(b))   (d) Pruned NN (after Step 1:2)

Fig. 3. Step 1 of NN-LFIT

CNF inputs. A prime implicant cover of F is then generated by duality from the prime implicate cover of $\tilde{F}$ generated by primer.

Example 3. Let us consider the neuron o_1 of the NN drawn in Fig. 3d that represents the system of Example 1. Due to the simplification of the network, o_1 only depends on i_1 and i_2. Then using $\mathcal{N}_1$ as a black box, we query all the different combinations of (i_1, i_2) inputs, keeping only the ones that activate o_2. In this example, o_1 is activated only in the following cases:

- i_1 is off and i_2 is on;
- i_1 is on and i_2 is on.

Then o_1 can be represented by the formula: $F_1 = (\neg i_1 \wedge i_2) \vee (i_1 \wedge i_2)$. Finally, the simplification of the formula F_1 is done by computing a prime implicant cover of F_1 as explained previously, resulting in the creation of the formula $F_1' = i_2$. Note how the term of F_1' subsumes the two terms of F_1 making F_1' equivalent to F_1. Going back to the original transition system, the rule describing the evolution of p extracted from the NN is thus: $p(t+1) \leftarrow q(t)$.

Now let us consider the neuron o_2 which, this time, depends on all the inputs i_1, i_2 and i_3. Then using $\mathcal{N}_2$ as a black box, we query all the different combinations of (i_1, i_2, i_3) inputs, keeping only the ones that activate o_2. In this example, o_2 is activated only in the following cases:

- i_1 is on, i_2 and i_3 are off;
- i_1 and i_2 are on and i_3 is off;
- i_1, i_2 and i_3 are on.

Then o_2 can be represented by the formula: $F_2 = (i_1 \wedge \neg i_2 \wedge \neg i_3) \vee (i_1 \wedge i_2 \wedge \neg i_3) \vee (i_1 \wedge i_2 \wedge i_3)$. Finally, the simplification of the formula F_2 is done by computing a prime implicant cover of F_2 as explained previously, resulting in the creation of the formula $F_2' = (i_1 \wedge \neg i_4) \vee (i_1 \wedge i_2)$. Note how the first term of F_2' subsumes the two first terms of F_2 and the second one subsumes the two last ones of F_2, making F_2' equivalent to F_2. Going back to the original transition system, the rule describing the evolution of q extracted from the NN is thus: $q(t+1) \leftarrow (p(t) \wedge \neg r(t)) \vee (p(t) \wedge q(t))$.

Finally, the neuron o_3 only depends on i_1. Then using $\mathcal{N}_3$ as a black box, we query the two different combinations of i_1. o_3 is activated only when i_1 is on. Then o_3 can be represented by the formula: $F_3 = i_1$. The only term is already a prime implicant of F_3, the rule describing the evolution of q extracted from the NN is thus: $r(t+1) \leftarrow p(t)$.

Note that extracting rules from the fully connected NN right after the steps 1(a) and (b) using the exact same method is possible. However, as shown in the experimental results, the performances of the NN are better after all the steps. In addition, thanks to the pruning (step 1.2), the rule extraction process is less time consuming because the number of input variables to consider for each output can be significantly smaller than before the pruning.

4   Experimental Results

The benchmarks used in the experiments are three Boolean networks from [20] also used for evaluating LFIT in [15]. They respectively describe the cell cycle regulation of budding yeast, fission yeast and mammalians. We randomly assign the $2^{n_{var}}$ transitions describing these networks into the test set and training set (that includes the validation set). Although it is standard to put around 80% of the available data in the training set, we want to simulate the fact that real world data are often incomplete especially in biology, hence we start by analyzing the influence of the size of the training set on the accuracy of the NN (see Fig. 4)[3]. It is measured by E_{test} and averaged over 30 random allocations of the data in the different sets. We observe that each successive sub-step of NN-LFIT improves the accuracy of the model and that, as expected, E_{test} decreases when the size of the training set increases. It reaches an error rate of only 1% while training only on 15% of the data and becomes negligible when the training covers 50% of the data. In comparison, LFIT [15] has a nearly constant error rate on the test set (resp. 36% and 33% on the mammalian and fission benchmarks) for all sizes of the training set. Obviously the accuracy of the NN varies depending on the system it models but still these results show that the generalization power of NNs

---

[3] The results for the budding benchmark are omitted due to space limitations.

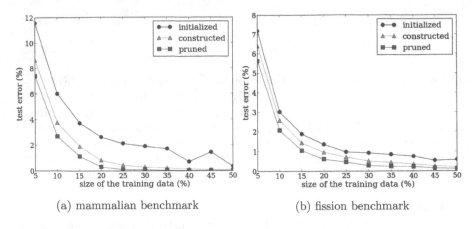

(a) mammalian benchmark   (b) fission benchmark

Fig. 4. Influence of the train size on E_{test} for every step of NN-LFIT.

is a real advantage over a purely symbolic approach. The following experiments are conducted allocating 15% of the data to the training set and the results are also averaged over 30 random allocations.

Table 1 shows the parameters of the NN architectures produced by NN-LFIT and their corresponding E_{test} as well as the error rate of LFIT on the test set, already mentioned in the previous experiment. The numbers of neurons and links decrease significantly during the pruning step (16% less hidden neurons and 65% less links) along with E_{test} (29% reduction) showing that the simplification step not only reduces the complexity of the NN but also improves the model performances through an efficient generalization. In addition, the accuracy of NN-LFIT clearly outperforms that of LFIT.

Table 1. Architecture and test error evolution during NN-LFIT steps.

Architecture	Mammalian, $n_{var} = 10$			Fission, $n_{var} = 10$			Budding, $n_{var} = 12$		
	Neurons	Links	E_{test} (%)	Neurons	Links	E_{test} (%)	Neurons	Links	E_{test} (%)
Initial	7.10	142	3.19	9.07	181	2.23	11.4	273	0.313
Constructed	13.5	270	1.92	13.73	275	1.61	14.4	346	0.237
Pruned	11.2	98.6	1.37	11.7	97.8	1.21	12.2	91	0.156
LFIT	-	-	36	-	-	33	-	-	-

Finally we evaluate the correctness and simplicity of the rules learned by NN-LFIT. For each variable x_i, we identify three categories: true positives, i.e. *valid* rules that output the same result as the original ones; false positives, i.e. *wrong* rules that contradict the original ones; and false negatives, i.e. *missing* rules that appear in the original model but are not present in the reconstructed one. Figure 5 shows the distribution of these categories after the construction and pruning steps of NN-LFIT for each variable[4]. The pruning step reduces

---

[4] Note that a rule of a logic program as defined in [15] is a term here, except for constant rules, e.g., x_1 in Fig. 5b which is always false and thus contains no term.

the number of terms (true and false positives) in almost all the rules which means they are simpler. Moreover the proportion of false positives and negatives diminishes after the pruning, reflecting the increase of the accuracy of the rules observed on Table 1.

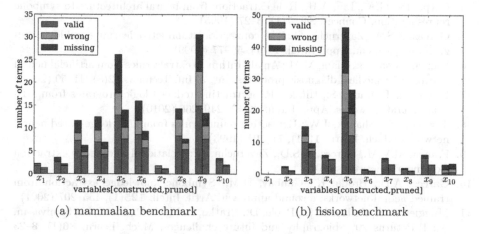

(a) mammalian benchmark   (b) fission benchmark

Fig. 5. Distributions of the categories of term on each variables.

5   Conclusion

In this paper, we present NN-LFIT, a method using feed-forward NNs to extract a logic program describing a dynamical system from the observation of its evolution. It includes a method to automatically tune a feed-forward NN to predict the evolution of the considered Boolean system and an original mechanism for the extraction of human-understandable rules from the NN. Experimental results indicate good overall performances in term of correctness and simplicity of the obtained rules, even when handling only as little as 15% of the data. Extensions of NN-LFIT exploiting more capacities of NNs are planned. One possibility is to extract the rules using a decompositional approach as in, e.g., [10] which details a sound but incomplete extraction algorithm improving the *complexity* × *quality* trade-off. Other extensions include the handling of noisy data and systems with continuous variables which can be naturally handled by feed-forward NNs. We are also considering how to use deep NNs to model systems with delays where $\mathbf{x}(t)$ depends not only on $\mathbf{x}(t-1)$ but also on some $\mathbf{x}(t-k)$ for k greater than one.

References

1. Cherkassky, V., Friedman, J.H., Wechsler, H.: From Statistics to Neural Networks: Theory and Pattern Recognition Applications, vol. 136. Springer Science & Business Media, Heidelberg (2012)

2. Svozil, D., Kvasnicka, V., Pospichal, J.: Introduction to multi-layer feed-forward neural networks. Chemom. Intell. Lab. Syst. **39**(1), 43–62 (1997)
3. Augasta, M.G., Kathirvalavakumar, T.: Rule extraction from neural networks - a comparative study. In: 2012 International Conference on Pattern Recognition, Informatics and Medical Engineering (PRIME), pp. 404–408. IEEE (2012)
4. Carpenter, G.A., Tan, A.H.: Rule extraction: from neural architecture to symbolic representation. Connect. Sci. **7**(1), 3–27 (1995)
5. Garcez, A.S.A., Zaverucha, G.: The connectionist inductive learning and logic programming system. Appl. Intell. **11**(1), 59–77 (1999)
6. Kamruzzaman, S., Islam, M.M.: An algorithm to extract rules from artificial neural networks for medical diagnosis problems. Int. J. Inf. Technol. **12**(8), 41–59 (2006)
7. Lehmann, J., Bader, S., Hitzler, P.: Extracting reduced logic programs from artificial neural networks. Appl. Intell. **32**(3), 249–266 (2010)
8. Towell, G.G., Shavlik, J.W.: Extracting refined rules from knowledge-based neural networks. Mach. Learn. **13**(1), 71–101 (1993)
9. França, M.V.M., Garcez, A.S.D., Zaverucha, G.: Relational knowledge extraction from neural networks (2015)
10. Garcez, A.D., Broda, K., Gabbay, D.M.: Symbolic knowledge extraction from trained neural networks: a sound approach. Artif. Intell. **125**(1), 155–207 (2001)
11. Muggleton, S., De Raedt, L., Poole, D., Bratko, I., Flach, P., Inoue, K., Srinivasan, A.: ILP turns 20 – biography and future challenges. Mach. Learn. **86**(1), 3–23 (2012)
12. Comet, J.-P., Fromentin, J., Bernot, G., Roux, O.: A formal model for gene regulatory networks with time delays. In: Chan, J.H., Ong, Y.-S., Cho, S.-B. (eds.) CSBio 2010. CCIS, vol. 115, pp. 1–13. Springer, Heidelberg (2010). doi:10.1007/978-3-642-16750-8_1
13. Ribeiro, T., Magnin, M., Inoue, K., Sakama, C.: Learning delayed influences of biological systems. Front. Bioeng. Biotechnol. (2014). doi:10.3389/fbioe.2014.00081
14. Ash, T.: Dynamic node creation in backpropagation networks. Connect. Sci. **1**(4), 365–375 (1989)
15. Inoue, K., Ribeiro, T., Sakama, C.: Learning from interpretation transition. Mach. Learn. **94**(1), 51–79 (2014)
16. Gentet, E., Tourret, S., Inoue, K.: Learning from interpretation transition using feed-forward neural networks. In: CEUR Workshop Proceedings of the 26th International Conference on Inductive Logic Programming (ILP 16 Short Papers) (2016)
17. Caferra, R.: Logic for Computer Science and Artificial Intelligence. Wiley, New York (2013)
18. Hornik, K., Stinchcombe, M., White, H.: Multilayer feedforward networks are universal approximators. Neural Netw. **2**(5), 359–366 (1989)
19. Previti, A., Ignatiev, A., Morgado, A., Marques-Silva, J.: Prime compilation of non-clausal formulae. In: Proceedings of the 24th International Conference on Artificial Intelligence, pp. 1980–1987. AAAI Press (2015)
20. Dubrova, E., Teslenko, M.: A SAT-based algorithm for finding attractors in synchronous Boolean networks. IEEE/ACM Trans. Comput. Biol. Bioinform. (TCBB) **8**(5), 1393–1399 (2011)

Stability and Stabilization of Time-Delayed Fractional Order Neural Networks via Matrix Measure

Fei Wang[1], Yongqing Yang[1($\boxtimes$)], Jianquan Lu[2], and Jinde Cao[2]

[1] School of Science, Jiangnan University, Wuxi 214122, China
fei_9206@163.com, yongqingyang@163.com
[2] School of Mathematics, Southeast University, Nanjing 210096, China
{jqluma,jdcao}@seu.edu.cn

Abstract. The stability problem of delayed neural networks with fractional order dynamics has been studied in this paper. Several criteria for the stability of the equilibrium point are derived via matrix measure method and fractional order differential inequality. All criteria are formed as matrix measure, which can be easy to verify in practice. Based on which, feedback controllers are designed to stabilize a kind of chaotic fractional order neural network. Finally, two simulations are given to check the theoretical results and compare with some exist results.

Keywords: Fractional-order · Matrix measure · Neural networks · Stability · Delay

1 Introduction

Neural networks have been widely investigated in the last decades, due to their successful applications in lots of areas, such as signal processing [1], automatic control [2], pattern recognition [3] and so on. Fractional-order derivatives has been receiving much attention recently by its advantages for the description of memory. The Hopfield neural network with fractional order dynamics was first studied in [4]. In the same year, authors of [5] found that fractional differentiation is more fit for describe neurons firing rate. There were some important results of stability analysis of systems with fractional order dynamics have been published, such as root locus method based on Laplace transform, theory of fractional order linear system, the second method of Lyapunov method [6,7].

Recently, a novel approach, named matrix measure strategies and Halanay inequality has been used to deal with stability of delayed neural networks [8]. Different from some exist methods, Lyapunov function need not to be constructed under this method, and the conditions of stability could be formed as matrix measure. Which means that the effects of both positive values and

Y. Yang—This work was jointly supported by the Natural Science Foundation of Jiangsu Province of China under Grant No. BK20161126, the Graduate Innovation Project of Jiangsu Province under Grant No. KYLX16_0778.

F. Cong et al. (Eds.): ISNN 2017, Part I, LNCS 10261, pp. 493–501, 2017.
DOI: 10.1007/978-3-319-59072-1_58

negative values of the matrix could be considered, while the most of the exists conditions of stability are formed as algebra or norm. Thus, many significant results about matrix measure are obtained in the past two years.

Noting that the above results about matrix measure are concerned with integer-order neural networks. Can the matrix measure method be extended to fractional order case? With the inspiration from this question and above discussions, this paper studied the matrix measure strategies for the neural networks with fractional order dynamics. Based on fractional order Dini-like derivative which has been introduced in [9,10], this paper investigated stability and stabilization of delayed neural networks with fractional order dynamics. All the stability conditions are formed as matrix measure, which can utilize the information of diagonal elements of parameter matrices more sufficiently.

This paper's frame would be as follows: In Sect. 2, Caputo fractional operator, fractional order Dini-like derivative, matrix measures and some lemmas will be introduced. The results about the stability of fractional order time-delayed neural networks will be presented in Sect. 3. Then, some examples are given to verify the main results in Sect. 4. Finally, conclusions are drawn in Sect. 5.

2 Preliminaries and Model Description

In this part, some preliminaries of Caputo fractional operator are presented at first. Then, fractional order Dini-like derivative would been introduced. The matrix measures and their properties will be shown later. Finally, the delayed fractional order neural networks model will be given.

2.1 Caputo Fractional Operator

The initial conditions of Caputo's type fractional differential operator are same as it of integer-order, which can describe physical meanings for the system of real world [11]. Therefore, the Caputo derivatives will be used in following. The definition of the Caputo derivative operator can be found in [11]. For simply, $D^\alpha x(t)$ will be denoted as the $_C D^\alpha_{0,t} x(t)$.

2.2 Fractional Order Dini-Like Derivative

In this subsection, fractional order Dini-like derivative would be introduced. Consider the fractional order functional system:

$$^C D^\alpha x(t) = \mathcal{F}(t, x_t) \tag{1}$$

where $\mathcal{F} : [t_0, \infty] \times \mathcal{PC}$, $\mathcal{PC} = \{\phi : [-\tau, 0] \to \mathbb{R}^n, \phi(t) \ is \ a \ continuous \ function\}$, we denote by x_t an element of $\mathcal{PC}$ defined by $x_t(s) = x(t+s)$, $-\tau \leqslant s \leqslant 0$. For a function $V \in \mathcal{C}_0$, where $\mathcal{C}_0 = \{V|V : [t_0, \infty) \times \mathbb{R}^n \to \mathbb{R}_+, \ V(t, 0) \equiv 0, \ V \ is \ Lipschitz \ continuous \ in \ x \in \mathbb{R}^n\}$, the fractional-order Dini-like derivative is defined as following:

Definition 1 [9,10]. *Given a function $V \in \mathcal{C}_0$. For $\phi \in \mathcal{PC}$, the upper right-hand derivative of V in Caputo's sense of order $\alpha(\alpha \in (0,1))$ with respect to the system (1) is defined by*

$$^C D_+^\alpha V(t, \phi(0)) = \lim_{h \to 0^+} \frac{V(t, \phi(0)) - V(t - h, \phi(0) - h^\alpha \mathcal{F}(t, \phi))}{h^\alpha}.$$

Based on the above Dini-like derivative, there are some stability theories of fractional order systems have been published [12–14]. The following Lemma has been obtained recently, which could deal with stability of delayed fractional order systems.

Lemma 1 [15]. *Assume that $\mathcal{F}(t, 0) \equiv 0$, $t \in [t_0, \infty)$, there exists a function $V \in \mathcal{C}_0$ such that*

$$\varphi_1(\|\, x \,\|) \leqslant V(t, x) \leqslant \varphi_2(\|\, x \,\|), \quad \varphi_1, \; \varphi_2 \in \mathcal{K},$$

and the inequality
$$D_+^\alpha V(t, \phi_0) \leqslant -cV(t, \phi_0)$$

whenever $V(t + \theta, \phi(\theta)) \leqslant p(V(t, \phi(0)))$ for $-\tau \leqslant \theta \leqslant 0$, $t \in [t_0, \infty)$, $\phi \in \mathcal{PC}$, where $c > 0$ is a const, $p(\cdot)$ is continuous and non-decreasing on $\mathbb{R}+$, and $p(u) > 0$ as $u > 0$, then asymptotical stability of the zero solution of system (1) can be dirived.

The definition of matrix measure for A can be found in [8], the following lemma about matrix measure will be used later.

Lemma 2. *If the matrix measure of A is satisfied $\mu_p(A) < 0$, then A is non-singular, i.e. $\mid A \mid \neq 0$, where $p = 1, 2, \infty$.*
Proof.
Case 1: $p = 1$
Noting that $\mu_1(A) < 0$ implies that $a_{jj} + \sum\limits_{i=1, i \neq j}^{n} \mid a_{ij} \mid < 0$, $j = 1, 2, ..., n$. Then,
$\mid a_{jj} \mid > \sum\limits_{i=1, i \neq j}^{n} \mid a_{ij} \mid$, *which means that A^T is a strictly diagonally dominant matrix, then, $\mid A \mid = \mid A^T \mid = \neq 0$, t hus, A is a non-singular matrix.*
Case 2: $p = 2$
When $\mu_2(A) < 0$, i.e. $\lambda_{max}(\frac{A+A^T}{2}) < 0$, which implies that $\lambda_A < 0$, thus, A is non-singular.
Case 3: $p = \infty$
$\mu_\infty(A) < 0$ *implies that $a_{ii} + \sum\limits_{j=1, i \neq j}^{n} \mid a_{ij} \mid < 0$, $i = 1, 2, ..., n$. Then, $\mid a_{ii} \mid >$*
$\sum\limits_{j=1, i \neq j}^{n} \mid a_{ij} \mid$, *which means that A is a strictly diagonally dominant matrix, thus, A is a non-singular matrix.*

2.3 Model Description

The following delayed neural network with fractional-order will be considered:

$$D^\alpha x_i(t) = -c_i x_i(t) + \sum_{j=1}^{n} a_{ij} f_j(x_j(t)) + \sum_{j=1}^{n} b_{ij} g_j(x_j(t - \tau_j(t))) + I_i(t), \quad (2)$$

or equivalently

$$D^\alpha x(t) = -Cx(t) + Af(x(t)) + Bg(x(t - \tau(t))) + I, \quad (3)$$

in which, $0 < \alpha < 1$, $c_i > 0$; $x_i(t)$ denotes the ith neuron's state at time t, $f_j(*), g_j(*)$ denote the activation functions, $\tau_j(t)$ are time-varying delays of the jth neuron, which is bounded in this paper, $0 \leqslant \tau_j(t) \leqslant \tau$; a_{ij} denotes the connection weight between the jth neuron and the ith neuron, I_i denotes an external input of ith neuron. It's easy to see that $x(t) = (x_1(t), x_2(t), x_3(t), ..., x_n(t))^T \in R^n$, $C = diag(c_1, c_2, c_3, ..., c_n)$, $A = (a_{ij})_{n \times n}$, $B = (b_{ij})_{n \times n}$, $f(x(t)) = (f_1(x_1(t)), f_2(x_2(t)), f_3(x_3(t)), ..., f_n(x_n(t)))^T$, $g(x(t)) = (g_1(x_1(t)), g_2(x_2(t)), ..., g_n(x_n(t)))^T$, $I = (I_1, I_2, ..., I_n)^T$, $\tau(t) = (\tau_1(t), \tau_2(t), ..., \tau_n(t))^T$. Let $x_i(t) = \phi_i(t) \in \mathcal{PC}$ are the initial conditions of system (6).

3 Main Results

The stability and stabilization of neural networks (3) will be studied in this section by applying the theoretical results which have been given in previous section.

3.1 Stability Analysis of Neural Network (3) via Matrix Measure Method

In this subsection, some sufficient conditions will be given, which is formed as matrix measure, to ensure the exist and stability of the equilibrium point of system (3).

Assumption 1.
The functions $f(\cdot), g(\cdot)$ are bounded and satisfied the following conditions: there exist positive constants l_p^f, l_p^g such that

$$\| f(x) - f(y) \|_p \leqslant l_p^f \| x - y \|_p, \| g(x) - g(y) \|_p \leqslant l_p^g \| x - y \|_p, \forall x, y \in \mathbb{R}^n. \quad (4)$$

Theorem 1. *Assuming that assumption 1 holds, if the parameters of the neural network (6) satisfied*

$$\mu_p(C) - l_f^p \| A \|_p > l_g^p \| B \|_p > 0.$$

The fractional order neural networks (3) has a unique equilibrium point, and the equilibrium point is asymptotically stable.

Proof.

Due to the active functions f_i and g_i are bounded, the existence of the equilibrium can be obtained easily by Brouwer's fixed point theorem, the proof is omitted here. Assume (6) has two equilibria u^* and v^*, one has:

$$-Cu^* + Af(u*) + Bg(u^*) + I = 0,$$

and

$$-Cv^* + Af(v*) + Bg(v^*) + I = 0.$$

By some simple calculations, there must be some constant m_i^f and m_i^g, $i = 1, 2, ..., n$ such that

$$(-C + AM^f + BM^g)(u^* - v^*) = 0_n, \tag{5}$$

where $M^f = diag(m_1^f, m_2^f, ..., m_n^f)$, $M^g = diag(m_1^g, m_2^g, ..., m_n^g)$. According to assumption 1, we have $\| M^p \|_p \leqslant l_p^f$ and $\| M^g \|_p \leqslant l_p^g$. Then,

$$
\begin{aligned}
\mu_p(-C + AM^f + BM^g) &\leqslant -\mu_p(C) + \mu_p(AM^f) + \mu_p(BM^g)\\
&\leqslant -\mu_p(C) + \| AM^f \|_p + \| BM^g \|_p\\
&\leqslant -\mu_p(C) + l_f^p \| A \|_p + l_g^p \| B \|_p\\
&< 0.
\end{aligned}
$$

Based on the Lemma 2, $(-C + AM^f + BM^g)$ is non-singular, then, (5) has a unique solution, thus, $u^* = v^*$. Therefore, Equilibrium points of the neural network (3) is unique. In the following, we will proof asymptotical stability of the equilibrium point. Let's assume that x^* is the equilibrium point of the system. Define that $e_i(t) = x_i(t) - x_i^*$, then equations (3) convert into

$$D^\alpha e_i(t) = -c_i e_i(t) + \sum_{j=1}^n a_{ij}[f_j(e_j(t) + x_j^*) - f_j(x_j^*)] + \sum_{j=1}^n b_{ij}[g_j(e_j(t - \tau_j(t)) + x_j^*) - g_j(x_j^*)], \tag{6}$$

or equivalently

$$D^\alpha e(t) = -Ce(t) + AF(e(t)) + BG(e(t - \tau(t))), \tag{7}$$

where $F(e(t)) = f(x(t)) - f(x^*)$, $G(e(t - \tau(t))) = g(x(t - \tau(t))) - g(x^*)$.

First, calculating the fractional order Dini-like derivative of $\| e(t) \|_p$ along the system (6) and using Assumption 1, one has:

$$
\begin{aligned}
&\varliminf_{h \to 0+} \frac{\| e(t) \|_p - \| e(t) - h^\alpha D^\alpha e(t) \|_p}{h^\alpha}\\
&= \lim_{h \to 0+} \frac{\| e(t) \|_p - \| e(t) - h^\alpha(-Ce(t) + AF(e(t)) + BG(e(t - \tau(t)))) \|_p}{h^\alpha}\\
&\leqslant \lim_{h \to 0+} \frac{\| e(t) \|_p - \| e(t) + h^\alpha Ce(t) \|_p + h^\alpha \| AF(e(t)) + BG(e(t - \tau(t)))) \|_p}{h^\alpha}\\
&\leqslant \lim_{h \to 0+} \frac{\| e(t) \|_p - \| e(t) + h^\alpha Ce(t) \|_p}{h^\alpha} + l_p^f \| A \|_p \| e(t) \|_p + l_p^g \| B \|_p \| e(t - \tau(t)) \|_p\\
&= \lim_{h \to 0+} \frac{1 - \| I + h^\alpha C \|_p}{h^\alpha} \| e(t) \|_p + l_p^f \| A \|_p \| e(t) \|_p + l_p^g \| B \|_p \| e(t - \tau(t)) \|_p\\
&= -(\mu_p(C) - l_p^f \| A \|_p) \| e(t) \|_p + l_p^g \| B \|_p \| e(t - \tau(t)) \|_p
\end{aligned}
$$

when $\| e(s) \|_p \leqslant \| e(t) \|_p$, $t - \tau \leqslant s \leqslant t$, *we have*

$$^C D_+^\alpha \| e(t) \|_p = \lim_{h \to 0^+} \frac{\| e(t) \|_p - \| e(t) - h^\alpha D^\alpha e(t) \|_p}{h^\alpha} \leqslant -(\mu_p(C) - l_p^f \| A \|_p - l_p^g \| B \|_p) \| e(t) \|_p .$$

Under Lemma 1, the system (6) can be asymptotical stability, which completes our proof.

Remark 1. Recently, there were lots results about stability of neural networks with fractional order dynamics, most of which are based on Lyapunov function. The conditions of stability are formed as the absolute value of elements coefficient matrices, or the eigenvalues of the coefficient matrices. However, a new criterion in the Theorem 1 are obtained without constructing Lyapunov function, and the conditions have include some exist results, which will be shown in the simulation part.

3.2   Stabilization of Neural Networks via Matrix Measure Method

In this subsection, we will give several criteria to stabilize a kind of chaotic fractional order systems. Noting that if A, B, C and $\tau(t)$ for neural networks are appropriately chosen, the dynamical of (2) may lead to chaos. Those neural networks are unstable, to stabilize them to their equilibrium points. The control model of (2) can be described as following:

$$D^\alpha x_i(t) = -c_i x_i(t) + \sum_{j=1}^n a_{ij} f_j(x_j(t)) + \sum_{j=1}^n b_{ij} g_j(x_j(t - \tau_j(t))) + I_i(t) + u_i(t), \quad (8)$$

where c_i, a_{ij}, b_{ij}, I_i are the same as defined in (2), and $u_i(t)$ is linear state feedback controller defined by $u_i(t) = d_i x_i(t)$. Let $U(t) = (u_1(t), u_2(t), ..., u_n(t))^T$, and $U(t) = Dx(t)$, where D is the feedback control gain to be determined. Consequently, the controlled network can be rewritten as:

$$D^\alpha x(t) = -(C - D)x(t) + Af(x(t)) + Bg(x(t - \tau(t))) + I, \quad (9)$$

where matrix C, A, B, I are the same as defined in (3).

Theorem 2. *Assuming that Assumption 1 holds, if the parameters of (9) satisfied*

$$\mu_p(C - D) - l_f^p \| A \|_p > l_g^p \| B \|_p > 0.$$

The (9) has an asymptotically stable unique equilibrium point under the feedback control.

Proof.
Let $C^* = C - D$, under the condition in this theorem, then Theorem 2 can be immediately derived from Theorem 1.

4   Numerical Simulations

Two simulations to illustrate the our results will be shown in this part.

Example 1. Considering the parameters of (2) as following: $f_i(\cdot) = tanh(\cdot)$, $g_i(\cdot) = sin(\cdot)$, $\tau_i(t) = \frac{0.5e^t}{1+e^t}$ $I_i = 0$, $i = 1, 2, ..., 5$, and

$$C = \begin{pmatrix} 2 & 0 & 0 & 0 & 0 \\ 0 & 4 & 0 & 0 & 0 \\ 0 & 0 & 4 & 0 & 0 \\ 0 & 0 & 0 & 5.5 & 0 \\ 0 & 0 & 0 & 0 & 3 \end{pmatrix}, A = B = \begin{pmatrix} 0.6 & -0.6 & 0.3 & 0.6 & 0.3 \\ 1.5 & -1.5 & -0.6 & 0.6 & 0.3 \\ 0 & -0.3 & 0.6 & 0.3 & 0 \\ 0 & 0.3 & 0.3 & -0.9 & 0.3 \\ 1.5 & 0 & 0 & -0.3 & -0.9 \end{pmatrix}.$$

Remark 2. In this simulation, let $p = 2$, then we have $l_f^2 = l_g^2 = 1$, and $\mu_2(C) = 5.5$, $\| A \|_2 = \| B \|_2 = 2.695$. It is obviously the condition in the Theorem 1 can be satisfied when $p = 2$. But, these coefficient matrices can not meet the conditions in [9]. Thus, there is more conservative in this paper. The state trajectories of this neural network is depicted in Fig. 1.

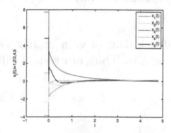

Fig. 1. Numerical solution of equation (1)

Example 2. Consider the parameters of the neural network (2) as following: $f_i(\cdot) = g_i(\cdot) = tanh(\cdot)$, $\tau_i = 0.5$ $I_i = 0$, $i = 1, 2, 3$, and

$$C = \begin{pmatrix} 1 & 0 & 0 \\ 0 & 1 & 0 \\ 0 & 0 & 1 \end{pmatrix}, A = B = \begin{pmatrix} 0.62 & -1.605 & -1.605 \\ -1.605 & 0.55 & -2.2 \\ -1.605 & 2.2 & 0.5 \end{pmatrix}.$$

The system with these parameters has a double-scrolling chaotic attractor with the initial $\phi(t) = (0.1, -0.5, -0.5)^T$, $\forall t \in [-0.5, 0]$. Without controller, the state trajectories of neural network (2) is depicted in Fig. 2. Let the feedback control gain $D = \begin{pmatrix} -2.5 & -7.5 & -7.5 \\ -5 & -2.5 & -10 \\ -7.5 & 10 & -2.5 \end{pmatrix}$, from Fig. 3, we can see the state variables $x_i(t)$ is converge to the $x^* = [0, 0, 0]^T$.

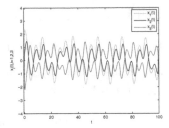

Fig. 2. State trajectories of variable in (1) without controller

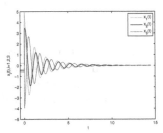

Fig. 3. State trajectories of variable in (1) with controller

Remark 3. Noting the feedback control gain matrix D is not symmetric. Furthermore, it is not positive definite. Thus, our results will have wider applications in practice.

5   Conclusion

The matrix measure method has been applied for dealing with the stability problem of neural networks with fractional order dynamics in this paper. Combined with fractional order differential inequality, some conditions related to matrix measure have been obtained to ensure the stability of the delayed fractional order models. According to the above results, feedback controllers have been designed for stabilizing the studied neural networks. In the simulation section, the efficiency and less conservatism for the derived criteria were shown by two examples.

References

1. Herault, J., Jutten, C.: Space or time adaptive signal processing by neural network models. In: Neural Networks for Computing, vol. 151, no. 1, pp. 206–211 (1986)
2. Hunt, K.J., Sbarbaro, D., Zbikowski, R., Gawthrop, P.J.: Neural networks for control systems survey. Automatica **28**(6), 1083–1112 (1992)
3. Carpenter, G.A.: Neural network models for pattern recognition and associative memory. Neural Netw. **2**(4), 243–257 (1989)

4. Boroomand, A., Menhaj, M.B.: Fractional-order hopfield neural networks. In: Köppen, M., Kasabov, N., Coghill, G. (eds.) ICONIP 2008. LNCS, vol. 5506, pp. 883–890. Springer, Heidelberg (2009). doi:10.1007/978-3-642-02490-0_108
5. Lundstrom, B.N., Higgs, M.H., Spain, W.J., Fairhall, A.L.: Fractional differentiation by neocortical pyramidal neurons. Nat. Neurosci. **11**(11), 1335–1342 (2008)
6. Kaslik, E., Sivasundaram, S.: Nonlinear dynamics and chaos in fractional-order neural networks. Neural Netw. **32**, 245–256 (2012)
7. Yu, J., Hu, C., Jiang, H.: α-stability and α-synchronization for fractional-order neural networks. Neural Netw. **35**, 82–87 (2012)
8. Cao, J., Wan, Y.: Matrix measure strategies for stability and synchronization of inertial BAM neural network with time delays. Neural Netw. **53**, 165–172 (2014)
9. Stamova, I.: Global Mittag-Leffler stability and synchronization of impulsive fractional-order neural networks with time-varying delays. Nonlinear Dyn. **77**(4), 1251–1260 (2014)
10. Stamova, I., Stamov, G.: Stability analysis of impulsive functional systems of fractional order. Commun. Nonlinear Sci. Numer. Simul. **19**(3), 702–709 (2014)
11. Podlubny, I.: Fractional Differential Equations: An Introduction to Fractional Derivatives, Fractional Differential Equations, to Methods of Their Solution and Some of Their Applications, vol. 198. Academic press, Cambridge (1998)
12. Yakar, C., Gücen, M.B., Cicek, M.: Strict stability of fractional perturbed systems in terms of two measures. In: Baleanu, D., Machado, J.A.T., Luo, A.C. (eds.) Fractional Dynamics and Control, pp. 119–132. Springer, New York (2012)
13. Cicek, M., Yakar, C., Gücen, M.B.: Practical stability in terms of two measures for fractional order dynamic systems in Caputo's sense with initial time difference. J. Frankl. Inst. **351**(2), 732–742 (2014)
14. Stamov, G., Stamova, I.: Second method of Lyapunov and almost periodic solutions for impulsive differential systems of fractional order. IMA J. Appl. Math. (2015). doi:10.1093/imamat/hxv008
15. Stamova, I.: On the Lyapunov theory for functional differential equations of fractional order. Proc. Am. Math. Soc. **144**(4), 1581–1593 (2016)

Metrics and the Cooperative Process of the Self-organizing Map Algorithm

William H. Wilson[✉]

UNSW, Sydney, Australia
billw@cse.unsw.edu.au

Abstract. This paper explores effects of using different the distance measures in the cooperative process of the Self-Organizing Map algorithm on the resulting map. In standard implementations of the algorithm, Euclidean distance is normally used. However, experimentation with non-Euclidean metrics shows that this is not the only metric that works. For example, versions of the SOM algorithm using the Manhattan metric, and metrics in the same family as the Euclidean metric, can converge, producing sets of weight vectors indistinguishable from the regular SOM algorithm. However, just being a metric is not enough: two examples of such are described. Being analogous to the Euclidean metric is not enough either, and we exhibit members of a family of such distance measures that do not produce satisfactory maps.

Keywords: Self-organizing map · Metric · Cooperative process

1 Introduction

The self-organizing map (SOM) algorithm, developed by Kohonen [3,4], is well known, and in practice standard versions normally converge reliably to a sensible map. Many studies of convergence and of convergence speed properties of SOMs exist; earlier ones are can be found in the extensive reference list in [4]. Messages from these studies include that the map converges to the probability distribution of the training data as the epoch number $n \to \infty$ (e.g. [6]).

What features of the algorithm lead to its convergence? Let's review the algorithm, for ease of reference. First, the weight vectors are given initial values: this can affect convergence speed ([4], p. 142) but does not seem to affect convergence as $n \to \infty$. In the first part of the algorithm loop, the *competitive* process, finds the weight vector **x** that is closest (in a certain sense) to the current training pattern. The sense of closeness used is a Euclidean geometric sense, using the metric derived from the l_m^2-norm (where m is in this case the dimension of the weight space); other metrics have been used (see e.g. [5]). The second process, the *cooperative* process, takes the "winning node", that is, the map node associated with the weight vector that was found in the competitive process, and in the variant we consider, computes the distance d, usually using the Euclidean (l_m^2) metric, of each map node from the winning map node c (but here m is

© Springer International Publishing AG 2017
F. Cong et al. (Eds.): ISNN 2017, Part I, LNCS 10261, pp. 502–510, 2017.
DOI: 10.1007/978-3-319-59072-1_59

the *map* dimension). This distance d is then used (see [4], p. 111) to compute a *neighborhood function* h used in the next process, (notation as in [1]):

$$h = \exp\left(-\frac{d^2}{2\sigma^2(n)}\right) \tag{1}$$

where $\sigma(n)$ is the "neighborhood width" term for epoch n. Another variant bases h on a *neighborhood set* N_c ([4], p. 111); we do not consider this here. The *adaptive* process changes each map node's weight vector $\mathbf{w}$ using

$$\Delta\mathbf{w} = \eta(n)h(\mathbf{x} - \mathbf{w}) \tag{2}$$

where $\eta(n)$ is the learning rate for epoch n, and $\mathbf{x}$ is as defined above.

Thus distance measures are used in the SOM algorithm in two places - in the competitive process, and in the cooperative process. These need not be the same. In this paper we are interested in the distance measure d used in the cooperative process. As noted above, in this version of the SOM algorithm, d is based on the l_m^2-norm (though [4], p. 111 mentions an alternative). We'll recall the definition of a metric (also found in [4] p. 4), in order to be able to refer to its parts later. A metric on a set X is a function $d : X \times X \to \mathbb{R}_0^+$ (non-negative real numbers) with the following 3 properties:

(a) $\forall x, y \in X, d(x, y) = 0 \Leftrightarrow x = y$
(b) (symmetry) $\forall x, y \in X, d(x, y) = d(y, x)$
(c) (triangle inequality): $\forall x, y, z \in X, d(x, z) \leq d(x, y) + d(y, z)$.

There are other metrics besides that derived from the l_m^2-norm, and we shall explore SOM algorithm performance when that metric is replaced by others, as described in the next section. Kohonen (e.g. [4]) and others consider metrics, perhaps mainly in relation to the *competitive* process of the SOM algorithm. Kohonen describes the Euclidean metric in discussing the neighborhood function (e.g. [4] p. 111), though he mentions an alternative (not based on an explicit metric). Section 2 describes some metric and non-metric distance measures, and Sect. 3 reports on experiments with their use in the cooperative process.

2   Distance Measures in the SOM Algorithm

There are many metrics available that might be applied in the cooperative process of the SOM algorithm; the ones described below by no means exhaust the possibilities, but they do allow us to find a range of different behaviours in the experimental results in the following section.

2.1   Standard Euclidean Metric

As is well known, when the standard SOM algorithm is used on clustered data, the algorithm places map node weight vectors in the middles of clusters (see Fig. 1), in a way that respects the map adjacency relationships. With uniformly distributed data, the algorithm distributes map node weight vectors across the data space, again in a way that respects the map adjacency relationships.

2.2   Metrics Based on the l_m^p-Norm $(p \geq 1)$

The l_m^p-norm is defined for a vector $\mathbf{x} \in \mathbb{R}^m$ by

$$||\mathbf{x}|| = \left(\sum_{i=1}^{m} |x_i|^p \right)^{\frac{1}{p}} \tag{3}$$

and then the associated metric, sometimes called the Minkowski metric, is defined by $d(\mathbf{x}, \mathbf{y}) = ||\mathbf{x} - \mathbf{y}||$. When $p = 2$ we have Euclidean distance, and when $p = 1$ we get the Manhattan, or city-block metric. p can be fractional.

2.3   Metric Based on the Max-Norm

Another metric (related to the l_m^p-family) is based on the max-norm, defined by

$$||\mathbf{x}|| = \max_i x_i \tag{4}$$

so that in this case $d(\mathbf{x}, \mathbf{y}) = ||\mathbf{x} - \mathbf{y}|| = \max_i (x_i - y_i)$.

2.4   Discrete Metric

The *discrete* metric is defined for $x, y \in X$ by $d(x, y) = 1$ unless $x = y$, when $d(x, y) = 0$. This is *a priori* the metric least likely to work in a SOM algorithm variant, as it gives no information to enable the algorithm to distinguish between map neurons that are near or far in terms of the map grid adjacency relations.

2.5   Distance Measures Based on the l_m^p-Formula $(p < 1)$

When $0 < p < 1$, the l_m^p-norm formula still makes sense, but the distance measure defined by $d(\mathbf{x}, \mathbf{y}) = ||\mathbf{x} - \mathbf{y}||$ is *not* a metric, because the triangle inequality (c) in the definition of metric will not (always) hold. Metric axioms (a) $d(x, y) = 0 \Leftrightarrow x = y$ and (b) (symmetry) do hold for these measures.

2.6   Post Office Metric

The "Post Office" metric is defined by analogy with physical mail systems, which frequently send all (non-local) items to a central location, such as a capital city, from which they are then distributed to their regional destinations. So in this distance measure, the distance between two distinct points is the sum of their distances from the central location. "Distance from the central location" needs a precise definition: possibilities include Euclidean distance and (rather reasonably for the post office analogy) Manhattan distance. The central location needs to be specified, too, and the results are likely to vary depending on which location is chosen. A possibility for a 2D map is for the central location to have the coordinates of the average x-value and average y-value (perhaps rounded).

3   Simulations with Non-standard Distance Measures

3.1   Data Sets

Data to which the SOM algorithm is applied are often of one of two types: (1) the points are spread out across the data space, uniformly or according to some gradation of density, or (2) the points belong to defined clusters. There are intermediate cases, e.g. with overlapping clusters. For the purpose of testing the performance of SOM variants, the experiments reported in this paper used two data sets, one clustered and one unclustered. The data sets were 2-dimensional ($m = 2$) so as to be easily visualised. The experiments used a regular square grid, rather than triangular or hexagonal grids, or grids with extra connectivity like those described in [2]. In one case the data (180 points) were distributed (pseudo-)uniformly in the square from $(1, 1) \rightarrow (8, 8)$, while in the other case, the data occupied 9 clusters each of 20 points inside the rectangle $(1, 0) \rightarrow (8, 8)$. These data distributions can be seen in the background of the figures.

3.2   Standard Euclidean Metric

Obviously, the SOM algorithm converges as described above when the original Euclidean metric is used as the distance measure. This is illustrated for clustered data in Fig. 1, and the weight vectors from the original SOM provide a standard for comparison for use with the other distance measures tried. This figure also represents the outcome for clustered data of all the other successful variants.

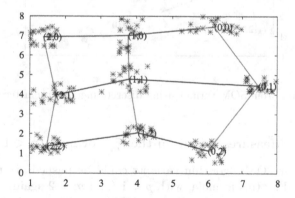

Fig. 1. Standard SOM clustering. Map nodes are at coordinate-labelled junctions of grid lines. This illustrates the map produced by all successful SOM variants (see text).

3.3   Distance Measures Based on the l_m^p-Norm ($p \geq 1$)

Variants of the SOM algorithm that use metrics based on the l_m^p-norm worked fine in practice. As noted in the introduction, $m = 2$, the dimension of the map for our 2D-map SOM simulations. For example, with $p = 3$, weight vectors found were identical to those for the standard SOM algorithm (to 3 decimal places, with the same random number generator seed). The same is true with $p = 1$ (Manhattan metric), and $p = 8$. Thus the weight diagrams in all three cases are identical to that in Fig. 1, except that coordinates may be rotated or reflected.

3.4 Metric Based on the Max-Norm

This SOM variant also performed perfectly well, with the same weight vectors (to 3 decimal places) as the standard SOM algorithm with the Euclidean metric.

3.5 Discrete Metric

Simulations using the discrete metric rshow that this variant of the SOM algorithm does *not* converge in such a way that the map nodes lie in the centers of clusters. Figure 2 visualizes what goes wrong for our clustered data: map points fall inside clusters, but grid topology is not reflected in the map.

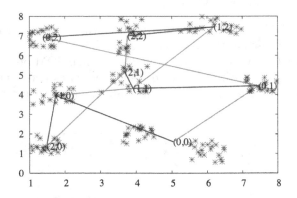

Fig. 2. Map from SOM variant using discrete metric for clustered data.

3.6 Distance Measures Based on the l_m^p-Formula $(p < 1)$

To probe how far SOM algorithm variants could be pushed, we tried distance measures defined by this formula, with $p < 1$ (and $m = 2$ again). In our experimental results, the performance of these SOM algorithm variants depends on how far below $p = 1$ they are. Thus, for the particular clustered data used, the weight diagram for $p = 0.9$ looks fine, $p = 0.5$ produces a clearly different weight diagram, and the diagram for $p = 0.25$ is very problematic. With the uniformly distributed data, $p = 0.9$ and $p = 0.5$ worked well, $p = 0.25$ looked a little strange, and for $p = 0.125$ the weights were very disturbed. Figures 3 and 4 illustrate this.

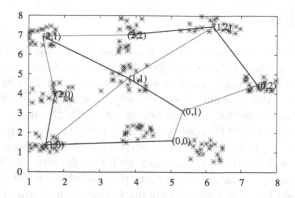

Fig. 3. Map from clustered data and SOM variant using l^p_m-style measure, $p = 0.5$.

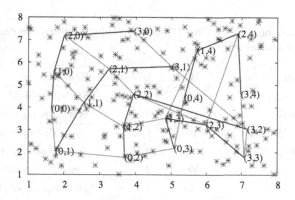

Fig. 4. Map from uniform data and discrete-metric-based SOM variant.

3.7 Post Office Distance Metric

The Post Office metric again fails to provide weight vectors that meet the intent of the SOM algorithm as we know it. Figure 5 illustrates what goes wrong.

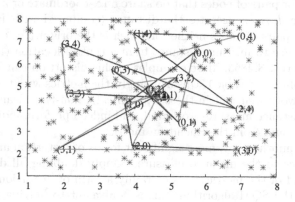

Fig. 5. Map from uniform data and SOM variant using Post Office metric.

4   Conclusion

The SOM algorithm works correctly with a range of non-standard distance measures in the cooperative process, not just Euclidean distance, for the data sets reported here. However, being a metric is not sufficient: the discrete and PO metrics do not work. Also in some cases (e,g, l_m^p-distance measure with p slightly less than 1) being a metric was not necessary. With the discrete metric, the problem is obvious: the SOM cooperative and adaptive processes must distinguish between map nodes close to the winning node and those further away, to change their weights differentially, but with the discrete metric, all pairs of distinct points are equi-distant. Other metrics for which SOM-like algorithms fail do distinguish between near and far map nodes, but distort the nearness if the two nodes do not share an x-coordinate or y-coordinate (in a 2-D map).

Table 1. Summary of Results: the "Distortion" column shows the distance from (2, 2) to (4, 4) for each measure.

Distance measure	Works?	Distortion	Remarks
Standard euclidean	Yes	2.83	Standard SOM algorithm
Manhattan metric	Yes	4	Some distance distortion
l_m^p-norm based ($p = 3$)	Yes	2.52	Some distance distortion
l_m^p-norm based ($p = 8$)	Yes	2.18	Some distance distortion
Based on max-norm	Yes	2	Limit of l_m^p-metric as $p \to \infty$
Discrete metric	No	1	No distance distinction
Post-office metric	No	8.49	Massive distance distortion
l_m^p-formula measure ($p = 0.91$)	Yes	5.08	Normal map
l_m^p-formula measure ($p = 0.25$)	No	32	Badly distorted

Most of the distance measures considered (e.g. the l_m^p-based ones) produce the same measure for pairs of nodes that do share an x-coordinate or a y-coordinate. The PO metric d_{PO} exaggerates the distance (compared with Euclidean distance d_{Euc}) for pairs of nodes neither of which is the origin. For example, using Euclidean distance to measure distance from a central location (namely (0,0)), $d_{PO}((2,2),(4,4)) \approx 8.4853$. More examples of this distortion can be seen in the "Distortion" column of Table 1. Unsurprisingly, the metrics with the greatest distortions don't work well, but in these experiments, non-Euclidean metrics with "reasonable" distance measures work well, and indeed perform indistinguishably from each other and from the Euclidean metric.

To put this another way, more significant forms of distortion arise from *non-uniform* changes to the distance measure. Simply doubling all distances - i.e. using $2d$ instead of the metric d - is certainly a form of distortion. However, in the context of the SOM algorithm, this is equivalent to halving neighborhood widths. If this interfered with map formation, it could be dealt with by increasing the initial neighborhood width and adjusting the neighborhood decay rate. The problems with the l^p_m based measures with $p < 1$ seem to arise because of mismatches between the degree of distortion along the constant-x and constant-y directions on the one hand, and the diagonal directions on the other hand (as indicated in the "Distortion" column of Table 1).

The notion of neighborhood distance in SOM algorithm is robust, since several variants function correctly. There is little to choose between the different metrics that do work, from a computational perspective, for the datasets used in the these experiments. Specific types of datasets may benefit from the use of a distance measure that somehow reflects the structure of the particular dataset.

In models of biological neural systems, one might prefer an intermediate between the Euclidean metric and the Manhattan metric, since in such systems, neurons that function as "neighbours" in the SOM algorithm sense will not be connected by straight-line axons but rather by one (or more) axons that need to divert from a straight line path because of obstacles, efficiency constraints, etc.

Many algorithms use Euclidean distance measures, and while in some cases there are good reasons to do so, in others they might also work fine with different metric-based distance measures. One could also try non-Euclidean distance measures in the competitive phase (this paper considers the cooperative phase).

We noted in the Introduction section that the metric axioms may not apply in certain real-life measures of distance-like phenomena like travel time and charging measures. While the non-metric measures explored in this paper did not perform impressively, in solving such problems, non-metric measures that model the particular problem, it may be profitable to explore the use of distance measures that violate metric axiom (a) $d(x, y) = 0 \Leftrightarrow x = y$ and/or axiom (b) $d(x, y) = d(y, x)$ (symmetry) for datasets where this matches reality.

Since the SOM algorithm works with a range of metrics, it may be appropriate, when creating SOMs, to consider which metric is most appropriate to the situation. Cases where a l^p_m-norm-based metric with $p = 8$, say, would be natural may be uncommon, but the Manhattan and max-norm metrics may well correspond better in some situations than the Euclidean norm.

Simulation Parameters: Initial learning rate was 0.1, and was multiplied by 0.999 after each epoch. The initial value of σ was the map radius of the map (i.e. $(mapHeight + mapWidth)/2$), and was multiplied by 0.999 after each epoch. Simulations ran for 5000 epochs.

References

1. Haykin, S.: Neural Networks and Learning Machines. Prentice-Hall, New York (2009)
2. Jiang, F., Berry, H., Schoenauer, M.: The impact of network topology on self-organizing maps. In: Proceedings of First ACM/SIGEVO Summit on Genetic and Evolutionary Computation, GEC-2009, pp. 247–254. ACM, New York (2009)
3. Kohonen, T.: Self-organized formation of topologically correct feature maps. Biol. Cybern. **43**, 59–69 (1982)

4. Kohonen, T.: Self-Organizing Maps, 3rd edn. Springer, Berlin (2001)
5. Płoński, P., Zaremba, K.: Improving performance of self-organising maps with distance metric learning method. In: Rutkowski, L., Korytkowski, M., Scherer, R., Tadeusiewicz, R., Zadeh, L.A., Zurada, J.M. (eds.) ICAISC 2012. LNCS, vol. 7267, pp. 169–177. Springer, Heidelberg (2012). doi:10.1007/978-3-642-29347-4_20
6. Yin, H., Allison, N.M.: On the distribution and convergence of feature space in self-organizing maps. Neural Comput. **7**, 1178–1187 (1995)

A Cooperative Projection Neural Network for Fast Solving Linear Reconstruction Problems

Youshen Xia[✉]

College of Mathematics and Computer Science, Fuzhou University, Fuzhou, China
ysxia@fzu.edu.cn

Abstract. This paper presents a new cooperative projection neural network for fast solving quadratic convex programming problems, including linear reconstruction problems. The proposed cooperative projection neural network consists of a weighted combination of two projection terms. Compared with conventional projection neural networks and numerical optimization methods, the proposed cooperative projection neural network has a small model size and no limit condition of initial points. Numerical results demonstrate that the proposed cooperative projection neural network has a faster speed than the existing projection neural networks. Therefore, the proposed cooperative projection neural network can fast solve linear reconstruction problems.

Keywords: Linear reconstruction problems · Continuous-time neural network · Low-dimensional model · Fast computation

1 Introduction

The linear reconstruction model is a basic model of many engineering applications, such as signal and image processing, medical imaging regression estimation, etc. [1–4,8,11]. The fundamental issue is to estimate the parameter vector of the linear reconstruction model in the presence of observation noise. The traditional least squares (LS) estimator is asymptotically unbiased when the noise distribution is white-Gaussian, but it could be very poor in non-Gaussian situations [5]. Since the assumption that the noise distribution is Gaussian is unrealistic, robust estimation methods were developed in [6–16]. Among these, the least absolute deviation (LAD) approach is a good choice since it is equivalent to the maximum likelihood method under a double exponential distribution, and thus the least absolute value norm estimation based on the LAD approach is efficient when the noise distribution is Laplace or Cauchy. Unfortunately, finding the *LAD* estimator is much more complex than finding the least square estimator. This is because the LAD estimation problem has a nonsmooth cost function and non-uniqueness solutions. In order to avoid the difficulty, the robust Huber

Y. Xia—This work is supported by the National Natural Science Foundation of China under Grant No. 61473330.

© Springer International Publishing AG 2017
F. Cong et al. (Eds.): ISNN 2017, Part I, LNCS 10261, pp. 511–520, 2017.
DOI: 10.1007/978-3-319-59072-1_60

M-estimator was proposed as an effective alternative to the LAD estimator. For robust sparse regression estimation, a robust sparse outlier estimator was developed. Because existing robust estimator problems include a non-smooth objective function, they are modeled exactly by a large convex quadratic programming problem. Therefore, the resulting estimator algorithms will have a slow speed in the case of large convex quadratic programming problems. Recently, a noise constraint-based least square (NCLS) method was presented for robust parameter estimation of autoregressive signals and data fusion, respectively. It was shown that the NCLS is robust against noise error, but the NCLS estimation problem may become a large convex quadratic programming problem when the input data is large.

This paper presents a new cooperative projection neural network for fast solving a class of quadratic convex programming problems, which can contain the robust NCLS estimation problem as its special case. The proposed cooperative projection neural network consists of two projection terms. Moreover, compared with existing projection neural networks and numerical optimization algorithms [17–25, 27, 28], the proposed cooperative projection neural network has a very smaller model size than the original size of the quadratic convex programming problems. Moreover, there is no limit condition of initial points. Numerical results demonstrate that the proposed neural network has faster speed than the existing projection neural networks. Because this type of quadratic convex programming problems generalize the robust reconstruction problems, the proposed cooperative projection neural network can be effectively used for fast computation of the robust linear reconstruction problems.

2   Linear Reconstruction Model and Estimation

Consider linear reconstruction model:

$$\mathbf{y} = A\mathbf{x} + \mathbf{n}, \tag{1}$$

where $\mathbf{x} \in R^m$ is the unknown parameter vector to be estimated, $\mathbf{y} \in R^n$ is a observation vector, $\mathbf{n} \in R^n$ is observation noise vector, and $A \in R^{n \times m}$ is a observation matrix. The linear reconstruction model is called underdetermined when $m > n$ and overdetermined when $m < n$. The linear reconstruction model can contain many application models as its special cases, such as the model of MR image reconstruction [3], CT image reconstruction [4], and signal and image restoration [8, 11]. The objective of the linear reconstruction model is to estimate the unknown parameter vector from the observation vector in the presence of observation noise.

Under an assumption of Gaussian noise, the traditional least squares (LS) estimator is given by:

$$\mathbf{x}_{LS} = \arg \min_{\mathbf{x}} \|\mathbf{y} - A\mathbf{x}\|_2^2 \tag{2}$$

or equivalently is the solution of one normal linear equation:

$$A^T A \mathbf{x} = A^T \mathbf{y} \tag{3}$$

where $\| \cdot \|$ denotes l_2 norm. In practices, matrix $A^T A$ is singular or very close to singular so that the precision of the LS solution becomes very poor. The LS estimator is thus replaced by the ridge estimator:

$$\mathbf{x}_R = \arg\min_{\mathbf{x}} \|\mathbf{y} - A\mathbf{x}\|_2^2 + \lambda\|\mathbf{x}\|_2^2 \tag{4}$$

or equivalently is the solution of another normal linear equation:

$$(A^T A + \lambda I)\mathbf{x} = A^T \mathbf{y} \tag{5}$$

where $\lambda > 0$ is a regularization parameter. The ridge estimator is thus given by:

$$\mathbf{x}_R = (A^T A + \lambda I)^{-1} A^T \mathbf{y},$$

which is also called the regularization solution. To achieve edge preservation in image information, the total variational (TV) regularization solution is presented by:

$$\mathbf{x}_{TV} = \arg\min_{\mathbf{x}} \|\mathbf{x}\|_{TV} + \|\mathbf{y} - A\mathbf{x}\|_2, \tag{6}$$

where $\| \cdot \|_{TV}$ is the discrete TV regularization operator term. Both the ridge solution and the TV regularization solution are in general biased since an optimal regularization parameter is usually difficult to be taken.

To deal with non-Gaussian noise, several robust estimators were presented. The traditional least absolute deviation (LAD) estimator is given by [1]:

$$\mathbf{x}_{LAD} = \arg\min_{\mathbf{x}} \|A\mathbf{x} - \mathbf{y}\|_1 \tag{7}$$

where $\| \cdot \|_1$ is l_1 norm. In order to deemphasize outliers, the robust Huber M-estimator is used as one abstracted alterative estimator to minimize the reconstruction problem [7]:

$$\text{minimize} \quad \varphi_1(\mathbf{x}, \mathbf{z}) = \gamma\|A\mathbf{x} - \mathbf{y} - \mathbf{z}\|_1 + \frac{1}{2}\|z\|_2^2$$
$$\text{subject to} \quad \mathbf{x} \in R^m, \ \mathbf{z} \in R^n \tag{8}$$

where $\gamma > 0$ is a design parameter. A generalized least absolute deviation estimator is another abstracted robust estimator to minimize the reconstruction problem [12–14]:

$$\text{minimize} \quad f_1(\mathbf{x}, \mathbf{z}) = \|A\mathbf{x} - \mathbf{b} - \mathbf{z}\|_1$$
$$\text{subject to} \quad \mathbf{z} \in X \tag{9}$$

where $X = \{\mathbf{z} \in R^n \mid \mathbf{l} \le \mathbf{z} \le \mathbf{h}\}$ is the error set. Recently, in order to avoid the nondifferentiability of the robust absolute value error residual, a noise constrained least square (NCLS) estimator was introduced to minimize the reconstruction problem [15, 16]:

$$\text{minimize} \quad f_1(\mathbf{x}, \mathbf{z}) = \|A\mathbf{x} - \mathbf{b} - \mathbf{z}\|_2$$
$$\text{subject to} \quad \mathbf{z} \in X \tag{10}$$

where $\|\cdot\|_2$ denotes l_2 norm. It was shown that the NCLS estimator is not only robust [16] but also fast obtained by using a low-dimensional recurrent neural network [20].

For general applications, this paper considers the following generalized reconstruction problem:

$$\text{minimize} \quad f_2(\mathbf{x}, \mathbf{z}) = \|A\mathbf{x} - Q\mathbf{z} - \mathbf{b}\|_2$$
$$\text{subject to} \quad \mathbf{x} \in \Omega_1, \mathbf{z} \in \Omega_1 \tag{11}$$

where Q is an orthogonal matrix, and $\Omega_1 \subset R^m$ and $\Omega_2 \subset R^n$ are two closed convex sets. It is seen that the robust reconstruction problem (10) is a special case of (11). The objective of this paper is to propose a new recurrent neural network for fast solving (11).

3   Cooperative Projection Neural Network

Recurrent neural networks have been used as computational models for solving computationally intensive problems. Because of the inherent nature of parallel and distributed information processing in neural networks, recurrent neural networks are promising computational models for real-time applications. Several popular projection neural networks (PRNNs) for effectively solving (11) were developed in [17–21]. Because the reconstruction problem (11) contains double unknown vectors, the PRNNs have model size being $n+m$ and model complexity being $O((n+m)^2)$. For fast computation, we here introduce a new cooperative projection neural network with both the model size being m and model complexity being $O(n)$ only.

3.1   Proposed Neural Network

For discussion convenience, we rewrite (11) as:

$$\min \quad f(x_I, x_{II}) = \frac{1}{2}\|Ax_I + Bx_{II} - c\|^2 \tag{12}$$
$$\text{s.t} \quad x_I \in \Omega_1, x_{II} \in \Omega_2$$

where $x_I \in \mathbb{R}^m$, $x_{II} \in \mathbb{R}^n$, $A \in \mathbb{R}^{n \times m}$, $B = -Q$ is an orthogonal matrix, and $c = \mathbf{b}$. By the optimality condition [24] we see that (x_I^*, x_{II}^*) is an optimal solution of (12) if and only if (x_I^*, x_{II}^*) satisfies for any $x_I \in \Omega_1$, $\forall x_{II} \in \Omega_2$

$$(x_I - x_I^*)^T \frac{\partial f(x_I^*, x_{II}^*)}{\partial x_I} + (x_{II} - x_{II}^*)^T \frac{\partial f(x_I^*, x_{II}^*)}{\partial x_I} \geq 0$$

where $\frac{\partial f(x_I, x_{II})}{\partial x_I}$ and $\frac{\partial f(x_I, x_{II})}{\partial x_{II}}$ denote the gradient of $f(x_I, x_{II})$ at x_I and x_{II}, respectively. Using the projection techniques [26] we can establish the following result:

Proposition 1. If $u^* = (u_I^*, u_{II}^*)$ satisfies equations:

$$\begin{cases} u_I = -A^T(AP_{\Omega_1}(u_I) + QP_{\Omega_2}(u_{II}) - c) + P_{\Omega_1}(u_I) \\ \qquad u_{II} = -Q^T(AP_{\Omega_1}(u_I) - c) \end{cases} \tag{13}$$

where $P_{\Omega_1}(u_I) = arg\ min_{v \in \Omega_1}\|u_I - v\|_2$, and $P_{\Omega_2}(u_{II}) = arg\ min_{\hat{v} \in \Omega_2}\|u_{II} - \hat{v}\|_2$. Then $(P_{\Omega_1}(u_I^*), P_{\Omega_2}(u_{II}^*))$ is an optimal solution of (12). Motivated by the reformulation (13), we propose a new cooperative projection neural network (CPNN) as follows:

State equation

$$\frac{du_I(t)}{dt} = \mu\{-A^T(AP_{\Omega_1}(u_I(t)) + QP_{\Omega_2}(w(t)) - c) + P_{\Omega_1}(u_I(t)) - u_I(t)\}$$

$$\tag{14}$$

Output equation

$$v(t) = (P_{\Omega_1}(u_I(t)), P_{\Omega_2}(w(t)))$$

where $w(t) = -Q^T(AP_{\Omega_1}(u_I(t)) - c)$, $u_I(t) \in R^n$ is state trajectory, $v(t) \in R^{n+m}$ is the output trajectory, $\mu > 0$ is a design constant. It is seen that the proposed neural network consists of two projection terms with a cooperative structure and thus we call it as the cooperative projection neural network. Because the proposed cooperative projection neural network has the number of neurons being n, the proposed neural network has the model size m only. As for the stability and convergence of the proposed neural network, using the analysis technique [19] we have the following results:

Theorem 1. The proposed CPNN in (14) is globally stable at its equilibrium point and the output trajectory of the proposed CPNN in (14) will converge globally to an optimal solution of (12).

3.2 Comparison with Related Works

To analyze the model complexity of the proposed neural network, let $W_1 = I - A^T A$, $W_2 = -A^T Q$, $q = A^T c$, and $p = Q^T c$ where $I \in R^{n \times n}$ is an unit matrix. Then (14) becomes:

$$\frac{du_I(t)}{dt} = \mu\{-u_I(t) + W_1 P_{\Omega_1}(u_I(t)) + W_2 P_{\Omega_2}(w(t)) + q\}$$

where $w(t) = W_2^T P_{\Omega_1}(u_I(t)) + p$. We see that the proposed neural network has the total number of multiplications being $2mn + m^2 + m$ per iteration. Thus its model complexity is $O(n)$ when $n << m$.

We now compare the proposed CPNN with existing projection neural networks for solving (12). First, one projection neural network, called the extended projection neural network (EPNN), was developed in [17], defined as:

State equation

$$\frac{d}{dt}\begin{pmatrix} x_I(t) \\ x_{II}(t) \end{pmatrix} = \mu\begin{pmatrix} P_{\Omega_1}(x_I(t) - A^T(Ax_I(t) + Qx_{II}(t) - c)) - x_I(t) \\ P_{\Omega_2}(Q^Tc - Q^T Ax_I(t)) - x_{II}(t) \end{pmatrix} \quad (15)$$

Output equation

$$x(t) = (x_I(t), x_{II}(t))$$

where $(x_I(t), x_{II}(t)) \in \mathbb{R}^n$ is the state trajectory, $\mu > 0$ is a design constant, $P_{\Omega_1}(x_I)$ and $P_{\Omega_2}(x_{II})$ are defined in Proposition 1. It is seen that EPNN has both the number of neurons being $n + m$ and model complexity being $O((n + m)^2)$.

Second, another projection neural network, called the cooperative projection neural network, was developed in [19], defined as:

State equation

$$\frac{d}{dt}\begin{pmatrix} u_I(t) \\ u_{II}(t) \end{pmatrix} = \mu\begin{pmatrix} -A^T(AP_{\Omega_1}(u_I) + QP_{\Omega_2}(u_{II}) - c) + P_{\Omega_1}(u_I) - u_I \\ -Q^T(AP_{\Omega_1}(u_I) + QP_{\Omega_2}(u_{II}) - c) + P_{\Omega_2}(u_{II}) - u_{II} \end{pmatrix} \quad (16)$$

Output equation

$$v(t) = (P_{\Omega_1}(u_I(t)), P_{\Omega_2}(u_{II}(t)))$$

where $u(t) \in R^{n+m}$ is state trajectory, $v(t) \in R^{n+m}$ is the output trajectory. It is seen that the projection neural network has both the number of neurons being $n + m$ and model complexity being $O((n + m)^2)$.

Recently, a bi-projection neural network (BPNN) was developed in [21], defined as:

State equation

$$\frac{dx_I(t)}{dt} = \mu\{P_{\Omega_1}(W_1x_I(t) + W_2P_{\Omega_2}(z(t)) + q) - x_I(t)\} \quad (17)$$

Output equation

$$v(t) = (x_I(t), P_{\Omega_2}(z(t)))$$

where $z(t) = p + W_2^T x_I(t)$, $x_I(t) \in R^n$ is the state trajectory, $v(t) \in R^n$ is the output trajectory, and W_1 and W_2 are defined in the CPNN. The BPNN has the number of neurons being n, but it assumes that initial point $x_I(t_0) \in \Omega_1$. By contrast, the proposed CPNN can be guaranteed to converge globally to the optimal solution of (1) without the initial point condition.

Finally, we compare the proposed CPNN with the gradient projection numerical algorithm [22, 23] and the disciplined convex programming algorithm [25]. Because the generalized reconstruction problem (11) has the solution space dimension being $n + m$, the solution algorithm has the algorithm complexity being $O((n + m)^2)$ at least.

4   Illustrative Examples

In this section, we give illustrative examples to demonstrate the effectiveness of the proposed CPNN algorithm. The simulation is conducted in MATLAB 7.0 platform where computation time unit is taken as second.

Example 1. Consider the generalized reconstruction problem (12):

$$\min \quad \frac{1}{2}\|Ax_I + Bx_{II} - c\|^2 \tag{18}$$
$$\text{s.t} \quad Cx_I = b, \ l \le x_{II} \le h$$

where A is the $n \times m$ random matrix, B is the $n \times n$ random orthogonal matrix, C is the $r \times m$ random matrix, c is the n dimensional random vector, and b is the r dimensional random vector. We choose two random integer vectors z_I and z_{II} to construct vector c contaminated by uniformly distributed random noise. We construct the vector b satisfying linear equality constraints, and construct the inequality constants by using two uniformly distributed random vectors l and h. All simulation results show the proposed CPNN is always convergent to the optimal solution. For a comparison, we also perform the EPNN defined in (15), the BPNN defined in (17), and the CVX numerical algorithm [25]. Computed results are listed in Table 1. From the Table 1 we see that the proposed CPNN has a faster speed than other three algorithms.

Table 1. Computed results of four algorithms in Example 1

Algorithm	Problem size	Error	CPU time (sec.)
Proposed CPNN	$n_1 = 200, \ n_2 = 400, \ m = 400, \ r = 100$	2.59	1.92
BPNN (17)	$n_1 = 200, \ n_2 = 400, \ m = 400, \ r = 100$	2.61	2.06
EPNN (15)	$n_1 = 200, \ n_2 = 400, \ m = 400, \ r = 100$	2.61	4.09
CVX numerical algorithm	$n_1 = 200, \ n_2 = 400, \ m = 400, \ r = 100$	4.53	19.57
Proposed CPNN	$n_1 = 400, \ n_2 = 600, \ m = 600, \ r = 200$	3.16	5.34
BPNN (17)	$n_1 = 400, \ n_2 = 600, \ m = 600, \ r = 200$	3.29	6.08
EPNN (15)	$n_1 = 400, \ n_2 = 600, \ m = 600, \ r = 200$	3.26	16.74
CVX numerical algorithm	$n_1 = 400, \ n_2 = 600, \ m = 600, \ r = 200$	4.76	71.21
Proposed CPNN	$n_1 = 600, \ n_2 = 800, \ m = 800, \ r = 400$	1.89	11.61
BPNN (17)	$n_1 = 600, \ n_2 = 800, \ m = 800, \ r = 400$	1.90	12.32
EPNN (15)	$n_1 = 600, \ n_2 = 800, \ m = 800, \ r = 400$	1.90	31.50
CVX numerical algorithm	$n_1 = 600, \ n_2 = 800, \ m = 800, \ r = 400$	1.98	302.57

Table 2. Computed results of four algorithms in Example 2

Algorithm	Problem size	Error	CPU time (sec.)
Proposed CPNN	$n_1 = 200$, $n_2 = 400$, $m = 400$, $r = 100$	1.97	1.45
BPNN (17)	$n_1 = 200$, $n_2 = 400$, $m = 400$, $r = 100$	2.03	1.48
EPNN (15)	$n_1 = 200$, $n_2 = 400$, $m = 400$, $r = 100$	1.98	2.81
CVX numerical algorithm	$n_1 = 200$, $n_2 = 400$, $m = 400$, $r = 100$	2.35	24.57
Proposed	$n_1 = 400$, $n_2 = 600$, $m = 600$, $r = 200$	2.17	4.46
BPNN (17)	$n_1 = 400$, $n_2 = 600$, $m = 600$, $r = 200$	2.19	4.92
EPNN (15)	$n_1 = 400$, $n_2 = 600$, $m = 600$, $r = 200$	2.18	10.66
CVX numerical algorithm	$n_1 = 400$, $n_2 = 600$, $m = 600$, $r = 200$	2.45	112.06
Proposed CPNN	$n_1 = 600$, $n_2 = 800$, $m = 800$, $r = 400$	3.69	10.16
BPNN (17)	$n_1 = 600$, $n_2 = 800$, $m = 800$, $r = 400$	3.78	13.69
EPNN (15)	$n_1 = 600$, $n_2 = 800$, $m = 800$, $r = 400$	3.74	39.84
CVX numerical algorithm	$n_1 = 600$, $n_2 = 800$, $m = 800$, $r = 400$	6.19	211.89

Example 2. Consider the generalized reconstruction problem (12):

$$\min \quad \frac{1}{2}\|Ax_I + Bx_{II} - c\|^2 \tag{19}$$
$$\text{s.t} \quad p \le x_I \le q, \quad Dx_{II} = d$$

where A is the $n \times m$ random matrix, B is the $n \times n$ random orthogonal matrix, D is the $r \times n$ random matrix, c is the n dimensional random vector, and d is the r dimensional random vector. We choose two random integer vectors z_I and z_{II} to construct vector c contaminated by uniformly distributed random noise. We construct vector d satisfying linear equality constraints, and construct the inequality constants by using two uniformly distributed random vectors p and q. All simulation results show the proposed CPNN is always convergent to the optimal solution. For a comparison, we also perform the EPNN defined in (15), the BPNN defined in (17), and the CVX numerical algorithm. Computed results are listed in Table 2. From the Table 2 we see that the proposed CPNN has a faster speed than other three algorithms.

5   Conclusion

This paper has presented a new cooperative projection neural network for fast solving generalized reconstruction problems. The proposed cooperative projection neural network has two projection terms. Compared with existing projection neural networks, the proposed cooperative projection neural network has not only a small model size but also no limit condition of initial points. Numerical results demonstrate that the proposed cooperative projection neural network has a faster speed than the existing projection neural networks. Therefore, the proposed cooperative projection neural network can be effectively used for fast computation of the robust linear reconstruction problems.

References

1. Dodge, Y., Jana, J.: Adaptive Regression. Springer, New York (2000)
2. Huber, P.J.: Robust Statistics, 2nd edn. Wiley, Hoboken (2008)
3. Angshul, M., Ward, R.K.: An algorithm for sparse MRI reconstruction by Schatten p-norm minimization. Magn. Reson. Imaging 29, 408–417 (2011)
4. Joost, B.K., Linda, P.: Fast approximation of algebraic reconstruction methods for tomography. IEEE Trans. Image Process. 21, 3648–3658 (2012)
5. Shalvi, O., Weinstein, E.: Maximum-likelihood and lower bounds in system-identification with non-Gaussian inputs. IEEE Trans. Inf. Theor. 40, 328–339 (1994)
6. Cadzow, J.A.: Minimum l_1, l_2, and l_∞ norm approximate solutions an overdetermined system of linear equations. Digit. Sig. Process. 12, 524–560 (2002)
7. Mangasarian, O.L., Musicant, D.R.: Robust linear and support vector regression. IEEE Trans. Pattern Anal. Mach. Intell. 22(9), 1–6 (2000)
8. Cichocki, A., Amari, S.: Adaptive Blind Signal and Image Processing: Learning Algorithms and Applications. Wiley, Hoboken (2002)
9. Kim, S.J., Koh, K., Lustig, M., Boyd, S., Gorinevsky, D.: An interior-point method for large-scale l_1-regularized least squares. IEEE J. Select. Topics Sig. Process. 1, 606–617 (2007)
10. Papageorgiou, G., Bouboulis, P., Theodoridis, S.: Obust linear regression analysis: a greedy approach. IEEE Trans. Sig. Process. 63, 3872–3887 (2015)
11. Campisi, P., Egiazarian, K.: Blind Image Deconvolution: Theory and Applications. CRC Press, Cambridge (2007)
12. Xia, Y.S., Kamel, M.S.: Novel cooperative neural fusion algorithms for image restoration and image fusion. IEEE Trans. Image Process. 16, 367–381 (2007)
13. Xia, Y.S., Kamel, M.S.: Cooperative learning algorithms for data fusion using novel L_1 estimation. IEEE Trans. Sig. Process. 56, 1083–1095 (2008)
14. Xia, Y.S., Kamel, M.S.: A generalized least absolute deviation method for parameter estimation of autoregressive signals. IEEE Trans. Neural Netw. 19, 107–118 (2008)
15. Xia, Y.S., Kamel, M.S., Henry, L.: A fast algorithm for AR parameter estimation using a novel noise-constrained least squares method. Neural Netw. 33, 396–405 (2010)
16. Xia, Y.S., Leung, H., Kamel, M.S.: A discrete-time learning algorithm for image restoration using a novel L-2-norm noise constrained estimation. Neurocomputing 198, 155–170 (2016)
17. Xia, Y.S.: An extended projection neural network for constrained optimization. Neural Comput. 16, 863–883 (2004)
18. Xia, Y.S.: Further results on global convergence and stability of globally projected dynamical systems. J. Optim. Theor. Appl. 122, 149–627 (2004)
19. Xia, Y.S., Feng, G., Wang, J.: A novel recurrent neural network for solving nonlinear optimization problems with inequality constraints. IEEE Trans. Neural Netw. 19, 1340–1353 (2008)

20. Xia, Y.S., Wang, J.: Low-dimensional recurrent neural network-based Kalman filter for speech enhancement. Neural Netw. **67**, 131–139 (2015)
21. Xia, Y.S., Wang, J.: A bi-projection neural network for solving constrained quadratic optimization problems. IEEE Trans. Neural Netw. Learn. Syst. **27**, 214–224 (2016)
22. Daubechies, I., Fornasier, M., Loris, I.: Accelerated projected gradient method for linear inverse problems with sparsity constraints. J. Fourier Anal. Appl. **14**, 764–792 (2008)
23. Malitsky, Y.: Projected reflected gradient methods for monotone variational inequalities. SIAM J. Optim. **25**, 502–520 (2015)
24. Boyd, S., Vandenberghe, L.: Convex Optimization. Cambridge University Press, Cambridge (2006)
25. Grant, M., Boyd, S., Ye, Y.Y.: Disciplined Convex Programming. Global Optimization: From Theory to Implementation. Kluwer, Dordrecht (2005)
26. Kinderlehrer, D., Stampacchia, G.: An Introduction to Variational Inequalities and Their Applications. Academic Press, New York (1980)
27. Cheng, L., Hou, Z.G., Lin, Y., et al.: Recurrent neural network for non-smooth convex optimization problems with application to the identification of genetic regulatory networks. IEEE Trans. Neural Netw. Learn. Syst. **22**, 714–726 (2011)
28. Liu, Q.S., Wang, J.: A one-layer projection neural network for nonsmooth optimization subject to linear equalities and bound constraints. IEEE Trans. Neural Netw. Learn. Syst. **24**, 812–824 (2013)

A Complex Gradient Neural Dynamics for Fast Complex Matrix Inversion

Lin Xiao[1]([✉]), Bolin Liao[1], Qinli Zeng[1,2], Lei Ding[1], and Rongbo Lu[1]

[1] College of Information Science and Engineering,
Jishou University, Jishou 416000, China
xiaolin860728@163.com

[2] School of Electronics and Information Technology, Sun Yat-sen University,
Guangzhou 510275, China

Abstract. Complex-valued matrix inversion problem is investigated by using the gradient-neural-dynamic method. Differing from the traditional processing method (only for real-valued matrix inversion), the proposed method develops a complex gradient neural dynamics for complex-valued matrix inversion in the complex domain. The advantages of the proposed method decrease the complexities in the aspects of computation, analysis, and computer simulations. Theoretical discussions and computer simulations demonstrate the efficacy and superiorness of the proposed method for online the complex-valued matrix inversion in the complex domain, as compared to the traditional processing method.

Keywords: Complex-valued matrix inversion · Theoretical analysis · Complex domain · Neural dynamic model

1 Introduction

The matrix inversion is often required in optimization [1], electromagnetic systems [2], and robot kinematics [3]. In addition, many physical phenomena can be depicted by matrix inversion. Therefore, the desired inverses of the matrices should be exploited to know the principle of physics about phenomena [4].

The approaches of the finding matrix inverses could be divided into two classes: one based on numerical algorithms and one based on neural networks. At first, much effort has been made towards rapid matrix inversion and various iterative algorithms were developed to find the matrix inverses [4]. However, since the computational operations of these algorithms are proportional to the dimension of the matrix [5], they might not be valid enough for real-time applications and large-scale datum computation.

In order to overcome the computational bottleneck of traditional iterative algorithms, recurrent neural network (RNN), as one kind of the key parallel-processing approaches, is developed for computing matrix inversion problem [6–9]. Compared with conventional numerical algorithms, RNNs have some prominent advantages in real-time applications (such as high fault tolerance, self-adaptation, distributed storage, and parallel processing) [10–17].

© Springer International Publishing AG 2017
F. Cong et al. (Eds.): ISNN 2017, Part I, LNCS 10261, pp. 521–528, 2017.
DOI: 10.1007/978-3-319-59072-1_61

Noting that the above aforementioned approaches are usually used to compute matrix inverse in the real domain, and few effort has been made towards the research in the complex domain for finding complex-valued matrix inversion. However, in some cases (for example, the input signals incorporate the phase and magnitude information), complex-valued matrix inversion problems can also appear [18]. Thus, the complex-valued problems have obtained a great of attention of researchers. For example, Liao et al. [19] extended the results on quadratic programming problems solving from the real to the complex domains. In that case the real and imaginary components of complex quadratic programming problems were solved respectively in the real domain, which increases the computational complexities in modeling, simulation, and some real-time applications. Obviously, it is more convenient for the analysis and solution to such a complex-valued problem in the complex field directly. This paper demonstrates that the complex gradient neural dynamics can be applied directly to complex-valued matrix inversion.

2   Problem Formulation and Equivalent Real-Valued GND Model

The complex matrix inversion equation is described as follows:

$$AZ(t) = I \in \mathbb{C}^{n \times n}, \text{ or } Z(t)A = I \in \mathbb{C}^{n \times n}, \tag{1}$$

where complex $Z(t) \in \mathbb{C}^{n \times n}$ stands for an unknown matrix, complex matrix $A \in \mathbb{C}^{n \times n}$ is a complex-valued coefficient of (1), and $I \in \mathbb{C}^{n \times n}$ denotes complex-valued identity matrix. Besides, $Z^* \in \mathbb{C}^{n \times n}$ is used to stand for the theoretical inversion of (1).

It is worth mentioning that, in most of past papers about complex matrix inversion, complex coefficient $A \in \mathbb{C}^{n \times n}$ is usually split into its imaginary and real parts, and handled separately. That is to say, $A \in \mathbb{C}^{n \times n}$ is considered to be the union of its imaginary and real parts, i.e., $A(t) = A_{\mathrm{re}} + jA_{\mathrm{im}}$ with $j = \sqrt{-1}$ denoting an imaginary unit $[Z(t) = Z_{\mathrm{re}}(t) + jZ_{\mathrm{im}}(t)$ as well]. Therefore, complex-valued matrix inversion Eq. (1) is equivalent to the following one:

$$[A_{\mathrm{re}} + jA_{\mathrm{im}}][Z_{re}(t) + jZ_{\mathrm{im}}(t)] = I_{\mathrm{re}} + jI_{\mathrm{im}}, \tag{2}$$

where $A_{\mathrm{re}} \in \mathbb{R}^{n \times n}$, $A_{\mathrm{im}} \in \mathbb{R}^{n \times n}$, $Z_{\mathrm{re}} \in \mathbb{R}^{n \times n}$, $Z_{\mathrm{im}} \in \mathbb{R}^{n \times n}$, $I_{\mathrm{re}} \in \mathbb{R}^{n \times n}$ and $I_{\mathrm{im}} \in \mathbb{R}^{n \times n}$.

Then, based on the traditional processing approach [19] and considering Eq. (2), the real-valued linear system is obtained as below:

$$\begin{cases} A_{\mathrm{re}}Z_{\mathrm{re}}(t) - A_{\mathrm{im}}Z_{\mathrm{im}}(t) = I_{\mathrm{re}} \in \mathbb{R}^{n \times n}, \\ A_{\mathrm{re}}Z_{\mathrm{im}}(t) + A_{\mathrm{im}}Z_{\mathrm{re}}(t) = I_{\mathrm{im}}(t) \in \mathbb{R}^{n \times n}, \end{cases}$$

which is further equivalent to the following expression:

$$\begin{bmatrix} A_{\mathrm{re}} & -A_{\mathrm{im}} \\ A_{\mathrm{im}} & A_{\mathrm{re}} \end{bmatrix} \begin{bmatrix} Z_{\mathrm{re}}(t) \\ Z_{\mathrm{im}}(t) \end{bmatrix} = \begin{bmatrix} I_{\mathrm{re}} \\ I_{\mathrm{im}} \end{bmatrix} \in \mathbb{R}^{2n \times n}. \tag{3}$$

Now, matrices $B \in \mathbb{R}^{2n \times 2n}$, $X(t) \in \mathbb{R}^{2n \times n}$ and $D \in \mathbb{R}^{2n \times n}$ is used to stand for the above matrices for presentation convenience, i.e.,

$$B = \begin{bmatrix} A_{\mathrm{re}} & -A_{\mathrm{im}} \\ A_{\mathrm{im}} & A_{\mathrm{re}} \end{bmatrix}, \quad X(t) = \begin{bmatrix} Z_{\mathrm{re}}(t) \\ Z_{\mathrm{im}}(t) \end{bmatrix}, \quad D = \begin{bmatrix} I_{\mathrm{re}} \\ I_{\mathrm{im}} \end{bmatrix}.$$

Thus, the complex matrix inversion problem can be changed into the following real matrix inversion problem:

$$BX(t) = D \in \mathbb{R}^{2n \times n}. \tag{4}$$

Then, based on the gradient neural dynamic approach [6,7], starting with the definition of energy function $\varepsilon(t) = \|BX(t) - D\|_F^2/2$, we have the following real-valued gradient neural dynamics (GND) for computing the real-valued matrix inversion Eq. (4):

$$\dot{X}(t) = -\gamma B^{\mathrm{T}}(BX(t) - D) \in \mathbb{R}^{2n \times n}, \tag{5}$$

where $\gamma > 0$ can adjust the convergence speed of the real-valued GND model (5), and matrix $X(t) \in \mathbb{R}^{2n \times n}$ is corresponding to $X^* = [Z_{\mathrm{re}}^*, Z_{\mathrm{im}}^*]^{\mathrm{T}} \in \mathbb{R}^{2n \times n}$ of (4), which makes up the complex inverse of (1), i.e., $Z(t) = Z_{\mathrm{re}}(t) + jZ_{\mathrm{im}}(t)$.

3 Fully Complex-Valued GND Model

Different from the conventional design processing, the complex gradient neural dynamics (GND) is constructed on account of the fully complex matrix inversion Eq. (1). To start with, an energy function can be defined as follows (instead of the equivalent real-valued matrix inversion problem solving):

$$\varepsilon(t) = \|AZ(t) - I\|_2^2/2. \tag{6}$$

Next, following the gradient neural dynamic method [6,7], we could design a complex-valued gradient algorithm to evolve towards the descent direction until the minimum value of this energy function is achieved. That is,

$$-\frac{\partial \|AZ(t) - I\|_F^2/2}{\partial Z(t)} = -A^{\mathrm{H}}(AZ(t) - I). \tag{7}$$

Third, the following complex GND model for the complex matrix inversion (3) can be derived by adopting the above negative gradient information:

$$\dot{Z}(t) = -\gamma A^{\mathrm{H}}(AZ(t) - I), \tag{8}$$

where $\gamma > 0$ is defined as before, and state matrix $Z(t) \in \mathbb{C}^{n \times n}$ is also corresponding to the $Z^* \in \mathbb{C}^{n \times n}$ of (3).

As we know, the convergence speed is of importance for a neural-dynamic model to be applied successfully. Therefore, we should study the convergence property of the complex GND model (8).

Theorem 1. *Considering the complex matrix inversion Eq. (2), complex matrix $Z(t)$ of the complex GND model (8), starting from a randomly-generated $Z(0) \in \mathbb{C}^{n \times n}$, can converge to the theoretical inverse Z^* of (2) exponentially. In addition, the convergence rate is the product of the value of γ and the minimum eigenvalue α of $A^H A$.*

Proof. Define $\tilde{Z}(t) = Z(t) - Z^*$ with Z^* being the theoretical inverse of Eq. (1) and $Z(t)$ generated by the complex GND model (8). Therefore, one can obtain:

$$\dot{Z}(t) = \dot{\tilde{Z}}(t) \in \mathbb{C}^{n \times n} \text{ and } Z(t) = \tilde{Z}(t) + Z^* \in \mathbb{C}^{n \times n}.$$

Let us substitute the above equations to (8), and based on $AZ^* - I = 0$; and the following dynamic system can be derived:

$$\dot{\tilde{Z}}(t) = -\gamma A^H A \tilde{Z}(t). \tag{9}$$

Thus, we could select a Lyapunov function $v(t)$ as below:

$$v(t) = \|A\tilde{Z}(t)\|_F^2 / 2 = \text{tr}\left((A\tilde{Z}(t))^H (A\tilde{Z}(t)) \right) / 2 \geqslant 0,$$

where $\text{tr}(\cdot)$ represents the trace of a matrix. Evidently, the Lyapunov function is positive definite, since $v(t) = 0$ only for $\tilde{Z}(t) = 0$, and $v(t) > 0$ for any $\tilde{Z}(t) \neq 0$.

On the other hand, the time derivative of $v(t)$ can be further estimated as follows:

$$\dot{v}(t) = \frac{dv}{dt} = \text{tr}\left((A\tilde{Z}(t))^H (A\dot{\tilde{Z}}(t)) \right) = -\gamma \|A^H A \tilde{Z}(t)\|_F^2 \leqslant 0, \tag{10}$$

which ensures the negative-definiteness of $\dot{v}(t)$. That is to say, $\dot{v}(t) = 0$ only for $\tilde{Z}(t) = 0$, and $\dot{v}(t) < 0$ for any $\tilde{Z}(t) \neq 0$. It can also be concluded that $v(t) \to +\infty$, when $\|\tilde{Z}\| \to +\infty$. In view of Lyapunov theory, complex matrix $\tilde{Z}$ converges to zero globally. Then, according to $\tilde{Z}(t) = Z(t) - Z^*$, complex matrix $Z(t)$ can converge to the theoretical inverse Z^* globally.

Now, let us consider the exponential convergence rate. Given $\alpha > 0$ as the minimum eigenvalue of $A^H A$, from (10), one can have:

$$\begin{aligned}
\dot{v}(t) &= -\gamma \|A^H A \tilde{Z}(t)\|_F^2 \\
&= -\gamma \text{tr}\{ (A\tilde{Z}(t))^H A A^H A \tilde{Z}(t) \} \\
&\leqslant -\alpha\gamma \text{tr}\{ (A\tilde{Z}(t))^H A \tilde{Z}(t) \} \\
&= -\alpha\gamma \|A\tilde{Z}(t)\|_F^2 \\
&= -2\alpha\gamma v(t).
\end{aligned} \tag{11}$$

According to Eq. (11), the analytic solution can be estimated as below:

$$v(t) \leqslant v(0) \exp(-2\alpha\gamma t).$$

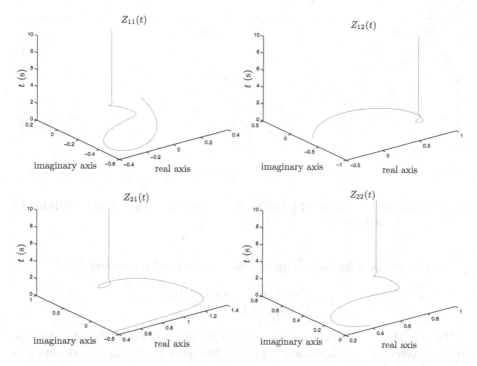

Fig. 1. Transient behavior of $Z(t)$ generated by the complex GND model (8) with $\gamma = 10$.

Moreover, $v(t) = \mathrm{tr}((A\tilde{Z}(t))^{\mathrm{H}} A\tilde{Z}(t))/2 \geqslant \alpha \mathrm{tr}(\tilde{Z}(t)^{\mathrm{H}}\tilde{Z}(t))/2 = \alpha\|\tilde{Z}(t)\|_{\mathrm{F}}^2/2$, and $v(0) = \|A\tilde{Z}(0)\|_{\mathrm{F}}^2/2 \leqslant \|A\|_{\mathrm{F}}^2\|\tilde{Z}(0)\|_{\mathrm{F}}^2/2$. In addition,

$$\alpha\|\tilde{Z}(t)\|_{\mathrm{F}}^2/2 \leqslant v(t) \leqslant v(0)\exp(-2\alpha\gamma t),$$

which is written further as follows:

$$\|Z(t) - Z^*\|_{\mathrm{F}} \leqslant \frac{\|A\|_{\mathrm{F}}\|\tilde{Z}(0)\|_{\mathrm{F}}}{\sqrt{\alpha/2}}\exp(-\alpha\gamma t).$$

It can be concluded that, as $t \to \infty$, $Z(t) \to Z^*$ exponentially with the convergence rate being $\alpha\gamma$, which indicates, complex $Z(t)$ of the complex GND model (8) can converge to the theoretical inverse Z^* of (1) with the convergence rate being $\alpha\gamma > 0$. This completes the proof. □

4 Illustrative Example

The real-valued GND model (5) and the fully complex-valued GND model (8) are provides in the previous two sections for computing online complex-valued matrix Eq. (1). In this part, one illustrative example is applied to demonstrate the efficacy of the complex GND model (8).

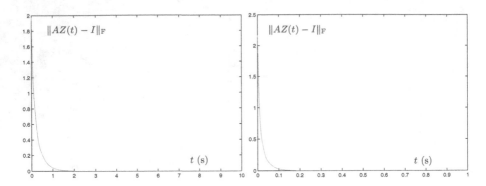

Fig. 2. Convergence behavior of $\|AZ(t) - I\|_F$ generated by the complex GND model (8) with $\gamma = 10$ and $\gamma = 100$.

We choose the following complex-valued matrix to be inverted:

$$A = \begin{bmatrix} \exp(j10) & -\exp(-j10) \\ -\exp(j10) & 0 \end{bmatrix}.$$

To validate the solution accuracy of the complex GND model (8), the theoretical inverse Z^* of the complex matrix inversion Eq. (1) in this situation is obtained as below:

$$Z^* = \begin{bmatrix} 0 & 0.8391 - j0.5440 \\ 0.8391 + j0.5440 & 0.8391 + j0.5440 \end{bmatrix}. \tag{12}$$

To begin with, starting from an arbitrary initial state $Z(0) \in \mathbb{C}^{2\times 2}$ and with design parameter $\gamma = 10$, the complex GND model (8) is used to calculate the above complex matrix inverse. The simulation results are described in Figs. 1 and 2. As observed from Fig. 1, complex neural-state matrix $Z(t) \in \mathbb{C}^{2\times 2}$ generated by the complex GND model (8) converges to a straight line within a little time. In addition, these values are the theoretical inverses of complex neural-state matrix $Z(t)$, as compared to the theoretical inverse in (12). This conclusion illustrates that the complex GND model (8) is valid for the complex matrix inversion.

To directly demonstrate the solution processing of the complex GND model (8), the convergence property of the residual error $\|AZ(t) - I\|_F$ is obtained in Fig. 2(a) under the same conditions. From Fig. 2(a), it can be concluded that $\|AZ(t) - I\|_F$ of the complex GND model (8) converges to zero after about 1.8 s. That is, the complex solution of the GND model (8) fits with the theoretical matrix inverse very well. The simulative observation agrees well with the results of Fig. 1.

Besides, as seen from Fig. 2(b), the convergence property of the complex GND model (8) is improved when the value of γ becomes larger. Specifically, the convergence time of the complex GND model (8) is decreased from 1.8 s to 0.18 s when the value of γ increases from 10 to 100. In a word, the complex GND model (8) is valid for solving complex matrix inverse depicted in Eq. (1).

5   Conclusions

A complex gradient neural dynamics (GND) is proposed and studied in this paper for finding complex matrix inversion in the complex domain. The proposed processing method does not change the complex-valued matrix inversion into the real-valued matrix inversion. The theoretical analysis and computer simulations show the validness and superiorness of the complex GND model for calculating the complex matrix inversion in the complex domain.

Acknowledgment. This work is supported by the Research Foundation of Education Bureau of Hunan Province, China (grant no. 15B192), the Natural Science Foundation of Hunan Province, China (grant no. 2016JJ2101), and the National Natural Science Foundation of China (grant nos. 61503152, 61563017, and 61363073).

References

1. Xiao, L.: A nonlinearly-activated neurodynamic model and its finite-time solution to equality-constrained quadratic optimization with nonstationary coefficients. Appl. Soft Comput. **40**, 252–259 (2016)
2. Chen, Q., Chakarothai, J., Sawaya, K.: Hybrid approach of SPM and matrix-inversion to estimate current distribution of electromagnetic radiation source. In: Proceedings of IEEE Electrical Design of Advanced Packaging and Systems Symposium, pp. 1–4 (2011)
3. Zhang, Z., Li, Z., Zhang, Y., Luo, Y., Li, Y.: Neural-dynamic-method-based dual-arm CMG scheme with time-varying constraints applied to humanoid robots. IEEE Trans. Neural Netw. Learn. Syst. **26**(12), 3251–3262 (2015)
4. Zhang, Y., Leithead, W.E., Leith, D.J.: Time-series Gaussian process regression based on Toeplitz computation of $O(N^2)$ operations and $O(N)$-level storage. In: Proceedings of the 44th IEEE Conference on Decision and Control, pp. 3711–3716 (2005)
5. Mathews, J.H., Fink, K.D.: Numerical Methods Using MATLAB. Prentice Hall, New Jersey (2004)
6. Xiao, L., Zhang, Y.: Zhang neural network versus gradient neural network for solving time-varying linear inequalities. IEEE Trans. Neural Netw. **22**, 1676–1684 (2011)
7. Zhang, Y., Shi, Y., Chen, K., Wang, C.: Global exponential convergence and stability of gradient-based neural network for online matrix inversion. Appl. Math. Comput. **215**, 1301–1306 (2009)
8. Zhang, Y., Ge, S.S.: Design and analysis of a general recurrent neural network model for time-varying matrix inversion. IEEE Trans. Neural Netw. **16**(6), 1477–1490 (2005)
9. Zhang, Y., Ma, W., Cai, B.: From Zhang neural networks to Newton iteration for matrix inversion. IEEE Trans. Circuits Syst. I **56**(7), 1405–1415 (2009)
10. Xiao, L., Lu, R.: Finite-time solution to nonlinear equation using recurrent neural dynamics with a specially-constructed activation function. Neurocomputing **151**, 246–251 (2015)
11. Li, S., Chen, S., Liu, B.: Accelerating a recurrent neural network to finite-time convergence for solving time-varying Sylvester equation by using a sign-bi-power activation function. Neural. Process. Lett. **37**, 189–205 (2013)

12. Xiao, L., Zhang, Y.: Two new types of Zhang neural networks solving systems of time-varying nonlinear inequalities. IEEE Trans. Circuits Syst. I **59**, 2363–2373 (2012)
13. Xiao, L.: A nonlinearly activated neural dynamics and its finite-time solution to time-varying nonlinear equation. Neurocomputing **173**, 1983–1988 (2016)
14. Xiao, L.: A new design formula exploited for accelerating Zhang neural network and its application to time-varying matrix inversion. Theor. Comput. Sci. **647**, 50–58 (2016)
15. Zhang, Y., Xiao, L., Xiao, Z., Mao, M.: Zeroing Dynamics, Gradient Dynamics, and Newton Iterations. CRC Press, Cambridge (2015)
16. Xiao, L., Liao, B.: A convergence-accelerated Zhang neural network and its solution application to Lyapunov equation. Neurocomputing **193**, 213–218 (2016)
17. Xiao, L., Zhang, Y.: Dynamic design, numerical solution and effective verification of acceleration-level obstacle avoidance scheme for robot manipulators. Int. J. Syst. Sci. **47**(4), 932–945 (2016)
18. Xiao, L.: A finite-time convergent neural dynamics for online solution of time-varying linear complex matrix equation. Neurocomputing **167**, 254–259 (2015)
19. Liao, W., Wang, J., Wang, J.: A recurrent neural network for solving complex-valued quadratic programming problems with equality constraints. In: Tan, Y., Shi, Y., Tan, K.C. (eds.) ICSI 2010. LNCS, vol. 6146, pp. 321–326. Springer, Heidelberg (2010). doi:10.1007/978-3-642-13498-2_42

Burst and Correlated Firing in Spiking Neural Network with Global Inhibitory Feedback

Jinli Xie[✉], Qinjun Zhao, and Jianyu Zhao

School of Electrical Engineering,
University of Jinan, Jinan 250022, Shandong, China
{cse_xiejl, cse_zjy, cse_zhaoqj}@ujn.edu.cn

Abstract. Burst and correlated firing activities are observed experimentally in a variety of brain areas, which transmit and communicate information predominantly through spikes. The firing mode of spiking neurons relies on specific network characteristics. The inhibitory feedback is thought to be crucial to the burst firing. However, the effects of inhibitory feedback, and in particular the resulting bursting, on neural correlations need further studies. In order to understand how inhibitory feedback circuit modulates correlations and burst, we carry out numerical simulations of spiking neural network with global inhibitory feedback. Owing to the feedback inhibition, the neurons fire correlated action potentials of a long time scale and exhibit bursting fire pattern. We also found that, with constant output firing rate, the burst firing enhanced network correlations. These results suggest that in the spiking neural network with globally inhibitory feedback the shifts in the feedback strength can induce changes in burst probability, and then effect the correlated firing activities.

Keywords: Burst · Correlation · Inhibitory feedback · Leaky integrate-and-fire neuron

1 Introduction

Neurons in many sensory systems tend to fire correlated action potentials [1–4]. Correlated firing activities have been participated in various neural functions, including attention, memory, olfaction, vision, and motor behavior [1, 5–8]. Evidence has collected that burst firing can also implicate in neural communication, which carries specific, stimulus-related information [9–12]. Recent research sheds light on the relationship between burst firing and network correlation, suggesting that any features affecting the prevalence of neural burst firing may play an important role in modulating the overall level of correlations in neural network [13]. Given that experimentally observed firing of bursts dependents on intrinsic cellular mechanisms that relate to feedback from upper centers [10, 14], it is important to investigate the feedback induced burst and correlated firing activities.

Our former studies suggested that network correlations can be modulated by inhibitory feedback via the interplay of mean firing rate and oscillatory activity [15, 16]. The correlated firing activity could be suppressed by reducing the output firing rate. While, the oscillation induced by the inhibitory feedback can modulate the firing

© Springer International Publishing AG 2017
F. Cong et al. (Eds.): ISNN 2017, Part I, LNCS 10261, pp. 529–535, 2017.
DOI: 10.1007/978-3-319-59072-1_62

activity of the post-synaptic neurons, leading to an improvement of the correlations of the network. However, little is known about the influence of changes in strength of feedback on bursts.

With the network of spiking neurons, we aimed to study the relationship between inhibitory feedback and bursts, and the effects of resulting bursts on correlated firing by numerical simulations in this work. The output firing rate is kept constant to avoid the influence of firing rate on correlated firing [17]. Thus the firing activities of the network are characterized by the level of input activities with various feedback gains. The bursts and network correlations, as quantified by the firing time of the spike trains, are obtained numerically. We demonstrate that bursting can be controlled by inhibitory feedback input, which successively contributes to the network correlated firing.

2   Spiking Neural Network Model

The structure of the network model includes N excitatory neurons and M inhibitory neurons. ALL the inhibitory neurons responded to the projection from the excitatory neurons provide inhibitory feedback to all the excitatory after a fixed time delay. The leaky integrate-and-fire (LIF) neuron models are used to simulate the neural dynamics. The dynamic of membrane potential of the spiking neurons is described by:

$$C\frac{dV_i(t)}{dt} = -g_l V_i(t) + \mu_E + \eta_i(t) + \xi_i(t) - I_g \tag{1}$$

where C is the membrane capacitance, $V_i(t)$ is the membrane potential, g_l is the membrane leak conductance, μ_E denotes the base current, $\eta_i(t)$ represents an internal zero-mean Gaussian white noise of intensity D_E, and $\xi_i(t)$ represents the external stimuli. The dynamics of membrane potential follow to a simple spike-and-reset rule, where the values of firing threshold V_{th} and reset potential V_r are set to be 1 and 0 respectively. On behalf of the absolute refractory state, a time constant τ_r is selected after firing. The spikes of the LIF neuron is modeled as δ-functions:

$$y(t) = \sum_{fire} \delta(t - t_{fire}) \tag{2}$$

where t_{fire} is the threshold crossing instants. I_g in Eq. 1 indicates the feedback loop which is computed by the spike trains of inhibitory neurons convoluted with an α function:

$$I_g(t) = G \int_{\tau_D}^{\infty} \alpha(\tau) \sum_{j=1}^{M} y_j(t - \tau) d\tau \tag{3}$$

with

$$\alpha(t) = \frac{t - \tau_D}{\tau_S^2} \exp\left[-\frac{t - \tau_D}{\tau_S}\right] \tag{4}$$

Here G is the feedback gain, t_j are the successive firing times, τ_D is the transmission delay of feedback, and τ_S is the synaptic time constant.

Each inhibitory neuron receives the convolution of the spike trains of excitatory neurons and another delayed α function:

$$I_f = \int_{\tau_D}^{\infty} \alpha(\tau) \sum_{i=1}^{N} y_i(t - \tau) d\tau \tag{5}$$

and it also has internal Gaussian white noise $\eta(t)$ with intensity D_I.

3 Characterizing Network Correlation and Burst Firing

Since the firing rate of the post-synaptic neurons is crucial to the correlated firing of the network [17], the simulations are achieved by adjusting the external stimuli μ_E such that the average output firing rate keeps constant (12 Hz) with varying values of parameter.

The probability distribution of the joint inter-spike interval (ISI) of an output spike train is used to gauge the prevalence of burst firing, which is defined by the probability of two successive spikes which fires compactly with interval time being less than 10 ms [13]:

$$P_{burst} = \frac{N_{spike}(t)(ISI < 10\,\text{ms})}{N_{spike}(t)} \tag{6}$$

where $N_{spike}(t)$ is the number of spikes of a spike train. Typically, according to Ref. [13], when the value of interval time is smaller than on tenth of their mean, the spikes are considered to be clustered.

To quantify the correlated firing of the network, we use spike train cross-correlograms (CCGs) [1, 17]. Since the network firing rate is set to be constant, we ignore it in the equation of CCGs.

$$CCG_{ij}(\tau) = \frac{\sum_{k=1}^{K}\sum_{t=0}^{L} y_i^p(t)y_j^p(t+\tau)}{K(L - |\tau|)} \tag{7}$$

There are K realizations of numerical simulation with time duration L for every realization. As figured by Ref. [17], we use $L - |\tau|$ as function of time lag τ to correct CCGs in order to eliminate the degree of overlap of two spike trains.

The auto-correlograms (ACGs) of the neurons, calculated similarly as the CCGs, are used to normalize the CCGs, but by letting $i = j$:

$$ACG_{jj}(\tau) = \frac{\sum_{k=1}^{K}\sum_{t=0}^{L} y_j^p(t)y_j^p(t+\tau)}{K(L - |\tau|)} \tag{8}$$

Here we need the signal correlations. By subtracting the shift predictor (SPT) from all CCGs and ACGs we can obtain the corrected values [1]. The SPT is calculated with different trials, k and k':

$$SPT_{ij}(\tau) = \frac{\sum_{k=1}^{K}\sum_{t=0}^{L} y_i^k(t)y_j^{k'}(t+\tau)}{K(L-|\tau|)} \tag{9}$$

Then the pairwise correlation is obtained as follow:

$$C_{ij}(T) = \frac{\sum_{\tau=-T}^{T} CCG_{ij}(\tau) - \sum_{\tau=-T}^{T} SPT_{ij}(\tau)}{\sqrt{[\sum_{\tau=-T}^{T} ACG_{ii}(\tau) - \sum_{\tau=-T}^{T} SPT_{ii}(\tau)][\sum_{\tau=-T}^{T} ACG_{jj}(\tau) - \sum_{\tau=-T}^{T} SPT_{jj}(\tau)]}} \tag{10}$$

where T represents the time window used to estimate correlations.

In order to obtain network correlations, we average the correlation over all pairs of neurons in the output layer:

$$Cor = \frac{1}{N(N-1)/2}\sum_{i=1}^{N}\sum_{j=i+1}^{N} C_{ij} \tag{11}$$

4 Results

The values of parameters are: $C = 1, D_E = D_I = 0.08, N = 80, M = 20, \tau_D = 4\,ms$, $g_l = 1, \tau_S = 0.5\,ms, V_{th} = 1, V_r = 0, \tau_r = 1\,ms, K = 100, L = 100\,s$. The time duration includes a transient period of $1s$ to remove the transient effects of initial conditions.

First, the responses of the model is clarified when the feedback loop is open and non-zero respectively. Figure 1 presents the inter-spike interval for one randomly chosen neuron. When simulated with global inhibition (Fig. 1 bottom), excitatory neurons produce two visible peaks. The first strong peak at short delays represents burst response, corresponding to spikes firing within time lags less than 10 ms. The second peak represents oscillatory component, which corresponds to larger probability of two spike firing within 20 ms. These two peaks are divided by a valley, indicating that the transmission delay of inhibitory feedback blocks spikes in the time interval around 10 ms after every spike. The patterns of the neural firing are shifted when the inhibitory feedback is blocked. The burst firing is weaken and the oscillatory activity is vanished (Fig. 1 top). The switch of the response mode of excitatory neurons indicates that the inhibitory feedback is of great significance as the neurons shift between firing modes. The generation of burst is thus proved to relay on feedback.

These results reveal that the burst firing of the network can be modified and identify a inhibitory feedback pathway as a basis for neurophysiology related to this

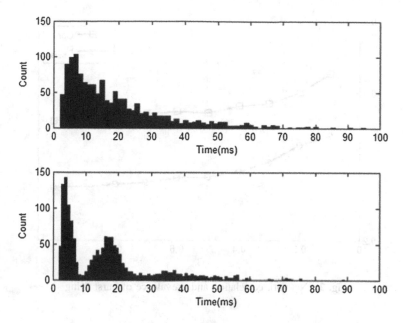

Fig. 1. ISI histograms

modification. The illustration of refined relationship between inhibitory feedback gain and the prevalence of bursts needs further studies. Correlated firing activities processed and transmitted from a layer of neurons to the next is proved to be associated with feedback. With constant firing rate and abolished oscillations, we further explore the contributions of feedback inhibition on the correlations and bursts of the network by changing the strength of the inhibitory feedback. Figure 2 reveals the results of numerical simulations.

It is anticipated that post-synaptic neurons are harder to fire again within a circle after every spike with increasing feedback strength. Hence, the prevalence of burst firing is reduced. This is indeed shown to be true as illustrate in Fig. 2. The P_{burst} decreases as G increase, suggesting that the burst firing which is induced by feedback inhibition could be suppressed when the strength of feedback loop is enhanced continuously. Thus the possibility of burst firing can be switched by changing feedback gain, corresponding to the predictions of Ref. [13].

In the absence of network oscillations, the firing rate of the network is remained constant by changing the intensity of external stimuli. Therefore, the relationship between bursting firing and network correlations could be identified in the network involving inhibitory feedback. As shown in Fig. 2, Cor decreases with G, which signals the reduction of P_{burst}. This suggests that the decreases in network correlations are resulted from the relatively weak burst firing activity. Thus the correlated firing are almost monotonic decreasing with burst.

In conclusion, in the spiking neural network, inhibitory feedback is responsible for bursts. While the induced bursting firing contributes to the increase of network correlations of a long time scale.

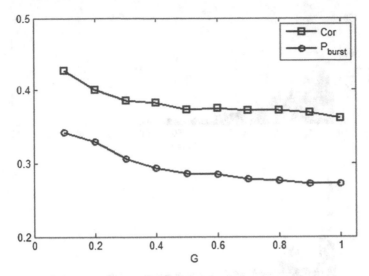

Fig. 2. Network correlation and prevalence of burst firing

5 Conclusions

The role of the variability of neural firing activity in sensory systems is a topic receiving increasing focus. One of the key questions is how network structure, formerly considered to indicate averaged trial phenomena, affects the properties of population activity. Especially, feedback is proved to be significant and participated in behaviors related to transmitting of sensory information. Here, we propose to identify the effects of inhibitory feedback on burst firing with interaction with network correlations by numerical simulations.

In this work, we first showed that the generation of burst firing relied on the feedback inhibition of the network. The excitatory neurons with inhibitory feedback input exhibited much stronger bursts than those without inhibitory feedback. We then found that bursting in response to inhibitory feedback could be reduced by increasing the strength of feedback loop. Furthermore, adjusting the amount of bursting spikes by the inhibitory feedback strength, we were able to show the relationship between burst firing and correlations in this network model. The neurons in the output layer showed corresponding shifts in burst and correlated firing. From the resulting analysis we conclude that the feedback resulting burst firing plays a dominant role in modulating the network correlated firing activities, which provides some enlightenments in understanding the correlated coding for external stimulus.

Acknowledgement. This work is supported by the National Natural Science Foundation of China (No. 61203375).

References

1. Kohn, A., Smith, M.A.: Stimulus dependence of neuronal correlation in primary visual cortex of the macaque. J. Neurosci. **25**, 3661–3673 (2005)
2. Averbeck, B.B., Latham, P.E., Pouget, A.: Neural correlations, population coding and computation. Nat. Rev. Neurosci. **7**, 358–366 (2006)
3. Okun, M., Lampl, I.: Instantaneous correlation of excitation and inhibition during ongoing and sensory-evoked activities. Nat. Neurosci. **11**, 535–537 (2008)
4. Gerkin, R.C., Tripathy, S.J., Urban, N.N.: Origins of correlated spiking in the mammalian olfactory bulb. Proc. Natl. Acad. Sci. U.S.A. **110**, 17083–17088 (2013)
5. Stephen, C., Andre, L., Leonard, M.: The neural dynamics of sensory focus. Nat. Commun. **6**, 8764 (2015)
6. Dipoppa, M., Gutkin, B.S.: Correlations in background activity control persistent state stability and allow execution of working memory tasks. Front. Comput. Neurosci. **7**, 139 (2013)
7. Smith, M.A., Kohn, A.: Spatial and temporal scales of neuronal correlation in primary visual cortex. J. Neurosci. **28**, 12591–12603 (2008)
8. Miller, K.J., Leuthardt, E.C., Schalk, G., Rao, R.P., Anderson, N.R., Moran, D.W., Miller, J.W., Ojemann, J.G.: Spectral changes in cortical surface potentials during motor movement. J. Neurosci. **27**, 2424–2432 (2007)
9. Kepecs, A., Lisman, J.: Information encoding and computation with spikes and bursts. Network **14**, 103–118 (2003)
10. Krahe, R., Gabbiani, F.: Burst firing in sensory systems. Nat. Rev. Neurosci. **5**, 13–23 (2004)
11. Sang-Yoon, K., Woochang, L.: Frequency-domain order parameters for the burst and spike synchronization transitions of bursting neurons. Cogn. Neurodyn. **9**(4), 411–421 (2015)
12. Sah, N., Sikdar, S.K.: Transition in subicular burst fring neurons from epileptiform activity to suppressed state by feedforward inhibition. Eur. J. Neurosci. **38**, 2542–2556 (2013)
13. Hoka, C., Dongping, Y., Changsong, Z., Thomas, N.: Burst firing enhances neural output correlation. Front. Comput. Neurosci. **10**, 42 (2016)
14. Sherman, S.M., Guillery, R.W.: The role of the thalamus in the flow of information to the cortex. Phil. Trans. R. Soc. Lond. B **357**, 1695–1708 (2002)
15. Jinli, X., Zhijie, W., Andre, L.: Correlated firing and oscillations in spiking networks with global delayed inhibition. Neurocomputing **83**, 146–157 (2012)
16. Jinli, X., Zhijie, W.: Effect of inhibitory feedback on correlated firing of spiking neural network. Cogn. Neurodyn. **7**(4), 325–331 (2013)
17. de La Rocha, J., Doiron, B., Shea-Brown, E., Josic, K., Reyes, A.: Correlation between neural spike trains increases with firing rate. Nature **448**, 802–806 (2007)

A Soft Computing Prefetcher to Mitigate Cache Degradation by Web Robots

Ning Xie, Kyle Brown, Nathan Rude, and Derek Doran[⊠]

Department of Computer Science and Engineering, Kno.e.sis Research Center,
Wright State University, Dayton, OH, USA
{xie.25,brown.718,howard.rude,derek.doran}@wright.edu

Abstract. This paper investigates the feasibility of a resource prefetcher able to predict future requests made by web robots, which are software programs rapidly overtaking human users as the dominant source of web server traffic. Such a prefetcher is a crucial first line of defense for web caches and content management systems that must service many requests while maintaining good performance. Our prefetcher marries a deep recurrent neural network with a Bayesian network to combine prior global data with local data about specific robots. Experiments with traffic logs from web servers across two universities demonstrate improved predictions over a traditional dependency graph approach. Finally, preliminary evaluation of a hypothetical caching system that incorporates our prefetching scheme is discussed.

Keywords: LSTM · Deep learning · Bayesian model · Web Caching · Resource prediction

1 Introduction

A Web robot is an autonomous agent that sends HTTP requests to web servers around the world. Recent studies, including our own [16], indicate that upwards of 60% of the traffic faced by web servers comes from robots [17], while only 20% of traffic came from web robots a decade ago [13]. This rise in traffic may come from the necessity for services to retrieve share-in-the-moment news and social data [15]. Moreover, internet-of-things devices will increase this proportion as more devices which operate autonomously are connected to the web.

Prefetching web resources for caching and content management systems is a common technique to anticipate and pre-load the resources likely to be requested next for fast, low latency access [3,8,14]. Prefetchers for human traffic are an essential component of web caches, but as robots exhibit different functionality [5], access patterns [7], and traffic characteristics, traditional prefetching strategies applied to traffic with high levels of web robot activity exhibit degraded performance. The degree of uncertainty and few restrictions on robot behavior requires powerful soft computing techniques to find and utilize the

---

N. Xie and K. Brown are joint first authors of this paper.

© Springer International Publishing AG 2017
F. Cong et al. (Eds.): ISNN 2017, Part I, LNCS 10261, pp. 536–546, 2017.
DOI: 10.1007/978-3-319-59072-1_63

possibly weak latent patterns existing in their requests. This paper proposes a prefetcher with such techniques for robot-laden web traffic. Evaluations show that the synergistic use of a deep recurrent neural network (RNN) and a Bayesian network greatly improves prefetching performance for robot traffic.

2 Related Work

There have been many previous studies which have examined the characteristics of web robot traffic. Doran et al. [7] study the distribution of request types (e.g. image files vs web pages) for human and robot traffic. The authors note that web robots have a strong penchant for larger resources; they are 10 times more likely to request resources larger than 10 MB when compared to human traffic. Rude and Doran [16] develop an Elman Neural Network to predict global trends of request types for robot traffic. Almeida et al. [1] investigate the impact of web robots on cache hit rate. The authors examined the referencing pattern of web robots and found out the pattern exhibits "round-robin" traits, disrupting locality assumptions which can leave an LRU cache useless. In [5], the authors classify web robots according to their functionality and request patterns. In [2], the authors suggest web servers should implement new interfaces for robot traffic that provide metadata archives describing the content. By querying the metadata instead of requesting all resources from some precomputed list, a robot is able to narrow its selection down to a more refined set of resources reducing the amount of bandwidth consumed. Li et al. [12] propose a hybrid cache design that is broken down into 3 sections: (i) an *index cache* which stores inverted indices for user search queries (ii) a *results cache* which caches a results page returned for a unique user search query and (iii) a *document cache* to cache documents on the server. However, the authors filter out robot traffic to better understand the user behavior and queries.

3 Prefetching Scheme

This section introduces the design of the request prefetcher. Its design and integration into a hypothetical caching architecture is illustrated in Fig. 1. After a request is processed, it is labeled as originating from a robot or a human through a real-time detection algorithm (of which many exist in the literature [6]). If the request is labeled as a robot, the proposed prefetcher will predict what subsequent robot requests will be made. To identify the more subtle patterns that exist in robot request sequences, prefetching is done by combining a deep recurrent neural network (RNN) with a Bayesian network model. The RNN is used to predict the 'orientation' of the robot request stream by predicting the subdirectories future robots will likely visit next given their past history. From the k subdirectories the RNN predicts robots will visit next, a Bayesian network selects the subset of resources within those directories to be prefetched. The Bayesian network incorporates prior information in the form of global web robot patterns, as well as the request patterns of the particular web robot who made

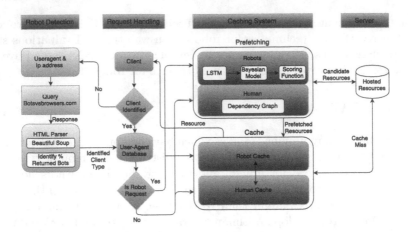

Fig. 1. A caching architecture with robot request prefetching

the last request, to tailor the prediction to future requests from that robot. This idea is motivated by the idea that robot request sequences often come from the same robot, who send a number of requests within a short period of time.

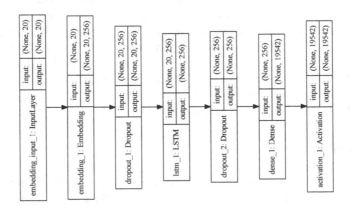

Fig. 2. RNN structure to predict subdirectory requests

RNN Specification. The RNN architecture is presented in Fig. 2. It has an embedding layer that translates ordered sequences of the subdirectories robots visited to an input, a long short-term memory (LSTM) layer that learns sequence patterns, dropout layers that control for overfitting, and a fully connected layer to compute the likelihood a subdirectory will be visited next. An LSTM layer is often used for sequence processing since LSTMs can hold "memory" over long periods [9,10]. The RNN was trained with dropout parameters set to 0.2. Validation loss was monitored with a patience of two. Training ended when no improvement in loss over a validation set was observed or after 128 epochs (passes over the training data).

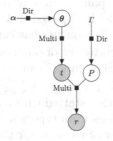

Fig. 3. Bayesian network of the simple model

Bayesian Networks. Two Bayesian models were developed to predict which resources a robot will request next in a given set of subdirectories. The first model, called the simple model (SM), first draws a resource type from a multinomial distribution and then an individual resource from another multinomial distribution corresponding to the set of resources of the drawn type. Request types are modeled based on past work that demonstrated request type preferences for Web robots [11,16]. Prior knowledge is given by hyper-parameter settings for the resource type and resource request distributions, which considers the number of times we observe *any* robot requesting resources in a subdirectory. This is useful if a specific robot has not made many requests to the web server in the past. The generative process and its hyper-parameters are shown in Fig. 3 using the notation in Dietz [4]. The data likelihood of the simple model is $\Pr(r_1, \cdots, r_M, t_1, t_M | \boldsymbol{\theta}, P) = \exp\{\sum_{j=1}^{K}(m_j \log(\theta_j) + \sum_{l=1}^{R_j} n_{j,l} \log(p_{j,l}))\}$ where M is the total number of observed requests by the robot in this subdirectory, K is the number of resource types, m_j is the number of requests for a resource of type j by the robot in this subdirectory, θ_j is the multinomial parameter for resource type j, R_j is the number of resources of type j in this subdirectory, $n_{j,l}$ is the number of times the l-th resource of type j in this subdirectory was requested by the robot, and $p_{j,l}$ is the l-th component of the multinomial parameter vector $\boldsymbol{p}_j$ for resources of type j in the subdirectory. The parameter vector $\boldsymbol{\theta}$ of resource types is drawn from $\boldsymbol{\theta} \sim \text{Dirichlet}(\boldsymbol{\alpha})$ and for each resource type j, the parameter vector is drawn from $\boldsymbol{p}_j \sim \text{Dirichlet}(\boldsymbol{\gamma}_j)$. The values of the hyper-parameters $\boldsymbol{\alpha}$ and $\Gamma = \{\boldsymbol{\gamma}_j\}_{j=1}^{K}$ are chosen using global statistics from all robots by $\alpha_j = \alpha m_j^{(g)} / M^{(g)}$, where α is the prior strength for $\boldsymbol{\alpha}$, so that $\sum_{j=1}^{K} \alpha_j = \alpha$, $m_j^{(g)}$ is the global number of requests for resources of type j, and $M^{(g)}$ is the global number of requests. $\boldsymbol{\gamma}_j$ is chosen as $\gamma_{j,k} = \gamma n_{j,k}^{(g)} / m_j^{(g)}$ where $n_{j,k}^{(g)}$ is the global number of requests for the k-th resource of type j, and γ is the prior strength for Γ. A second model is an extension of the simple model, where resource types are now generated by a Markov process instead of draws from a multinomial distribution. It attempts to consider patterns in the request type sequence of a robot, which may be helpful when robots are only interested in specific kinds of resources (e.g. an image scraper) [16]. For the Markov model, an "observation" is not just a single resource-type pair, but a sequence of resource-type pairs. The generative process for this model is depicted in Fig. 4. We assume there are L such observations, and that the ith sequence has length M_i and is represented by $(r_1^{(i)}, t_1^{(i)}), \cdots, (r_{M_i}^{(i)}, t_{M_i}^{(i)})$. The entire set of observations is denoted by $\mathcal{R}$. Then the data likelihood for the Markov model can be written as: $\Pr(\mathcal{R}|\boldsymbol{\theta}, P, A) = \exp\{\sum_{j=1}^{K}(m_j \log(\theta_j) + \sum_{k=1}^{K} T_{j,k} \log(a_{j,k}) + \sum_{l=1}^{R_j} n_{j,l} \log(P_{j,l}))\}$ where m_j is the number of times an observation started with a request for a resource of type j, $T_{j,k}$ is the number of transitions from type j to k within an observation were observed, and $n_{j,l}$ is the number of requests for the l-th resource of type j over all observations. The other symbols are the same as in the simple model. The parameters $\boldsymbol{\theta}$ and P in the Markov model are assumed to be generated from the same distributions as the simple model. Each row $\boldsymbol{a}_j$ of the transition matrix A is drawn from $\boldsymbol{a}_j \sim \text{Dirichlet}(\boldsymbol{\lambda}_j)$ for $1 \leq j \leq K$. Here $\boldsymbol{\lambda}_j$ is the j-th row of the hyper-parameter matrix Λ. As for the simple model, hyper-parameters for the Markov model are computed from global statistics for all robots. $\boldsymbol{\alpha}$ and Γ are computed the same way as the simple model. Each element $a_{j,k}$ of A is computed as $a_{j,k} = aT_{j,k}^{(g)}/T_j^{(g)}$ where $T_{j,k}^{(g)}$ is the global number of transitions from a resource of type j to type k, $T_j^{(g)} = \sum_{k=1}^{K} T_{j,k}^{(g)}$ is the global transition count from a resource of type j, and a is the prior strength of A.

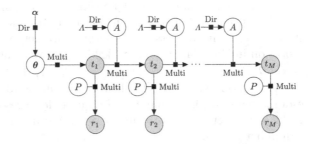

Fig. 4. Bayesian network where request types are generated by a Markov process

Parameter estimation for all models was done using maximum *a posteriori* estimation (MAP) because of the ability to obtain a closed-form solution for parameters for the simple and Markov models. Since the models need to be updated as requests come in to the web server, this approach was used to enable an efficient implementation. The MAP parameter values for the simple model are $\tilde{\theta}_j = (\alpha_j + m_j - 1)/(\alpha + M - 1)$ for $1 \leq j \leq K$ and $\tilde{p}_{j,l} = (\gamma_{j,l} + n_{j,l} - 1)/(\Gamma_j + m_j - 1)$ for $1 \leq j \leq K$ and $1 \leq l \leq R_j$, where $\Gamma_j = \sum_{l=1}^{R_j} \gamma_{j,l}$.

4   Prefetching Evaluation

To evaluate the models, Web logs across all web servers at Wright State University (WSU) and the University of Pavia were collected. The WSU logs represent a 3 month period in 2016 while the Pavia logs span a 5 month period over 2014. Robot traffic was extracted from the logs by following a simple heuristic procedure using the crowdsourced database *botsvsbrowsers*[1]. Checking the user-agent of each request in the log against the nearly 1.5 million agents recorded on this database, a user-agent match to a known robot is used to mark the session as a robot. While many probabilistic detection methods exist [6], this heuristic approach guarantees that we only evaluate the prefetcher on true robot traffic while still giving us a sizable number of sessions. Specifically, 221,683 and 14,401 robot sessions were extracted from WSU and the University of Pavia logs, respectively. The models were trained on two-thirds of earlier arriving sessions from each dataset, while the remaining third was used for evaluation.

As a baseline for comparison of the system, a multinomial distribution was fit over all resources in the directory. In the plots below, this is called a *base model*. To show the improvement of the Markov model when using prior information it was fit using both MLE and maximum a-posteriori estimation (MAP). Evaluation was carried out by: (i) examining the top-k accuracy of the RNN for subdirectory prediction; (ii) comparing the performance of the Bayesian models for representative subdirectories on both servers; and (iii) testing the prefetchers capability to accurately identify the next request a robot in a caching system.

RNN Evaluation. RNN evaluation was carried out by checking its top-k accuracy, which is defined as the percentage of time the true subdirectory visited next by a robot is among the k most probable subdirectories predicted by the RNN (called a *hit*). These accuracies are compared against a simple predictor that always predicts the most commonly requested subdirectory as the next one a robot will visit. the empirical probability of the most frequently requested subdirectory in the in Fig. 5[2]. They both show promising results as the RNN is able to identify the subdirectory of the next resource to be requested 68% of the time on WSU and 42% of the time over the Univ. of Pavia data. If we allow the RNN to suggest $k = 2$ subdirectories to the Bayesian model, the accuracy substantially improves to 74% and 54%, respectively. It is interesting to note that the accuracy of subdirectory predictions taper off as we let $k > 5$, where the RNN sports accuracies of 77% and 66% on the two servers. This is a desirable property for cache prefetching; the RNN should submit a minimum number of subdirectories to the Bayesian model, thus minimizing the number of resources it needs to predict requests for.

Bayesian Model Evaluation. Bayesian models were evaluated by checking its average top-k accuracy for predicting resources across each subdirectory of the WSU and Univ. of Pavia logs. These averages were taken over models for all

---

[1] www.botsvsbrowsers.com.
[2] The RNN is labeled *LSTM* in the figures.

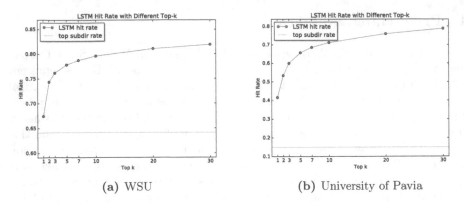

(a) WSU (b) University of Pavia

Fig. 5. RNN subdirectory prediction performance

robots seen making at least one request to a given subdirectory. Comparisons of the model within subdirectories of the same web server and between the different web servers provides information about when the models perform well or poorly.

Figure 6 compares the performance of the Simple Model to the Markov Model (with parameters fitted using both MAP estimation and MLE) and to a 'base model', which is defined as the prediction by drawing from a multinomial distribution fitted to the proportion of times each resource was requested in the training data, to two directories on WSU. The 'base model' represents a basic empirical approach for predicting requests made by a single robot with only information about that robot. The WSU root directory is extraordinarily popular, with robots requesting resources from it over 60% of the time, and it contains around 5% of the resources on the server. Despite the fact that about a thousand resources are located in this directory, Fig. 6a shows how the large number of requests provide many observations from various robots to yield a rich prior distribution that allows the Simple Model to make better predictions compared to the base model. It is also interesting to note how poorly the Markov Model performs, even compared to the Base Model. This may be because the Markov Model seeks to fit resource type patterns that are common to both global and local robot traffic when no such pattern exists. For example, if the root subdirectory contains an even distribution of resource types, it may be the case that the Markov Model's attempt to find patterns in resource type sequences may add noise that reduces performance.

Figure 6b shows a comparison of the models against a different, less popular subdirectory on the website. In contrast to the root directory, we find that both Bayesian models perform almost identically to the Base Model. Such a pattern may emerge when very few requests from all robots are made to the subdirectory, since the Simple Model reduces to the Base Model when using MLE, which could occur when the prior information is the same as the observations for the current robot, or if there is no prior information. This suggests that the power of the Bayesian models relies on observing a large number of requests from a diverse

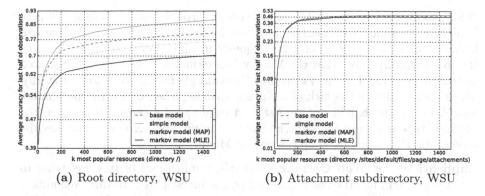

(a) Root directory, WSU   (b) Attachment subdirectory, WSU

Fig. 6. Bayesian model performance, WSU subdirectories

set of robots in a subdirectory. It may also imply that, out of a risk for admitting a number of resources that have low probability of being requested, prefetching resources from some directories should not be done at all.

Figure 7 examines the performance of the Bayesian model over subdirectories at the University of Pavia. This web server represents the interesting condition where the simple MLE-based Base Model for request prediction performs astonishingly well (with 96% prediction accuracy on the root and on a seldom requested image directory). This may be due to the different structure of the University of Pavia's website along with the smaller size of the dataset. As the top-k plots tend to plateau quickly, this might indicate that there are fewer resources commonly requested by robots for each subdirectory. Having a smaller dataset size would provide less prior information, reducing the ability of the Bayesian models to outperform the Base model.

Despite this, the Simple Model is still able to over-perform the Base Model by fusing together data about global and specific-robot patterns. As with the

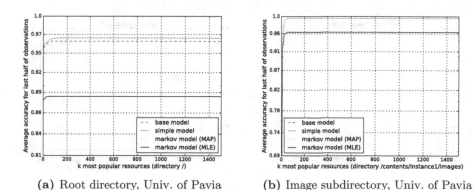

(a) Root directory, Univ. of Pavia   (b) Image subdirectory, Univ. of Pavia

Fig. 7. Bayesian model performance, Univ. of Pavia

WSU subdirectories, the Markov Model still underperforms the Base and Simple Models in the root directory. This probably indicates that there is little pattern between the resource types of robot requests in high-traffic directories such as the root directory. Note that the Markov Model fit with MAP performs as well as the Simple Model in the /contents/instance1/images directory (Fig. 7b), which could indicate the lack of a pattern that is picked up on by the Markov Model, i.e. the probability of requesting a resource of a certain type given the previous type is always the same no matter the previous type and matches the type probabilities predicted by the Simple Model.

Implications for Web Caching. Next, the implications of this prefetcher to a caching system like the one shown in Fig. 1 is investigated. In this preliminary study, the actions of this architecture were simulated over the later 1/3 of the WSU logs (the same data used for testing the RNN and Bayesian models). The cache limit was set to 100 MB, of which 80 MB was reserved to cache human traffic and 20 MB was reserved to store requests predicted to be requested by web robot traffic by the proposed prefetcher. Experiments were ran with varying levels of robot sessions in the data by interweaving a random oversampling of human sessions that were extracted by the *botsvsbrowsers* identification scheme.

The performance of the prefetcher is compared to a dependency graph, a popular and often used Markov model for prefetching resources in caching systems. Figure 8 compares the precision and recall of these two prefetching approaches over robot traffic. Here, precision is defined as the number of times the prefetcher successfully prefetched the resource requested next divided by the total number of prefetches that were made. Recall is defined as the percent of time that a prefetched resource still in the cache was requested by *any* robot. Recall measures instances where, for example, a resource may be requested multiple times after it is added to the cache, or a resource prefetched but not requested next by a robot was eventually requested in the future while it was still in the cache. The top-k subdirectories considered and the top-n resources chosen by the soft prefetched were set to $k = 3$ and $n = 2$ following a parameter sweep of different

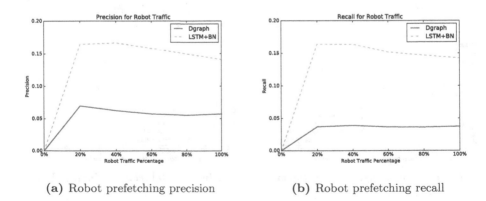

(a) Robot prefetching precision   (b) Robot prefetching recall

Fig. 8. Cache prefetching performance: soft prefetcher vs. dependency graph

values, choosing the pair that yielded the highest F_1 score. In these settings, Fig. 8 shows how the precision of a dependency graph is ~5.5–7% and its recall is 4% regardless of how many robots are visiting the cache. The soft prefetcher, however, enjoys a 2-4x gain in performance, leading to significant improvements in the efficiency of a web cache. Moreover, the soft prefetcher is able to maintain these strong values even in the face of extraordinarily high levels of Web robot traffic, which the web may experience in the future.

5 Conclusions and Future Work

This paper introduced a novel soft computing prefetcher for web caches tailored to mitigate the degradation of web system caches by web robots. The approach is rooted in the idea that a deep recurrent neural network would be able to find patterns in the ordering of subdirectories visited by web robots, and that Bayesian models can incorporate observations about all robot traffic with data about a particular robot to formulate accurate request predictions within subdirectories. Evaluation results indicate significant gains over MLE-based approaches for predicting web robot request patterns, and significantly better prefetching performance for a web cache compared to the popular dependency graph approach. Future work will further evaluate the dual caching approach and will be exercised over datasets from different universities. Alternative Bayesian models, incorporating patterns besides request type patterns, will also be explored for robot resource request prediction.

Acknowledgment. The authors thank Logan Rickert for data processing support, Maria-Carla Calzarossa for data from the University of Pavia, and Mark Anderson for data from Wright State University. This paper is based on work supported by the National Science Foundation (NSF) under Grant No. 1464104. Any opinions, findings, and conclusions or recommendations expressed in this material are those of the author(s) and do not necessarily reflect the views of the NSF.

References

1. Almeida, V., Menascé, D., Riedi, R., Peligrinelli, F., Fonseca, R., Meira Jr., W.: Analyzing web robots and their impact on caching. In: Proceedings of Sixth Workshop on Web Caching and Content Distribution, pp. 20–22 (2001)
2. Brandman, O., Cho, J., Garcia-Molina, H., Shivakumar, S.: Crawler-friendly web servers. In: Proceedings of Performance and Architecture of Web Servers Conference (2000)
3. Chen, X., Zhang, X.: A popularity-based prediction model for web prefetching. Computer **36**(3), 63–70 (2003)
4. Dietz, L.: Directed factor graph notation for generative models. Technical report, Max Planck Institute for Informatics (2010)
5. Doran, D., Gokhale, S.: A classification framework for web robots. J. Am. Soc. Inf. Sci. Technol. **63**, 2549–2554 (2012)
6. Doran, D., Gokhale, S.S.: Web robot detection techniques: overview and limitations. Data Mining Knowl. Discov. **22**(1–2), 183–210 (2011)

7. Doran, D., Morillo, K., Gokhale, S.: A comparison of web robot and human requests. In: Proceedings of ACM/IEEE Conference on Advances in Social Network Analysis and Mining, pp. 1374–1380 (2013)
8. Gellert, A., Florea, A.: Web prefetching through efficient prediction by partial matching. World Wide Web **19**(5), 921–932 (2016)
9. Graves, A.: Neural networks. In: Graves, A. (ed.) Supervised Sequence Labelling with Recurrent Neural Networks, pp. 15–35. Springer, Heidelberg (2012)
10. Hochreiter, S., Schmidhuber, J.: Long short-term memory. Neural Comput. **9**(8), 1735–1780 (1997)
11. Lee, J., Cha, S., Lee, D., Lee, H.: Classification of web robots: an empirical study based on over one billion requests. Comput. Secur. **28**(8), 795–802 (2009)
12. Li, H., Lee, W.-C., Sivasubramaniam, A., Giles, C.L.: A hybrid cache and prefetch mechanism for scientific literature search engines. In: Baresi, L., Fraternali, P., Houben, G.-J. (eds.) ICWE 2007. LNCS, vol. 4607, pp. 121–136. Springer, Heidelberg (2007). doi:10.1007/978-3-540-73597-7_10
13. Menascé, D., Almeida, V., Riedi, R., Ribeiro, F., Fonseca, R., Meira Jr., W.: In search of invariants for e-business workloads. In: Proceedings of the 2nd ACM Conference on Electronic Commerce, pp. 56–65 (2000)
14. Pallis, G., Vakali, A., Pokorny, J.: A clustering-based prefetching scheme on a web cache environment. Comput. Electr. Eng. **34**(4), 309–323 (2008)
15. Qualman, E.: Socialnomics: How Social Media Transforms the Way We Live and Do Business. Wiley, Hoboken (2012)
16. Rude, H.N., Doran, D.: Request type prediction for web robot and internet of things traffic. In: Proceedings of IEEE International Conference on Machine Learning and Applications, pp. 995–1000 (2015)
17. Zeifman, I.: Report: Bot traffic is up to 61.5% of all website traffic. bit.ly/MoMRxE

A Caputo-Type Fractional-Order Gradient Descent Learning of BP Neural Networks

Guoling Yang, Bingjie Zhang, Zhaoyang Sang, Jian Wang$^{(\boxtimes)}$, and Hua Chen

College of Science, China University of Petroleum, Qingdao 266580, China
yangguolingfwz@163.com, bingjie_zhang_1993@163.com, sun1410@163.com,
{wangjiann1,chenhua}@upc.edu.cn

Abstract. Fractional calculus has been found to be a promising area of research for information processing and modeling of some physical systems. In this paper, we propose a fractional gradient descent method for the backpropagation (BP) training of neural networks. In particular, the Caputo derivative is employed to evaluate the fractional-order gradient of the error defined as the traditional quadratic energy function. Simulation has been implemented to illustrate the performance of presented fractional-order BP algorithm on large dataset: MNIST.

Keywords: Fractional calculus · BP · Caputo derivative · MNIST

1 Introduction

Fractional differential calculus has been a classical notion in mathematics for several hundreds of years. It is based on differentiation and integration of arbitrary fractional order, and as such it is a generalization of the popular integer calculus. Yet only recently it has been applied to the successful modeling of certain physical phenomena.

The fractional differential calculus has been successfully adopted also in the field of neural networks. Some remarkable research of fractional-order neural networks has been presented in [1–5]. In [1], fractional calculus was used for the Backpropagation (BP) [15] algorithm for feedforward neural networks (FNNs). By extending the second method of Lyapunov, the Mittag-Leffler stability analysis was performed for fractional-order Hopfield neural networks [2].

However, most research findings for fractional-order systems have been limited to studies of fully coupled recurrent networks of Hopfield type [2–9]. The vast majority of papers have been focused on studying properties of fixed points

J. Wang—This work was supported in part by the National Natural Science Foundation of China (No. 61305075), the China Postdoctoral Science Foundation (No. 2012M520624), the Natural Science Foundation of Shandong Province (Nos. ZR2013FQ004, ZR2013DM015), the Specialized Research Fund for the Doctoral Program of Higher Education of China (No. 20130133120014) and the Fundamental Research Funds for the Central Universities (Nos. 14CX05042A, 15CX05053A, 15CX02079A, 15CX08011A).

© Springer International Publishing AG 2017
F. Cong et al. (Eds.): ISNN 2017, Part I, LNCS 10261, pp. 547–554, 2017.
DOI: 10.1007/978-3-319-59072-1_64

for non-integer order differential equations that describe such networks. The researched networks vary in their properties: they are with or without delay in the feedback loop, while other extensions have provided generalizations to complex-valued neurons. In contrast, this work concerns fractional-order error BP in FNNs.

Gradient descent method is commonly used to train FNNs by minimizing the error function, being the norm of a distance between the actual network output and the desired output. There exist other optional methods to implement the BP algorithm for FNNs, such as conjugate gradient, Gauss-Newton and Levenberg-Marquardt. We note that all of the above optimal methods are typically employed to train integer-order FNNs.

Inspired by [1, 10], we apply the fractional steepest descent algorithm to train FNNs. In particular, we employ the Caputo derivative formula to compute the fractional-order gradient of the error function with respect to the weights and obtain the deterministic convergence.

The structure of the paper is as follows: in Sect. 2, the definitions of the commonly used fractional-order derivative is introduced. The traditional BP algorithm and our novel algorithm of fractional-order BP neural networks training based on Caputo derivative are presented in Sect. 3. Numerical simulation is presented to illustrate the effectiveness of our results in Sect. 4. Finally, the paper is concluded in the last Section.

2   Fractional-Order Derivative

Unlike the situation with integer-order derivatives, several definitions are used for fractional-order derivatives. The three most common definitions are referred to Grunvald-Letnikov (GL), Riemann-Liouiville (RL), and Caputo [11–14]. In this paper, We mainly focus on the Caputo definition.

Definition 1 *(Caputo fractional-order derivative). The definition of the Caputo fractional-order derivative of order α is defined as follows*

$$^{Caputo}_{\quad a}D_t^\alpha f(t) = \frac{1}{\Gamma(n-\alpha)} \int_a^t (t-\tau)^{n-\alpha-1} f^{(n)}(\tau)d\tau, \tag{1}$$

where $^{Caputo}_{\quad a}D_t^\alpha$ is the Caputo derivative operator, α is the fractional order.

Particularly, when $\alpha \in (0, 1)$, the expression for Caputo derivative is as follows

$$^{Caputo}_{\quad a}D_t^\alpha f(t) = \frac{1}{\Gamma(1-\alpha)} \int_a^t (t-\tau)^{-\alpha} f'(\tau)d\tau. \tag{2}$$

3   Algorithm Description

We think about a network with three layers and the numbers of neurons for the input, hidden and output layers are p, n and 1, respectively. The training sample set that we suppose in this paper is $\{\mathbf{x}^j, O^j\}_{j=1}^J \subset \mathbb{R}^p \times \mathbb{R}$, in which $\mathbf{x}^j$ is the input of the j-th sample and O^j is the corresponding desired output. Let $g, f : \mathbb{R} \to \mathbb{R}$ be given activation functions for the hidden and the output layers, separately. Let $\mathbf{V} = (v_{ij})_{n \times p}$ be the weight matrix connecting the input and the hidden layers, and write $\mathbf{v}_i = (v_{i1}, v_{i2}, \cdots, v_{ip})^T \in \mathbb{R}^p$ for $i = 1, 2, \cdots, n$. The weight vector connecting the hidden and output layers is denoted by $\mathbf{u} = (u_1, u_2, \cdots, u_n)^T \in \mathbb{R}^n$. Aimed to simplify the statement, we integrate the weight matrix $\mathbf{V}$ with the weight vector $\mathbf{u}$, namely, $\mathbf{w} = (\mathbf{u}^T, \mathbf{v}_1^T, \cdots, \mathbf{v}_n^T)^T \in \mathbb{R}^{n(p+1)}$. For the sake of convenience, we present the following vector valued function

$$G(\mathbf{z}) = (g(z_1), g(z_2), \cdots, g(z_n))^T, \quad \forall \mathbf{z} \in \mathbb{R}^n. \tag{3}$$

3.1   BP Algorithm Based on Gradient Descent Method

For any given j-th input $\mathbf{x}^j \in \mathbb{R}^p$, the final actual output is

$$y = f(\mathbf{u} \cdot G(\mathbf{V}\mathbf{x}^j)). \tag{4}$$

For any fixed weights $\mathbf{w}$, the error of the neural networks is defined as

$$E(\mathbf{w}) = \frac{1}{2} \sum_{j=1}^J \left(O^j - f\left(\mathbf{u} \cdot G\left(\mathbf{V}\mathbf{x}^j\right)\right)\right)^2$$
$$= \sum_{j=1}^J f_j(\mathbf{u} \cdot G(\mathbf{V}\mathbf{x}^j)), \tag{5}$$

where $f_j(t) = \frac{1}{2}(O^j - f(t))^2, j = 1, 2, \cdots, J, t \in \mathbb{R}$. We note that the constructed function $f_j(\cdot)$ is a composite function of f and G in terms of the j-th sample.

Given an initial weight $\mathbf{w}^0$, the batch learning of standard BP [15] updates the weights iteratively by

$$u_i^{k+1} = u_i^k - \eta E_{u_i^k}(\mathbf{w}), \tag{6}$$

$$v_{im}^{k+1} = v_{im}^k - \eta E_{v_{im}^k}(\mathbf{w}), \tag{7}$$

where $k \in \mathbb{N}$, $i = 1, 2, \cdots, n$, $m = 1, 2, \cdots, p$; $\eta > 0$ is the learning rate; and

$$E_{u_i^k}(\mathbf{w}) = \sum_{j=1}^J f_j'(\mathbf{u} \cdot G(\mathbf{V}\mathbf{x}^j))g(v_i^k \cdot \mathbf{x}^j), \tag{8}$$

$$E_{v_{im}^k}(\mathbf{w}) = \sum_{j=1}^J f_j'(\mathbf{u} \cdot G(\mathbf{V}\mathbf{x}^j))u_i g'(v_i^k \cdot \mathbf{x}^j)x_m^j, \tag{9}$$

where $k \in \mathbb{N}$, $i = 1, 2, \cdots, n$, $j = 1, 2, \cdots, J$.

3.2   BP Algorithm Based on Caputo Fractional-Order Derivative

For the given j-th input sample $\mathbf{x}^j \in \mathbb{R}^p$, $\theta^{i,j} = v_{i1}x_1{}^j + v_{i2}x_2{}^j + \cdots + v_{ip}x_p{}^j = \mathbf{v}_i \cdot \mathbf{x}^j (i = 1, 2, \cdots, n)$ is the j-th input value of i-th hidden neuron. The input of output layer is represented by $\zeta^j = u_1 g(\mathbf{v}_1 \cdot \mathbf{x}^j) + u_2 g(\mathbf{v}_2 \cdot \mathbf{x}^j) + \cdots + u_n g(\mathbf{v}_n \cdot \mathbf{x}^j) = \mathbf{u} \cdot G(\mathbf{V}\mathbf{x}^j)$ and $y^j = f(\mathbf{u} \cdot G(\mathbf{V}\mathbf{x}^j))$ after activation as the actual output. For any fixed weights $\mathbf{w}$, a conventional square error function is the same as (5).

Given an initial weight $\mathbf{w}^0 = (\mathbf{u}^0, \mathbf{V}^0)$, without loss of generality, assume that $c = \min\{u_i{}^k, v_{im}{}^k\}(k \in \mathbb{N}, i = 1, \cdots, n, m = 1, \cdots, p)$. The BP network with Caputo α-order derivative updates the weights $\{\mathbf{w}^k\}$ iteratively by

$$u_i^{k+1} = u_i^k - \eta_c D_{u_i^k}^\alpha E(\mathbf{w}), \tag{10}$$

$$v_{im}^{k+1} = v_{im}^k - \eta_c D_{v_{im}^k}^\alpha E(\mathbf{w}), \tag{11}$$

where $\eta > 0$ is the learning rate, $0 < \alpha < 1$ is the fractional order, $i = 1, 2, \cdots, n$ and $m = 1, 2, \cdots, p$.

According to the definitions of Caputo fractional derivative and the fractional-order differential of a composite function, we have

$$_c D_{u_i^k}^\alpha E(\mathbf{w}) = \frac{1}{(1-\alpha)\Gamma(1-\alpha)} \sum_{j=1}^J f_j'(\mathbf{u} \cdot G(\mathbf{V}\mathbf{x}^j)) g(v_i^k \cdot \mathbf{x}^j)(u_i^k - c)^{1-\alpha}. \tag{12}$$

$$_c D_{v_{im}^k}^\alpha E(\mathbf{w}) = \frac{1}{(1-\alpha)\Gamma(1-\alpha)} \sum_{j=1}^J f_j'(\mathbf{u} \cdot G(\mathbf{V}\mathbf{x}^j)) u_i^k g'(\mathbf{v}_i \cdot \mathbf{x}^j) x_m^j (v_{im}^k - c)^{1-\alpha}. \tag{13}$$

Write

$$_c D_{\mathbf{w}}^\alpha E(\mathbf{w}) = (_c D_{u_1}^\alpha E(\mathbf{w}), \cdots, _c D_{u_n}^\alpha E(\mathbf{w}), _c D_{v_{11}}^\alpha E(\mathbf{w}), \cdots, _c D_{v_{np}}^\alpha E(\mathbf{w})). \tag{14}$$

4   Experiment

To verify the convergence of the proposed factional-order BP algorithm, simulation has been done on the MNIST handwritten digital dataset.

The MNIST handwritten digital dataset is collected from the NIST database of the National Institute of Standards and Technology. Digital images are normalized to an image 28×28, and represented as 784×1 vectors. The value of each element in the vector is between 0 and 255 on behalf of the gray levels of each pixel. There have 60,000 training samples and 10,000 testing samples in the MNIST database.

To display the performances of differences of different fractional-order BP neural networks, we employed different learning parameters: learning rates and

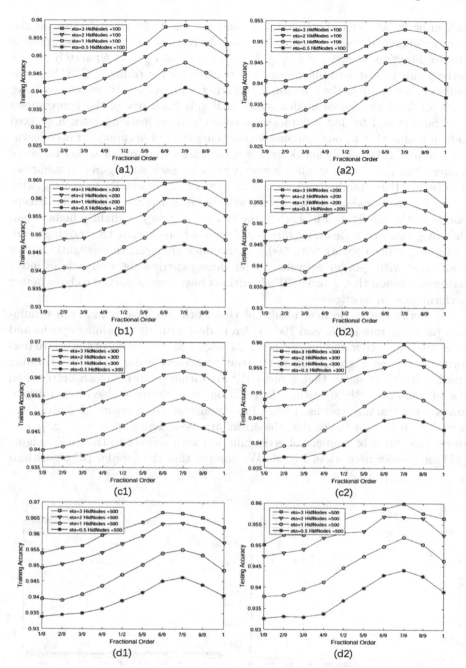

Fig. 1. The comparison of different fractional and integer order BP algorithms for various learning rates with fixed numbers of hidden nodes.

hidden nodes are set to be $\{0.5, 1, 2, 3\}$ and $\{100, 200, 300, 500\}$. In this simulation, we employ different fractional α-order derivatives to compute the gradient of error function, where $\alpha = \frac{1}{9}, \frac{2}{9}, \frac{3}{9}, \frac{4}{9}, \frac{1}{2}, \frac{5}{9}, \frac{6}{9}, \frac{7}{9}, \frac{8}{9}$ and $\frac{9}{9} = 1$, separately ($\alpha = 1$ corresponds to standard integer-order derivative for the common BP). For convenience, we graphed the training and testing accuracies with various learning rates for fixed number of hidden nodes in Fig. 1. Each row of the figure shows the training and testing accuracies based on various learning rates and fixed hidden nodes, that is, one can observe the tendencies of accuracy for each learning parameter. Generally speaking, the accuracies with larger learning rates are higher than those with smaller learning rates in each sub-figure. In addition, we observe that training and testing accuracies are gradually increasing with increasing fractional-orders and then reach the peak around $\frac{7}{9}$-order. One exception is in the sub-figure Fig. 1. ($b2$), the testing accuracy is slightly higher than that in $\frac{8}{9}$-order case (learning rate is 3). Another one is that the training accuracies of $\frac{6}{9}$-order network with 500 hidden nodes are equal to or slightly higher than those with $\frac{7}{9}$-order network. Under these experiments, we reach an interesting conclusion that $\frac{7}{9}$-order BP algorithms have been observed to have better performances in most cases.

To verify the theoretical results of this work, we have redone the simulation (learning rate is 0.5, and 100 hidden nodes) with 1000 training epochs and compare the $\frac{7}{9}$-order and integer-order BP algorithms in terms of the above performance figures. Figure 2a shows the training accuracies for up to 1000 training epochs. It clearly shows that $\frac{7}{9}$-order BP algorithm performs much better than the integer-order BP algorithm. In addition, it performs stable after approximately 400 training epochs. Figure 2b demonstrates the errors of $\frac{7}{9}$-order BP networks. It is clear to see that the errors are monotonically decreasing. It also shows the norms of gradient of error function with respect to the total weights (14) for $\frac{7}{9}$-order network in Fig. 2b. We observe that the $\frac{7}{9}$-order BP algorithms

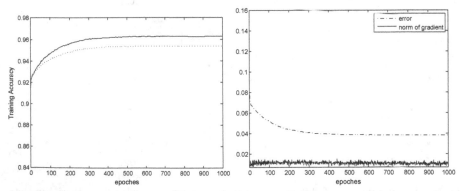

(a) The training accuracies of $\frac{7}{9}$-order and common integer BP algorithms.

(b) The training error and the norm of gradient with respect to weights.

Fig. 2. Performance of fractional-order BP algorithm and its convergent behavior.

perform stably and tend to converge to zero with increasing iterations. These observations effectively verify the theoretical results of the presented algorithm in this paper, such as monotonicity and convergence.

5 Conclusions

In this paper, we have extended the fractional steepest descent approach to BP training of FNNs. The Caputo derivative is employed to evaluate the fractional-order gradient of the error defined as the traditional quadratic energy function. Numerical simulation is reported to illustrate the effectiveness of the proposed factional-order neural networks. The illustrated simulation shows that there exists one specific fractional-order ($\frac{7}{9}$-order) BP network that often performs the best for the example problem. Fractional orders $[\frac{6}{9}, \frac{7}{9}, \frac{8}{9}]$ usually perform better than the integer-order BP, for which, of course, the fractional order is one. A $\frac{5}{9}$ fractional order BP performs often at a comparable training level to the integer-order BP.

References

1. Chen, X.: Application of fractional calculus in BP neural networks (in Chinese). Master thesis. Nanjing Forestry University, Nanjing, Jiangsu (2013)
2. Zhang, S., Yu, Y., Wang, H.: Mittag-Leffler stability of fractional-order Hopfield neural networks. Nonlinear Anal. Hybri. **16**, 104–121 (2015)
3. Chen, B., Chen, J.: Global $O(t^{-\alpha})$ stability and global asymptotical periodicity for a non-autonomous fractional-order neural networks with time varying delays. Neural Netw. **73**, 47–57 (2016)
4. Rakkiyappan, R., Sivaranjani, R., Velmurugan, G., Cao, J.: Analysis of global $O(t^{-\alpha})$ stability and global asymptotical periodicity for a class of fractional-order complex-valued neural networks with time varying dela. Neural Netw. **77**, 51–69 (2016)
5. Rakkiyappan, R., Cao, J., Velmurugan, G.: Existence and uniform stability analysis of fractional-order complex-valued neural networks with time delays. IEEE Trans. Neural Netw. Learn. **26**, 84–97 (2015)
6. Xiao, M., Zheng, W., Jiang, G., Cao, J.: Undamped oscillations generated by Hopf Bifurcations in fractional-order recurrent neural networks with Caputo derivative. IEEE Trans. Neural Netw. Learn. **26**, 3201–3214 (2015)
7. Wang, H., Yu, Y., Wen, G.: Stability analysis of fractional-order Hopfield neural networks with time delays. Neural Netw. **55**, 98–109 (2014)
8. Wu, A., Zhang, J., Zen, Z.: Dynamic behaviors of a class of memristor-based Hopfield networks. Phys. Lett. A. **375**, 1661–1665 (2011)
9. Wu, A., Wen, S., Zen, Z.: Anti-synchronization control of a class of memristive recurrent neural networks. Commun. Nonlinear Sci. **18**, 373–385 (2013)
10. Pu, Y., Zhou, J., Zhang, Y., Zhang, N., Huang, G., Siarry, P.: Fractional extreme value adaptive training method: fractional steepest descent approach. IEEE Trans. Neural Netw. Learn. **26**, 653–662 (2015)
11. Love, E.R.: Fractional derivatives of imaginary order. J. Lond. Math. Soc. **3**, 241–259 (1971)

12. Oldham, K.B., Spanier, J.: The Fractional Calculus: Theory and Applications of Differentiation and Integration to Arbitrary Order. Academic, Cambridge (1974)
13. Mcbride, A.C.: Fractional Calculus. Halsted, USA (1986)
14. Nishimoto, K.: Fractional Calculus: Integrations and Differentiations of Arbitrary Order. New Haven University Press, New Haven (1989)
15. Rumelhart, D.E., Hinton, G.E., Williams, R.J.: Learning representations by back propagating errors. Nature **323**, 533–536 (1986)

Attracting Sets of Non-autonomous Complex-Valued Neural Networks with both Distributed and Time-Varying Delays

Zhao Yang and Xiaofeng Liao[⊠]

College of Electronic and Information Engineering,
Southwest University, Chongqing 400715, China
xfliao@swu.edu.cn

Abstract. In this paper, we investigate the attracting sets of a class of complex-valued neural networks by using integro-difference inequality and properties of the M-matrix. To be specific, we consider both time-varying and infinite distributed delays in complex-valued neural networks and establish some sufficient conditions by setting up integro-differential inequality and applying conjugate system of complex-valued neural networks. Some new results on attracting sets of neural networks are obtained. Simulation verifies our results.

Keywords: Complex-valued neural networks · Attracting sets · Distributed delays · Integro-differential inequality

1 Introduction

The complex-valued neural network (CVNN) is an extension of the neural network in real domain and it uses complex-valued states, connection weights, or activation functions. In recent years, CVNNs have attracted a public concern due to their practical application in computer vision, image processing, optoelectronics, filtering, speech synthesis, remote sensing, electromagnetics, ultrasonic, and quantum waves [1–4]. Therefore, it is important and necessary to investigate complex-valued differential systems. Besides, considering stability is an important performance criterion commonly used to quantize the quality of CVNNs, many existed literature have discussed the stability analysis in CVNNs [6–11]. In [9], the equilibrium point of the exponential stability of CVNNs with time-varying delays has been derived by using conjugate system of the CVNNs works, the theorem of Brouwer's fixed point and a delay differential inequality. In addition, some researchers have established the sufficient condition to verify the existence of equilibrium and show its uniqueness. In the meanwhile, CVNNs' asymptotic stability has been derived. Zhang studied global asymptotic stability of equilibrium based on a Lynapunov function and mathematical analysis and derived the sufficient condition by separating CVNNs into real and imaginary parts [10].

© Springer International Publishing AG 2017
F. Cong et al. (Eds.): ISNN 2017, Part I, LNCS 10261, pp. 555–563, 2017.
DOI: 10.1007/978-3-319-59072-1_65

Although these aforementioned works studied the CVNNs, few literatures have discussed non-autonomous CVNNs, and there are few results about the attracting set of non-autonomous CVNNs by using differential inequality. Motivated by the above discussions, we will use the integro-differential established in [5] and conjugate system in [9] to investigate the attracting set of the CVNNs.

Motivated by the above considerations, we consider a CVNN and analyze its global attracting set. To be more specific, this paper has the following contributions: (1) Using a different method in [9], we do not separate the activation functions into their real and imaginary parts. (2) Taking advantage of CVNNs' conjugate system, integro-differential inequality and M-matrix theory, we obtain some sufficient conditions for checking the attracting set of non-autonomous CVNNs with both time-varying delays and infinite distributed delays. We organize the rest of this paper as follows. In Sect. 2, we illustrate the system model of distributed delay and time-varing CVNNs and describe some definitions and lemma. Furthermore, we discuss a class of intego-differential inequality and present the main results in Sect. 3. In Sect. 4, we simulate one instance by computers and give an numerical example to verify the effectiveness of our results. In the end, we give a summary of this paper in Sect. 5.

Notations: Unless otherwise specified in this article, notations we use in this paper can be summarized as follows. To be specific, E, $\mathbb{R}$ denotes the unit matrix and the set of real numbers, $\mathbb{R}_+ = [0, +\infty)$. $\mathbb{R}^n$, $\mathbb{C}^n$ denote the n-dimensional Euclidean space and the n-dimensional complex vector space, respectively. $\mathbb{R}^{m \times n}$, $\mathbb{C}^{m \times n}$ denote the sets of $m \times n$ real and complex matrices, respectively. Let $y = a + ib$ be a complex number, $a, b \in \mathbb{R}$, $|y| = \sqrt{a^2 + b^2}$. $\bar{y} = a + i(-b)$ is the conjugate complex number of y. For $y, \phi \in \mathbb{C}^n$, we define $[|y(t)|] = (|y_1(t)|, \ldots, |y_n(t)|)^T$, $\|y(t)\| = \sqrt{\Sigma_{k=1}^n |y_k(t)|^2}$. Let $C \triangleq ((-\infty, 0], \mathbb{C}^n)$, $PC \triangleq ((-\infty, 0], \mathbb{R}^n)$. $\hbar = \{v(t) : \mathbb{R} \longrightarrow \mathbb{R} | (t) \text{ is a positive integral continuous function and } \lim_{t \to \infty} \int_a^t v(s)ds = \infty, \sup_{a \le t < b} \int_{t-\tau}^t v(s)ds = \sigma_1 < \infty \text{ and } \sup_{a \le t < b} \{ \sup_{-\infty < s < 0} \{v(t+s)\} \} = \sigma_2 < \infty \}$. For $C, D \in \mathbb{R}^{m \times n}$ or $\in \mathbb{C}^{m \times n}$, we define the Hadamard product or Schur product $C \circ D = (c_{ij}d_{ij})_{m \times n}$, $|C| = (|c_{ij}|)_{m \times n}$.

2   Problem Formulation and Preliminaries

In our paper, we investigate a model of time-delay CVNNs:

$$\dot{y}(t) = -D(t)y(t) + A(t)f(y(t)) + C(t)g(y(t - \tau(t)))$$

$$+ \int_{-\infty}^t P(t - s)w(y(t))ds + J(t), \tag{1}$$

the initial conditions of (1) can be given by $y(t) = \phi(t)$, $-\infty \le t \le t_0$. Where $y(t) = (y_1(t), \ldots, y_n(t))^T \in \mathbb{C}^n$ is the state vector of the neural network with n neurons, $\dot{y}(t) = (\dot{y}_1(t), \ldots, \dot{y}_n(t))^T \in \mathbb{C}^n$. $D(t) = diag\{d_1(t), \ldots, d_n(t)\} \in \mathbb{R}^{n \times n}$ with $d_i(t) > 0$ is the self-feedback connection weight matrix.

$A(t) = (a_{ij}(t))_{n \times n}, C(t) = (c_{ij}(t))_{n \times n}, P(t) = (p_{ij}(t))_{n \times n} \in \mathbb{C}^{n \times n}$ is the connection weight matrix. $J(t) = (J_1(t), \ldots, J_n(t))^T \in \mathbb{C}^n$ is the external input vector. $f(\cdot), g(\cdot), w(\cdot) : \mathbb{C}^n \to \mathbb{C}^n$, are the activation functions of the neurons and $f(y(t)) = (f_1(y_1(t)), \ldots, f_n(y_n(t)))^T$, $w(y(t)) = (w_1(y_1(t)), \ldots, w_n(y_n(t)))^T$, $g(y(t - \tau(t))) = (g_1(y_1(t - \tau_1(t))), g_2(y_2(t - \tau_2(t))), \ldots, g_n(y_n(t - \tau_n(t))))^T$ ($0 \leq \tau_i(t) \leq \tau$ is the transmission delays).

A conjugate system of model (1) is represented by

$$\dot{\overline{y}}(t) = -D(t)\overline{y}(t) + \overline{A}(t)\overline{f}(\overline{y}(t)) + \overline{C}(t)\overline{g}(\overline{y}(t - \tau(t)))$$
$$+ \int_{-\infty}^{t} \overline{P}(t - s)\overline{w}(\overline{y}(t))ds + \overline{J}(t), \tag{2}$$

where $\overline{y}(t) = (\overline{y}_1(t), \ldots, \overline{y}_n(t))^T$, $\overline{C}(t) = (\overline{c}_{ij}(t))_{n \times n}$, $\overline{C}(t) = (\overline{c}_{ij}(t))_{n \times n}$, $\overline{P}(t) = (\overline{p}_{ij}(t))_{n \times n}$, $\overline{f}(\overline{y}(t)) = (\overline{f}_1(\overline{y}_1(t)), \ldots, \overline{f}_n(\overline{y}_n(t)))^T$, $\overline{w}(\overline{y}(t)) = (\overline{w}_1(\overline{y}_1(t)), \ldots, \overline{w}_n(\overline{y}_n(t)))^T$, $\overline{g}(\overline{y}(t - \tau(t))) = (\overline{g}_1(\overline{y}_1(t - \tau_1(t))), \ldots, \overline{g}_n(\overline{y}_n(t - \tau_n(t))))^T$.

Moreover, there are some necessary definition and lemma need to be illustrated in detail.

Definition 1. If for any initial value $\phi \in C$, the set $S \subset C$ is called a global attracting set of (1), the solution $y(t, t_0, \phi)$ converges to S as $t \longrightarrow +\infty$; that is,

$$dist(y(t, t_0, \phi)), S) \longrightarrow 0, \qquad t \longrightarrow +\infty,$$

where $dist(y, S) = \inf_{\psi \in S} dist(y, \psi)$ and $dist(y, \psi) = \sup_{-\tau \leqslant s \leqslant 0} |y(s) - \psi(s)|$.

The following Lemma has basically the same proof in [9].

Lemma 1. Let $c(t), d(t) \in \mathbb{C}^n$, $c(t) \circ d(t) = (c_1(t)d_1(t), \ldots, c_n(t)d_n(t))^T$, $[|c(t)|]^2 = (|c_1(t)|^2, \ldots, |c_n(t)|^2)^T$, and $Re(c(t))$ is the real part of $c(t)$, then:

(1) $c(t) + \overline{c}(t) = 2Re(c(t))$,
(2) $[|c(t)|] = [|\overline{c}(t)|]$,
(3) $[|c(t)|^2] = c(t) \circ \overline{c}(t)$,
(4) $[c(t) \circ d(t)]' = c(t) \circ d'(t) + d(t) \circ c'(t)$.

If D is a M matrix, according to the characteristic of M matrix, we can obtain

$$\Omega_M(D) \triangleq \{\xi \in \mathbb{R}^{n \times n} | D\xi > 0, \xi > 0\}.$$

3   Main Results

For network (1), we give the following hypotheses.

(H1) $\frac{1}{\rho_2} \leq h(t) \leq \frac{1}{\rho_1}, |D(t)| \geq \tilde{D}h(t)$, and $|J| \leq \tilde{J}h(t)$, where $h(t) \in \hbar$, constants $\rho_1, \rho_2 > 0$, nonnegative constant matrices $\tilde{D} = diag\{\tilde{d}_{11}, \ldots, \tilde{d}_{nn}\} \in \mathbb{R}^{n \times n}$ and nonnegative vector $\tilde{J} = [\tilde{J}_1, \ldots, \tilde{J}_n]^T \in \mathbb{R}^n$.

(H2) There exist nonnegative constant matrices $\tilde{A} = (\tilde{a}_{ij})_{n \times n}$, $\tilde{C} = (\tilde{c}_{ij})_{n \times n} \in \mathbb{R}^{n \times n}$ make $|A(t)| \leq \tilde{A}h(t)$, $|C(t)| \leq \tilde{C}h(t)$ established.

(H3) There exist nonnegative constant diagonal matrices $F = diag\{f_1, \ldots, f_n\}$, $G = diag\{g_1, \ldots, g_n\}$ and $W = diag\{w_1, \ldots, w_n\} \in \mathbb{R}^{n \times n}$ make $|f(y(t))| \leq F|y(t)|$, $|g(y(t))| \leq G|y(t)|$ and $|w(y(t))| \leq W|y(t)|$ established.

(H4) Let $-\Pi = (\tilde{D} - \tilde{A}F - \tilde{C}G - PW)$ be an M matrix, where $P = (p_{ij})_{n \times n} \triangleq (\int_0^\infty p_{ij}(s)ds)_{n \times n}$.

First, we propose a integro-difference inequality as follows.

Theorem 1. Let $x(t) \in PC([t_0, \infty), \mathbb{R}^n)$ satisfy the following integro-differential inequality

$$
\begin{cases}
D^+x(t) \leq u(t)\{Wx(t) + V[x(t)]_\tau + \int_0^\infty R(s)x(t-s)ds + L\}, & t \geq t_0 \\
x(t) = \phi(t), & -\infty < t \leq t_0
\end{cases}
$$
$$(3)$$

where $0 \leq x(t) = (x_1(t), \ldots, x_n(t))^T, \phi(t) \in PC([t_0 - \tau, t_0], \mathbb{R}^n)$, $W = (w_{ij})_{n \times n}$ and $w_{ij} \geq 0$ for $i \neq j$, $V = (v_{ij})_{n \times n} \geq 0$, $L = (L_1, \ldots, L_n)^T \geq 0$, $u(t) \in \hbar$, $R = (r_{ij})_{n \times n} \triangleq (\int_0^\infty r_{ij}(s)ds)_{n \times n}$.

If $K = -(W + V + R)$ is a nonsingular M matrix, there exists a positive vector $\xi = (\xi_1, \ldots, \xi_n)^T \in \Omega_M(K)$, then

$$
x(t) \leq \kappa \xi e^{-\lambda \int_{t_0}^t u(s)ds} - (W + V + R)^{-1}L, \qquad t \geq t_0, \tag{4}
$$

where $\kappa \geq 0$, and the inequality determines the constant λ and λ is always more than 0:

$$
[\lambda E + W + Ve^{\lambda \sigma_1} + \int_0^\infty R(s)e^{\lambda \sigma_2 s}ds]\xi < 0, \tag{5}
$$

for the given $\xi \in \Omega_M(K)$.

Proof. The proof is similar to [5, Theorem 3], so omitted.

Theorem 2. Suppose that (H1)–(H4) hold, then the global attracting set of (1) can be represented as $S = \{\phi \in C | \|\phi\| \leq -\Pi^{-1}\tilde{J}\}$.

Proof. On the basis of the property of Lemma 1 with systems (1) and (2), we can achieve

$$
\begin{aligned}
[|y(t)|^2]' &= (\overline{y}(t) \circ y(t))' \\
&= \overline{y}(t) \circ \dot{y}(t) + y(t) \circ \dot{\overline{y}}(t) \\
&= -2[\overline{y}(t) \circ D(t)y(t)] \\
&\quad + \overline{y}(t) \circ A(t)f(y(t)) + y(t) \circ \overline{A}(t)\overline{f}(\overline{y}(t))
\end{aligned}
$$

$$+ \overline{y}(t) \circ C(t)g(y(t - \tau(t))) + y(t) \circ \overline{C}(t)\overline{g}(\overline{y}(t - \tau(t)))$$

$$+ \overline{y}(t) \circ \int_{-\infty}^{t} P(t - s)w(y(s))ds + \overline{y}(t) \circ J(t) + y(t) \circ \overline{J}(t)$$

$$+ y(t) \circ \int_{-\infty}^{t} \overline{P}(t - s)\overline{w}(\overline{y}(s))ds. \tag{6}$$

From hypotheses (H1) and (H2), we can derive

$$\overline{y}(t) \circ A(t)f(y(t)) + y(t) \circ \overline{A}(t)\overline{f}(\overline{y}(t))$$
$$= \overline{y}(t) \circ A(t)f(y(t)) + \overline{\overline{y}(t) \circ A(t)f(y(t))}$$
$$\leq 2[|\overline{y}(t)|] \circ (|A(t)|[|f(y(t))|])$$
$$\leq 2[|\overline{y}(t)|] \circ (\tilde{A}h(t)[F|y(t)|])$$
$$= 2[|y(t)|] \circ (\tilde{A}h(t)[F|y(t)|]). \tag{7}$$

Similarly,

$$-\overline{y}(t) \circ D(t)y(t) - y(t) \circ D(t)\overline{y}(t)$$
$$= -2[|y(t)|] \circ (\tilde{D}h(t)[|y(t)|]), \tag{8}$$

$$\overline{y}(t) \circ C(t)g(y(t - \tau(t))) + y(t) \circ \overline{C}(t)\overline{g}(\overline{y}(t - \tau(t)))$$
$$= 2[|y(t)|] \circ (\tilde{C}h(t)G[|y(t)|]_\tau), \tag{9}$$

$$\overline{y}(t) \circ \int_{-\infty}^{t} P(t - s)w(y(s))ds + y(t) \circ \int_{-\infty}^{t} \overline{P}(t - s)\overline{w}(\overline{y}(s))ds$$
$$= 2[|y(t)|] \circ \int_{0}^{\infty} |P(s)|W[|y(t - s)|]ds, \tag{10}$$

$$y(t) \circ \overline{J}(t) + \overline{y}(t) \circ J(t)$$
$$\leq 2[y(t)] \circ [\tilde{J}v(t)]. \tag{11}$$

Taking (7)–(11) into (6), with the following equation

$$[|y(t)|^2]' = 2[|y(t)|] \circ [|y(t)|]',$$

we can get

$$2\,[|y(t)|] \circ [|y(t)|]'$$
$$\leq -2[|y(t)|] \circ \{\tilde{D}h(t)[|y(t)|] - \tilde{A}h(t)F[|y(t)|]$$
$$- \tilde{C}h(t)G[|y(t)|]_\tau - \int_{0}^{\infty} |P(s)|W[|y(t - s)|]ds - \tilde{J}v(t)\}$$
$$\leq -2[|y(t)|] \circ h(t)\{\tilde{D}[|y(t)|] - \tilde{A}F[|y(t)|]$$
$$- \tilde{C}G[|y(t)|]_\tau - \int_{0}^{\infty} \rho_1|P(s)|W[|y(t - s)|]ds - \tilde{J}\}. \tag{12}$$

Finally, we can obtain

$$[|y(t)|]' \leq h(t)\{-\tilde{D}[|y(t)|] + \tilde{A}F[|y(t)|]$$

$$+ \tilde{C}G[|y(t)|]_\tau + \int_0^\infty \rho_1|P(s)|W[|y(t-s)|]ds + \tilde{J}\}. \quad (13)$$

On the basis of Theorem 1 and (H4), we can get that $-\Pi = (\tilde{D} - \tilde{A}F - \tilde{C}G - PH)$ is a M matrix. Moreover, according to the property of M matrix, there must exists a vector ξ, $-\Pi\xi > 0$, where $\xi = (\xi_1, \ldots, \xi_n)^T \in \Omega_M(-\Pi)$, and $-\Pi^{-1}\tilde{J} \geq 0$.

Based on the initial conditions mentioned above in (1), $y(t) = \phi(t)$, where $t \in (-\infty, t_0]$ and $\phi(t) \in C$, we can get

$$[|y(t)|] \leq \kappa_0\xi[|\phi|], \qquad \kappa_0 = \frac{1}{min_{1 \leq i \leq n}\{\xi_i\}}, \quad -\infty < t < t_0,$$

and

$$[|y(t)|] \leq \kappa_0\xi[|\phi|]e^{-\lambda \int_{t_0}^t u(s)ds} - \Pi^{-1}\tilde{J}, \qquad t \geq t_0.$$

This implies

$$\|y(t)\| \leq \kappa_0\xi\|\phi\|e^{-\lambda \int_{t_0}^t u(s)ds} - \Pi^{-1}\tilde{J}, \qquad t \geq t_0,$$

where λ is determined by

$$[\lambda E + (-\tilde{D} + \tilde{A}F) + \tilde{C}Ge^{\lambda\sigma_1} + \int_0^\infty \rho_1|P(s)|We^{\lambda\sigma_2 s}ds]\xi < 0.$$

The proof is completed.

Remark 1. With conditions (H1)–(H4) and $\tilde{J} = 0$, we can find that CVNNs (1) is globally exponentially stable at zero point.

Remark 2. If we put the system (1) into the real domain, then we can obtain $y(t), J(t) \in \mathbb{R}^n$, $A(t) \in \mathbb{R}^{n\times n}, C(t) \in \mathbb{R}^{n\times n}, P(t) \in \mathbb{R}^{n\times n}$ and $f(\cdot), g(\cdot), w(\cdot) : \mathbb{R}^n \to \mathbb{R}^n$, that is to say, the complex-valued system (1) become to a real valued system. Then we can count the upper right derivative of $[|y(t)|]'$ under equation (1) as

$$[|y(t)|]' \leq -|D(t)||y(t)| + |A(t)||f(y(t))| + |C(t)||g(y(t-\tau(t)))|$$

$$+ \int_{-\infty}^t |P(t-s)||w(y(t))|ds + |J(t)|,$$

Using (H1)–(H3), we can obtain Eq. (13). Next, repeating the proof below Eq. (13) in Theorem 2, we can get the global attracting set S as $S = \{\phi \in PC|\|\phi\| \leq -\Pi^{-1}\tilde{J}\}$ for the real-valued system which is transformed by the complex-valued system (1).

4   Numerical Example

Example 1. Consider a two-neuron CVNNs (1) with the following parameters:
where $d_{11} = 8$, $d_{22} = 7$, $a_{11} = \arctan(t) + \frac{\sqrt{3}}{10} \cos(t)i$, $a_{22} = e^{-t} + \arctan(t)i$,
$c_{11} = \arctan(t)i$, $c_{22} = p_{22} = e^{-2t}$, $p_{11} = e^{-t}$, $J_1 = \frac{\sqrt{3}}{10} \sin(t) + \frac{\sqrt{5}}{10} \cos(t)i$,
$J_2 = \frac{\sqrt{5}}{10} \cos(t) + \frac{\sqrt{3}}{10} \sin(t)i$, $d_{12} = d_{21} = b_{12} = b_{21} = c_{12} = c_{21} = p_{12} = p_{21} = 0$.
and the activation functions $f_i(y_i) = g_i(y_i) = w_i(y_i) = 0.5|x_i| + 0.5i|y_i|$, with
$y_i = x_i + iy_i \in \mathbb{C}$ and $i = 1, 2$.

The parameters of hypothesis (H1)–(H3) are as follows:

$$\tilde{D} = \begin{bmatrix} 8 & 0 \\ 0 & 7 \end{bmatrix}, \qquad \tilde{A} = \begin{bmatrix} 3.8 & 0 \\ 0 & 3.8 \end{bmatrix}, \qquad \tilde{C} = \begin{bmatrix} 2.461 & 0 \\ 0 & 2.227 \end{bmatrix}, \qquad P = \begin{bmatrix} 1 & 0 \\ 0 & 0.5 \end{bmatrix},$$

$$\tilde{J} = \begin{bmatrix} 0.282 \\ 0.282 \end{bmatrix}, \qquad F = W = H = \begin{bmatrix} \sqrt{5} & 0 \\ 0 & \sqrt{5} \end{bmatrix},$$

we set $h(t) = \frac{1}{1+e^{-t}} \in \hbar, \frac{1}{2} \le h(t) \le 1, t \ge 0$.

It is easy to compute that $-\Pi = \tilde{D} - \tilde{A}F - \tilde{C}G - PW = \begin{bmatrix} 5.5437 & 0 \\ 0 & 4.6226 \end{bmatrix}$
is a nonsingular M-matrix. Therefore, by Theorem 2, we can obtain the global attracting sets of non-autonomous complex-valued system (1) that is
$S = \{\phi \in C \mid \|\phi\| \le -\Pi^{-1}\tilde{J} = (0.0510, 0.0612)^T\}$.

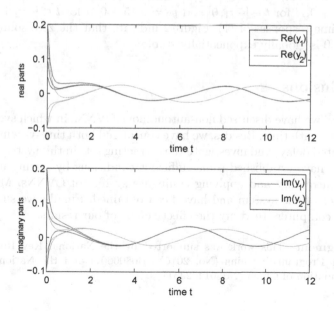

Fig. 1. Real and imaginary parts of state trajectories.

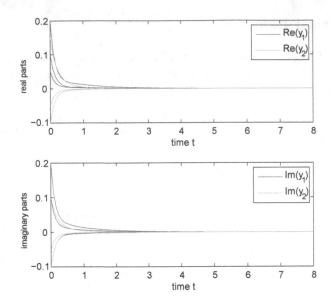

Fig. 2. Real and imaginary parts of state trajectories with $J = 0$.

Then we consider three cases. Case 1: $\tau_1 = 0.5$ and $\tau_2 = 0.2$, the initial sate $y_1(t) = 0.1 + 0.2i$ for $t \in [-\tau_1, 0]$ and $y_2 = -0.05 - 0.05i$ for $t \in [-\tau_2, 0]$. Case 2: $\tau_1 = 0.3$ and $\tau_2 = 0.4$, the initial sate $y_1(t) = 0.2 - 0.1i$ for $t \in [-\tau_1, 0]$ and $y_2 = -0.1 - 0.1i$ for $t \in [-\tau_2, 0]$. Case 3: $\tau_1 = \tau_2 = 0.2$, the initial sate $y_1(t) = 0.05 + 0.1i$ for $t \in [-\tau_1, 0]$ and $y_2 = 0.15 + 0.1i$ for $t \in [-\tau_2, 0]$. Figure 1 shows the time responses of (1). Figure 2 indicate that the zero solution of (1) with $J(t) = 0$ is globally exponentially stable.

5   Conclusions

In this paper, we have discussed non-autonomous CVNNs in which system parameters change with time. Besides, we have considered both time-varying and infinite distributed delays and investigate the attracting sets in this system model. In addition, we have established some sufficient conditions by setting up integro-differential inequality and applying conjugate system of CVNNs. Moreover, a attracting set of the system and have been obtained. Finally, we simulate one instance by computers to verify the effectiveness of our results.

Acknowledgment. This work was supported by the National Key Research and Development Program of China (No. 2016YFB0800601) and the National Natural Science Foundation of China (No. 61472331).

References

1. Hirose, A.: Complex-Valued Neural Networks: Theories and Applications. World Scientific, Singapore (2003)
2. Bohner, M., Sanyal, S.: Global fof complex-valued neural networks on time case. Differ. Equ. Dyn. Syst. **19**, 3–11 (2011)
3. Liu, X., Fang, K., Liu, B.: A synthesis method based on stability analysis for complex-valued Hopfield neural network. In: 2009 Asian Control Conference, pp. 1245–1250 (2009)
4. Hirose, A.: Recent progress in applications of complex-valued neural networks. In: Rutkowski, L., Scherer, R., Tadeusiewicz, R., Zadeh, L.A., Zurada, J.M. (eds.) ICAISC 2010. LNCS, vol. 6114, pp. 42–46. Springer, Heidelberg (2010). doi:10.1007/978-3-642-13232-2_6
5. Xu, L.G., Xu, D.Y.: Exponential stability of nonlinear impulsive neutral integro-differential equation. Nonlinear Anal. **69**, 2910–2923 (2008)
6. Zhou, B., Song, Q.K.: Boundedness and complete stability of complex-valued neural networks with time delay. IEEE Trans. Neural Netw. Learn. Syst. **24**, 1227–1238 (2013)
7. Xu, X.H., Zhang, J.Y., Shi, J.Z.: Exponential stability of complex-valued neural networks with mixed delays. Neurocomputing **128**, 483–490 (2014)
8. Song, Q.K., Yan, H., Zhao, Z.J., Liu, Y.R.: Global exponential stability of complex-valued neural networks with both time-varying delays and impulsive effects. Neural Netw. **79**, 108–116 (2016)
9. Pan, J., Liu, X.Z.: Exponential stability of a class of complex-valued neural networks with time-varying delays. Neurocomputing **164**, 293–299 (2015)
10. Zhang, Z.Y., Lin, C., Chen, B.: Global stability criterion for delayed complex-valued recurrent neural networks. IEEE Trans. Neural Netw. Learn. Syst. **25**, 1704–1708 (2014)
11. Hu, J., Wang, J.: Global stability of complex-valued recurrent neural networks with time-delays. IEEE Trans. Neural Netw. Learn. Syst. **23**, 853–865 (2012)

Stability of Complex-Valued Neural Networks with Two Additive Time-Varying Delay Components

Zhenjiang Zhao[1](✉), Qiankun Song[2](✉), and Yuchen Zhao[3](✉)

[1] College of Science, Huzhou University, Huzhou 313000, China
zhaozjcn@163.com
[2] Department of Mathematics, Chongqing Jiaotong University,
Chongqing 400074, China
qiankunsong@163.com
[3] Deloitte Consulting (Shanghai) Co. Ltd., Shanghai 200002, China
yuczhao@deloitte.com.cn

Abstract. In this paper, a class of complex-valued neural networks including two additive time-varying delay components has been discussed. By making use of the combinational Lyapunov-Krasovskii functional and free weighting matrix method, as well as matrix inequality technique, a delay-dependent criterion of stability is derived.

Keywords: Complex-valued neural networks · Additive delay components · Stability · Lyapunov-Krasovskii functional · Time-varying delays

1 Introduction

With the application of artificial intelligence technology, the research of artificial neural networks (NNs) is becoming more and more important. Especially, in the past twenty years, neural networks have been applied in many fields, such as optimization problem, associative memory, model identification, pattern recognition, signal processing, and other engineering and scientific areas, so the neural network is attracting more and more attention [1]. As we all know, in the process of the realization of neural networks, time delays often occur, which may lead to the performance of neural networks to reduce, or even induce instability [2]. Therefore, the research of delayed NNs have attracted great interest, also many stability criterion have been obtained [1–5].

Meanwhile, in the literature [6], the authors introduced a new type of time-varying delay with two additive components in the state of neural networks. In many practical applications, such as remote control, network control system, etc., we may encounter such a system. For example, in networked controlled systems, the signal transmitted from one location to another location may experience some segments of networks, which can possibly cause a series of delays. Due to the uncertain network transmission conditions, the delay of the sensor to the

© Springer International Publishing AG 2017
F. Cong et al. (Eds.): ISNN 2017, Part I, LNCS 10261, pp. 564–571, 2017.
DOI: 10.1007/978-3-319-59072-1_66

controller has different characteristics with the controller to actuator delay. This means that the study of the system having time-vary delays will become more complicated and more meaningful. Therefore, the stability of the NNs having additive time-varying delays have been widely investigated [6–9].

Complex-valued NNs is an extension of real-valued NNs, due to its practical applications of complex-valued neural networks in physical systems for processing quantum waves, electromagnetic, ultrasonic and light, complex-valued NNs consisting of complex-valued states, outputs, connection weights, and activation functions has also become a research hot spot [10]. In addition, some problems can be only solved by complex-valued NNs, but cannot be solved by real-valued NNs [11]. At present, some achievements have been obtained in the study of stability of various complex-valued neural networks, relevant results are available in [12–20]. In [12–16], the methods used to examine the stability of complex-valued NNs were still using the method of analyzing the stability of real-valued NNs. It is a method of dividing complex-valued NNs into two parts, real parts and imaginary parts. However, this method will bring two questions. One problem is that the dimension of the real-valued NNs is twice as much as the complex-valued neural network. This will result in the difficulty of analysis. The other is that the real and imaginary parts of the activation function need to be distinct, but there is lack of a analytical form to express such difference. In [17–20], under the condition that both the real and imaginary portions of the complex-valued neural network were not split, the stability of the system was explored. Several criteria for the stability of the system were obtained.

In the above complex-valued NNs, the time delay in a state was only a singular form. However, as far as we know, few scholars have studied the stability of the complex-valued NNs having additive time-varying delays so far. Based on the above analysis, we study the problem. In the second section of this article, we described the problem and made some preparatory work. In the third section, we did the analysis of delay dependent stability of complex-valued NNs having additive time-varying delay. Then by making hybrid use of the Lyapunov-Krasovskii functional and the free weighting matrix approach, innovative delay dependent stability criteria are derived.

2   Problem Description and Preliminaries

The stability analysis of complex-valued NNs having two additive time-varying delays components is examined.

$$\dot{z}(s) = -Cz(s) + Af(z(s)) + Bf(z(s - \tau_1(s) - \tau_2(s))) + J \qquad (1)$$

$s \geq 0$, $z(s) = (z_1(s), z_2(s), \cdots, z_n(s))^T \in \mathbb{C}^n$, where $z_i(s)$ is the state of the ith neuron at time t, $i = 1, 2, \cdots, n$; $f(z(s)) = (f_1(z_1(s)), f_2(z_2(s)), \cdots, f_n(z_n(s)))^T \in \mathbb{C}^n$ is the vector-valued activation function; $A = (\mathbf{a}_{ij})_{n \times n} \in \mathbb{C}^{n \times n}$ and $B = (\mathbf{b}_{ij})_{n \times n} \in \mathbb{C}^{n \times n}$ are the connection weight matrices; $C = \mathrm{diag}\{c_1, c_2, \cdots, c_n\} \in \mathbb{R}^{n \times n}$ is the self-feedback connection weight matrix, where

$c_i > 0$; the input vector is $J = (J_1, J_2, \cdots, J_n)^T \in \mathbb{C}^n$; $\tau_1(s)$ and $\tau_2(s)$ are two time-varying delays.

The following assumptions are made:

(H1). Two time-varying delays should meet the following conditions

$$0 \leq \tau_1(s) \leq \tau_1, \quad 0 \leq \tau_2(s) \leq \tau_2, \quad \dot\tau_1(s) \leq \mu_1, \quad \dot\tau_2(s) \leq \mu_2,$$

where $\tau_1, \tau_2, \mu_1, \mu_2$ are constants, and we denote

$$\tau(s) = \tau_1(s) + \tau_2(s), \quad \tau = \tau_1 + \tau_2, \quad \mu = \mu_1 + \mu_2.$$

(H2). There exists a diagonal matrix $L = \mathrm{diag}\{l_1, l_2, \cdots, l_n\}$ such that

$$|f_i(\gamma_1) - f_i(\gamma_2)| \leq l_i|\gamma_1 - \gamma_2|, i = 1, 2, \cdots, n$$

for all $\gamma_1, \gamma_2 \in \mathbb{C}$, where $l_i > 0$.

The initial conditions of model (1) are $z_i(u) = \phi_i(u)$, $u \in [-\tau, 0]$, where ϕ_i is continuous and bounded on $[-\tau, 0]$, $i = 1, 2, \cdots, n$.

To simplify the model, suppose that $\tilde{z}$ is an equilibrium point of system (1), using the transform $y(s) = z(s) - \tilde{z}$, (1) is converted to the following system

$$\dot{y}(s) = -Cy(s) + Ag(y(s)) + Bg(y(s - \tau_1(s) - \tau_2(s))), \tag{2}$$

where $g(y(s)) = f(y(s) + \tilde{z}) - f(\tilde{z})$.

3 Main Result

Theorem 1. Under the assumptions of **(H1)** and **(H2)**, system (2) is globally asymptotically stable given the seven positive-definite Hermite matrices P_i ($i = 1, 2, \cdots, 7$), two real-valued positive diagonal matrices R and S, four complex-valued matrices $Q_i(i = 1, 2, 3, 4)$ such that the following complex-valued LMI holds:

$$\Pi = (\Pi_{ij})_{11 \times 11} < 0, \tag{3}$$

where $\Pi_{11} = -P_1C - CP_1 + P_2 + P_3 + P_4 + P_5 + C(\tau_1 P_6 + \tau_2 P_7)C + Q_2 + Q_2^* + KRK$, $\Pi_{12} = -Q_2$, $\Pi_{16} = P_1A - C(\tau_1 P_6 + \tau_2 P_7)A$, $\Pi_{17} = P_1B - C(\tau_1 P_6 + \tau_2 P_7)B$, $\Pi_{18} = Q_2$, $\Pi_{22} = -(1 - \mu_1)P_2 + Q_1 + Q_1^*$, $\Pi_{23} = -Q_1$, $\Pi_{29} = Q_1$, $\Pi_{33} = -P_3 + Q_4 + Q_4^*$, $\Pi_{34} = -Q_4$, $\Pi_{3,10} = Q_4$, $\Pi_{44} = -(1 - \mu)P_4 + Q_3 + Q_3^* + KSK$, $\Pi_{45} = -Q_3$, $\Pi_{4,11} = Q_3$, $\Pi_{55} = -P_5$, $\Pi_{66} = A^*(\tau_1 P_6 + \tau_2 P_7)A - R$, $\Pi_{67} = A^*(\tau_1 P_6 + \tau_2 P_7)B$, $\Pi_{77} = B^*(\tau_1 P_6 + \tau_2 P_7)B - S$. $\Pi_{88} = -\frac{1}{\tau_1}P_6$, $\Pi_{99} = -\frac{1}{\tau_1}P_6$, $\Pi_{10,10} = -\frac{1}{\tau_2}P_7$, $\Pi_{11,11} = -\frac{1}{\tau}P_7$, others are zero.

Proof. The Lyapunov-Krasovskii functional candidate for model (2) is constructed as follows

$$V(s) = V_1(s) + V_2(s) + V_3(s) + V_4(s), \tag{4}$$

where

$$V_1(s) = y^*(s)P_1y(s), \tag{5}$$

$$V_2(s) = \int_{s-\tau_1(s)}^{s} y^*(\theta)P_2y(\theta)d\theta + \int_{s-\tau_1}^{s} y^*(\theta)P_3y(\theta)d\theta, \tag{6}$$

$$V_3(s) = \int_{s-\tau_1(s)-\tau_2(s)}^{s} y^*(\theta)P_4y(\theta)d\theta + \int_{s-\tau_1-\tau_2}^{s} y^*(\theta)P_5y(\theta)d\theta. \tag{7}$$

$$V_4(s) = \int_{-\tau_1}^{0} \int_{s+\xi}^{s} \dot{y}^*(\theta)P_6\dot{y}(\theta)d\theta d\xi + \int_{-\tau_1-\tau_2}^{\tau_1} \int_{s+\xi}^{s} \dot{y}^*(\theta)P_7\dot{y}(\theta)d\theta d\xi. \tag{8}$$

From assumption (H1), calculating the time derivative of $V_1(s)$, $V_2(s)$, $V_3(s)$ and $V_4(s)$, we get that

$$\begin{aligned}
\dot{V}_1(s) &= y^*(s)P_1\dot{y}(s) + \dot{y}^*(s)P_1y(s) \\
&= -y^*(s)(P_1C + CP_1)y(s) + y^*(s)P_1Ag(y(s)) + g^*(y(s))A^*P_1y(s) \\
&\quad + y^*(s)P_1Bg(y(s-\tau(s))) + g^*(y(s-\tau(s)))B^*P_1y(s), \tag{9}
\end{aligned}$$

$$\begin{aligned}
\dot{V}_2(s) &\le y^*(s)(P_2 + P_3)y(s) - (1-\mu_1)y^*(s-\tau_1(s))P_2y(s-\tau_1(s)) \\
&\quad -y^*(s-\tau_1)P_3y(s-\tau_1), \tag{10}
\end{aligned}$$

$$\begin{aligned}
\dot{V}_3(s) &\le y^*(s)(P_4 + P_5)y(s) - (1-\mu)y^*(s-\tau(s))P_4y(s-\tau(s)) \\
&\quad -y^*(s-\tau)P_5y(s-\tau), \tag{11}
\end{aligned}$$

$$\begin{aligned}
\dot{V}_4(s) &= \dot{y}^*(s)(\tau_1P_6 + \tau_2P_7)\dot{y}(s) - \int_{s-\tau_1}^{s} \dot{y}^*(\theta)P_6\dot{y}(\theta)d\theta \\
&\quad - \int_{s-\tau}^{s-\tau_1} \dot{y}^*(\theta)P_7\dot{y}(\theta)d\theta \\
&= y^*(s)C(\tau_1P_6 + \tau_2P_7)Cy(s) \\
&\quad -y^*(s)C(\tau_1P_6 + \tau_2P_7)Ag(y(s)) - g^*(y(s))A^*(\tau_1P_6 + \tau_2P_7)Cy(s) \\
&\quad -y^*(s)C(\tau_1P_6 + \tau_2P_7)Bg(y(s-\tau(s))) \\
&\quad -g^*((s-\tau(s)))B^*(\tau_1P_6 + \tau_2P_7)Cy(s) \\
&\quad +g^*(y(s))A^*(\tau_1P_6 + \tau_2P_7)Ag(y(s)) \\
&\quad +g^*(y(s))A^*(\tau_1P_6 + \tau_2P_7)Bg(y(s-\tau(s))) \\
&\quad +g^*(y(s-\tau(s)))B^*(\tau_1P_6 + \tau_2P_7)Ag(y(s)) \\
&\quad +g^*(y(s-\tau(s)))B^*(\tau_1P_6 + \tau_2P_7)Bg(y(s-\tau(s))) \\
&\quad -\int_{s-\tau_1}^{s-\tau_1(s)} \dot{y}^*(\theta)P_6\dot{y}(\theta)d\theta - \int_{s-\tau_1(s)}^{s} \dot{y}^*(\theta)P_6\dot{y}(\theta)d\theta \\
&\quad -\int_{s-\tau}^{s-\tau(s)} \dot{y}^*(\theta)P_7\dot{y}(\theta)d\theta - \int_{s-\tau(s)}^{s-\tau_1} \dot{y}^*(\theta)P_7\dot{y}(\theta)d\theta. \tag{12}
\end{aligned}$$

From assumption **(H2)**, we can get

$$0 \leq y^*(s)LRLy(s) - g^*(y(s))Rg(y(s)), \tag{13}$$

$$0 \leq y^*(s - \tau(s))LSLy(s - \tau(s)) - g^*(y(s - \tau(s)))Sg(y(s - \tau(s))). \tag{14}$$

By Newton-Leibniz formula, we have

$$0 = y^*(s - \tau_1(s))Q_1\left(y(s - \tau_1(s)) - y(s - \tau_1) - \int_{s-\tau_1}^{s-\tau_1(s)} \dot{y}(\theta)d\theta\right)$$
$$+\left(y(s - \tau_1(s)) - y(s - \tau_1) - \int_{s-\tau_1}^{s-\tau_1(s)} \dot{y}(\theta)d\theta\right)^* Q_1^* y(s - \tau_1(s)), \tag{15}$$

$$0 = y^*(s)Q_2\left(y(s) - y(s - \tau_1(s)) - \int_{s-\tau_1(s)}^{s} \dot{y}(\theta)d\theta\right)$$
$$+\left(y(s) - y(s - \tau_1(s)) - \int_{s-\tau_1(s)}^{s} \dot{y}(\theta)d\theta\right)^* Q_2^* y(s), \tag{16}$$

$$0 = \left(y(s - \tau(s)) - y(s - \tau) - \int_{s-\tau}^{s-\tau(s)} \dot{y}(\theta)d\theta\right)^* Q_3^* y(s - \tau(s))$$
$$+y^*(s - \tau(s))Q_3\left(y(s - \tau(s)) - y(s - \tau) - \int_{s-\tau}^{s-\tau(s)} \dot{y}(\theta)d\theta\right), \tag{17}$$

$$0 = \left((s - \tau_1) - y(s - \tau(s)) - \int_{s-\tau(s)}^{s-\tau_1} \dot{y}(\theta)d\theta\right)*Q_4^* y(s - \tau_1)$$
$$+y^*(s - \tau_1)Q_4\left((s - \tau_1) - y(s - \tau(s)) - \int_{s-\tau(s)}^{s-\tau_1} \dot{y}(\theta)d\theta\right). \tag{18}$$

By Eqs. (12), (15), (16), (17), (18) and inequality (13), (14), we can get

$$\dot{V}(t) \leq y^*(s)\left(-P_1C - CP_1 + P_2 + P_3 + P_4 + P_5 + C(\tau_1 P_6 + \tau_2 P_7)C\right.$$
$$\left.+Q_2 + Q_2^* + LRL\right)y(s) - y^*(s)Q_2 y(s - \tau_1(s)) - y^*(s - \tau_1(s))Q_2^* y(s)$$
$$-y^*(s)Q_2\int_{s-\tau_1(s)}^{s} \dot{y}(\theta)d\theta - \left(\int_{s-\tau_1(s)}^{s} \dot{y}(\theta)d\theta\right)^* Q_2^* y(s)$$
$$+y^*(s)\left(P_1A - C(\tau_1 P_6 + \tau_2 P_7)A\right)g(y(s))$$
$$+g^*(y(s))\left(A^*P_1 - A^*(\tau_1 P_6 + \tau_2 P_7)C\right)y(s)$$
$$+y^*(s)\left(P_1B - C(\tau_1 P_6 + \tau_2 P_7)B\right)g(y(s - \tau(s)))$$

$$+g^*(y(s-\tau(s)))\Big(B^*P_1 - B^*(\tau_1 P_6 + \tau_2 P_7)C\Big)y(s)$$

$$+y^*(s-\tau_1(s))\Big(Q_1 - (1-\mu_1)P_2 + Q_1^*\Big)y(s-\tau_1(s))$$

$$-y^*(s-\tau_1(s))Q_1 y(s-\tau_1) - y^*(s-\tau_1)Q_1^* y(s-\tau_1(s))$$

$$-y^*(s-\tau_1(s))Q_1 \int_{s-\tau_1}^{s-\tau_1(s)} \dot{y}(\theta)d\theta$$

$$-\Big(\int_{s-\tau_1}^{s-\tau_1(s)} \dot{y}(\theta)d\theta\Big)^* Q_1^* y(s-\tau_1(s))$$

$$+y^*(s-\tau_1)\Big(Q_4 - P_3 + Q_4^*\Big)y(s-\tau_1)$$

$$-y^*(s-\tau_1)Q_4 y(s-\tau(s)) - y^*(s-\tau(s))Q_4^* y(s-\tau_1)$$

$$-y^*(s-\tau_1)Q_4 \int_{s-\tau(s)}^{s-\tau_1} \dot{y}(\theta)d\theta - \Big(\int_{s-\tau(s)}^{s-\tau_1} \dot{y}(\theta)d\theta\Big)^* Q_4^* y(s-\tau_1)$$

$$+y^*(s-\tau(s))\Big(Q_3 - (1-\mu)P_4 + Q_3^* + LSL\Big)y(s-\tau(s))$$

$$-y^*(s-\tau)Q_3^* y(s-\tau(s)) - y^*(s-\tau(s))Q_3 y(s-\tau)$$

$$-\Big(\int_{s-\tau}^{s-\tau(s)} \dot{y}(\theta)d\theta\Big)^* Q_3^* y(s-\tau(s)) - y^*(s-\tau(s))Q_3 \int_{s-\tau}^{s-\tau(s)} \dot{y}(\theta)d\theta$$

$$-y^*(s-\tau)P_5 y(s-\tau)$$

$$+g^*(y(s))\Big(A^*(\tau_1 P_6 + \tau_2 P_7)A - R\Big)g(y(s))$$

$$+g^*(y(s))A^*(\tau_1 P_6 + \tau_2 P_7)Bg(y(s-\tau(s)))$$

$$+g^*(y(s-\tau(s)))B^*(\tau_1 P_6 + \tau_2 P_7)Ag(y(s))$$

$$+g^*(y(s-\tau(s)))\Big(B^*(\tau_1 P_6 + \tau_2 P_7)B - S\Big)g(y(s-\tau(s)))$$

$$-\int_{s-\tau_1(s)}^{s} \dot{y}^*(\theta)P_6 \dot{y}(\theta)d\theta - \int_{s-\tau_1}^{s-\tau_1(s)} \dot{y}^*(\theta)P_6 \dot{y}(\theta)d\theta$$

$$-\int_{s-\tau(s)}^{s-\tau_1} \dot{y}^*(\theta)P_7 \dot{y}(\theta)d\theta - \int_{s-\tau}^{s-\tau(s)} \dot{y}^*(\theta)P_7 \dot{y}(\theta)d\theta$$

$$= \eta^*(s)\tilde{\Pi}\eta(s) + \tau_1(s)y^*(s)Q_2 P_6^{-1}Q_2^* y(s)$$

$$+(\tau_1 - \tau_1(s))y^*(s-\tau_1(s))Q_1 P_6^{-1}Q_1^* y(s-\tau_1(s))$$

$$+(\tau - \tau(s))y^*(s-\tau(s))Q_3 P_7^{-1}Q_3^* y(s-\tau(s))$$

$$+(\tau(s) - \tau_1)y^*(s-\tau_1)Q_4 P_7^{-1}Q_4^* y(s-\tau_1)$$

$$-\int_{s-\tau_1(s)}^{s} [y^*(s)Q_2 + \dot{y}^*(\theta)P_6]P_6^{-1}[Q_2^* y(s) + P_6 \dot{y}(\theta)]d\theta$$

$$-\int_{s-\tau_1}^{s-\tau_1(s)} [y^*(s-\tau_1(s))Q_1 + \dot{y}^*(\theta)P_6]P_6^{-1}[Q_1^* y(s-\tau_1(s)) + P_6 \dot{y}(\theta)]d\theta$$

$$-\int_{s-\tau}^{s-\tau(s)} [y^*(s-\tau(s))Q_3 + \dot{y}^*(\theta)P_7]P_7^{-1}[Q_3^* y(s-\tau(s)) + P_7 \dot{y}(\theta)]d\theta$$

$$- \int_{s-\tau(s)}^{s-\tau_1} [y^*(s-\tau_1)Q_4 + \dot{y}^*(\theta)P_7]P_7^{-1}[Q_4^* y(s-\tau_1) + P_7\dot{y}(\theta)]d\theta$$

$$\leq \eta^*(s)\tilde{\Pi}\eta(s) + \tau_1 y^*(s)Q_2 P_6^{-1} Q_2^* y(s)$$
$$+\tau_1 y^*(s-\tau_1(s))Q_1 P_6^{-1} Q_1^* y(s-\tau_1(s))$$
$$+\tau y^*(s-\tau(s))Q_3 P_7^{-1} Q_3^* y(s-\tau(s))$$
$$+\tau_2 y^*(s-\tau_1)Q_4 P_7^{-1} Q_4^* y(s-\tau_1)$$
$$= \eta^*(s)\Omega\eta(s), \tag{19}$$

where

$$\eta(s) = (y^*(s), y^*(s-\tau_1(s)), y^*(s-\tau_1), y^*(s-\tau(s)), y^*(s-\tau), g(y(s)), g(y(s-\tau(s))))^*,$$

$\tilde{\Pi} = (\Pi_{ij})_{7\times7}$, $\Pi_{ji} = \Pi_{ij}^*$, $\Omega = (\Omega_{ij})_{7\times7}$, $\Omega_{ji} = \Omega_{ij}^*$, $\Pi_{11} = -P_1 C - C P_1 + P_2 + P_3 + P_4 + P_5 + C(\tau_1 P_6 + \tau_2 P_7)C + Q_2 + Q_2^* + LRL$, $\Pi_{12} = -Q_2$, $\Pi_{16} = P_1 A - C(\tau_1 P_6 + \tau_2 P_7)A$, $\Pi_{17} = P_1 B - C(\tau_1 P_6 + \tau_2 P_7)B$, $\Pi_{22} = -(1-\mu_1)P_2 + Q_1 + Q_1^*$, $\Pi_{23} = -Q_1$, $\Pi_{33} = -P_3 + Q_4 + Q_4^*$, $\Pi_{34} = -Q_4$, $\Pi_{44} = -(1-\mu)P_4 + Q_3 + Q_3^* + LSL$, $\Pi_{45} = -Q_3$, $\Pi_{55} = -P_5$, $\Pi_{66} = A^*(\tau_1 P_6 + \tau_2 P_7)A - R$, $\Pi_{67} = A^*(\tau_1 P_6 + \tau_2 P_7)B$, $\Pi_{77} = B^*(\tau_1 P_6 + \tau_2 P_7)B - S$. $\Omega_{11} = \Pi_{11} + \tau_1 Q_2 P_6^{-1} Q_2^*$, $\Omega_{22} = \Pi_{22} + \tau_1 Q_1 P_6^{-1} Q_1^*$, $\Omega_{33} = \Pi_{33} + \tau_2 Q_4 P_7^{-1} Q_4^*$, $\Omega_{44} = \Pi_{44} + \tau Q_3 P_7^{-1} Q_3^*$, and the rest of Π_{ij} and Ω_{ij} are zero.

By Schur complement, we know that $\Pi < 0$ in (3) be equivalent to $\Omega < 0$, which means the system (2) is globally asymptotically stable. The proof is ended.

4   Conclusions

In this article, the stability for the complex-valued NNs model having two additive time-varying delay components has been studied. A delay-dependent stability criterion has been obtained by making use of the Lyapunov-Krasovskii functional and free weighting matrix method, and using matrix in-equality technique. The obtained result in this paper is to generalize some well-known research.

Acknowledgments. This work was supported by the National Natural Science Foundation of China under Grants 61473332, 11402214, and 61673169 and the Program of Chongqing Innovation Team Project in University under Grant CXTDX201601022.

References

1. Chen, T.: Global exponential stability of delayed Hopfield neural networks. Neural Netw. **14**, 977–980 (2001)
2. Arik, S., Orman, Z.: Global stability analysis of Cohen-Grossberg neural networks with time varying delays. Phys. Lett. A **341**, 410–421 (2005)
3. Song, Q., Cao, J.: Impulsive effects on stability of fuzzy Cohen-Grossberg neural networks with time-varying delays. IEEE Trans. Syst. Man Cybern. **37**, 733–741 (2007)

4. Kwon, O.M., Park, J.H.: New delay-dependent robust stability criterion for uncertain neural networks with time-varying delays. Appl. Math. Comput. **205**, 417–427 (2008)
5. Weera, W., Niamsup, P.: Novel delay-dependent exponential stability criteria for neutral-type neural networks with non-differentiable time-varying discrete and neutral delays. Neurocomputing **173**, 886–898 (2016)
6. Zhao, Y., Gao, H., Mou, S.: Asymptotic stability analysis of neural networks with successive time delay components. Neurocomputing **71**, 2848–2856 (2008)
7. Shao, H., Han, Q.: New delay-dependent stability criteria for neural networks with two additive time-varying delay components. IEEE Trans. Neural Netw. **22**, 812–818 (2011)
8. Xiao, N., Jia, Y.: New approaches on stability criteria for neural networks with two additive time-varying delay components. Neurocomputing **118**, 150–156 (2013)
9. Liu, Y., Lee, S.M., Lee, H.G.: Robust delay-depent stability criteria for uncertain neural networks with two additive time-varying delay components. Neurocomputing **151**, 770–775 (2015)
10. Hirose, A.: Dynamics of fully complex-valued neural networks. Electron. Lett. **28**, 1492–1494 (1992)
11. Lee, D.: Relaxation of the stability condition of the complex-valued neural networks. IEEE Trans. Neural Netw. **12**, 1260–1262 (2001)
12. Hu, J., Wang, J.: Global stability of complex-valued recurrent neural networks with time-delays. IEEE Trans. Neural Netw. Learn. Syst. **23**, 853–865 (2012)
13. Zhou, B., Song, Q.: Boundedness and complete stability of complex-valued neural networks with time delay. IEEE Trans. Neural Netw. Learn. Syst. **24**, 1227–1238 (2013)
14. Chen, X., Song, Q.: Global stability of complex-valued neural networks with both leakage time delay and discrete time delay on time scales. Neurocomputing **121**, 254–264 (2013)
15. Zhang, Z., Lin, C., Chen, B.: Global stability criterion for delayed complex-valued recurrent neural networks. IEEE Trans. Neural Netw. Learn. Syst. **25**, 1704–1708 (2014)
16. Liu, X., Chen, T.: Global exponential stability for complex-valued recurrent neural networks with asynchronous time delays. IEEE Trans. Neural Netw. Learn. Syst. **27**, 593–606 (2016)
17. Bohner, M., Sree Hari Rao, V., Sanyal, S.: Global stability of complex-valued neural networks on time scales. Differ. Equ. Dyn. Syst. **19**, 3–11 (2011)
18. Fang, T., Sun, J.: Further investigate the stability of complex-valued recurrent neural networks with time-delays. IEEE Trans. Neural Netw. Learn. Syst. **25**, 1709–1713 (2014)
19. Song, Q., Zhao, Z., Liu, Y.: Impulsive effects on stability of discrete-time complex-valued neural networks with both discrete and distributed time-varying delays. Neurocomputing **168**, 1044–1050 (2015)
20. Song, Q., Yan, H., Zhao, Z., Liu, Y.: Global exponential stability of complex-valued neural networks with both time-varying delays and impulsive effects. Neural Netw. **79**, 108–116 (2016)

Alpine Plants Recognition with Deep Convolutional Neural Network

Tomoaki Negishi[✉] and Motonobu Hattori

Interdisciplinary Graduate School of Medicine, Engineering and Agriculture,
University of Yamanashi, Kofu, Yamanashi, Japan
{g16tk010,m-hattori}@yamanashi.ac.jp

Abstract. In this study, the ultimate goal is to build a system which identifies species of alpine plants from pictures. In this paper, in order to build such a system, its fundamental recognition part is constructed using a Deep Convolutional Neural Network (DCNN). A lot of recent studies reveal that DCNNs show excellent performance for recognition tasks by acquisition of feature representations from raw data. However, it is necessary to prepare sufficient number of data for obtaining good feature representations. Especially for alpine plants, this is rather difficult because of their habitat. In this paper, we add images of plants other than alpine ones, and examine how such supplementary data have influence on the recognition accuracy for the target domain, i.e., alpine plants. Experimental results show the effectiveness of using supplementary images for alpine plants recognition.

Keywords: Deep Convolutional Neural Network · Alpine plants · Image recognition

1 Introduction

Alpine plants are plants that grow in the alpine climate. The total number of alpine plants species is about 440 in Japan where subspecies, varieties, and breeds are excluded [1,2]. In general, a species of an alpine plant is identified by its color of the flower, shape of the leaf, height and so on. However, it is very tired of taking a picture book to high mountains and looking into it page by page in order to identify an alpine plant among a large number of species.

The goal of this study is to develop an application software in hand-held terminal such as a smart phone, which can identify the type of alpine plants from pictures taken by the hand-held itself. Here, for this purpose, a recognition system for images of alpine plants is constructed using feature representations extracted from a Deep Convolutional Neural Network (DCNN).

In deep learning of a multilayer Neural Network (NN), it has been generally recognized that it is necessary to use large-scale training data to obtain good generalization ability [3,4].

However, it is not very likely to collect enough training data for alpine plants recognition due to an abundance of species and their habitat. On the other hand,

F. Cong et al. (Eds.): ISNN 2017, Part I, LNCS 10261, pp. 572–577, 2017.
DOI: 10.1007/978-3-319-59072-1_67

if plants are not limited to *alpine plants*, there are a large number of images available in many databases. Therefore, in this paper, we try to supplement the lack of training data with images of plants other than alpine plants, and examine how such supplementary data improves recognition accuracy for the target domain.

The rest of this paper is organized as follows. In Sect. 2, the experimental design is explained. In Sect. 3, we will show experimental results. Finally, conclusions are given in Sect. 4.

2   Experimental Design

2.1   Deep Convolutional Neural Network

In this study, the deep learning framework, Caffe was used [5]. The structure of a DCNN is shown in Fig. 1 and Table 1. It consists of 10 hidden layers and 3 fully-connected (fc) layers. In DCNNs, the max pooling, Rectified Linear Units (ReLUs), and Local Response Normalization (LRN) were employed [3]. LRN divides each input value x_i by

$$\left(k + \alpha \sum_{j=max(0,i-n/2)}^{min(N-1,i+n/2)} x_j^2 \right)^{\beta} \tag{1}$$

where N is the total number of kernels in the layer, and k, n, α, and β are hyper-parameters; we used $k = 1$, $n = 5$, $\alpha = 10^{-4}$, and $\beta = 0.75$. Each of the first and second layers in the fully-connected layers have 4,096 units. The output of a DCNN was calculated by the following softmax function,

$$f(u_n) = \frac{\exp(u_n)}{\sum_i^C \exp(u_i)} \tag{2}$$

where u_i denotes the internal state of the ith unit in the output layer, and C denotes the number of classes to recognize.

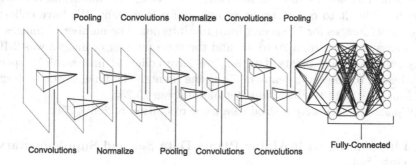

Fig. 1. Structure of a DCNN

Table 1. Details of the structure of the DCNN used in the experiments

Layer	Patch	Map size
Data	-	$227 \times 227 \times 3$
Convolution 1	11×11	$55 \times 55 \times 96$
Pooling 1	3×3	$27 \times 27 \times 96$
Normalization 1	5×5	$27 \times 27 \times 96$
Convolution 2	5×5	$27 \times 27 \times 256$
Pooling 2	3×3	$13 \times 13 \times 256$
Normalization 2	5×5	$13 \times 13 \times 256$
Convolution 3	3×3	$13 \times 13 \times 384$
Convolution 4	3×3	$13 \times 13 \times 384$
Convolution 5	3×3	$13 \times 13 \times 256$
Pooling 5	3×3	$6 \times 6 \times 256$
fc 6	-	$1 \times 1 \times 4096$
fc 7	-	$1 \times 1 \times 4096$
fc 8	-	$1 \times 1 \times 1000$

We used the DCNN which had learned roughly 1.2 million natural images in advance. The data set was provided by ImageNet Large Scale Visual Recognition Challenge 2012 (ILSVRC2012), and consisted of 1,000 classes. We have fine-tuned this prelearned DCNN by alpine plants images and other plants images.

In the previous research [6], Suzuki et al. have shown the effectiveness of feature representations extracted from the DCNN which has prelearned a large number of natural images and then has been fine-tuned by a small number of target images. Therefore, we can expect that using the prelearned DCNN for the feature extraction of alpine plants may be also effective.

2.2   Plants Data Set

As alpine plants, we chose 71 species out of 72 recorded in the alpine plant picture book [2]. Although this book covers 72 typical alpine plants which inhabit the alpine belt in Japan, we have omitted one species called *Senjo-azami* because it was quite difficult to collect accurate images for this plant. We have collected alpine plants images for 71 species from the Internet. The number of images for each class was varied from 20 to 36, and the total number of images was 2,104.

As supplementary plants data set, we have collected images of 55 species which belong to the same genus as 71 alpine plants. The images were collected from ImageNet, and the total number of data was 25,255.

All plants images were resized to a fixed resolution of 256 × 256.

2.3   Fine-Tuning with Alpine Plants Data Set and Supplementary Data Set

In order to examine how supplementary plants data set effects the recognition accuracy, we have compared the following four methods to fine-tune DCNNs,

(a) Fine-tuned with only alpine plants images.
(b) Fine-tuned with only supplementary plants images.
(c) Fine-tuned with supplementary plants images first, and followed by fine-tuning with the alpine plants images.
(d) Fine-tuned by the alpine plants and supplementary images at the same time.

In addition, we examined the performance of DCNNs without any fine-tuning as the baseline.

After fine-tuning, we used DCNNs as a feature extractor. That is, 4,096 dimensional output of the last hidden layer of the DCNN was used as feature representations for input images. These feature vectors were learned by support vector machines (SVMs) to classify plant images.

As for the methods (a) and (c), since the fully-connected layers in DCNNs were also able to be used as classifiers, we examined their performance, too.

2.4   Details of Learning

We trained each DCNN using stochastic gradient descent with a batch size of 256 examples, momentum of 0.9, and weight decay of 0.0005 [3]. The learning rate was initialized at 0.01 in prelearning and 0.001 in fine-tuning. In prelearning, the learning rate was divided by 10 every 100,000 iterations. Dropout was applied on the fully connected layer with ratio 0.5 [7].

To serve as data augmentation, we extracted random 227×227 patches from the 256×256 images at each input and trained each DCNN with these extracted patches.

3   Experimental Results

Table 2 shows the result when SVMs were used as classifiers. We used the Gaussian kernel in SVMs and set the parameters: the soft margin parameter, $C = 100$ [8] and the kernel parameter, $\gamma = 0.001$. The parameter pair was selected by a grid search on $C \in \{1.0 \times 10^1, 1.0 \times 10^2, 1.0 \times 10^3\}$ and $\gamma \in \{1.0 \times 10^{-2}, \cdots, 1.0 \times 10^{-6}\}$.

As shown in the table, the fine-tuning even only with alpine plants images, (a) improves the recognition accuracy in comparison with the result without it. Interestingly, the fine-tuning only with supplementary plants images other than alpine plants, (b) also improves the recognition accuracy. Collecting supplementary images of 55 species belonging to the same genus as 71 alpine plants may result in this improvement. Moreover, we can see that the fine-tuning with supplementary plants images first and followed by that with domain images, (c) improves recognition rate further. Statistically, the difference of the recognition rates between (a) and (c) was at significance level of 5%. This result shows that the feature representations of plants which could not be obtained by alpine plants images were extracted by using supplementary plants images. The difference of recognition rates between (b) and (c) was also at significance level of

5%. On the other hand, the performance for the method (d) was as same as the one without any fine-tuning. In (d), since both alpine plants images and other plants ones were applied simultaneously, the resultant feature representations for alpine plants images might be blurred due to the large number of supplementary images.

Table 2. Recognition rates based on 10-fold cross validation (Classifier: SVMs)

Fine-tuning method	Accuracy (%)	Std. dev
Without fine-tuning	73.56	4.87
(a) Alpine plants	78.37	4.55
(b) Supplementary plants	78.54	4.25
(c) Supplementary plants → alpine plants	80.21	5.13
(d) Alpine plants + supplementary plants	73.02	3.70

Table 3 shows recognition results when we used the fully-connected layers in DCNNs as classifiers. In comparison with the results using SVMs in Table 2, the recognition rates are much improved. Since the hidden layers in a DCNN learn training data in cooperation with the fully-connected layers, feature representations extracted by the hidden layers should come to suit to the fully-connected layers through supervised learning. This may contribute to the better performance than SVMs.

Moreover, in Table 3, the recognition rate by (c) was better than that by (a) ($p < 0.05$). Therefore, we can say that the supplementary images are also effective in this case.

Table 3. Recognition rates based on 10-fold cross validation (Classifiers: fully-connected layers in DCNNs)

Fine-tuning method	Accuracy (%)	Std. dev
(a) Alpine plants	83.62	4.66
(c) Supplementary plants → alpine plants	85.73	4.06

4   Conclusions

In this paper, we have constructed a recognition system for Japanese alpine plants by using features extracted from a Deep Convolutional Neural Network (DCNN) which prelearned a large number of natural images in advance. Since it is very difficult to prepare enough number of alpine plants images for fine-tuning in DCNNs, we collected images of plants other than alpine ones which belong to the same genus as the alpine plants, and used them for fine-tuning to supplement the lack of the number of training data.

Experimental results show that the fine-tuning with supplementary plants images first followed by that with the alpine plants images significantly improves the recognition accuracy. That is, we have shown the effectiveness of using supplementary plants images for alpine plants recognition.

In the future research, we will expand species of alpine plants to recognize, and consider employing other data augmentation techniques for better performance. Moreover, we will implement our recognition system in a hand-held terminal.

References

1. Shimizu, T., Kaota, Y., Kihara, H.: Flowers bloom in the alpine (in Japanese), p. 512, Yama to keikokusya (2014)
2. Arai, K.: Basics of alpine plants (in Japanese), p. 125, Eisyuppan (2012)
3. Krizhevsky, A., Sutskever, I., Hinton, G.E.: ImageNet classification with deep convolutional neural networks. Adv. Neural Inf. Process. Syst. **25**, 1097–1105 (2012)
4. Donahue, J., Jia, Y., Vinyals, O., Hoffman, J., Zhang, N., Tzeng, E., Darrell, T.: DeCAF: a deep convolutional activation feature for generic visual recognition, arXiv preprint arXiv:1310.1531v1 (2013)
5. Yangqing, J., Evan, S.: Caffe | Fine-tuning for style recognition, Caffe | Deep Learning Framework. 8 December 2015. http://caffe.berkeleyvision.org/gathered/examples/finetune_flickr_style.html
6. Suzuki, S., Shouno, H., Kido, S.: Feature analysis for diffuse lung disease with deep convolutional neural network (in Japanese), IEICE Technical report, NC2014-114, pp. 259–264 (2015)
7. Hinton, G.E., Srivastava, N., Krizhevsky, A., Sutskever, I., Salakhutdinov, R.R.: Improving neural networks by preventing co-adaptation of feature detectors, arXiv preprint arXiv:1207.0580 (2012)
8. Chang, C., Lin, C.: LIBSVM: a library for support vector machines. ACM Trans. Intell. Syst. Technol. (TIST) **2**(3), 27 (2011)

Author Index